W0246007

PHYSICAL PROCESSES IN FRAGMENTATION AND STAR FORMATION

ASTROPHYSICS AND SPACE SCIENCE LIBRARY

A SERIES OF BOOKS ON THE RECENT DEVELOPMENTS OF SPACE SCIENCE AND OF GENERAL GEOPHYSICS AND ASTROPHYSICS PUBLISHED IN CONNECTION WITH THE JOURNAL SPACE SCIENCE REVIEWS

VOLUME 162

PROCEEDINGS

PHYSICAL PROCESSES IN FRAGMENTATION AND STAR FORMATION

PROCEEDINGS OF THE WORKSHOP ON
'PHYSICAL PROCESSES IN FRAGMENTATION AND STAR FORMATION',
HELD IN MONTEPORZIO CATONE (ROME), ITALY, JUNE 5–11, 1989

Edited by

ROBERTO CAPUZZO-DOLCETTA
Institute of Astronomy, University of Rome I, Italy

CESARE CHIOSI
Department of Astronomy, University of Padova, Italy

and

ALBERTO DI FAZIO
Observatory of Rome, Italy

KLUWER ACADEMIC PUBLISHERS
DORDRECHT / BOSTON / LONDON

Library of Congress Cataloging in Publication Data

Workshop on Physical Processes in Fragmentation and Star Formation (1989 : Rome, Italy)
Physical processes in fragmentation and star formation : proceedings of the Workshop in Physical Processes in Fragmentation and Star Formation, held in Monteporzio Catone (Rome), Italy, June 5-11, 1989 edited by Roberto Capuzzo-Dolcetta, Cesare Chiosi, Alberto Di Fazio.
p. cm. -- (Astrophysics and space science library ; 162)
ISBN 0-7923-0769-0 (alk. paper)
1. Stars--Formation--Congresses. 2. Astrophysics--Congresses. I. Capuzzo-Dolcetta, Roberto. II. Chiosi, C. (Cesare) III. Di Fazio, Alberto. IV. Title. V. Title: Fragmentation and star formation. VI. Series.
QB806.W65 1989
523.8--dc20

90-4436
CIP

ISBN-13: 978-94-010-6760-7 e-ISBN-13: 978-94-009-0605-1
DOI: 10.1007/978-94-009-0605-1

Published by Kluwer Academic Publishers,
P.O. Box 17, 3300 AA Dordrecht, The Netherlands.

Kluwer Academic Publishers incorporates
the publishing programmes of
D. Reidel, Martinus Nijhoff, Dr W. Junk and MTP Press.

Sold and distributed in the U.S.A. and Canada
by Kluwer Academic Publishers,
101 Philip Drive, Norwell, MA 02061, U.S.A.

In all other countries, sold and distributed
by Kluwer Academic Publishers Group,
P.O. Box 322, 3300 AH Dordrecht, The Netherlands.

Printed on acid-free paper

TABLE OF CONTENTS

1. ATOMIC, MOLECULAR PROCESSES, TURBULENCE, AND MAGNETIC FIELD IN FRAGMENTATION.

2. STAR FORMATION IN DIFFERENT ASTROPHYSICAL ENVIRONMENTS.

3. HYDRODYNAMICAL AND DYNAMICAL MODELS.

4. THE INITIAL MASS FUNCTION.

5. STAR CLUSTERS: FRAGMENTATION AND STAR FORMATION.

6. DISCUSSION ON THE LOW MASS END OF THE IMF AND CONCLUDING REMARKS.

PREFACE

In the recent years, we have witnessed the expansion and multiplication of the observations regarding star formation and fragmentation, and the consequent growth of the problematics concerning the underlying physical processes, the chemistry, the sites, the times, etc.. At the same time, we had a correspondingly increasing demand of meetings dedicated to these subjects. It became particularly stimulating to have specific meetings on the underlying physical processes (both in the microphysics and not) and on the observations directly aimed at which of the physical mechanisms are relevant, and how do they interact and concur to give the observed data. Moreover, the studies of the latest 10 years have shown that formation of stars is likely to share many features (and many physical driving mechanisms, maybe even the main instabilities causing its onset) with the formation of other self-gravitating objects (dust and molecular clouds, star systems, etc.). Therefore, it appeared to be very useful and interesting to discuss the formation of the mentioned objects not separately, but in a more comparative way. Being aware of the presence of the latter concept in the recent literature, it has been fairly natural to include in a meeting on the above subjects the more general key-word "fragmentation".

For all the above reasons, we decided to organize an interactive workshop with the title *Physical Processes in Fragmentation and Star Formation.* This initiative was taken while attending the IAU symposium 115 "Star Forming Regions". The idea to have such a workshop met the enthusiastic support of many specialists, and led to the formation of the Scientific Organizing Committee. The latter was composed of R. Capuzzo- Dolcetta, C. Chiosi, A. Di Fazio, F. Ferrini, R. B. Larson, F. Palla, J. M. Scalo, D. A. Varshalovich, S. V. Vereshchagin. The decision taken by the SOC was that of organizing a specialistic workshop limited to a maximum number of 45 participants. The chosen site for the workshop has been the Monteporzio Catone branch of the Astronomical Observatory of Rome. A relevant part of the world specialists joined in, and the workshop experienced a lively, highly interactive participation, thanks to the high number of new contributions presented and to the open discussions organized.

We would like to express our gratitude to all the colleagues, for their great help. We enjoyed the presence of a large group of Soviet colleagues, which guaranteed a large representativity of the scientific community, and for this we thank the Academy of Sciences of the USSR and the Italian National Council of Researches (CNR).

On behalf of the Scientific Organizing Committee we thank very warmly all the participants for their contributed papers and for their lively collaboration. We also would like to thank the sponsors, namely the CNR, the University of Rome "La Sapienza", the Astronomical Observatory of Rome, and the Astrophysical Observatory of Arcetri (Florence) for their financial support. Finally, special thanks are due to the director, Prof. A. Cavaliere, to dr. P. Battinelli and to the staff of the Astronomical Observatory of Rome for their unvaluable help.

The Editors,

R. Capuzzo–Dolcetta, C. Chiosi, A. Di Fazio

Scientific Organizing Committee

R. Capuzzo–Dolcetta, University of Roma I, Italy
C. Chiosi, University of Padova, Italy (chairman)
A. Di Fazio, Observatory of Roma, Italy
F. Ferrini, University of Pisa, Italy
R.B. Larson, Yale University, New Haven, U.S.A.
F. Palla, Arcetri Observatory, Firenze, Italy
J.M. Scalo, University of Texas, Austin, U.S.A.
D.A. Varshalovich, Ioffe Institute, Leningrad, U.S.S.R.
S.V. Vereshchagin, Astron. Counc. Acad. of Sciences, Moscow, U.S.S.R.

Local Organizing Committee

P. Battinelli, Observatory of Roma, Italy
R. Capuzzo–Dolcetta, University of Roma I, Italy
A. Di Fazio, Observatory of Roma, Italy
with the help of A. Lattanzi, A. Restante, F. Rosati (Observatory of Roma, Italy).

LIST OF PARTICIPANTS

P. Battinelli
Osservatorio Astronomico
via del Parco Mellini 84
I-00136 Roma, ITALY

R. Bedogni
Dip. di Astronomia
via Zamboni 33
I-40126 Bologna, ITALY

A. Boss
Dept. of Terrestrial Magnetism, Carnegie Inst. of Washington
5241 Broad Branch RD. N.W.
Washington D.C. 20015, USA

R. Capuzzo–Dolcetta
Institute of Astronomy, University of Roma I
via G.M. Lancisi 29
I-00161 Roma, ITALY

R. Cayrel
Observatoire de Paris
61 Avenue de l'Observatoire
F-75014 Paris, FRANCE

T. Celandroni
Istituto di Astronomia
University of Pisa
Piazza Torricelli 2
I-56100 Pisa, ITALY

J. P. Chièze
Centre d'Etudes de Bruyères le Châtel
Service P.T.N., B. P. 12,
F-91680 Bruyères le Chatel, FRANCE

C. Chiosi
Dipartimento di Astronomia
Vicolo Osservatorio 5
I-35122 Padova, ITALY

C. Clarke
Institute of Astronomy
Madingley road
Cambridge CB3 OHA, England U.K.

F. D' Antona
Astronomical Observatory of Roma
I-00040 Monteporzio Catone (Roma), ITALY

C. de Boisanger
Centre d'Etudes de Bruyères le Châtel
Service P.T.N., B. P. 12,
F-91680 Bruyères le Chatel, FRANCE

A. Di Fazio
Astronomical Observatory of Roma
viale del Parco Mellini 84
I-00136 Roma, ITALY

M. Ferrari–Toniolo
Istituto di Astrofisica Spaziale del CNR
C.P. 67
00044 Frascati, ITALY

F. Ferrini
Istituto di Astronomia, Universita‘ di Pisa
Piazza Torricelli 2
I-56100 Pisa, ITALY

P. Giannone
Institute of Astronomy, University of Roma I
via G.M. Lancisi 29
I-00161 Roma, ITALY

M. Giroux
Department of Astronomy, University of Texas at Austin
Austin, TX 78712, USA

M. Hawkins
Royal Observatory
Blackford Hill
Edinburgh 3H9 3HJ, Scotland U.K.

E. Herbst
Erstes Physikalisches Institut Univ. Zu Köln
Univesitätsstrasse 14
D-5000, Köln 41, F.R. Germany

Yu.I. Izotov
Central Astronomical Observatory
Ukrainian Academy of Sciences
Goloseevo
SU-252127 Kiev, USSR

M. Kiguchi
Research Institute for Science and Technology Kowakae
Higashiosaka
Osaka 577, JAPAN

I. Kolesnik
Central Astronomical Observatory
Ukrainian Academy of Sciences
Goloseevo
SU-252127 Kiev, USSR

E. Krügel
Max Planck Institut Für Radioastronomie
Auf Dem Hügel 69
D-5300 Bonn 1, F.R. GERMANY

E. Lada
Department of Astronomy, University of Texas at Austin
Austin, TX 78712, USA

R. B. Larson
Dept. of Astronomy, Yale University
P. O. Box 6666
New Haven, CT 06511, USA

P. Lenzuni
Astrophysical Observatory of Arcetri
Largo E. Fermi 5
I-50125 Firenze, ITALY

A. Lioure
Centre d'Etudes de Bruyères le Châtel
Service P.T.N., B. P. 12,
F-91680 Bruyères le Chatel, FRANCE

S. Lizano
Inst. de Astron. Univ. Nacional Autonoma de Mexico
Apartado postal 70-264
DF 04510, MEXICO

M. Mateo
The Observatories, Carnegie Institution of Washington
813 Santa Barbara St
Pasadena CA 91101, USA

A.B. Men'shchikov
Astronomical Council of the Acad. of Sciences
Pyatnitskaya 48
SU-109017 Moscow, USSR

T. Mouschovias
Dept. of Physics and Astron., Univ. of Illinois
Urbana, IL 61801, U. S. A.

F. Palla
Astrophysical Observatory of Arcetri
Largo E. Fermi 5
I-50125 Florence, ITALY

A. Parravano
Universidad de Los Andes, Departmento de Fisica
Merida 5101, VENEZUELA

M. Perault
Lab. de Physique, Radioastronomie mill.,École Normale Supérieure
24 rue Lhomond
F-75005 Paris, FRANCE

P. Persi
Istituto di Astrofisica Spaziale del CNR
C.P. 67
I-00044 Frascati, ITALY

H. Pongracic
University College Cardiff, Dept. of Physics
P.O. Box 913,
Cardiff, CF1 3TH Wales, U.K.

N. C. Rana
Tata Inst. of Fundamental Research
Homi Bhabha Road
Bombay 400 005, INDIA

B. Rocca Volmerange
Institut d'Astrophysique
98 Bis Bd. Arago
F-75014 Paris, FRANCE

Prof. P. Saraceno
Istituto di Fisica dello Spazio Interplanetario del CNR
C.P. 27
I-00044 Frascati (Roma). ITALY

P. Shapiro
Department of Astronomy, University of Texas at Austin
Austin, TX 78712, USA

J. M. Scalo
Department of Astronomy, University of Texas at Austin
Austin, TX 78712, USA

W. Tscharnuter
Institut für Theoretische Astrophysik, Universität Heidelberg
Im Neuenheimer Feld 294
D-6900, Heidelberg, F.R. GERMANY

D.A. Varshalovich
Ioffe Physico-Technical Institute of the USSR Acad. of Sciences
SU-194021 Leningrad, USSR

E. Vazquez
Department of Astronomy, University of Texas at Austin
Austin, TX 78712, USA

S.V. Vereshchagin
Astronomical Council of the Acad. of Sciences
Pyatnitskaya 48
SU-109017 Moscow, USSR

K. Yorke
Max Planck Institut für Radioastronomie
Auf Dem Hügel 69
D-5300 Bonn 1, F.R. GERMANY

H. Zinnecker
Max Planck Institut für Astrophysik, Institut für Extraterrestrische Physik
D-8046 Garching bei München, F.R. GERMANY

IDENTIFICATION LIST

1 A. Di Fazio
2 T. Celandroni
3 A. Parravano
4 F. Ferrini
5 R. Capuzzo–Dolcetta
6 C. Chiosi
7 P. Giannone
8 R. Cayrel
9 H. Zinnecker
10 P. Lenzuni
11 D.A. Varshalovich
12 A. Lioure
13 M. Hawkins
14 F. D'Antona
15 E. Krügel
16 E. Lada
17 C. Clarke
18 J.P. Chièze
19 A. Men'shchikov
20 M. Mateo
21 S. Lizano
22 B. Rocca–Volmerange
23 P. Shapiro
24 M. Giroux
25 I. Kolesnik
26 E. Vazquez
27 C. de Boisanger
28 Yu. Izotov
29 A. Restante
30 K. Yorke
31 R. Bedogni
32 W. Tscharnuter
33 R. Porreca
34 H. Pongracic
35 R.B. Larson
36 J.M. Scalo
37 M. Kiguchi
38 P. Battinelli
39 S.V. Vereshchagin
40 N.C. Rana
41 A.P. Boss
42 F. Rosati
43 L. Bontempi

1. ATOMIC, MOLECULAR PROCESSES, TURBULENCE, AND MAGNETIC FIELD IN FRAGMENTATION.

E.HERBST - Chemistry and star formation.

J. P. CHIÈZE , G. PINEAU DE FORETS - Molecular clouds chemistry with mixing.

M. KIGUCHI - Collapse and fragmentation of molecular clouds.

Yu. I. IZOTOV - On the role of molecular hydrogen in the formation and evolution of blue compact dwarf galaxies and giant HII regions.

R. CAPUZZO–DOLCETTA, A. DI FAZIO, F. PALLA - NLTE H_2 cooling function and protogalactic evolution.

P.R. SHAPIRO, H. KANG - Radiative shocks and nonequilibrium chemistry in the early universe: Galaxy and primordial star formation.

M.L. GIROUX, P.R. SHAPIRO - The intergalactic medium: initial and boundary conditions for Galaxy and primeval star formation.

I. G. KOLESNIK, Y. Yu. OHUL'CHANSKY - Supersonic turbulent fragmentation of giant molecular clouds.

P. BATTINELLI, R. CAPUZZO–DOLCETTA, A. DI FAZIO, V. A.URPIN, S.V. VERESHCHAGIN - Turbulence in the frame of the evolution of a self-gravitating protocloud.

A.DI FAZIO, A. SOLOVIEV, V.A. URPIN, S.V. VERESHCHAGIN - Fragmentation and supersonic turbulence in self-gravitating gas clouds.

T. Ch. MOUSCHOVIAS - Fragmentation and collapse in magnetic molecular clouds: natural length scales and protostellar masses.

CHEMISTRY AND STAR FORMATION

Eric Herbst
I. Physikalisches Institut
Universitaet zu Koeln
D-5000 Koeln 41
Federal Republic of Germany

ABSTRACT. The normal gas phase chemistry of dense interstellar clouds is strongly affected by the process of star formation. Three mechanisms in which star formation impacts the chemistry of the surrounding material have been discussed in the literature. These are: evaporation of grain mantles due to heating, mixing of material driven by stellar winds into the ambient interstellar gas, and interstellar shock waves. There is evidence for each of these mechanisms in the Orion Molecular Cloud and it also appears that the mechanisms operate in physically distinct sections of the cloud.

1. Ambient Chemistry of Dense Interstellar Clouds

Dense interstellar clouds, the birthplace of future generations of stars, are large agglomerations of gas and dust. Since these are regions of high visual extinction, they cannot be studied by the standard optical techniques, but instead must be studied by long wavelength techniques, especially radioastronomy. A large number of molecules have been discovered in the gas phase of such regions, chiefly by observations of discrete rotational spectral features at millimeter wavelengths ; these molecules range up to thirteen atoms in size. In dense interstellar clouds, all species other than molecular hydrogen are trace species, with carbon monoxide (the second most abundant molecule) having a fractional abundance with respect to H_2 of 10^{-4}. The molecular spectra have been utilized to characterize the physical conditions of these clouds; typical densities of 10^4 cm^{-3} and temperatures of 10 - 70 K have been derived although both higher densities and temperatures are seen in star-forming regions (Blake *et al.* 1987). In addition, there is evidence from broad infrared features seen in absorption against continua emitted from hot dust in star-forming regions that molecules such as water, ammonia, and carbon monoxide are present on the surfaces of dust particles, either because they are formed there or because they are adsorbed from the gas (Eiroa and Hodapp 1989) . Recent observational work has shown that the distribution of molecules within dense interstellar clouds is far from uniform (Blake *et al.* 1987; Stutzki *et al.* 1988). The most dramatic chemical effects are related to star-forming regions, in which higher temperatures, higher densities, turbulent motions, and shock excitation change the character of the chemistry In this review, we shall first discuss the normal chemistry of dense interstellar clouds and then discuss how the physical conditions in star-forming regions affect this chemistry and lead to dramatic observational differences in molecular abundances.

R. Capuzzo-Dolcetta et al. (eds.), Physical Processes in Fragmentation and Star Formation, 3–16.

1.1 CHEMICAL PROCESSES ON DUST SURFACES

The chemistry of the ambient dense interstellar cloud medium, a term by which we shall refer to non-star-forming regions, has been studied extensively in recent years. Both chemical processes occurring on the surfaces of dust particles and in the gas phase have been considered; however, there is a much larger laboratory data base for gas phase processes (see, viz., Anicich and Huntress 1986) and there is much less uncertainty concerning them. Although gas phase processes can produce many if not most of the molecules seen under ambient conditions in dense clouds (Winnewisser and Herbst 1987; Herbst and Leung 1989), they cannot produce the most abundant molecule - molecular hydrogen - in sufficient quantity (Gould and Salpeter 1963; Watson 1976). Molecular hydrogen must be produced predominantly by recombination of two hydrogen atoms on dust surfaces - this process has been most recently discussed by Tielens and Allamandola (1987). In the process, two hydrogen atoms first stick to cold grains with reasonable efficiency and, because of their lightness, are then able to migrate from one active site to another until they are on adjacent sites and can form a chemical bond. In both the sticking to the grain surface and the formation of the chemical bond, energy must be transferred to the dust particle, with a consequent rise in its temperature. Once formed, molecular hydrogen evaporates from the surface. The process of converting a gas initially assumed to be atomic hydrogen into its molecular form takes somewhat more than 10^5 yr., which is a minimum lifetime for all dense clouds since these objects all contain molecular hydrogen as the overwhelmingly abundant species (Watson 1976).

The question of whether any other molecules that are observed in the gas phase, are formed on the surfaces of dust particles has remained unanswered for many years. One basic problem is that it is not clear how molecules heavier than molecular hydrogen can be removed from the surface once they are formed since thermal evaporation appears to be impossible under ambient conditions. Among the suggestions for desorption made over the years have been: photodesorption, utilization of the exothermicity of chemical reactions, cosmic-ray bombardment, and grain explosion triggered by radical buildup (Léger *et al.* 1985). These suggestions, which involve continuous mechanisms, are less convincing than intermittent mechanisms based on star formation such as thermal heating of grain mantles and interstellar shock waves. We will therefore assume for simplicity that, outside of star formation regions, molecules other than H_2 that are formed on grain surfaces remain there to a large extent.

Since the exact nature of interstellar grains is still uncertain, the chemical processes that form species on them cannot be identified unambiguously. Still, certain dominant processes can be identified (Watson 1976; Tielens and Allamandola 1987). Since hydrogen atoms can migrate freely, they can combine with species other than with themselves, for example heavy atoms such as oxygen, nitrogen, and carbon. The reactive radicals formed by such reactions (OH, NH, and CH) will also react with hydrogen atoms and the process will continue until the so-called saturated species H_2O, NH_3, and CH_4 are formed. These species will not be very reactive since, unlike radicals, stable neutral species react slowly at low temperatures due to the presence of activation (potential) energy barriers (see below). Unless the grains are so-called active catalysts, which remove activation energy barriers, it would appear that saturated species will not react further on grain surfaces. Catalysts are normally found to be specific rather than universal in scope. Although reactive radicals (species such as OH, NH, and CH) can presumably react with other radicals as well as H atoms without activation energy, these reactions will occur less efficiently due to slower migration. It therefore seems a good first approximation to regard the dominant products of grain surface chemistry as the simple saturated molecules above, although there is no unanimity of opinion on this point (Tielens and Allamandola 1987). These molecules will then be retained to a large extent on the grain surfaces in the ambient medium. It is possible however that photochemical processes can convert some of these simple species to more complex organic forms (Greenberg 1987).

1.2 GAS PHASE CHEMICAL PROCESSES

Once H_2 is formed on grain surfaces and is released into the gas phase, it can initiate a chain of important gas phase reactions. As on the surfaces of non-catalytic grains, the only gas phase reactions that can occur rapidly at low temperatures are those that possess no activation energy. The reason for this assertion can be seen from the Arrhenius form of the rate coefficient k(T), which is given by the expression (Levine and Bernstein 1974)

$$k(T) = A(T) \exp(-E_a/kT) \quad (1)$$

where A(T) is the pre-exponential factor, which is at most weakly temperature dependent, E_a is the activation energy barrier, and k is the Boltzmann constant. Since activation energy barriers for stable neutral species range from 1-5 eV, the exponential term at temperatures below 100 K is normally so large and negative that rate coefficients are close to zero. Although there are radical species (e.g. OH) that do not possess activation energy barriers in many of their studied reactions, the more important exceptions to this bleak assessment concern reactions involving at least one charged species. To be rapid at low temperatures, these reactions must first of all be exothermic (give off energy) rather than endothermic (take in energy) since only in the former case can the activation energy be zero. Unlike the case of neutral-neutral reactions, the study of ion-molecule reactions, which involve one charged species and one neutral species as reactants, shows these systems to possess no exponential temperature dependence of the type shown in eq. (1) in almost all cases (Anicich and Huntress 1986). In other words, for the vast majority of ion-molecule reactions, $E_a = 0$, although there are some important exceptions. A simple reason for this aberrant behavior is that ion-molecule systems possess strong long-range attractive forces which lead to attractive rather than repulsive potentials at short range. Furthermore, for reactions in which the neutral reactant is nonpolar, A(T), the pre-exponential factor, is completely independent of temperature and given by a simple expression first derived by Langevin (Herbst 1987a). Its value lies in the vicinity of 10^{-9} $cm^3\ s^{-1}$. If the neutral reactant is polar, A(T) actually possesses an inverse dependence on temperature, varying something like $T^{-1/2}$ (Marquette *et al.* 1989). Consequently, exothermic ion-molecule reactions will occur rapidly at even the lowest temperatures of dense interstellar clouds.

Ion-molecule synthetic schemes start mainly with the cosmic-ray induced ionization of molecular hydrogen (Herbst and Klemperer 1973):

$$H_2 + \text{Cosmic Ray} \longrightarrow H_2^+ + e + \text{cosmic ray}, \quad (2)$$

which is the dominant producer of positive ions and electrons, followed by the rapid reaction

$$H_2^+ + H_2 \longrightarrow H_3^+ + H \quad (3)$$

which produces the simplest polyatomic species. A subsequent simple and representative reaction pathway leading to methane (CH_4) is the reaction sequence

$$C + H_3^+ \text{ -----> } CH^+ + H_2 \quad (4)$$

$$CH^+ + H_2 \text{ -----> } CH_2^+ + H \quad (5)$$

$$CH_2^+ + H_2 \text{ -----> } CH_3^+ + H \quad (6)$$

$$CH_3^+ + H_2 \text{ -----> } CH_5^+ + h\nu \quad (7)$$

$$CH_5^+ + e \text{ -----> } CH_4 + H . \quad (8)$$

Reactions (7) and (8) require special comment. The normal ion-molecule reaction

$$CH_3^+ + H_2 \text{ -----> } CH_4^+ + H \quad (9)$$

does not occur since it is endothermic. Reaction (7) is an example of a radiative association reaction, in which the two reactants manage to stick together by emitting a photon. Such processes are quite inefficient for small reactants and have been studied in the laboratory in only a few instances. Theoretical determinations of the rates of these processes have been carried out (Bates and Herbst 1988). Reaction (7) itself has been studied in the laboratory and found to proceed at 10 K on one out of every 10^4 collisions (Barlow, Dunn, and Schauer 1984). Although inefficient, processes such as (7) can be competitive because of the large abundance of H_2. It is interesting to note that the radiative association reaction between two hydrogen atoms, which is a gas phase synthesis for H_2, is much less efficient than (7), occurring on fewer than 1 in 10^{10} collisions (Watson 1976; Gould and Salpeter 1963). The place of radiative association reactions in other ion-molecule syntheses has been previously discussed (Winnewisser and Herbst 1987).

Reaction (8) is an example of a dissociative recombination reaction. Recombination reactions between positive molecular ions and electrons are known to be very rapid ($k[300\ K] \approx 10^{-6}\ cm^3\ s^{-1}$), to increase in rate somewhat as temperature decreases, and to be dissociative in nature (Adams and Smith 1988). Until recently, the exact neutral products formed in such reactions had not been studied in the laboratory and modellers using ion-molecule rate processes were forced to estimate the relative amounts of the assorted possible products (see e.g. Green and Herbst 1979). The situation has begun to change, however, as laboratory studies become feasible. For example, Herd *et al.* (1989) have now measured that in the dissociative recombination reaction between H_3O^+ and e, the dominant heavy molecule produced is OH, which is made on 65 ± 10 % of collisions, whereas H_2O is produced on at most 35 % of collisions. The simple estimate used most frequently by modellers in the past has been that each species was produced on 50% of collisions. Preliminary experimental results on the products of reaction (8) show methane to be a major product (Smith 1989).

Ion-molecule syntheses have been written down for a large number of interstellar molecules up to ten and more atoms in size (Herbst and Leung 1989; Winnewisser and Herbst 1987; Herbst 1988); these will not be reviewed here. It is important to note however that the reaction sequence leading to methane shown above is unusual in that it leads to a fully saturated (hydrogen-rich) ion (CH_5^+). For hydrocarbon ions with more than one carbon atom, it appears that neither normal nor radiative association reactions can produce saturated species (Herbst and Leung 1989) and the neutral hydrocarbons produced in the ghas phase are then quite unsaturated (hydrogen-deficient). Thus, for example, ion-molecule reactions do not lead to the production of the saturated C_2H_6

(ethane) molecule but do produce large amounts of C_2H_2 (acetylene).

Once the assorted ionic and neutral species are produced via gas phase reactions, they are destroyed by such processes as well. In addition, there is the possibility of destruction by photon-initiated processes. The question of the extent of the penetration by hard (visible and ultra-violet) photons into the interiors of dense clouds has been debated and depends on the geometry of the cloud, the grain albedos and scattering functions, and the degree to which the cloud material is inhomogeneous and clumped (Boissé 1989). Even with the assumption that little in the way of external radiation penetrates dense clouds, there is still the problem of cosmic-ray-induced ultra-violet radiation since cosmic rays do indeed penetrate and can cause the emission of radiation (Gredel *et al.* 1989).

2. Gas Phase Models of Ambient Interstellar Clouds

Since the pioneering study of Herbst and Klemperer (1973), a large number of investigators have constructed gas phase chemical models of dense clouds in which only H_2 is formed on the surfaces of dust particles. A partial list of models is contained in Herbst (1988). Originally, models were steady-state in nature; i.e., the chemistry was assumed to be time independent. Now however, it is more frequently assumed that the chemistry evolves from initial conditions. If this evolution is assumed to occur under fixed physical conditions, the model is referred to as pseudo-time-dependent whereas if the time dependence (e.g. gravitational collapse) of the physical conditions is treated as well, the model is referred to as truly time dependent. The most frequently attempted models are those of the pseudo-time-dependent variety (Herbst and Leung 1989) and these are overall quite sucessful in producing a large number of molecules in reasonable agreement with observation in representative dense interstellar clouds. Rather than give a detailed comparison between theory and observation here, we note the following basic predictions of these pseudo-time-dependent gas phase models:

(a) the most abundant organic molecules tend to be unsaturated or hydrogen-poor,
(b) the abundances of organic molecules peak at an early time before declining at steady state,
(c) the peak abundances are normally sufficiently large to account for observations,
(d) isotopic fractionation can be large, especially for deuterium.

Point (a) has been previously discussed. Point (b) refers to the abundances of complex molecules only. Starting from zero, these abundances increase with time and appear to peak at times ($\approx 10^5$-10^6 yr) well before steady-state is reached ($\geq 10^7$ yr). The peak abundances are typically only somewhat greater than those observed in the "molecule factory" TMC-1. Steady-state abundances of complex molecules are very low and in poor agreement with observation because the carbon needed to produce organic species has all been consumed in the formation of CO (Herbst and Leung 1989 and references therein.) Since interstellar clouds are thought by astronomers to be older than 10^5-10^6 yr, the comparison of chemical abundances calculated for such an age with observation is somewhat suspect. Chieze and Pineau des Forets (1989) have just shown that if the customary simple homogeneous picture of a cloud is replaced by a more realistic two-phase model in which high density clumps exist in a lower density gas (Stutzki *et al.* 1988), and if there is rapid mixing between the low and high density phases, the early-time solutions of Herbst and Leung (1989) can be preserved to steady state for at least some molecules. For times greater than 10^5-10^6 yr, however, it is not clear how in the absence of star formation a gas phase can continue to exist. Perhaps grain mantles in high density clumps are photodesorbed.

Point (d) has not previously been discussed in this article. Ion-molecule reactions lead to a large fractionation effect for deuterated molecules such that these species can reach their observed abundances of ≈ 1-10% of their hydrogen analogs despite the fact that the overall deuterium to hydrogen abundance ratio is only a few x 10^{-5}. A detailed study of gas phase deuterium fractionation has just appeared (Millar *et al.* 1989) and reviews the chemical processes responsible for it. Overall, fractionation is strongest at low temperature but weakens at higher temperatures. For most molecules, there is little time dependence of the *ratio* of deuterated to normal molecular abundances. Although the gas phase theory accounts for fractionation in ambient sources such as TMC-1 and the extended ridge in Orion, it fails to account for large fractionation effects seen in certain star-forming regions. This is discussed in Section 4.

3. The Chemical Consequences of Star Formation

As interstellar gas and dust collapse in the process of star formation, the density and eventually the temperature increase. The increasing density and temperature allow a greater diversity of chemical processes to occur than in the ambient interstellar medium. Higher gas densities permit three-body association reactions to occur (Bates and Herbst 1988); these are processes in which two species, say A^+ and B, come together to form a short-lived "complex" AB^{+*}, which will come apart within a short time unless it is stabilized. However, at high enough densities a third body C can collide inelastically with AB^{+*} to produce stable AB^+. The gas density at which three-body association is more rapid than radiative association, in which the complex is stabilized by radiative emission, is normally in the range 10^{11} - 10^{12} cm^{-3} (Bates and Herbst 1988). Three-body reactions are presumably important in the initial stages of condensation processes, such as the formation of dust particles in stellar atmospheres.

In addition to the increase in gas density, which renders three-body processes possible, the increase in temperature affects the chemistry of star formation regions strongly. As can be seen in eq. (1), the rate coefficients of processes with activation energy barriers depend on temperature in an exponential manner so that an increase in temperature from say 10 K (the ambient interstellar value) to 300 K or 1000 K can have a dramatic effect on rate. For a system with an activation energy of 0.1 eV, the exponential term in eq. (1) is exp(-116) at 10 K but only exp(-1.16) at 1000 K. Thus, as the temperature rises, reactions that are weakly endothermic or exothermic with small potential barriers must be considered. The constraint, operative for ambient interstellar chemistry, that most neutral-neutral reactions can be ignored vanishes. However, one must be careful in utilizing reaction rates measured in the laboratory at high temperatures for models of interstellar gas since interstellar gas may not be in thermodynamic equilibrium (Wagner and Graff 1987) It would appear that the need to include far more chemical processes in models of star formation regions than in models of ambient regions makes the former problem by far the more difficult. In general this is so, but one must remember that eventually, as temperature and density increase sufficiently, the chemistry approaches the limit of thermodynamic equilibrium and the kinetic differential equations can be replaced by thermodynamic equations which are far simpler, as is the case for stellar atmospheres.

As temperature and density increase, not only the gas phase is affected. Given the relative weakness of adsorption forces, molecules will be desorbed from the surfaces of dust particles. Any grain chemistry that might have transpired during the low temperature phase will now affect the composition of the gas. Indeed, at sufficiently high temperatures, the refractory cores of dust particles will vaporize. High temperatures need not be limited to the immediate vicinity of the star or protostar since shock waves can carry heat over large distances.

These mainly qualitative statements show that there should be severe observational chemical consequences of star formation and that the temperature, density profile of material in the immediate vicinity of a protostar should lead to a chemical profile. Unfortunately, astronomers have not yet succeeded in observing the change in chemistry as a function of increasing temperature and density, or proximity to the core, in the protostar or very early stellar stage. What has been undertaken up to now have been lower spatial resolution studies of larger regions in areas known to be sites of active star formation. These studies show that the chemistry is indeed altered from ambient gas phase chemistry along some of the lines discussed above. The *chemical* alterations have been ascribed to three specific *physical* causes: interstellar shocks, desorption of grain mantles, and mixing of winds with quiescent gas.

Grain desorption will affect the chemistry in two ways. First, molecules formed in the gas but condensed onto the grains, will once more be released into the gas phase. Secondly, as discussed above, molecules formed on grain surfaces but unable to come off the surfaces under ambient conditions, will be desorbed and give an indication of grain chemistry under ambient conditions.

The effects of interstellar shocks and of interaction between stellar winds and the ambient gas are more difficult to ascertain. A significant number of investigators have studied the physics and chemistry of interstellar shocks (Mitchell 1987; Shull and Draine 1987; Flower *et al.* 1988). The properties of a shock depend on such parameters as the velocity of the shock front and the size of the magnetic field perpendicular to the shock front. If the magnetic field can be ignored, then the shock is termed J-type, and is characterized by a rapid jump in temperature as the shock front passes through the medium followed by a much slower cooling. Temperatures in excess of several thousand K can be achieved with moderate shock velocities. A simple J-type shock model was first used by Elitzur and Watson (1978, 1980) to account for the abundance of the diatomic radical CH^+ in diffuse interstellar clouds. Unlike most other diatomic species in such sources, which are well accounted for by ambient gas phase models, CH^+ cannot be produced under low temperature conditions because the only important reaction forming it:

$$C^+ + H_2 \text{ -----> } CH^+ + H \qquad (10)$$

is endothermic by 0.39 eV. At a peak temperature of 4500 K, however, the reaction can proceed rapidly. The role of magnetic fields in moderating the effects of shocks was not appreciated by Elitzur and Watson (1978, 1980). As is now well known, once the magnetic field perpendicular to the shock front is larger than a certain critical value, the neutral gas shows a smooth variation in temperature rather than a rapid jump, but there is a net velocity between the ionic and neutral phases (C-type shock). It is this velocity rather than a high kinetic temperature which can power endothermic reactions between ions and neutrals such as reaction (10). The latest C-type shock models show however that even with a variety of parameters, shock models do not quite account for the CH^+ abundance (Flower *et al.* 1988). Besides diffuse clouds, shock models have been applied to star forming regions in dense clouds. In general, the dense cloud shock models do not contain as detailed a treatment of the role of the magnetic field. These will be discussed below in the particular context of the Orion Nebula.

The third physical cause of chemical aberrations is mixing between material blown out of star formation regions and the ambient medium. It is clear that if alien material is mixed into the quiescent gas, its chemistry will be changed. It has been suggested that such a mixing is required to form the oxygen-containing organic molecules such as dimethyl ether, methyl formate, and ethanol in giant molecular clouds since normal gas phase syntheses, while successful at predicting the abundances of these species in cold sources such as TMC-1, do not account for their abundances in giant clouds (Blake *et al.* 1987; Herbst 1987b; Herbst and Leung 1989).

All three of these physical causes of chemical effects have been proposed to explain detailed observations of molecular abundances in the Orion Nebula. In particular, three physically distinct regions - the Hot Core, the Plateau, and the Compact Ridge - have been identified (Blake *et al.* 1987) and their chemistries explained by grain mantle depletion, shock waves, and mixing, respectively. In the next section, these regions and chemical models to explain them are discussed.

4. Case Study: The Orion Nebula

Throughout much of the core of the Orion Nebula, called the "Extended Ridge" source, the physical conditions pertaining are a gas density n of 10^5 cm^{-3} and a temperature of 70 K. However, as noted above, there are three distinct regions observed in the direction of Orion-KL with different physical conditions (Blake *et al.* 1987). The Hot Core source is thought to have a kinetic temperature in the range 150 - 300 K and a density in excess of 10^6 - 10^7 cm^{-3}. The Plateau source, in which large line widths are often observed, probably consists of shock excited material with rotational temperatures in the range 100 - 150 K and densities in excess of 10^6 cm^{-3}. Finally, the Compact Ridge source is thought to have a kinetic temperature in the range 80 - 140 K with a density in excess of 10^6 cm^{-3}. None of these densities is sufficiently large for three-body association reactions to be important so that chemical effects can only be caused by the high temperatures, both directly and indirectly through grain mantle depletion, and by the mixing of stellar-driven material.

4.1 THE HOT CORE

The chemistry of the so-called Hot Core of Orion is distinguished from the ambient (Extended Ridge) source in several aspects. First, chemically saturated (hydrogen-rich) molecules are more abundant than in the Extended Ridge. Some saturated species - e.g. C_2H_5CN - are not found at all outside of the Hot Core whereas others - NH_3, H_2O - have abundances in the Hot Core that are several orders of magnitude larger than in the ambient source (Irvine *et al.* 1987; Blake *et al.* 1987). Secondly, the deuterium fractionation of the Hot Core is extreme, implying a much lower temperature than the standard estimates of 150 - 300 K (Millar *et al.* 1989). Both of these aspects can be accounted for by a chemical model in which grain chemistry followed by mantle desorption is evoked.

Although the ideas had been discussed to some extent by previous authors (Tielens and Allamandola 1987; Walmesley *et al.* 1987; Mauersberger *et al.* 1988), detailed models of the chemistry of the Hot Core are associated with Millar and co-workers (Brown *et al.* 1988 a,b; Brown and Millar 1989 a,b). In their models, the chemistry is divided into three phases. In phase (i), gas phase chemistry occurs with an intial temperature of 10 K and gas density of 3 x 10^3 cm^{-3}. As the gas phase chemistry occurs, two other events occur: the cloud undergoes isothermal free-fall collapse and gas phase species stick to the dust particles. The dust mantle composition is followed and surface reactions between H atoms and other atoms and radicals are allowed to occur. The result is that saturated species such as ammonia and water build up on the surfaces of the dust particles. Phase (i) is halted at a time of 8 x 10^5 yr when the gas density reaches 10^7 cm^{-3}. By this time, condensible (heavy) species are all on the grains. Other than some saturated molecules, most of these species have been produced by gas phase chemistry and then deposited on the grains. Phase (ii) is a rapid heating caused by star formation in the vicinity of the source. The heating causes the grain mantles to be released into the gas phase. In phase (iii), gas phase chemistry once again occurs, but now at an enhanced density of 10^7 cm^{-3} and a temperature of 200 K. Adsorption onto dust particles does not occur efficiently at this elevated temperature. The high density reduces the ion content of the gas sufficiently that the chemistry is very slow. Consequently, the abundances produced in phase (i) are preserved for long periods of time and can be compared with observation.

The results of the model calculations (Brown *et al.* 1988a) show that saturated species such as ammonia and water are quite enhanced compared with ambient gas phase models due to the extra formation mechanism on the grain surfaces. The result for ammonia can be compared with observation and is somewhat too high. Abundances of other species such as HC_3N resemble those of ambient models. The Hot Core also shows a high degree of deuteration which is not understandable in terms of normal gas phase chemistry since the temperature of the source is too high. As discussed by Walmsley *et al.* (1987), the size of the deuterium fractionation suggests that it occurred in a previous low temperature phase. In the model of Brown and Millar (1989a) the deuteration occurs during phase (i) in which both low temperature deuteration in the gas and on the surfaces of dust particles occurs. The dust mantle deuteration occurs via reactions between D atoms and heavy species since low temperature gas phase chemistry produces a rather high atomic D/H abundance ratio. The high D/H abundance ratio at 10 K ($\approx$ 0.005) leads to a comparably large NH_2D/NH_3 and HDO/H_2O abundance ratios, in good agreement with observation (Walmsley *et al.* 1987). In addition, the latest model (Brown and Millar 1989b) predicts rather large abundances for double deuterated species such as NHD_2, a species tentatively detected in Orion by Turner (private communication). The confirmation of this species in the Hot Core will strengthen the case that deuteration occurs on the surfaces of dust particles.

Overall then, despite the simplicity of the chemical model of the Hot Core, it would appear that it is certainly a reasonable picture of the complex interaction between physical conditions and chemical mechanisms that must be occurring in this source. The only skepticism that can be voiced is why certain segments of the process (grain adsorption, cloud collapse) are not more general in scope, leading to complete deposition of the gas onto dust particles in a relatively short time in the absence of star formation in other regions of the cloud.

4.2 THE PLATEAU SOURCE

Shock models directed at this source have attempted to explain the large abundance of certain sulphur-bearing species found here (e.g. Hartquist *et al.* 1980; Mitchell 1984). A recent work by Leen and Graff (1988) contains results for these species which can be compared both with observation and with normal ambient models of sulphur chemistry. A comparison of older shock and ambient models has been undertaken by Blake *et al.* 1987). One must note that the shock models do not seem to have a large enough column density of H_2 to be compared directly with the Plateau source. The problem appears to be that it is difficult to follow the abundances as the temperature cools completely down to ambient levels. As noted by Leen and Graf (1988), their column densities are still increasing as their calculation terminates at a temperature of 300 K and a total gaseous column density we estimate to be 10^{21} cm^{-2}. It is probably better to compare abundance ratios and utilize different theoretical and actual H_2 column densities. In the comparison below (Table 1), the recent ambient-phase sulphur-chemistry model of Millar and Herbst

TABLE 1. A comparison of shock model, ambient model, and observed abundances in the Plateau source.

	Fractional Abundances		
Species	Observed	Shocked	Ambient
HS	---------	4(-10)	9.5(-14)
CS	2.2(-08)	2(-09)	2.6(-08)
SO	5.2(-07)	2(-08)	1.0(-08)
H_2S	9.8(-08)	4(-09)	1.9(-11)
OCS	5.2(-08)	7(-10)	1.6(-10)
SO_2	5.2(-07)	3(-08)	4.1(-10)

Note: a(-b) signifies a x 10^{-b}.

(1989) is utilized at "early time" with the sulphur elemental abundance ratio $S/H_2 = 6 \times 10^{-8}$ and with physical conditions resembling those of the Extended Ridge source. Its results are compared with the shock model of Leen and Graff (1988) in which $v = 10$ km s^{-1}, the initial gas density is 10^5 cm^{-3}, and the H_2/H ratio is set at 10^5. The observed abundances are from Blake *et al.* (1987). It can be seen that for HS, H_2S, and SO_2 the shock model shows a noticeable enhancement. The ambient SO_2 abundance is a strong function of time, however. At slightly longer times than the early time used here, the SO_2 abundance rises strongly and is not noticeably different from the shocked value (Millar and Herbst 1989). Still, the ambient model is clearly the more deficient for H_2S, which is observed to be two orders of magnitude more abundant in the plateau source than in the ambient medium. Observations of HS would help to confirm the superiority of the shock model. Neither the shock model of Leen and Graff (1988) nor the ambient model of Millar and Herbst (1989) comes close to predicting the large abundance of OCS. The older shock model of Hartquist *et al.* (1980) does show a strong enhancement for OCS, based on differing initial conditions and a somewhat different chemistry. In general, the older shock model shows higher abundances of sulphur-bearing molecules.

4.3 THE COMPACT RIDGE SOURCE

Notable for its large abundances of oxygen-containing organic molecules, this source cannot be understood in terms of ambient gas phase chemistry (Herbst 1987b). It has been suggested (Blake *et al.* 1987) that large amounts of water are mixed into the gas of this source from the nearby outflow region IRC 2. The large abundance of water can be utilized in gas phase reaction schemes to produce the oxygen-containing organic molecules. This idea has not been tested in detailed model calculations as yet although we are undertaking such calculations. The idea of Blake *et al.* (1987) is that the initial reaction involves water and methyl ion to form the protonated precursor of methanol:

$$CH_3^+ + H_2O \text{ -----> } CH_3OH_2^+ + h\nu \qquad (11)$$

which then reacts with electrons dissociatively to produce methanol. Once methanol and its protonated ion are produced, they can be utilized to generate the more complex organic species found in this source. For example, methyl formate can be formed by the reaction

$$CH_3OH_2^+ + H_2CO \text{ -----> } H_2COOCH_3^+ + H_2 \qquad (12)$$

followed by

$$H_2COOCH_3^+ + e \text{ -----> } HCOOCH_3 + H, \qquad (13)$$

although it must be mentioned that reaction (12) has not yet been studied in the laboratory. As another example, dimethyl ether can be formed by the studied reaction

$$CH_3OH_2^+ + CH_3OH \text{ -----> } (CH_3)_2OH^+ + H_2O \qquad (14)$$

followed by

$$(CH_3)_2OH^+ + e \text{ -----> } CH_3OCH_3 + H. \qquad (15)$$

The efficiency of the syntheses of Blake *et al.* (1987) depends upon the rate of reaction (11) which consumes the large amounts of water mixed into the quiescent medium. Radiative association reactions such as (11) are calculated to be most rapid at low temperatures and the relatively high temperature of the Compact Ridge source may make this reaction too slow to accomplish the production of large amounts of methanol. Since, however, large amounts of methanol are detected in this source, it is entirely possible that even if methanol is formed by some other mechanism, the more complex molecules are formed in the manner suggested by Blake *et al.* (1987). If so, another mechanism for the production of methanol must be determined, perhaps one involving the surfaces of dust particles.

5. Conclusions

Standard gas-phase, ion-molecule chemistry explains the syntheses of most gaseous molecules observed up to now in cool regions of dense interstellar clouds. However, the chemistry of star-forming regions is quite different because the physical conditions differ drastically from the low density, low temperature conditions present in the ambient interstellar medium. Observational evidence from sources such as the Orion Nebula indicate that three effects operate to change the chemistry from that observed in non-star-forming regions. These effects - evaporation of grain mantles, interstellar shock waves, and mixing of pre-stellar and stellar winds with the quiescent medium - may all occur in Orion. The evidence for grain mantle desorption is especially strong since in the so-called Hot Core source, the observed high abundances of simple saturated molecules such as ammonia and water as well as more complex saturated species such as ethyl cyanide (C_2H_5CN) can only be explained by hydrogenation on the surfaces of dust particles. In addition, the large abundance of deuterated species such as NH_2D as well as the possible identification of NHD_2 argue strongly for deuteration by deuterium atoms on the surfaces of the dust particles.

The chemical evidence for shock excitation in dense clouds is somewhat weaker since different shock models yield very different results because of their extreme dependence on uncertain physical parameters. In the Orion Nebula, shock excitation in the Plateau source is often indicated by the large line widths of the spectral features. The major chemical effect explainable by chemical shock models is the enhancement in the abundance of selected sulfur-bearing species. This enhancement is not as pronounced in the most recent shock model of Leen and Graff (1988) but is still evident for a few molecules, one of which (HS) has not yet been detected in the interstellar medium.

Finally, the evidence for the mixing of heated gas with the quiescent medium has not yet been subjected to a detailed quantitative test. Calculations to determine whether or not the richness of the Compact Ridge source in Orion in oxygen-containing organic molecules is due to the infusion of large amounts of water are currently being undertaken.

ACKNOWLEDGMENTS

The National Science Foundation (U. S.) is kindly acknowledged for support of this work via grant AST-8713151.

REFERENCES

Adams, N. G. and Smith, D. 1988, in *Rate Coefficients in Astrochemistry*, ed. T. J. Millar and D. A. Williams (Dordrecht: Kluwer), p. 173.

Anicich, V. G. and Huntress, W. T., Jr. 1986, *Ap. J. Suppl.*, **62**, 553.

Barlow, S. E., Dunn, G. H., and Schauer, K. 1984, *Phys. Rev. Letters* , **52**, 902 and **53**, 1610.

Bates, D. R. and Herbst, E. 1988, in *Rate Coefficients in Astrochemistry*, ed. T. J. Millar and D. A. Williams (Dordrecht: Kluwer), p. 17.

Blake, G. A., Sutton, E. C., Masson, C. R., and Phillips, T. G. 1987, *Ap. J.*, **315**, 621.

Brown, P. D., Charnley, S. B., and Millar, T. J. 1988a, *M. N. R. A. S.*, **231**, 409.

Brown, P. D., Charnley, S. B., and Millar, T. J. 1988b, in *Rate Coefficients in Astrochemistry* , ed. T. J. Millar and D. A. Williams (Dordrecht: Kluwer), p. 263.

Brown, P. D. and Millar, T. J. 1989a, *M. N. R. A. S.* , **237**, 661.

Brown, P. D. and Millar, T. J. 1989b, *M. N. R. A. S.* , submitted.

Boissé, P. 1989, *Astr. Ap.*, submitted.

Chieze, J. P. and Pineau des Forets, G. 1989, *Astr. Ap.*, in press.

Eiroa, C. and Hodapp, K.-W. 1989, *Astr. Ap.*, **210**, 345.

Elitzur, M. and Watson, W. D. 1978, *Ap. J. Letters* , **222**, L141 and **226**, L157.

Elitzur, M. and Watson, W. D. 1980, *Ap. J.*, **236**, 172.

Flower, D. R., Monteiro, T. S., Pineau des Forets, G., and Roueff, E. 1988, in *Rate Coefficients in Astrochemistry* , ed. T. J. Millar and D. A. Williams (Dordrecht: Kluwer), p. 271.

Gould, R. J. and Salpeter, E. E. 1963, *Ap. J.*, **138**, 393.

Gredel, R., Lepp, S., Dalgarno, A., and Herbst, E. 1989, *Ap. J.*, in press.

Green, S. and Herbst, E. 1979, *Ap. J.*, **229**, 121.

Greenberg, J. M. 1987, in *Astrochemistry*, ed. M. S. Vardya and S. P. Tarafdar (Dordrecht: Reidel), p. 501.

Hartquist, T. W., Oppenheimer, M., and Dalgarno, A. 1980, *Ap. J.*, **236**, 182.

Herbst, E. 1987a, in *Interstellar Processes*, ed. D. J. Hollenbach and H. A. Thronson, Jr. (Dordrecht: Reidel), p. 611.

Herbst, E. 1987b, *Ap. J.*, **313**, 867.
Herbst, E. 1988, in *Reviews in Modern Astronomy* , ed. G. Klare (New York: Springer-Verlag), p. 114.
Herbst, E. and Klemperer, W. 1973, *Ap. J.*, **185**, 505.
Herbst, E. and Leung, C. M. 1989, *Ap. J. Suppl.*, **69**, 271.
Herd, C. R., Adams, N. G., and Smith, D. 1989, *Ap. J. Letters*, in press.
Irvine, W. M., Goldsmith, P. F., and Hjalmarson, Å. 1987, in *Interstellar Processes* , ed. D. J. Hollenbach and H. A. Thronson, Jr. (Dordrecht: Reidel), p.561.
Leen, T. M. and Graff, M. M. 1988, *Ap. J.*, **325**, 411.
Léger, A., Jura, M., and Omont, A. 1985, *Astr. Ap.*, **144**, 147.
Levine, R. D. and Bernstein, R. B. 1974, *Molecular Reaction Dynamics* (New York: Oxford).
Marquette, J. B., Rebrion, C., and Rowe, B. R. 1989, *Astr. Ap.*, **213**, L29.
Mauersberger, R., Henkel, C., Jacq, T., and Walmsley, C. M. 1988, *Astr. Ap.*, **194**, L1.
Millar, T. J., Bennett, A., and Herbst, E. 1989, *Ap. J.*, **340**, 906.
Millar, T. J. and Herbst, E. 1989, *Astr. Ap.*, submitted.
Mitchell, G. F. 1984, *Ap. J.*, **287**, 665.
Mitchell, G. F. 1987, in *Astrochemistry* , ed. M. S. Vardya and S. P. Tarafdar (Dordrecht: Reidel), p. 275.
Shull, J. M. and Draine, B. T. 1987, in *Interstellar Processes* , ed. D. J. Hollenbach and H. A. Thronson, Jr. (Dordrecht: Reidel), p. 283.
Smith, D. 1989, private communication.
Stutzki, J., Stacey, G. J., Genzel, R., Harris, A. I., Jaffe, D. T., and Lugten, J. B. 1988, *Ap. J.*, **332**, 379.
Tielens, A. G. G. M. and Allamandola, L. J. 1987, in *Interstellar Processes*, ed. D. J. Hollenbach and H. A. Thronson, Jr. (Dordrecht: Reidel), p. 397.
Wagner, A. F. and Graff, M. M. 1987, *Ap. J.*, **317**, 423.
Walmsley, C. M., Hermsen, W., Henkel, C., Mauersberger, R., and Wilson, T. L. 1987, *Astr. Ap.*, **172**, 311.
Watson, W. D. 1976, *Rev. Mod. Phys.*, **48**, 513.
Winnewisser, G. and Herbst, E. 1987, *Topics Current Chem.*, **139**, 119.

Discussion

ZINNECKE: In some sense you discussed the effect of star formation on chemistry, but what is the effect of chemistry on star formation and fragmentation? In other words: is there chemically-induced fragmentation? For example, the fractional degree of ionization may depend on the chemistry, and this may affect clump formation.
HERBST: Your question poses an interesting topic for future research.

MATEO: If you need metals to form grains and you need grains to form H_2, then how do you cool zero metal clouds?
HERBST: You do not need grains to form H_2. Reasonable amounts of H_2 can be formed from processes starting from the recombination of H and e to form H^- and from the radiative association of H and H^+ to form H_2^+. Grains are required to form a gas that is *predominantly* H_2.

BOSS: Your model included gas-phase reactions to convert CO to organic molecules in dense clouds. Have you also considered other mechanisms such as Fischer-Tropsch-type reactions catalyzed by iron grains?
HERBST: The uncertainty in the exact make-up of the grain particles makes an understanding of the chemistry occurring on them difficult. Rather than include an uncertain grain chemistry, we have chosen to utilize gas phase models based on an extensive laboratory data base. The existence of grain surface chemistry can then, under certain circumstances, be inferred from deviations between our predictions and observations. An example of a region where ambient gas phase chemistry is inadequate is the Hot Core Region in Orion discussed in this review.

MOLECULAR CLOUDS CHEMISTRY WITH MIXING

J.P. CHIEZE
Commissariat à l'Energie Atomique
Centre d'Etudes de Bruyères-le-Châtel
Service P.T.N.
BP 12, F-91680 Bruyères-le-Châtel
France

G. PINEAU DES FORETS
DAMAp
Observatoire de Paris
F-92195 Meudon Principal Cedex
France

ABSTRACT. Turbulent motions are observed in molecular clouds which may prevent the achievment of equilibrium chemical abundances. The possibility of mass exchanges between the envelope and the dense cores of a cloud can lead to significant injection of ionized species in dense and UV shielded regions, opening new chemical chanels. We have calculated the steady state abundances of various molecules as a function of the mixing rate. High C/CO ratios and low H_2O abundances are obtained over a wide range of the mixing rate, pointing to a characteristic mixing time of the order of the free-fall time in cloud cores. This leads to put constraints on the efficiency of clump condensation by gravity alone.
As an alternative, preliminary results regarding clump formation and dynamics triggered by radiation field inhomogeneities are also presented.

1. Introduction

Molecular cloud chemistry, involving large reaction networks, is often devoted to dark clouds interiors, where the harvest of complex molecules is the richest. In these dense regions, photoionization and photodestruction processes are unimportant. As a result, calculated abundances of carbon are low, with C/CO ratios lower than 10^{-3}. This is in conflict with much higher values observed in various places at large optical depths (Phillips and Huggins 1981,Keene et al. 1985, Zmuidzinas, Betz and Goldhaber 1986, Genzel et al. 1988). Cosmic rays induced photodissociation of CO and other molecules has only limited effects (Sternberg et al. 1987, Gredel et al. 1987) but the suggestion of a clumpy medium pervaded by UV radiation (Stutzki et al. 1988, Boissé 1989) may account for the observations. Perhaps a less avoidable pitfall of closed box chemistry is that the calculated abundances of complex molecules is not a monotonic function of time, but reach a maximum at early times ($t \sim 3\ 10^5 yr$, Herbst and Leung 1989), and then decrease often by orders of magnitude below the values compatible with the

R. Capuzzo-Dolcetta et al. (eds.), Physical Processes in Fragmentation and Star Formation, 17–28.

observations. One is then forced to suppose that molecular clouds are less than one million years objects. Some difficulties further arise from the dependence of the early time abundances on their assumed initial values.

Another approach attempts to link the chemistry and the dynamical properties of clouds. Boland and de Jong (1982) examined the turbulent circulation of gas and dust between the core and the outer layers of a cloud, focusing on the formation and destruction of dust mantles by accretion and evaporation of molecules. It is commonly believed that gas cycling between dense clumps and their low density surroundings hangs heavily upon chemistry, preventing it from attaining equilibrium (Williams and Hartquist 1984, Goldsmith *et al.* 1986, Charnley *et al.* 1988a,b). Compared to classical chemical equilibrium, the mixing models result essentially in the injection of UV–exposed, and thus ionized gas in dense and dark regions.

The question is then twofold: what is the nature of the mixing processes, and what is, qualitatively and quantitatively, the influence of the related mixing rates on the chemical evolution. To some extent, the last question can be answered even in the absence of a precise picture of the actual mixing scenario. The single extra parameter entering chemistry will be the mixing time defined as:

$$t_{mix} = M / \mid \dot{M} \mid \tag{1}$$

where $\dot{M}$ is the mass injection rate in the region of mass M.

We have shown that large departures from equilibrium abundances require a mixing time smaller than 1 Myr. Thus, steady state non-equilibrium chemical abundances will be attained in regions with evolutionary time scales *larger* than this. The mixing model and results are presented in Section 2. The comparison of the molecular production of mixing models with observed abundances may single out a definite relation between the mixing rate and the physical properties (such as the density) of various regions, a precious indication regarding the nature of the mixing process(es). Any mixing scenario – connecting regions of substantially different densities – will be closely related both to the nature of the source of the turbulent energy which supports clouds and clumps against collapse, and to the dominant processes at the origin of the formation of molecular clumps.

Supposing that the dense clump phase is due to gravity which drives the collapse of the more diffuse material, we examine in Section 3 the constraints imposed by the chemical evolution on the gas condensation rate (see also Lioure and Chièze, this volume, for a more general discussion of gas condensation processes). The point in favour of this gravitational option is the fact that non-equilibrium chemistry at various densities brings out mixing times always of the order of the gravitational free-fall time. However, we present an alternative mechanism, where clump formation and "turbulent" motions are the gas response to inhomogeneities in the local radiation field intensity.

2. Mixing Model

We present in this section the simplest mixing model coupling the time dependent chemistry of different regions of a molecular clouds. We consider two media with total number densities $n^{(1)}$ and $n^{(2)}$ and temperatures $T^{(1)}$ and $T^{(2)}$ respectively. The chemical coupling is induced by a mass flow $\mid dM/dt \mid$, comming in and out the two regions which have then a constant mass. These opposit mass flows carry the chemical composition of the region from which they originate. A two zone model is characterized by two mixing times $t_{mix}^{(1)}$ and $t_{mix}^{(2)}$. (In a n–zones model, each zone may be connected to k adjacent ones, with k mixing times per zone). The time dependent thermodynamical evolution may be choosen isobaric in each region or isochoric. The following expressions are written for the later case, but we have checked that it differs from isobaric conditions only in very low density envelopes, near the threshold of thermal instability. The *chemical* net production rate of the specy X, with number density $n_X^{(i=1,2)}$, is:

$$\left(\frac{d}{dt} n_X^{(i)}\right)_{chem} = N_X^{(i)}, \tag{2}$$

where $N_X^{(i)}$ includes the production and destruction rates of species X in each region. The abundances $n_X^{(i)}$ evolve according to the mixing rate and local chemistry:

$$\left(\frac{d}{dt} n_X^{(i)}\right) = N_X^{(i)} - \left[n_X^{(i)} - \frac{\rho^{(i)}}{\rho^{(j)}} n_X^{(j)}\right] \frac{1}{M^{(i)}} \left|\frac{dM}{dt}\right| \tag{3}$$

with $i \neq j$. The temperature evolution is followed according to:

$$\frac{dT^{(i)}}{dt} = \frac{2}{3n^{(i)} k_B} \left[B^{(i)} - N^{(i)} k_B T^{(i)}\right] \tag{4}$$

where $N^{(i)} = \sum_X N_X^{(i)}$ is the total particle number density variation rate due to chemical reactions only, and $B^{(i)}$ the net heating rate per unit volume in the region (i). The heating and cooling rates we have adopted are described in Chièze and Pineau des Forêts (1987).

2.1 THE CHEMICAL NETWORK

The chemical networks are identical in each zone. Besides the modifications introduced by the mixing, the different regions evolve at a rate which depends upon the density and the local extinction A_v. We have chosen as initial composition in a given zone the equilibrium abundances calculated without mixing. But in fact, with mixing switched on, the final steady state abundances are *independent* of the initial ones, which may be just the elemental composition.

Most of the results presented here have been obtained with a chemical network including about 250 gas-phase reactions involving 50 atomic and molecular species, based on the dark cloud chemistry described in Pineau des Forêts, Flower and Dalgarno (1988). Photoreactions are listed in the Appendix of Pineau des Forêts et al. (1986, 1987). We have adopted the branching ratios of dissociative recombination reactions of Millar et al. (1988). The cosmic ray induced photodissociation processes are also included (Sternberg, Dalgarno and Lepp 1987, Gredel, Lepp and Dalgarno 1987), with a dust grain albedo of 0.5 and a cosmic ray ionization rate $\zeta = 10^{-17} s^{-1}$. We supposed that the CO photodissociation lines are completely shielded by H_2 molecules at optical depths $A_v \geq 0.3$ *mag* (van Dishoeck and Black, 1987). At smaller depths, we used the photodissociation rates given by van Dishoeck and Black (1988) in terms of the extinction. We have adopted a composition $n_C/n_H = 3.3\ 10^{-5}$ and $n_O/n_H = 4.25\ 10^{-5}$, which corresponds to depletion factors $\delta_C = \delta_O = 0.1$. The metal abundance is 1.5 10^{-8} (low metallicity gas).

2.2 THE CHEMICAL EVOLUTION

As an illustration, we first consider mass circulation between two regions. Their characteristics are listed in Table 1, along with the *equilibrium* C^0/CO and C^+/CO ratios. Two cases are considered, one for which the extinction in the dense region is moderate, the other for which the radiation field is negligible ($A_v = \infty$).

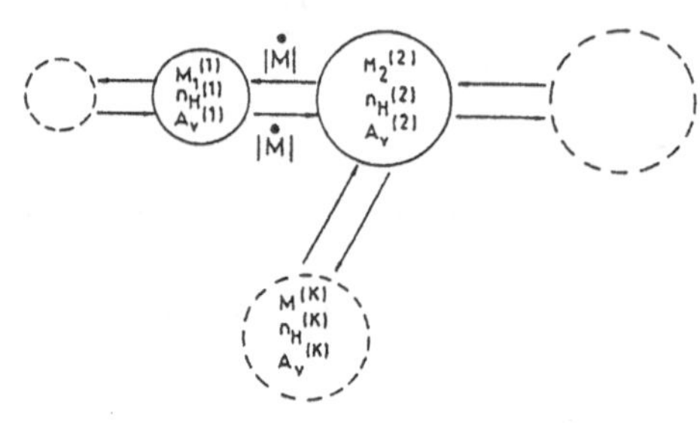

		(1) Core	(2) Envelope
Model (a)	n_H (cm^{-3})	10^4	10^2
	A_v	3.	0.5
	$\{C^0/CO\}_{eq}$	1.6 (−3)	9.2 (−4)
	$\{C^+/CO\}_{eq}$	5.2 (−4)	1.8
Model (b)	n_H (cm^{-3})	10^4	10^2
	A_v	∞	0.5
	$\{C^0/CO\}_{eq}$	1.4 (−3)	9.2 (−4)
	$\{C^+/CO\}_{eq}$	8.6 (−6)	1.8

Table 1: densities and extinctions of the two models (a) and (b) constituted by two interacting regions (1) and (2). The equilibrium C/CO ratios without mixing are also indicated.

The two regions (1) and (2) represent a clump embeded in the low density interclump phase. In the situation examined, the characteristic mixing time of the densest region is much shorter than that of the envelope. In other words, the later is viewed as a large reservoir of ionized low density gas.

A set of models is calculated for various mixing times in the range $10^{-2} Myr < t^{(1)}_{mix} < 1\ Myr$. In each case, the chemical evolution is followed until stationary

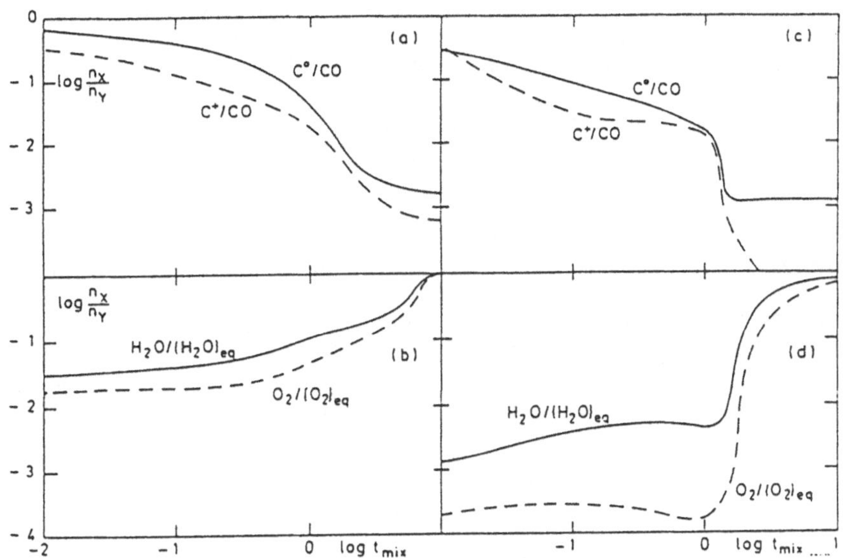

Figure 1 $\begin{smallmatrix} a & c \\ b & d \end{smallmatrix}$ *: calculated C/CO ratios in the "core" region of models (a) and (b) as functions of the mixing time* t_{mix} *in the core, expressed in units of* $10^6 yr$*. Model (a) on the left (1a,1b) and model (b) on the right (1c,1d). The corresponding abundances of* H_2O *and* O_2 *relative to their equilibrium values without mixing are also presented.*

state is attained. We report only on some of the most abundant species, C, CO, O_2, H_2O and some representative "complex" molecules, which feature the trends of the chemical evolution.

Fig. 1(a,c) show the stationary values of the C/CO ratio as a function of the mixing time. For $t_{mix}^{(1)} \leq 1\ Myr$, this ratio is one to two orders of magnitude greater than its equilibrium value without mixing. In moderately shielded clumps (model "a"), values $C^0/CO \geq 0.1$ and an enhencement of C^+ by more than a factor 50 are obtained for mixing times shorter than 6 $10^5 yr$. This is in agreement with the high C/CO ratios observed in various environments (Keene et al. 1985, Genzel et al. 1988, Zmuidzinas et al. 1988). It is interesting to notice that the required mixing time, in a clump with density $n_H \sim 10^4\ cm^{-3}$ is just of the order of the free-fall time.

Because O_2 and H_2O abundances are very sensitive to the ions circulation, they are strongly reduced for $t_{mix} \leq 10^6 yr$ (see Fig. 1b). These molecules are destroyed through ion–neutral reactions involving the principal ions, C^+, H^+ and

He^+, formed primarily in the low density gas. This behaviour is enhanced in model (*b*), with O_2 and H_2O stationary abundances reduced by three or four orders of magnitude relative to their equilibrium values. Thus, dynamical mixing with time scales shorter than $10^6 yr$ is in good position to reconcile the models and the stringent upper limits deduced from the non–detection of HDO (Olofson 1984, Moore et al. 1986) and $^{16}O^{18}O$ (Goldsmith et al. 1985, Liszt and Vanden Bout 1985).

More complex molecules, such as C_2H, C_3H_2, are exemples of "early time" molecules, which attain peak abundances after $t \sim 3\ 10^5 yr$ and then fall to much lower equilibrium values (Herbst and Leung 1989). In the present model, and contrary to O_2 and H_2O, the abundances of such molecules are greatly enhanced, depending on the mixing time. For exemple, the C_2H abundance is increased (relative to equilibrium) by a factor of 200 for $t_{mix} = 10^6 yr$ and more than 10^3 for $t_{mix} = 5\ 10^5 yr$. This is partly due to the presence of non negligible amounts of ionized carbon, injected by mixing, which combine with the abundant CH_4:

$$\begin{aligned} C^+ + CH_4 &\rightarrow C_2H_2^+ + H_2 \\ &\rightarrow C_2H_3^+ + H \end{aligned}$$

Therefore, mixing within clump and interclump material can reduce the difficulties of reproducing the observations. In particular, mixing models account *simultaneously* for:

(i) the large quantities of atomic carbon observed at large optical depths in molecular clouds. Appreciable amounts of C^+ still remain in dense clumps with an abundance relative to CO often larger 1%;
(ii) the low abundance of O_2 and H_2O predicted by measured upper limits: the ion flux formed in the UV–exposed external regions rapidly dissociates these two molecules;
(iii) the relatively high observed column density of "complex" molecules formed through ion–neutral reactions involving C^+.

The main point of this study is that coupling chemistry with mixing processes bring chemical models significantly closer to the observations. The required turn over time scales are less than 1 Myr. This may be relatively short, if one assumes that mixing is due to diffusive processes, as turbulent transport. The results presented so far concern a region of given density, $n_H = 10^4 cm^{-3}$. The same analysis has been performed for various densities in the range $10^3 cm^{-3} \leq n_H \leq 10^6 cm^{-3}$. Examination of the results show that the optimum mixing time is always close to the free-fall time of the region under consideration. It is thus tempting to adopt the point of view by which clump formation is driven by gravity, and examine the constraints imposed by the chemical model.

3. Mixing Dynamics

We propose in this section a model of the gravitational condensation of molecular gas from the mean density in a cloud towards the much higher densities of dense cores and clumps. We retain only the main features of the condensation model, which is described in Lioure and Chièze (this volume).

3.1 MIXING DRIVEN BY GRAVITY

We assume that gravity drives the condensation of cloud material at a rate $\omega = dLog\rho/dt$ equal to the asymptotic density growth rate ω_J of Jeans unstable modes of density ρ. Assuming a stationary state, the density distribution, defined as the mass of gas contained per unit logarithmic interval of the density, would be given by:

$$\frac{\partial M}{\partial Log\, n_H} = \frac{\Phi_0}{\omega_J} \tag{6}$$

where Φ_0 is the condensation mass flow (in $M_\odot Myr^{-1}$). This mass flow can be estimated from the total mass of molecular gas in the Galaxy, $M(H_2)$, the star formation rate $R_{SF} \sim (3.-10.) \times 10^6\ M_\odot Myr^{-1}$, and the star formation efficiency $r \sim 0.1$. This implies a destruction rate of molecular clouds equal to:

$$\begin{aligned} \left|\frac{dM(H_2)}{dt}\right| &= \frac{R_{SF}}{r} \\ &= \Phi_0 \\ &\sim (3. - 10.) \times 10^7 M_\odot Myr^{-1} \end{aligned} \tag{7}$$

We argue that a stationary state is achieved, among the molecular cloud population, through their disruption by massive star formation, which brings the cloud material to the warm phase of the interstellar medium. The cycle of the interstellar gas is completed by the efficient cooling of this warm phase which eventually condenses back to molecular clouds.

However, the simple choice of ω_J would picture a pure gravitational collapse of the molecular cloud population, which is not observed at the cloud scale, and further would overestimate the star formation rate by about *one* order of magnitude. This lead to a reduction of the condensation rate by the same factor, possibly due to the intrinsically 3–D nature of the collapse followed by virialization, or due to magnetic fields (Mouschovias, this volume). Thus we adopt $\omega = 0.1\ \omega_J$. We have seen previously that chemistry points to mixing rates precisely of the order of ω_J. This conflict is solved if the gravitational collapse of each clump, which starts at the rate ω_J is incomplete and that a fraction $z \sim 90\%$ of the collapsing flow is returned to its initial density.

According to this picture, condensations of molecular gas are continuously formed and destroyed, on the whole density scale inside a cloud. The small difference ($\sim 10\%$) between the condensation and destruction rates allows for the mass flow Φ_0, which drains the interstellar gas towards the formation of stars. The rate of the gravitational *condensation* flow:

$$\begin{aligned} \phi_G &= \omega_J \frac{\partial M}{\partial Log\, n_H} \\ &= \frac{1}{1-z}\Phi_0 \end{aligned} \tag{8}$$

is constant along the increasing density scale, anf, for the present purpose, practically balanced by the reverse flow. Following Eq.(6) and (8), the mass distribution along the density scale in a molecular cloud would be proportional to $n_H^{-1/2}$:

$$\frac{\partial M}{\partial Log\ n_H} \approx 250 \frac{\Phi_0}{\sqrt{n_H}} \tag{9}$$

where Φ_0 is expressed in $M_\odot Myr^{-1}$ and n_H in cm^{-3}.
According to this model, the dynamical scheme we adopt in the followings is characterized by a constant mass flux through each region, and a dynamical time scale $\omega_J^{-1} \approx 10/\sqrt{n_H}\ Myr$ depending upon the local density. These regions are represented by discrete density bins of mass:

$$\Delta M(n_H) = 250 \frac{\varphi_0}{\sqrt{n_H}} \tag{10}$$

where φ_0 is now the value of the condensation flux *per molecular cloud*. The explored density range, $10^2 cm^{-3} \leq n_H \leq 10^4 cm^{-3}$, is sampled by three interacting regions (I, II and III) presented in Table 2. It represents about 90% of the total mass of a 100 $M_\odot$ isolated molecular cloud with mean density $< n_H > \approx 100\ cm^{-3}$. Note that such a cloud may be the prototype of the most massive *homogeneous* clouds which are the building blocks of more developed molecular complexes (Falgarone and Pérault 1987, Chièze and Pineau des Forêts 1987).

Region	I	II	III
n_H (cm^{-3})	10^4	10^3	10^2
A_v	10	3	0.1
$M/M(I)$	1	3	10
T_{eq} (K)	15	32	125
$\{C/CO\}_{eq}$	1.4 (−3)	2.8 (−3)	12
$\{C^+/CO\}_{eq}$	8.5 (−6)	5.1 (−3)	4.6 (+4)

Table 2: Initial conditions in the three regions I, II and III discussed in the text. The equilibrium temperatures and C/CO ratios without mixing are also indicated.

According to Eq. (10), their respective masses are inversely proportional to the square root of their density. The mass flow model adopted here is a straightforward extension of the two zone mixing model already described. The results are presented in Table 3. Observed molecules such as C_2H, C_3H and C_3H_2 are

Region	I	II	III
n_H (cm^{-3})	10^4	10^3	10^2
C^0/CO	5.9 (−2)	1.4 (−1)	3.9
C^+/CO	1.4 (−4)	1.1 (−1)	1.1 (+4)
O_2/H_2	1.5 (−7)	1.7 (−9)	4.4 (−12)
H_2O/H_2	1.7 (−6)	4.8 (−8)	1.3 (−8)
C_2H/H_2	9.3 (−8)	6.4 (−8)	4.9 (−12)
C_3H/H_2	7.9 (−9)	5.1 (−8)	1.1 (−12)
C_3H_2/H_2	2.9 (−9)	1.9 (−8)	1.4 (−13)
$O_2/\{O_2\}_{eq}$	4.8 (−2)	2.8 (−1)	1.0
$H_2O/\{H_2O\}_{eq}$	2.8 (−1)	2.7 (−1)	1.0
$C_2H/\{C_2H\}_{eq}$	1.3 (+2)	2.6 (+2)	—
$C_3H/\{C_3H\}_{eq}$	1.8 (+3)	4.3 (+3)	—
$C_3H_2/\{C_3H_2\}_{eq}$	1.6 (+3)	4.6 (+3)	—

Table 3: Steady state C/CO ratios in the three regions of the model presented in Table 2. The abundances of some other molecules are also presented, normalized to H_2, and compared to their equilibrium abundances without mass exchange between the three regions

significantly enhanced, while H_2O and O_2 have low abundances. Large amounts of neutral carbon are also present, even in the densest and darkest region.

3.1 MIXING DRIVEN BY INHOMOGENEITIES OF THE RADIATION FIELD

Molecular clouds are clumpy objects. Density inhomogeneities may result in local fluctuations of the incident interstellar radiation field intensity, which in turn drive the formation of new clumps. Through photoelectric effect on dust grains, the UV radiation field dominates the gas heating sources at low and intermediate densities, up to a few $10^3 cm^{-3}$. In this density range, the gas pressure is thus a sensitive decreasing function of the extinction A_v of the radiation field. However, one expects that gas layers at a given level in a cloud are approximately isobaric. This results in local condensations of the gas in obscured regions. We have examined the response of the molecular gas to such extinction variations (de Boisanger et al. 1989). In this preliminary calculations, the idealized model consists in a two dimensional, slab geometry, molecular gas layer initially in hydrostatic equilibrium in a constant gravitational field. A finite perturbation of the extinction (ΔA_v of the order of +1 mag) is applied in a finite region. The gas response is followed by a 2-D hydrodynamical code, with the gas heating and cooling terms included.

Quite naturally, the shielded gas condenses by at least one order of magnitude. The characteristic cooling time is quite short, of the order of $10^{-3} Myr$. The perturbation of the pressure field sets matter in motion, with sonic velocities. This tends to smooth the disturbed pressure field, but gas condensations are swept along the (uniform) gravitational field by the Archimedes buoyant force. Thus, the formation of gas condensations is bound to the appearance of a sonic velocity field, which may ultimately decay to turbulence. In the numerical model, the gas is confined in a box with an upper free surface, maintained at the constant interstellar mean pressure. In the stationary regime, the gas is somewhat artificially forced to convection–like bulk motions. In more realistic conditions, clumps formed in a shielded region may be destroyed as they leave it, unless they are gravitationally bound. This may account for the gas cycling introduced in Section 2. However, an estimation of the characteristic cycling and mixing times requires more elaborated models of molecular clouds.

4. Conclusion

Mixing of fresh, ionized, envelope material with denser regions of a molecular cloud modifies the chemical flow diagram pertinent for closed–box chemistry, leading to high C/CO abundance ratios, high abundances of some multi–carbon molecules, together with low abundances of H_2O and O_2. The time scale introduced by the mixing processes also limits the efficiency of chemistry. In this respect, the chemical evolution bears some resemblance with the so–called "early time" chemistry. However, mixing models open new chemical channels, which may improve the comparison with the observed abundances. They lead to stationary models, independent of the initial abundances, which do not introduce a limitation on the age of molecular clouds, of the order of:

$$t_{cloud} = \frac{M(H_2)}{\Phi_0} \sim 70\ Myr \tag{11}$$

High C/CO values are obtained even in high density ($n_H = 10^4 cm^{-3}$), shielded regions ($A_v \geq 10\ mag$), provided that the time spent by the gas at any density ρ is of the order of the gravitational time scale $t_G = (4\pi G\rho)^{-1/2}$. This suggest that the cloud material is pushed to collapse by gravity. However, in order to match the star formation rate in the Galaxy, the collapse should be incomplete, with the bulk of the material returning to lower densities, where mixing occurs. Moreover, fluctuations of the UV field intensity in the envelope of molecular clouds may trigger the formation and the destruction of gas condensations, leading again to the mixing of material chemically processed at different densities and extinctions.

References

de Boisanger, C., Chièze, J.P., Besnard, D., Gambart, J.: 1989, (in preparation)
Boissé, P.: 1989, *Astron. Astrophys.* (in press)
Boland, W., de Jong, T.: 1982, *Astrophys. J.* **261**, 110
Charnley, S.B., Dyson, J.E., Hartquist, T.W., Williams, D.A.: 1988a, *Monthly Notices Roy. Astron. Soc.* **231**, 269
Charnley, S.B., Dyson, J.E., Hartquist, T.W., Williams, D.A.: 1988b, *Monthly Notices Roy. Astron. Soc.* **235**, 1257
Chièze, J.P., Pineau des Forêts, G: 1987, *Astron. Astrophys.* **183**, 98

Falgarone, E., Pérault, M.: 1987, in *Protostars and Molecular clouds*, eds. T. Montmerle, C. Bertout, Edition Doc Cen Saclay 87–257, p.15
Genzel, R., Harris, A.I., Jaffe, D.T., Stutzki, J.: 1988, *Astrophys. J.* **332**, 1049
Goldsmith, P.F., Langer, W.D., Wilson, R.W.: 1986, *Astrophys. J.* **303**, L11
Goldsmith, P.F., Snell, R.L., Erickson, N.R., Dickman, R.L., Schloerb, F.P., Irvine, W.M.: 1985, *Astrophys. J.* **289**, 613
Gredel, R., Lepp, S., Dalgarno, A.: 1987, *Astrophys. J.* **323**, L137
Herbst, E., Leung, C.M.: 1989, *Astophys. J. Suppl.* **69**, 271
Keene, J., Blake, G.A., Phillips, T.G., Huggins, P.J., Beichman, C.A.: 1985, *Astrophys. J.* **299**, 967
Liszt, H.S., Vanden Bout, P.A.: 1985, *Astrophys. J.* **291**, 178
Millar, T.J., DeFrees, D.J., McLean, A.D., Herbst, E.: 1988, *Astron. Astrophys.* **194**, 250
Moore, E.L., Langer, W.D., Huguenin, G.R.: 1986, *Astrophys. J.* **306**, 682
Olofsson,H.: 1984, *Astron. Astrophys.* **134**, 36
Phillips, T.G., Huggins, P.J.: 1981, *Astrophys. J.* **251**, 533
Pineau des Forêts, G., Flower, D.R., Dalgarno, A.: 1988, *Monthly Notices Roy. Astron. Soc.* **235**, 621
Pineau des Forêts, G., Flower, D.R., Hartquist, T.W., Dalgarno, A.: 1986, *Monthly Notices Roy. Astron. Soc.* **220**, 801
Pineau des Forêts, G., Flower, D.R., Hartquist, T.W., Millar, T.J.: 1987, *Monthly Notices Roy. Astron. Soc.* **277**, 562
Sternberg, A., Dalgarno, A., Lepp S.: 1987, *Astrophys. J.* **320**, 676
Stutzki, J., Stacey, G.J., Genzel, R., Harris, A.I., Jaffe, D.T., Lugten, J.B.: 1988, *Astrophys. J.* **332**, 379
van Dishoeck, E.F., Black, J.H.: 1987, in *Physical Processes in Interstellar Clouds*, eds. G.E. Morfill, M. Schoeler, Reidel, Dordrecht, p. 241
van Dishoeck, E.F., Black, J.H.: 1988, *Astrophys. J.* **334**, 771
Williams, D.A., Hartquist, T.W.: 1984, *Monthly Notices Roy. Astron. Soc.* **210**, 141
Zmuidzinas, J., Betz, A.L., Boreiko, R.T., Goldhaber, D.M.: 1988, *Astrophys. J.* **335**, 774
Zmuidzinas, J., Betz, A.L., Goldhaber, D.M.: 1986, *Astrophys. J.* **307**, L75

Questions

Palla: The depletion of H_2O and O_2 that you find in the models of chemical mixing would imply an enhancement of the OH abundance: is that in agreement with observations? (It is my feeling that the fact that H_2O is "depleted" results from the branching ratio between OH and H_2O that you used in the chemical network).

Chieze: The branching ratio of H_3O^+ dissociative recombination to OH recently measured by Herd, Adams and Smith (1989, *Astrophys. J.* (in press)) is 0.65. In the present calculation it was taken to be 0, according to the prediction of Bates (1986, *Astrophys. J.* **307**, L45). Of course, this branching ratio has a direct influence on H_2 abundance calculated at chemical equilibrium as well as in the mixing models.

Nevertheless, it has a very small influence on OH abundances, which is also formed through other channels (dissociation of H_2O reacting with He^+ or through secondary photons induced by cosmic rays) and is strongly destroyed owing to the high rates of the neutral–neutral reactions $OH + O \rightarrow O_2 + H$ and $OH + C \rightarrow CO + H$ (see Smith 1988, in *Rate Coefficients in Astrochemistry*, eds. T.J. Millar and D.A. Williams, (Dordrecht:Kluwer Academic Publishers)). Incidentally, the chemical mixing has practically *no* influence on the OH abundance whatever the mixing time is.

Lizano: What is the magnitude of the UV radiation flux you used?
Chieze: The mean interstellar radiation field we used is:
$u_\nu/h\nu = 2 \cdot 10^{-8}\ photon\ cm^{-2}\ s^{-1}\ Hz^{-1}$ at $1000A$ (Habing, 1968, *Bull. Astron. Inst. Neth.* **19**, 421). The mean ionizing radiation field of carbon adopted in the present calculation is: $\Phi_0 = 3 \cdot 10^7\ cm^{-2}s^{-1}$ in the range $10 - 13.6\ eV$ (de Boer *et al.*, 1973, *Astron. Astrophys.* **28**, 145).

Kiguchi: What determines the extinction, grain or molecular lines? How do you treat the radiative transfer? I think that the molecular process in mixing region is very complex and the radiative transfer in molecular lines modifies indeed the molecular abundance.
Chieze: The UV extinction is taken as a parameter characteristic of each zone of the model, based upon spherical, hydrostatic self–gravitating cloud models calculated in Chieze and Pineau des Forets (1987, *Astron. Astrophys.* **183**, 98). In spherical geometry, the extinction is calculated as: $A_v = 6.7x10^{-22} \int_r^{R_s} n_H dr$ (Bohlin, Savage, and Drake, 1978, *Astrophys. J.* **224**, 132). Regarding photodissociation of CO, we have supposed that it is completely shielded by H_2 for $A_v \geq 0.3\ mag$. For smaller extinctions, we used the extinction dependent photodissociation rates given by van Dishoeck and Black (1988, *Astrophys. J.* **334**, 771).

COLLAPSE AND FRAGMENTATION OF MOLECULAR CLOUDS

M.KIGUCHI
Research Institute for Science and Technology
Kinki University
3-4-1 Kowakae, Higashi-Osaka-city, 577 Osaka
Japan

ABSTRACT. When the geometry of a cloud configuration is not spherical, long range property of gravity induces varies unexpected behavior. In many cases, therefore, simple consideration leads us to a wrong result on the problems of star formation. To obtain detailed feature of gravity in star formation, we studied numerically the behavior of gravity for rotating isothermal clouds. We constructed equilibrium models extensively, and studied the evolution of non-axisymmetric perturbation.

1. Introduction

To make clear the physical process essential in star formation, we have studied numerically physical characteristics of rotating isothermal clouds embedded in external media, forming project team led by C. Hayashi of Kyoto University, with S. Narita of Doshisha University and M.S. Miyama of Kyoto University. This project points firstly to clarify the equilibrium (Kiguchi, Narita, Miyama and Hayashi (1987)) and stability (Miyama, Narita, Kiguchi and Hayashi (1989)) of rotating clouds, secondly to clarify the time scale of angular momentum transport in protostellar disks by magnetic field (as for the qualitative discussion, see Hayashi(1981)), and finally to form a total view of star formation. In this paper, I review the equilibrium and stability of rotating clouds.

2. Gravity in various configurations

In the process of star formation, various phenomena such as collapse and fragmentation occur in a cloud. These varieties are induced by the long range force of gravity coupled with their geometry. To analyze star forming process, therefore, we should know the properties of gravity in various geometrical configurations.

Usually, these properties are described in terms of a dispersion relation (for a review, see Larson (1985)). From these linear analyses, we know that when a thin disk or fine filament is formed, it soon fragments and reduces the gravitational energy. The scale of fragment is 2π times the scale height of the disk or the typical radius of the cylinder.

Here, according to Hayashi (1989, private communication), we analyze the gravity of a cloud in various configuration using virial technique. We assume that the cloud is embedded in media with uniform pressure P_{ext}.

According to cloud geometry, the integrated mass, which is the source term of gravity, has various forms. For the *spherical* configuration, it is given by

$$dm = \rho d(\frac{4}{3}\pi r^3), \tag{1}$$

R. Capuzzo-Dolcetta et al. (eds.), Physical Processes in Fragmentation and Star Formation, 29–34.

where ρ is the mass density, r is the distance from the center of the cloud. For the *cylindrical* configuration, it is the line mass m_l given by

$$dm_l = \rho d(\pi r^2), \tag{2}$$

where r is the radial distance from the axis of the cylinder. For the *disk* configuration, it is the surface mass m_s given by

$$dm_s = \rho dz, \tag{3}$$

where z is the height from equatorial plane of the cloud. From these equations, it seems adequate to describe the geometry by the parameter x, where $x = 4\pi r^3/3$ for the sphere, $x = \pi r^2$ for the cylinder, and $x = z$ for the disk. Using this parameter, gravitational equilibrium is described for the case of spherical configuration by

$$\frac{1}{\rho}\frac{dP}{dx} = -\frac{1}{3}\sqrt[3]{\frac{4\pi}{3}}\frac{Gm}{x^{4/3}}, \tag{4}$$

for the case of cylindrical configuration by

$$\frac{1}{\rho}\frac{dP}{dx} = -\frac{Gm_l}{x^1}, \tag{5}$$

and for the case of disk configuration by

$$\frac{1}{\rho}\frac{dP}{dx} = -4\pi\frac{Gm_s}{x^0}. \tag{6}$$

From these equation, especially from the power of the variable x, we see that the gravity of a cylinder is similar to a sphere rather than a disk. The M - P_c/P_{ext} relation for the isothermal cylinder just corresponds to the M - P_c/P_{ext} relation for the polytropic sphere with the polytropic index $n = 5$, where M is the total line mass and P_c is the pressure at the center of the cloud.

The equation of motion for a cylinder is given by

$$\pi r^2\frac{\partial r}{\partial t} = x\frac{\partial P}{\partial m} - Gm_l. \tag{7}$$

Integrating this equation over m, we get a virial relation for the equilibrium of a cylinder:

$$R = \frac{\bar{c}^2 M}{\pi p_{ext}}\left(1 - \frac{GM}{2\bar{c}^2}\right), \tag{8}$$

where c is the average sound velocity given by $\bar{c}^2 = (1/M)\int_0^M (p/\rho)dm$ and R is the radius of the cylinder. From this equation, we see clearly that the line mass of a cylinder is limited. The critical line mass is given by $M_{crit} = 2\bar{c}^2/G$.

3. Equilibrium of rotating isothermal clouds

In the 1970's, numerical experiments of collapse of rotating clouds are carried out extensively. Validity of the results obtained by numerical experiments was, regretfully, inconclusive. At that time, the knowledge about an equilibrium configuration was insufficient, and we couldn't judge the validity of numerical experiments. Under these circumstances, Hayashi, Narita and Miyama (1982) found analytic solutions for equilibrium of rotating isothermal clouds. Thus, we have a concrete ground for discussion about the collapse.

The analytic solutions are solutions for a infinite system but a realistic isothermal cloud has a finite volume which is embedded in external media such as an HII region. The range of gravity is long, so that it is possible that the property of realistic rotating isothermal clouds differs completely. In these analytic solutions, average mass density enclosed in a spherical shell is not so large, and there are two configurations with same mass and different flatness. We wanted to know how change these properties in a realistic cloud.

In 1983, Stahler constructed models of isothermal rotating clouds. The essence of his method was in the fact that he fixed the central density during iterative calculation to kill the overall collapse mode. We know, thus, how we can construct unstable configurations, which is essential for the study of stability.

The basic equation we solved is

$$\frac{1}{\rho}\mathrm{grad}P + \mathrm{grad}\phi_g - \mathrm{grad}\phi_c = 0, \tag{9}$$

where ρ is the mass density, P is the pressure, ϕ_g is gravitational potential which is given by Poisson equation

$$\Delta\phi_c = 0, \tag{10}$$

and ϕ_c is the centrifugal potential given by

$$\phi_c(\varpi) = \left(\frac{J}{M}\right)\int_0^{\varpi} d\varpi' \frac{j^2(\varpi')}{\varpi'^3}, \tag{11}$$

where M is the total mass and J is the total angular momentum and ϖ is the distance from rotation axis and j is the angular momentum distribution.

We constructed equilibrium configurations for various angular momentum distributions and obtains following result (Kiguchi, Narita , Miyama and Hayashi 1987):

1. The clouds with central density ρ_c greater than $800P_{ext}/c^2$ are unstable to global contraction or expansion, where P_{ext} is the pressure of surrounding media and c is the sound velocity of the isothermal gas.
2. The clouds with rotation energy greater than 0.44 times the gravitational energy are unstable to ring formation
3. The maximum mass of axially stable clouds is 31 times that of nonrotating clouds.
4. The maximum mean rotation velocity of clouds is $2.7c$.
5. The maximum mean density of clouds is $6P_{ext}/c^2$.
6. For each cloud, the maximum height of the boundary surface from the equatorial plane is nearly given by $0.3c^2/\sqrt{GP_{ext}}$.

4. Rotation-supported equilibrium

If we could assume that an extremely rotating isothermal cloud is an ellipsoid in a good approximation, we could get equilibrium properties of the cloud rather simply. In this approximation, total energy of the cloud is described in terms of two nondimensional parameters,

$$A = \frac{4\pi}{3}\frac{a^3G^3M^2P_{ext}}{c^8} \quad \text{and} \quad B = \left(\frac{Jc}{kGM^2}\right)^2, \tag{12}$$

where a is the length of major axis and k is a parameter of order unity which describes the mass density distribution in the cloud. The equilibrium and the stability of the cloud is found from the minimum of the total energy, and this approximation shows that there are two stable equilibria with same values of A and B, one of which is an extremely flattened configuration.

Our calculation, which is described in the previous section, does not show any evidence of a cloud in which rotational energy is dominated more than ten times over the thermal energy. This is contrary to the result obtained under the assumption of ellipsoid. In rotating clouds, the density distribution is never similar to the ellipsoid, and the property of gravity in an extremely flattened cloud changes drastically.

Hachisu and Eriguchi (1985a,b), however, claims the existence of two equilibria with the same mass and the different angular momentum, and, moreover, Tohline (1985) claims the phase transition between these two configurations. Then, we constructed carefully extremely flattened configuration of a cloud (Narita, Kiguchi, Miyama and Hayashi, 1989).

We searched equilibrium configurations using another method from the self-consistent field method. In the usual self-consistent field method, pressure is variated in the iteration. It is, therefore, easy to find pressure-supported configuration but difficult to find rotation-supported configuration. Thus, we variated gravity to make easy to find rotation-supported configuration.

As the configuration we searched is extremely flattened, we used thin-disk approximation. The validity of this assumption is checked by the calculation described in the previous section. In this approximation, we assume that the density distribution pararell to rotation axis at distance ϖ from the axis is given by

$$\rho(\varpi, z) = \rho_0(\varpi)\mathrm{sech}^2\left[\frac{\sqrt{2\pi G\rho_0(\varpi)}z}{c}\right], \tag{13}$$

and assume that the gravity in the ϖ direction is given by

$$F_g(\varpi) = -\frac{2G}{\varpi}\int_0^R d\varpi'\sigma(\varpi')\varpi'\left[\frac{K(\frac{2\sqrt{\varpi\varpi'}}{\varpi+\varpi'})}{\varpi+\varpi'} + \frac{E(\frac{2\sqrt{\varpi\varpi'}}{\varpi+\varpi'})}{\varpi-\varpi'}\right], \tag{14}$$

where a is the height from the equatorial plane, ρ_0 is the density on equatorial plane, σ is the surface density, and K and E is the complete elliptic integrals of the 1-st and 2-nd kinds, respectively.

To summarize the calculation, we must discriminate concepts which have been confused in the work by Hachisu-Eriguchi and Tohline. The equilibria of rotating clouds are classified into two state, i.e. pressure-supported and rotation-supported state. Tohline's classification does not correspond to this. He classifies the equilibrium into two state, i.e. diffuse and compact state. We will define these concepts as follows: The pressure-supported configuration is defined by $E_{rotation} < 10 \times E_{thermal}$, where E is the energy. The rotation-supported configuration is defined by $E_{rotation} >> 10 \times E_{thermal}$. The diffuse configuration is defined by $\bar{P} < 10 \times P_{ext}$, and the compact configuration is defined by $\bar{P} >> 10 \times P_{ext}$.

Using these concepts, we can summarize our calculation as follows:

1. For a value of J and P_c, there are at least three types of configurations other than a stable configuration.
2. There is only one stable configuration for a value of J and P_c.
3. Stable configuration is a pressure-supported configuration.
4. Pressure-supported configuration is necessary diffuse configuration.
5. Phase transition between diffuse and compact configuration is impossible, because the masses of the clouds with same angular momentum in these two configurations are very much different.

5. Non-axisymmetric instability and angular momentum transfer

In the previous sections, we assumed that the cloud is axisymmetric. In reality, when rotation energy is large compared with gravitational energy, we know that the non-axisymmetric instability is induced. It is interesting to know how the non-axisymmetric instability occurs. We calculated dynamical evolution starting from an axisymmetric equilibrium using three dimensional hydrodynamic code (Miyama, Narita, Kiguchi and Hayashi 1989).

From these calculation, we found that the evolution is classified by the β value, where β is the ratio of rotation energy to gravitation energy. The critical values of classification just correspond to the critical values of instability of the Maclaurin spheroid:

1. A cloud initially with $\beta < 0.27$ is stable against non-axisymmetric perturbation. The maximum mass of stable clouds against this perturbation is $4c^4/\sqrt{G^3 P_{ext}}$
2. In a cloud with $0.27 < \beta < 0.30$, two-armed spiral appears.
3. In a cloud with $0.30 < \beta < 0.46$, a large number of arms appears in the course of evolution.

For a typical case, we will explain the evolution. The cloud we take has initially β value just above the onset of dynamical bar instability of Maclaurin spheroid. This evolution proceeds as follows:

1. At $t = 15.8t_*$, where t_* is the free fall time at boundary of the cloud, bar-like pattern appears. The bar begins to be winded.
2. At $t = 23.6t_*$, two spiral arms appear. Then, spirals extend. Maximum density of the cloud oscillates just as in equilibrium.
3. At $t = 31.2t_*$, maximum density increases rapidly, i.e., begins to collapse.

We can interpret this as follows:

1. Collapse begins because angular momentum is transferred from center to envelope.
2. Angular momentum transfer begins after the spiral appears. The cause of transfer is the gravitational torque. The time scale is about 2.2 times the pattern rotation period.

References

Hachisu, I. and Eriguchi, Y. (1985), 'Equilibrium structures of rotating isothermal gas clouds. I', Astron. & Astrophys. **143**, 355.

Hachisu, I. and Eriguchi, Y. (1985), 'Equilibrium structures of rotating isothermal gas clouds. II. Dependence on the angular momentum distribution', Astron. & Astrophys. **147**, 13.

Hayashi, C. (1981), 'Structure of the solar nebula, growth and decay of magnetic fields and effects of magnetic and turbulent viscosities on the nebula', Prog. Theor. Phys. Suppl. **70**, 35.

Hayashi, C.,Narita, S. and Miyama, S.M. (1982), 'Analytic solutions for equilibrium of rotating isothermal clouds. One-parameter family of axisymmetric and conformal configurations', Prog. Theor. Phys. **68**, 1949.

Kiguchi, M., Narita, S., Miyama, S.M. and Hayashi, C. (1987), 'The equilibria of rotating isothermal clouds', Astrophys. J. **317**, 830.

Larson, R.B. (1985), 'Cloud fragmentation and stellar masses', Mon. Not. R. astr. Soc. **214**, 379.

Miyama, S.M., Narita, S., Kiguchi, M. and Hayashi, C. (1989), 'Non-axisymmetric instability of rotating isothermal equilibria', submitted to Astrophys. J..

Narita, S., Kiguchi, M., Miyama, S.M. and Hayashi, C. (1989), 'Rotation-dominant equilibria of isothermal clouds', submitted to Mon. Not. R. astr. Soc..

Stahler, S.W. (1983), 'The equilibria of rotating, isothermal clouds. I. Method of solution', Astrophys. J. **268**, 155.

Stahler, S.W. (1983), 'The equilibria of rotating, isothermal clouds. II. Structure and dynamical stability', Astrophys. J. **268**, 165.

Tohline, J.E. (1985), 'Star formation: phase transition, not Jeans instability', Astrophys. J. **292**, 181.

Discussion

A. Boss: After Steven Stahler found the pressure-supported diffuse equilibrium for rotating isothermal clouds, and Joel Tohline used the virial method to hypothesized the existence of compact, rotationally-supported equilibria (supported by detailed models by Eriguchi and Hachisu), Tohline hypothesized that phase transitions between these two equilibrium might occur. Could you explain why you have found that these phase transitions are impossible.

M. Kiguchi: As we doubted the validity of the models constructed by Hachisu and Eriguchi, we have also constructed extremely rotating models carefully.

Gravity is determined by the global mass distribution (See Eq.(14)). Except rare cases such as the spherical and the spheroidal mass distribution, gravity at a point on a shell is determined by the mass lying outer than the shell as well as lying inner than the shell. The joining of solutions obtained in the inner region with the one obtained in the outer region is, therefore, not allowed in principle in general case. Because there is a few case when the joining of solutions is allowed, and the simple virial approximation gives qualitatively same result as the one obtained by the joining procedure, the possibility is left that the joining procedure is allowed practically. To get correct answer, therefore, it is necessary to construct models, taking into account the mathematical properties of gravity. We have carried out this.

T. Ch. Mouschovias: You used an integral equation for the computation of the gravitational field, if I understand you correctly. What did you do at the cloud boundary, where that approach can cause numerical problems depending an how precisely the column density of the cloud varies?

M. Kiguchi: We did not use an integral equation. We used differential equation. At first stage of our project, we have used the integral method, but the integral method causes difficult problems as Mouschovias claimed. We,therefore, abandoned the integral method finally.

In original self-consistent field method by Ostriker and Mark, they calculated gravity from a given mass distribution by integral, but in our method, we solved Poisson differential equation by SOR method. The boundary condition is given on a sphere far from the cloud.

Even when the mass distribution is far from the spherical symmetry, gravity is almost spherical. This is an important property of gravity. Therefore, it is sufficient to impose the boundary condition that the gravity is given by $-GM/r$ at a point far from the mass.

T. Ch. Mouschovias: You stated that the time scale for redistribution of angular momentum by gravitational torque, once a spiral pattern develops, is a few times longer than the rotation period. Since a cloud (or filament) must lose several orders of magnitude of angular momentum for stars to form, you would wait for many of your e-folding times for gravitational redistribution of angular momentum to resolve this problem. Could you not then conclude that gravitational torques cannot resolve the angular momentum problem of star formation?

M. Kiguchi: I admit you. We are interested just in the mechanism how the non-axisymmetric instability is induced.

We are now calculating the angular momentum transfer rate by the magnetic field in a partially ionized rotating disk.

ON THE ROLE OF MOLECULAR HYDROGEN IN FORMATION AND EVOLUTION OF BLUE COMPACT DWARF GALAXIES AND GIANT HII-REGIONS

YU. I. IZOTOV
Main Astronomical Observatory of the Academy of Sciences of the Ukrainian SSR, Kiev, USSR

ABSTRACT. The universal mechanism controlling the thermal evolution of the gas clouds, namely the molecular hydrogen formation, is found. The cloud is effectively cooled from temperature $T \simeq 10^4$ K down to $T < 10^3$ K. The principal possibility has been shown to exist for the blue compact dwarf galaxies formation from matter with primeval chemical composition. In the galaxies with large heavy element abundances $Z > 0.2\ Z_\odot$ the star formation regions fragment on the clouds with masses $M \simeq 10^3 - 10^5\ M_\odot$ due to thermal instability. The star formation regions in the galaxies with $Z < 0.2\ Z_\odot$ are more homogeneous and have the masses $\simeq 10^6 - 10^7\ M_\odot$.

1. INTRODUCTION

It is established from observations that there are no giant galaxies at the formation stages in the vicinity of Galaxy. Probably, the young giant galaxies with redshifts $z \simeq 3$-5 may successfully be observed with cosmic Hubble telescope.

However, one class of galaxies, i.e. blue compact dwarf galaxies (BCDGs) are characterised with active star formation, low luminosities ($M_B \simeq -14^m \div -17^m$), large ultraviolet excesses ($U-B \simeq -0.6^m$), compactness (with linear sizes about 1 kpc), large gas content (10 ÷ 75% of total mass), and low content of heavy elements (1/40 ÷ 1/2 of solar value (Kunth (1987), Dufour (1986), Kunth and Sargent (1986), Hunter and Gallagher (1986)). Surface photometry of the blue compact dwarf galaxies in optical and near infrared ranges (Kunth et al. (1986), Loose and Thuan (1986)) has revealed that in the most cases they contain stars of the late spectral types and, probably, they have been subjected to star formation processes more than once. However, some galaxies are known to be observed with lack of the old star population and they may really be the young galaxies which undergo the first burst of star formation.

R. Capuzzo-Dolcetta et al. (eds.), Physical Processes in Fragmentation and Star Formation, 35–47.

To promote progress in solving the problem of the origin of the blue compact dwarf galaxies and giant HII-regions, the study of the thermal and dynamical evolution of the contracting clouds is necessary.

In this paper we describe some results of the thermal, dynamical, and chemical evolution of the gas clouds with chemical composition varying from initial ($Z = 0$) to the solar value ($Z = Z_\odot$).

It is a matter of common observations that the thermal evolution of the matter with primeval chemical composition at $T < 10^4$ K is determined by molecular hydrogen. In present paper in contrast to the papers published earlier, the study of the thermal and chemical evolution of the free contracting cloud, which undergoes the effects of radiation from already formed galaxies is carried out. The presence of the heavy elements may essentially influence the cloud contraction because of additional processes, such as gas cooling on heavy elements,molecular hydrogen formation on the dust grains, gas heating by cosmic rays which are the products of the previous star generations. Finally, the possibility of the contracting cloud fragmentation due to thermal instability is examined.

2. THERMAL AND CHEMICAL EVOLUTION OF THE CLOUD WITH PRIMEVAL CHEMICAL COMPOSITION

To begin with we consider thermal and chemical evolution of the clouds in absence of heavy elements. The cloud model which is determined in such a way is applied to solve the problem of the BCDG formation. At the time $t = 0$ the cloud temperature and number density are equal to $T \simeq 10^4$K and $N = 10^{-2}$ cm^{-3} respectively. Under such conditions the Jeans mass equals $10^8 - 10^9$ $M_\odot$, corresponding to neutral hydrogen mass and total mass of BCDGs. The main reactions of the molecular hydrogen formation in primeval matter are as follows:

$$H + e \longrightarrow H^- + \gamma \qquad\qquad H + H^+ \longrightarrow H_2^+ + \gamma$$
$$\text{and} \qquad\qquad (1)$$
$$H^- + H \longrightarrow H_2 + e \qquad\qquad H_2^+ + H \longrightarrow H_2 + H^+ .$$

The external radiation hinders the molecular hydrogen formation in two ways:

1) by destruction of ions H^- and H_2^+ ;

2) the photons at $\lambda \simeq 912 - 1100$ Å make the H_2 molecules to dissociate.

In the latter case the radiation line absorption by H_2 electronic transitions has been taken into account. In the present paper the homogeneous cloud contraction is under consideration. But the cloud collapse is not really homogeneous and is followed by more fast density growth in the cloud centre. Therefore, the bulk of the radiation is suggested to be absorbed in central dense part of the cloud, and in calculation of

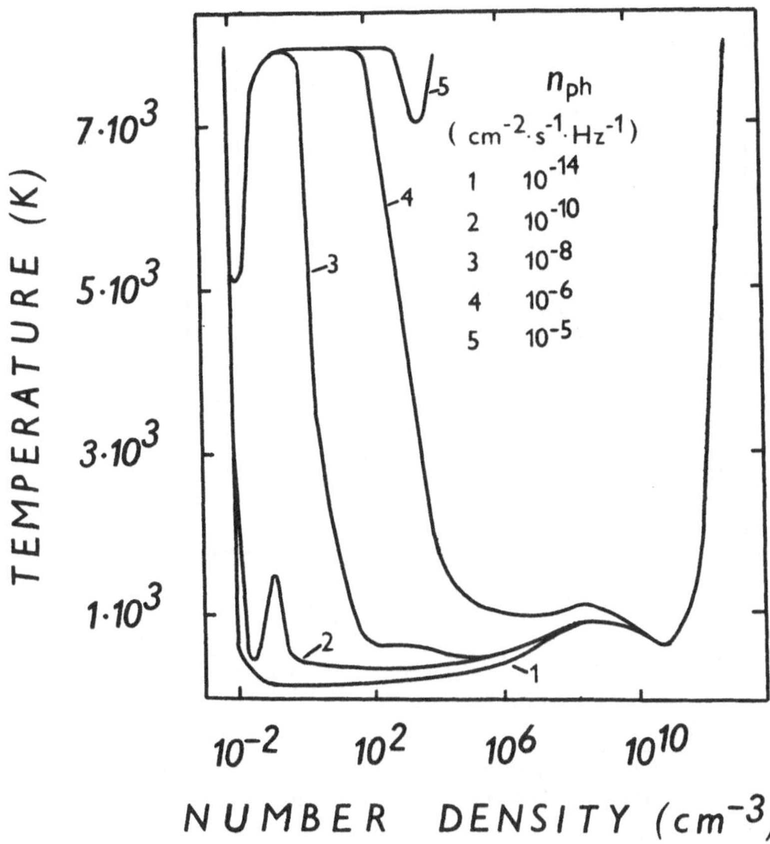

Figure 1. Gas temperature T as a function of number density N in homogeneously contracting cloud with primeval chemical composition. The curves for different values of the external radiation intensity are labeled with numbers.

the optical depth the Jeans radius R_J for N(t) and T(t) is taken for each time moment. It is lower than the cloud radius.

Fig.1 - 2 show the variations of the temperature and H_2 relative concentration as functions of number density N for different values of external radiation flux n_{ph}.

At low $n_{ph} < 10^{-10}$ $cm^{-2}s^{-1}Hz^{-1}$ the radiation makes no essential influence on the H_2 formation process. At $n_{ph} > 10^{-10}$ $cm^{-2}s^{-1}Hz^{-1}$ being typical for galaxies and clusters of galaxies, the effective H_2 molecules formation is possible starting from some limiting cloud density

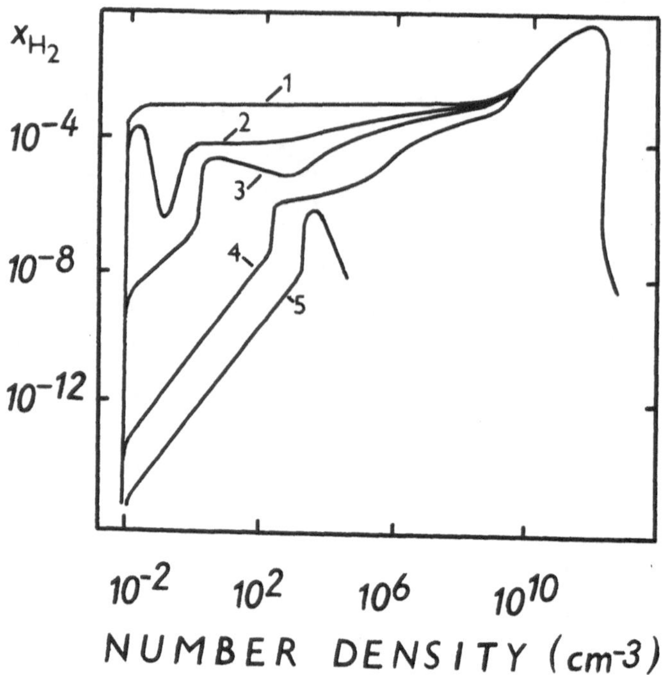

Figure 2. Variations of relative molecular hydrogen concentration as a function of number density N. Designations of curves are the same as on Figure 1.

(Fig.2). As it follows from Fig.2, the small increase of the number density ($\Delta N/N < 2$) causes the relative molecular hydrogen concentration to rise fast by a factor 10^2 , which is followed by gas cooling down to temperature $T \simeq 10^2 - 10^3$ K. The growth of the relative H_2 concentration is connected with increase of the optical depth in the electronic transition lines of the molecular hydrogen.

At $n_{ph} \simeq 10^{-9}$ $cm^{-2}s^{-1}Hz^{-1}$ being typical for intergalactic medium, the mass of the cloud central part, which has already cooled down, equals $\simeq 10^6 M_\odot$. This value is consistent with mass of BCDG central part experiencing the burst of active star formation. The typical densities of the cooled gas are about 1 cm^{-3}. Consequently, the size of the central cloud part being subjected to active star formation is about 300 pc, it corresponds to appropriate values for BCDGs.

At large external radiation fluxes $n_{ph} > 10^{-8}$ $cm^{-2}s^{-1}Hz^{-1}$ the central cloud part has the opportunity for molecular hydrogen formation and gas cooling under $N = 10^2 - 10^4$ cm^{-3}. In this case the mass of the cooled region is equal to $M \simeq 10^5$ $M_\odot$ and the typical sizes are about $r \simeq 5$ -

10 pc; it corresponds to the globular cluster parameters.

Finally, at $n_{ph} > 10^{-5}$ $cm^{-2} \cdot s^{-1} \cdot Hz^{-1}$ the molecular hydrogen is not formed and the cloud evolution takes place at $T \simeq 10^4$ K in accordance with results of Hasegawa et al. (1981) and Silk (1977).

3. BLUE COMPACT DWARF GALAXIES AND GLOBULAR CLUSTER FORMATION

We examine the evolution of the central cloud part consisting of primeval matter when the star formation starts. As it has been shown by Izotov (1987a, b), the massive stars with masses about 10^2 $M_\odot$ and luminosities about 10^6 $L_\odot$ can form from the matter with primeval chemical composition. As a result, an additional source of the radiation arises which causes molecular hydrogen to photodissociate and protostars to "evaporate". However, the rate of the "evaporation" depends on a particular model.

Now consider in detail the protostar "evaporation". As it follows from Izotov (1987a), in the part of the protostar with primeval chemical composition where molecular hydrogen formation takes place the density varies as a function of radius according to the law $\rho \propto r^{-2.9}$. The time of the molecular hydrogen photodissociation in layer with optical depth $\tau \simeq 1$ in electronic transition lines is equal to

$$\Delta t_{pd} = (\alpha_o \cdot n_{ph})^{-1}, \qquad (2)$$

where $\alpha_o = 1.9 \cdot 10^{-9}$ (Izotov and Kolesnik (1984)), and linear size of the protostar layer with $\tau \simeq 1$ can be obtained from

$$\Delta r = \frac{m_H}{k_o \cdot \rho_o \cdot r_o^{2.9} \cdot x_{H_2}} r^{2.9} , \qquad (3)$$

where ρ_o is the density on protostar surface, r_o is its radius, x_{H_2} is relative H_2 concentration, m_H is the hydrogen atom mass.

Having taken (2) and (3) one can obtain

$$\frac{\Delta r}{\Delta t} = \frac{\alpha_o \cdot n_{ph} \cdot m_H}{k_o \cdot \rho_o \cdot r_o^{2.9} \cdot x_{H_2}} r^{2.9} , \qquad (4)$$

Integrating the equation (4) we get the following expression for the H_2 dissociation time in protostar

$$t_{pd} = \frac{k_o \cdot \rho_o \cdot r_o^{2.3}}{\alpha_o \cdot n_{ph} \cdot m_H \cdot r^{1.3}} \; . \tag{5}$$

Compare t_{pd} with free-fall time

$$t_{ff} = (3\pi \; / \; 32G\rho \;)^{1/2} \; . \tag{6}$$

That part of protostar turns into a stellar core for which the condition $t_{ff} < t_{pd}$ is satisfied. Otherwise, the molecular hydrogen dissotiation occurs and it puts the end to protostar collapse (Izotov (1987b)). Now for different models we evaluate the mass of protostar, which turns into the star.

If the massive star with luminosity about 10^{39} $erg \cdot s^{-1}$ has been formed, the additional radiation flux density of 10^{-8} $cm^{-2} s^{-1} Hz^{-1}$ arises in the central part of the cloud with $M \simeq 10^6 M_\odot$ and $N \simeq 1$ cm^{-3} (model 2). From (5) we find $t_{pd} = 10^{13}$ s. Further, having used (6) one can get the density of that protostar part which has already turned into the star. It equals $N = 10^4$ cm^{-3} and corresponds to protostar mass about 10^3 – 10^4 $M_\odot$. For the model 3 ($n_{ph} = 10^{-8}$ $cm^{-2} s^{-1} Hz^{-1}$) the following values were obtained: the typical photodissociation time $t_{pd} = 10^{11}$ s, the critical number density $N_{cr} = 10^8$ cm^{-3} and $M_{cr} = 10^2$ $M_\odot$.

Finally, for the model 4 we determined $t_{pd} \simeq 10^9$ s, $N_{cr} \simeq 10^{11}$ cm^{-3} and $M_{cr} \simeq 2 - 3$ $M_\odot$.

The model 2 is consistent with cloud evolution scheme under the conditions typical for clusters of galaxies. In this case the star formation in the central part of the cloud does not affect substantially another protostar evolution and one would expect the formation of the large number of massive stars in the region with mass about 10^6 $M_\odot$ and size about 10^2 – 10^3 pc. Such values are consistent with the parameters of blue compact dwarf galaxies.

The model 4 is in agreement with conditions for periphery of the young giant galaxy which has quasar in its centre. Just these conditions were feasible for our Galaxy formation stage. In this case, the massive star, being already formed in the cloud, affects substantially the protostar evolution, setting a limit on stars mass at the level 2 – 3 $M_\odot$. The typical cloud central part mass equals 10^5 $M_\odot$, the size r is about $2 \cdot 10^{19}$ cm. These are characteristics of the globular cluster.

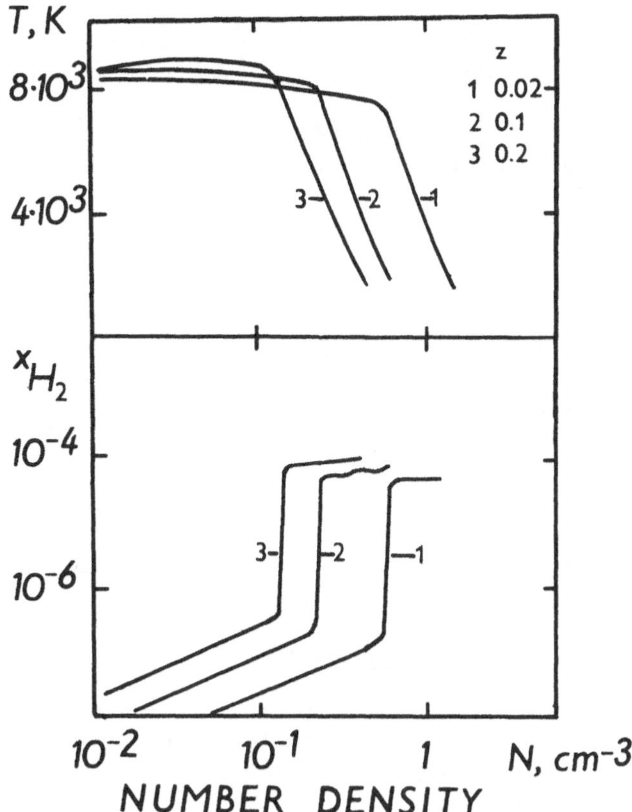

Figure 3. Gas temperature T and relative molecular hydrogen concentration x_{H_2} plotted versus the total number density for clouds with different content of the heavy elements z. It has been adopted that n_{ph}/n^{o}_{ph} = 1.0, n_{CR}/n^{o}_{CR} = 0.1.

4. THE THERMAL AND CHEMICAL EVOLUTION OF CLOUDS ENRICHED WITH HEAVY ELEMENTS

The presence of the heavy elements, characterised by value $z = Z/Z_{\odot}$ causes the additional mechanisms to be taken into account: the cooling on the heavy elements; the heating due to the photoelectronic emission from dust surface and heating by cosmic rays; the molecular hydrogen formation on the dust grains; the extinction of external ultraviolet radiation by dust. Nearly the half of the heavy elements is assumed to be incorporated into dust.

The approach used in this section is suitable to solve the problem of star forming region origin during recurrent starburst in dwarf galaxy

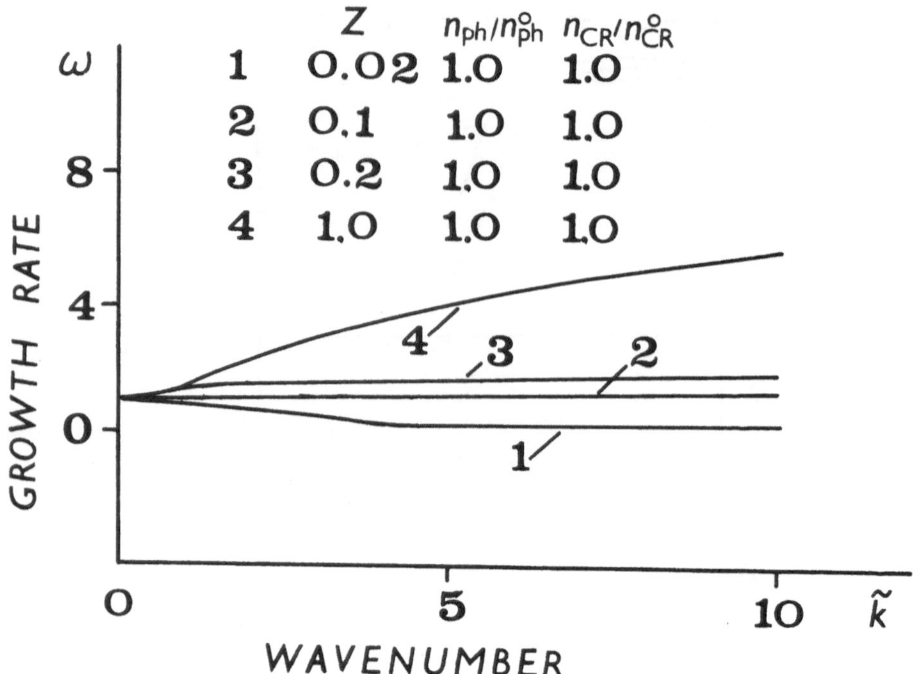

Figure 4. Variations of the rate of the thermal instability growth as a function of wave number $k = k/\sqrt{\gamma}\ k_J$ for models with different values of the heavy elements content z.

as well in spiral galaxy.

Thus, the thermal evolution of the contracting cloud enriched with heavy elements depends on three parameters, such as $z = Z/Z_\odot$, n_{ph}, and flux density n_{CR} of the soft cosmic rays which heat gas.

The H_2 content rises abruptly when the total number density reaches the limiting value $N_{cr} \simeq 0.1 - 1\ cm^{-3}$ (Fig.3). As a result, the cooling rate on the molecular hydrogen Λ_{H_2} exceeds that on heavy elements and atomic hydrogen.

The expressions for cloud Jeans mass and column density along the line of sight at the initial moment of molecular hydrogen cooling and of formation of the cold central part are:

$$M_J / M_\odot = 2 \cdot 10^8 \cdot (z/2)^{1/3} \cdot (n_{ph} / n^0_{ph})^{-1/3} , \qquad (7)$$

$$N_{HI} = 6 \cdot 10^{20} \cdot (z/2)^{-1/3} \cdot (n_{ph} / n^0_{ph})^{1/3}\ cm^{-2} , \qquad (8)$$

where $n^0_{ph} = 10^{-8}\ cm^{-2} s^{-1} Hz^{-1}$ is the ultraviolet radiation flux density in the solar vicinity. The mass $M_J / M_\odot$ and N_{HI} vary in the following range: $M_J / M_\odot = (4 \div 20) \cdot 10^7$, $N_{HI} = (7 \div 30) \cdot 10^{20} cm^{-2}$.

Thus independently of external conditions and chemical cloud composition the universal mechanism, namely molecular hydrogen formation, exists which is a decisive factor in the thermal cloud evolution and formation of the cold nucleus of the star forming region. The proposed mechanism has not been taken into account in papers published early by Silk (1977), Hasegawa et al. (1981).

5. THE CLOUD FRAGMENTATION AND STAR FORMATION REGION PROPERTIES IN GALAXIES OF DIFFERENT TYPES

In this section we briefly consider the possibility of the fragmentation of collapsing cloud due to thermal instability in the way developed by Kondo (1970) and Yoshii and Sabano (1979). The thermal instability evolution was studied for all contracting cloud models calculated in present paper.

The equation system describing the dynamical, thermal and chemical evolution takes the form:

$$\frac{d\rho}{dt} + \rho \nabla\vec{v} = 0, \qquad (9)$$

$$\frac{d\vec{v}}{dt} + \frac{1}{\rho}\nabla p + \nabla\phi = 0, \qquad (10)$$

$$\frac{dU}{dt} + \frac{p}{\rho^2}\frac{d\rho}{dt} + L = 0, \qquad (11)$$

$$\frac{dx_{H_2}}{dt} - F = 0, \qquad (12)$$

$$\frac{dx_e}{dt} - E = 0, \qquad (13)$$

$$\nabla^2\phi - 4\pi G\rho = 0, \qquad (14)$$

$$p - \frac{R\rho T}{\mu} = 0, \qquad (15)$$

$$U - \frac{1}{\gamma-1}\frac{RT}{\mu} = 0. \qquad (16)$$

In (9) – (16) ρ, T, p, $\vec{v}$ are the gas density, temperature, pressure and velocity, respectively; x_{H_2}, x_e are the relative concentrations of the H_2 molecules and free electrons; L is the cooling rate which takes into account all heating and cooling mechanisms mentioned above; G and R are gravitational and general gas constants respectively; F and E are the right parts of the kinetic equations for the molecular hydrogen and free electrons. These equations take into account the formation and destruction processes for the corresponding types of particles.

We express the solution of the equation system in the form of perturbed and unperturbed components. Having eliminated the unperturbed component from equations according to Kondo(1970) and Yoshii and Sabano (1979), one can get the system of equations for perturbed components in the form:

$$\frac{d(\ln \rho^*)}{d\tau^*} + \nabla^* \vec{v}^* = 0, \qquad (17)$$

$$\frac{d\vec{v}^*}{d\tau^*} + \alpha\vec{v}^* + \frac{1}{\gamma k_J^2}\frac{\nabla^* p^*}{\rho^*} + \nabla^*\phi^* = 0, \qquad (18)$$

$$\frac{d(\ln T^*)}{d\tau^*} + (\gamma - 1)\nabla^*\vec{v}^* + (P^* - P_0^*) = 0, \qquad (19)$$

$$\frac{d(\ln x_{H_2}^*)}{d\tau^*} - (\bar{F} - \bar{F}_0) = 0, \qquad (20)$$

$$\frac{d(\ln x_e^*)}{d\tau^*} - (\bar{E} - \bar{E}_0) = 0, \qquad (21)$$

$$\nabla^{*2}\phi^* - (\rho^* - 1) = 0, \qquad (22)$$

where notations are taken in the form:

$$\rho^* = \rho/\rho_0,\ T^* = T/T_0,\ p^* = p/p_0,\ x_{H_2}^* = x_{H_2}/x_{H_2}^0,\ x_e^* = x_e/x_e^0,\ \vec{v}^* = \frac{\tau_{ff_0}}{a}(\vec{v}-\vec{v}_0),$$

$$\phi^* = \frac{\tau^2_{ff_o}}{a^2}(\phi - \phi_o),\ k^2_J = \frac{4\pi G\rho^2_o a^2}{\gamma P_o},\ \tau_{ff_o} = (4\pi G\rho_o)^{-1/2},\ \alpha = d(\ln a)/d\tau^*,$$

$$a = (\rho_{oi}/\rho_o)^{1/3},\ \frac{d}{d\tau^*} = \frac{\partial}{\partial\tau^*} + \vec{v}^*\nabla^*,\ \frac{\partial}{\partial\tau^*} = \tau_{ff_o}\left(\frac{\partial}{\partial\tau} + \vec{v}\nabla\right),\ \nabla^* = a\nabla,\ \vec{v}_o = \alpha\vec{x}/\tau_{ff_o},$$

$$\tau^* = \int_o^t \frac{dt}{\tau_{ff_o}},\ \vec{x}^* = \frac{1}{a}\vec{x},\ \bar{F} = \frac{\tau_{ff_o}F_o}{x_{H_2}},\ \bar{E} = \frac{\tau_{ff_o}E_o}{x_e},\ P^* = (\gamma-1)\frac{\tau_{ff_o}\mu_o}{RT}L + \mu\tau_{ff_o}(E-F),$$

$\vec{x}$ is the space coordinate, ρ_{oi} is unperturbed density at the initial time moment.

Our calculations show that thermal instability doesn't arise in clouds with $z < 0.02$. The rate of the thermal instability growth ω is the bigger the larger heavy elements abundance z and ω exceeds the rate of the growth of the gravitational instability when $z > 0.2$ (Fig.4). Therefore, the parameters of the star formation regions are to differ substantially for galaxies of different types. In spiral galaxies where $z \simeq 1$ the star formation regions are to have inhomogeneous structure with typical fragments mass about $10^3 - 10^5\ M_\odot$. In the dwarf systems, just as blue compact dwarf galaxies where $z \simeq 0.1$, the fragmentation due to thermal instability doesn't arise. As a result, the star and gas distributions are to be more homogeneous and star formation regions in the dwarf galaxies are $10^2 - 10^3$ times as large as the separate star forming regions in the giant galaxies.

The observational data do not permit us to make definitive conclusions about differences in the star forming regions structure of the irregular and spiral galaxies. But in accordance with paper by Kunth (1987) there is the basis to believe that star formation process in the BCDGs is to spread over a large area about $10^6 - 10^7\ M_\odot$ of the galaxy while the star formation regions in spiral galaxies have masses about $10^5\ M_\odot$. Besides that, the observational data (Hodge (1983, 1987)) show that regions of the ionized hydrogen in the irregular galaxies have systematically larger sizes than HII-regions in spiral galaxies.

6. CONCLUSION

Our principal results are as follows:

1). The universal mechanism, controlling the thermal evolution of the gas clouds which have initially the temperature $T \simeq 10^4$ K and mass about $10^8 - 10^9\ M_\odot$, has been found. The molecular hydrogen gives the principal contribution into the gas cooling process at $T < 10^4$ K independently of the content of the heavy elements and of the external physical conditions. The effective formation of the H_2 molecules takes place when the cloud optical depth in lines of the H_2 electronic transitions reaches the value $\tau \simeq 1$.

2). The principal possibility for the blue compact dwarf galaxies formation from matter with primeval chemical composition is shown to exist. In this case the formation of the large number of massive stars is shown to be expected in the cloud central parts with sizes and masses which are typical for the regions of intense star formation in BCDGs.

3). The possibility for globular clusters formation on the periphery of the young galaxies is shown to exist. In this case, the massive stars which have already been formed in the cloud are to establish the upper limit about 2 - 3 $M_\odot$ on the masses of the stars which form later.

4). In the contracting cloud the presence of the heavy elements causes the thermal instability to progress. The rate of the thermal instability growth is the higher the larger content of heavy elements z, and it reaches the maximum value at $T < 4000$ K when the cooling rate on the heavy elements equates with the cooling rate on the H_2 molecules.

5). Within the adopted models, the star formation regions have been found out to differ substantially for galaxies of different types. In spiral galaxies enriched with heavy elements the star formation regions break down into fragments with masses $M \simeq 10^3 - 10^5\ M_\odot$. On the other hand, in dwarf galaxies with $z = Z / Z_\odot < 0.1$ the star formation regions must be more homogeneous complexes with masses about $10^7 - 10^8\ M_\odot$.

REFERENCES

Dufour, R.J. (1986) ' Abundances in dwarf irregular galaxies ', Publications of Astronomical Society of Pacific 98, 1025 - 1031.

Hasegawa, T., Yoshii, Y., and Sabano, Y. (1981) ' Thermal evolution of a contracting hydrogen gas cloud ', Astronomy and Astrophysics 98, 186 - 194.

Hodge, P.W. (1983) ' Size distribution of HII regions in galaxies. I. Irregular galaxies ', Astronomical Journal 88, 1323 - 1329.

Hodge, P.W. (1987) ' Size distribution of HII regions in galaxies. II. Spiral galaxies ', Publications of Astronomical Society of Pacific 99, 915 - 920.

Hunter, D. and Gallagher, J.S. (1986) ' Stellar populations and star formation in irregular galaxies ', Publications of Astronomical Society of Pacific 98, 5 - 28.

Izotov, Yu.I. (1987a) ' The formation of the first generation stars. I. The collapse regularities of protostars with primordial chemical composition ', Kinematics and Physics of Celestial Bodies 3, No. 3, 61 - 66 (in Russian).

Izotov, Yu.I. (1987b) ' The formation of the first generation stars. II. Maximum masses of stars ', Kinematics and Physics of Celestial Bodies 3, No. 3, 30 - 39 (in Russian).

Izotov, Yu.I. and Kolesnik, I.G. (1984) ' Kinetics of molecular hydrogen formation in primeval gas of the Universe ', Soviet Astro-

nomical Journal 61, 24 - 34.
Kondo, M. (1970) ' Thermal instability in an expanding medium ', Publications of Astronomical Society of Japan 22, 13 - 40.
Kunth, D. (1987) ' Star formation in dwarf galaxies ', Science Reports of Tohoku University, Eigth Series 7, 353 - 363.
Kunth, D., Maurogordato, S. and Vigroux, L. (1988) ' Blue compact galaxies: a mixing bag ', Astronomy and Astrophysics 204, 10 - 20.
Kunth, D. and Sargent, W.L.W. (1986) ' I Zw 18 and existence of very metal-poor blue compact dwarf galaxies ', Astrophysical Journal 300, 496 - 499.
Loose, H.H. and Thuan, T.X.T. (1986) ' The morfology and structure of blue compact dwarf galaxies from CCD observations ', in J. Tran Thanh Van (eds.), Star-forming dwarf galaxies and related objects, Frontieres, Paris, pp. 73 - 88.
Silk, J. (1977) ' On the fragmentation of cosmic gas clouds. I. The formation of galaxies and first generation of stars ', Astrophysical Journal 211, 633 - 648.
Yoshii, Y. and Sabano, Y. (1979) ' Stability of a contracting pregalactic gas cloud ', Publications of Astronomical Society of Japan 31, 505 - 521.

DISCUSSION

MATEO: You showed that the masses of star-forming regions from Z = 0 gas are about $10^7 - 10^8\ M_\odot$. However, the globular cluster masses are about 100 times smaller than this, even the metal poor ones. How do you reconcile this ?

IZOTOV: We obtained the mass of star-forming regions of $10^7 - 10^8\ M_\odot$ in the case, when the flux of ultraviolet radiation consistent with that for intergalactic medium. If we consider the model of young giant galaxy containing the quasar in the centre, then the masses and dimensions of star-forming regions will be about $10^5\ M_\odot$ and 5 - 10 pc.

SHAPIRO: Can you explain why you find that there is no thermal instability in the collapsing cloud if only H_2 cooling occurs, in the absence of metals ?

IZOTOV: The main reason of absence of thermal instability in collapsing cloud with primeval chemical composition is that the cooling function on H_2 molecules is steeper function of temperature rather than density.

MOUSCHOVIAS: If I understand you correctly, your calculation included heating by cosmic rays. Did you include cosmic rays with energies larger than 100 MeV, which can penetrate clouds with column densities smaller than about 60 gm/cm^2 and therefore ionize and heat the cloud ? And could these cosmic rays prevent the thermal instability at some stage ?

IZOTOV: I included in calculations only the soft cosmic rays with energies of a few MeV. It's difficult to say what are flux densities of cosmic rays outside Galaxy.

NLTE H_2 COOLING FUNCTION AND PROTOGALACTIC EVOLUTION

R. CAPUZZO-DOLCETTA
Istituto Astronomico Universitá 'La Sapienza'
via G.M. Lancisi 29
I-00161, Roma, Italy

A. DI FAZIO
Osservatorio Astronomico
viale del Parco Mellini 84
I-00136, Roma, Italy

F. PALLA
Osservatorio Astrofisico di Arcetri
Largo E. Fermi 5
I-50125, Firenze, Italy

ABSTRACT. We performed detailed NLTE calculation of the absorption (due to bound-bound transitions among the roto–vibrational levels of the molecule and due to scattering on bound electrons) and emission (due to the roto–vibrational transitions) functions of the H_2 molecule in a low temperature and density gas typical of protogalactic conditions. All the details of the computations together with the presentation of fitting functions to the results, suitable to be used in heavy numerical codes which have the aim to follow the protogalaxy collapse, will be published elsewhere (Capuzzo–Dolcetta, Di Fazio and Palla, 1989); here we report the results and conclusions on H_2 roto-vibrational cooling.

1. MOLECULAR HYDROGEN IN PRIMORDIAL GALAXIES.

The important role played by H_2 molecules in the first phases of galaxy evolution is well known. H_2 is the most abundant molecular species and the most efficient coolant in a primordial gas mixture (Saslaw and Zipoy, 1967). If galaxies formed after recombination, with a residual ionization fraction around 10^{-6}, the most efficient reactions to produce H_2 are:

$$\mathrm{H} + e^- \rightarrow \mathrm{H}^- + \gamma$$
$$\mathrm{H}^- + \mathrm{H} \rightarrow \mathrm{H}_2 + e^-$$

(McDowell, 1961). In order to avoid photo–destruction of H^-, a radiation temperature $T_{rad} \lesssim 300$ K is needed. At higher temperatures, when H^- is photo–destroyed or destroyed by collisions, the reactions effective to produce H_2 are:

$$\mathrm{H} + \mathrm{H}^+ \rightarrow \mathrm{H}_2^+ + \gamma$$
$$\mathrm{H}_2^+ + \mathrm{H} \rightarrow \mathrm{H}_2 + \mathrm{H}^+$$

R. Capuzzo-Dolcetta et al. (eds.), Physical Processes in Fragmentation and Star Formation, 49–54.

(Saslaw and Zipoy, 1967).

The importance of H_2 molecule is furtherly stressed by recent work of Mac Low and Shull (1986) and Shapiro and Kang (1987) who showed that a production of H_2 up to $n_{H_2}/n_H \simeq 10^{-3}$ is expected behind shocks in the primordial intergalactic medium. This significant abundance can survive to UV flux at the epoch of galaxy formation.

All these things indicate how crucial is the importance of a correct treatment of the radiative properties of the H_2 component of the protogalactic gas, when dealing with the task of describing the first phases of galactic evolution in the attempt of understanding the characteristics of the first stellar population in the halo.

2. THE EMISSION FUNCTION.

2.1 THE MODEL MOLECULE AND ITS EMISSIVITY.

In this report we deal with the roto-vibrational (rv) emissivity of H_2; the rv absorption and scattering on bound electrons coefficients are treated in Capuzzo–Dolcetta, Di Fazio and Palla (1989), and their values suggest that the optical depth due to molecules in a protogalaxy is less than 1.

The emission function Λ (defined as the power per unit volume released by H_2 molecules) due to the de–excitation of roto–vibrational (rv) levels populated by radiation and collisions was computed as function of density (ρ), matter and radiation temperatures (T, T_{rad}), and fractional H_2 mass density $f_{H_2} = \rho_{H_2}/(X\rho)$, where X is the hydrogen fractional abundance by mass.
The expression of Λ is:

$$\Lambda = \sum_{v,j} \Lambda_{v,j} = \sum_{v} \sum_{j} n_{v,j} A_{vj;v'j'} \Delta E_{vj;v'j'} [1 + J_{vj;v'j'}]$$

where $v \geq v'$, $j \geq j'$ define the vibrational and rotational levels of the transition from (v, j) to (v', j'); $A_{vj;v'j'}$ are the radiative spontaneous transition probabilities which were taken from Turner, Kirby–Docken and Dalgarno (1977); $\Delta E_{vj;v'j'}$ is the energy difference between the rv levels; $n_{v,j}$ is the level population, normalized in such a way that: $n_{H_2} = \sum_{v,j} n_{v,j} = X\rho f_{H_2}/m_{H_2}$; $J_{vj;v'j'}$ accounts for stimulated emission.

To compute Λ, we considered the first 3 vibrational levels of the H_2 molecule, each one splitted in its first 20 rotational levels. The choice of $v = 0, 1, 2$ is justified by the fact that the energy difference between the ground level and the $v = 2$ level is $\Delta E_{20;00}/k \simeq 12{,}000$ K, and the energy difference between the $j = 20$ and the $j = 0$ levels is up to $\approx 30{,}000$ K; at protogalactic densities, the hydrogen molecule is dissociated at such high temperatures.

The level population is computed at given $\rho, T, T_{rad}, f_{H_2}, X$ by solving the steady state system of the rate equations describing particle conservation in each rv level, following a procedure similar to that by Hutchins (1976). The analytic fits of Hollenbach and McKee (1979) for the collisional de-excitation rates were adopted. The importance of keeping separate the matter from the radiation temperature is easily understood when thinking that the collapse of a protocloud of galactic size and zero metals is, in a large part of the structure, in the transparent regime so that T_{rad} is constantly equal to the temperature of the cosmological background radiation field while T is governed by gravitation and radiative losses.

2.1 MAIN RESULTS.

The emission function was evaluated in the ranges:

$$10^{-26} \leq \rho(\text{gcm}^{-3}) \leq 10^{-18}, 100 \leq T(\text{K}) \leq 4000, 10 \leq T_{rad}(\text{K}) \leq 1000.$$

Figure 1 shows the dependence of $\Lambda/n_{H_2}^2$ (i.e. the emission per unit density) for three different T_{rad}. The most interesting thing to note is how different (particularly at the low densities typical of the first phases of a protogalactic collapse) are the values of the emission function at a given T when $T_{rad} \simeq T$ and when $T_{rad} \neq T$ (this last situation is a likely one during a protogalactic collapse) confirming the importance of keeping separate the two temperatures. Another relevant feature to note is how Λ gets independent of T at lower gas density when increasing T_{rad} due to the efficiency of the radiation field in determining the population of the excited states.

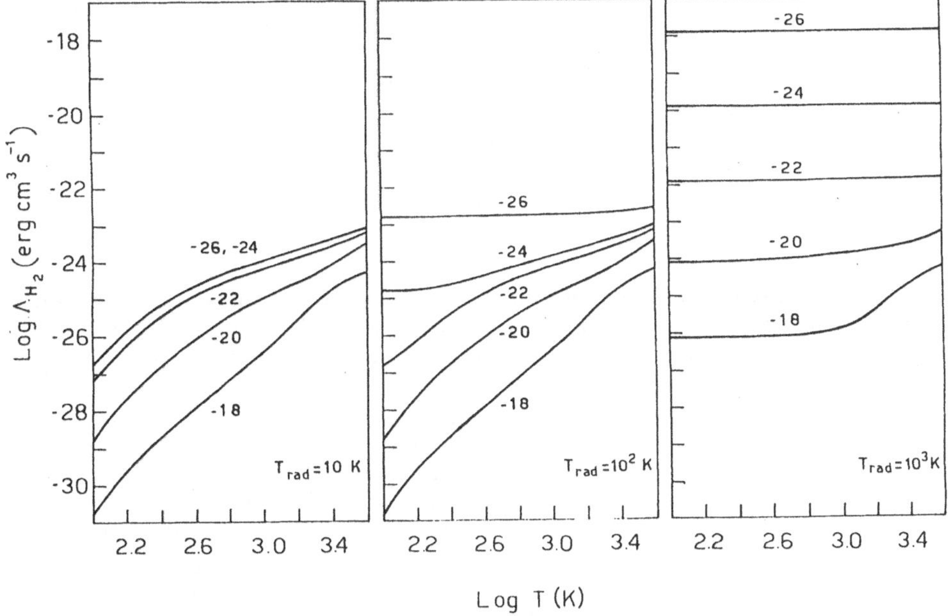

Figure 1. The emission function per unit density, $\Lambda_{H_2} = \Lambda/n_{H_2}^2$, vs. T for three values of T_{rad}. The Logρ value labels each curve.

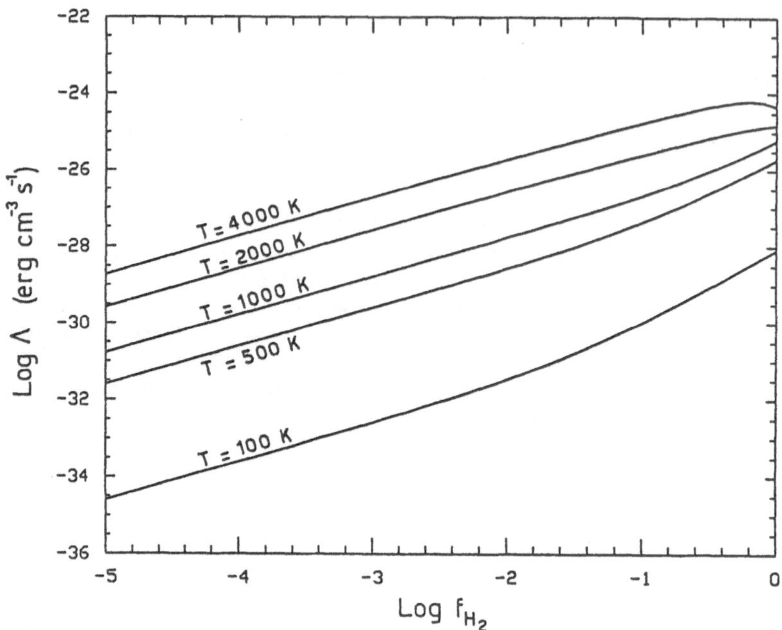

Figure 2. Λ as function of f_{H_2} at various gas temperature, for $T_{rad} = 10K$.

In Figure 2 the behaviour of Λ as function of f_{H_2} is shown: the linearity is kept until values of f_{H_2} such that $C_{H_2-H_2} n_{H_2} \gtrsim C_{H_2-H} n_H$ (where $C_{H_2-H_2}$ and C_{H_2-H} are the collisional coefficients) are reached. Of course this typical value of f_{H_2} depends on T.

A comparison of our results with the formula 6.36 given by Hollenbach and McKee (1979) as a fit to their approximated Λ is in agreement within a factor 1.6 (note that their fitting formula is accurate to a factor of 2). Another check is the comparison, letting $T = T_{rad}$ in our computations, with the fitting formula to H_2 cooling given by Smith (1980) which is valid for $\rho \gtrsim 2 \times 10^{-19}$ and does not take into account radiation field. The largest percentage relative difference is $\approx 40\%$ around $T = 100$ K decreasing with temperature to values less than 10 %.
To have a quantitative indication on the importance of differences in the cooling function on the evolution of the Jeans' mass M_J in a self-gravitating cloud we followed the dynamics and fragmentation of a cloud with $M = 10^7$ $M_\odot$, $R = 300$ pc, $T = 100$ K, X=0.75, Z=0, with the code described in Battinelli *et al.* (this volume) once with our fitted Λ and once with Smith's fitting formula. Typical values of the fragments mass is 1300 $M_\odot$ and 1500 $M_\odot$, respectively. This corresponds to a 15 % percentage difference.

Figure 3 shows the ratio between the cooling time t_c and the free-fall time t_{ff} (which is an indicator of how much radiative dissipation is relevant during the collapse) for a protogalaxy with mass 5×10^{11} $M_\odot$ and radius 100 kpc, in the assumption $f_{H_2} = 10^{-3}$ (panel a), and $f_{H_2} = 1$ (panel b). The figure illustrates how different may be t_c/t_{ff} as a function of T in the T_{rad} region corresponding to a red-shift of galaxy formation $z_f \lesssim 30$, which is a likely range for z_f. In the case $f_{H_2} = 1$, $t_c \simeq t_{ff}$ at $T_{rad} \simeq 100$ K for any matter temperature; this implies that, under the hypothesis that hydrogen is all in molecular form, if galaxies formed at $z_f \gtrsim 36$ their first dynamics was strongly dominated by radiative dissipation. In the case of $f_{H_2} = 10^{-3}$, the first galaxy evolution is radiative dissipative only if the galaxy formed at $z_f > 100$.

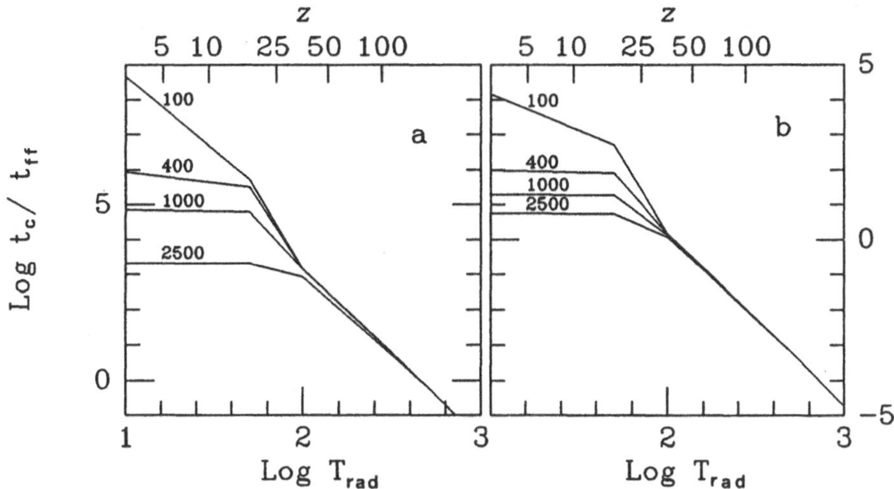

Figure 3. The cooling- to free fall-time ratio as a function of T_{rad} and red-shift (low and up abscissas). Panel a refers to the case $f_{H_2} = 10^{-3}$; panel b to $f_{H2} = 1$. Each curve is labelled by the corresponding value of T.

REFERENCES

Battinelli, P., Capuzzo–Dolcetta, R., Di Fazio, A., Urpin, V., Vereshchagin, S.: 1989, *this volume.*

Capuzzo–Dolcetta, R., Di Fazio, A. and Palla, F.: 1989, *in preparation.*

Hollenbach, D. and McKee, C.F.: 1979, *Astrophys. J. Suppl.* **41**, 555.

Hutchins, J.B.: 1976, *Astrophys. J.* **205**, 103.

Mac Low, M.M. and Shull, J.M.: 1986, *Astrophys. J.* **302**, 585.

McDowell, M.R.C.: 1961, *Observatory* **81**, 240.

Saslaw, C. and Zipoy, D.: 1967, *Nature* **216**, 976.

Shapiro, P.R. and Kang, H.: 1987, *Astrophys. J.* **318**, 32.

Smith, J.: 1980, *Astrophys. J.* **238**, 842.

Turner, J., Kirby–Docken, K. and Dalgarno, A.: 1977, *Astrophys. J. Suppl.* **35**, 281.

Discussion

Herbst: Where did you get your collisional cross sections from?
Capuzzo–Dolcetta: From Hollenbach and McKee (1979).
Shapiro: Is the radiation field in your computation modelled with a black body?

Capuzzo–Dolcetta: Yes, at a temperature T_{rad} which is an input parameter in our computations.

Rocca-Volmerange: To compare your galaxy formation models to observations of high-z galaxies, you need calculations from a much higher mass than you do. What would become your results if the initial mass is 10^{12} $M_{\odot}$?

Capuzzo–Dolcetta: We performed a test evolutionary computation for a cloud with mass 10^7 $M_{\odot}$ and not 10^{11} $M_{\odot}$ or even more just to save computer time; anyway this is not a limitation, because we were actually interested in testing the dependence of the Jean's mass on different cooling functions in a self-consistent evolution model in a transparent case, and so the choice of the total mass of the system is not crucial.

RADIATIVE SHOCKS AND NONEQUILIBRIUM CHEMISTRY IN THE EARLY UNIVERSE: GALAXY AND PRIMORDIAL STAR FORMATION

Paul R. Shapiro
Dept. of Astronomy
University of Texas
Austin, TX 78712
U.S.A.

Hyesung Kang
Dept. of Astronomy
University of Minnesota
116 Church St., S.E.
Minneapolis, MN 55455
U.S.A.

ABSTRACT. Shock waves in primordial composition gas occur in a wide range of circumstances in the theory of galaxy and pregalactic star formation. The radiative cooling of the postshock gas is crucial to the successful production of gravitationally bound fragments. Without such fragmentation, the models do not form stars and galaxies. We have studied in detail the nonequilibrium radiative cooling, recombination, and molecule formation behind steady-state shock waves in primordial composition gas. We have solved the hydrodynamical conservation equations, along with the rate equations fo nonequilibrium ionization, recombination, and molecule formation and the equation of radiative transfer. We find that the shocked gas cools faster than it can recombine and, as a result, is able to form an H_2 concentration as high as 10^{-3} or higher via the formation of H^- and H_2^+ intermediaries due to the enhanced nonequilibrium ionization at 10^4 K. With such an H_2 concentration, the gas cools by rotational-vibrational line excitation of H_2 molecules to well below the canonical final temperature of 10^4 K for a molecule-free gas without metals. This cooling below 10^4 K significantly lowers the characteristic gravitationally unstable mass estimated for shocks relative to the value if the gas cooling stops at 10^4 K. We show that, as the level of external ionizing and dissociating radiation flux is increased, the formation of and cooling by H_2 molecules can be inhibited and delayed. In addition to their relevance to the pregalactic and intergalactic medium, shocks such as these may be responsible for the formation of globular clusters inside protogalaxies. The implications of our shock calculations for this model of the origin of globular clusters will also be discussed.

R. Capuzzo-Dolcetta et al. (eds.), Physical Processes in Fragmentation and Star Formation, 55–70.

1. Introduction

Shock waves in a metal-free gas are predicted to have occurred in the pregalactic and intergalactic medium (IGM) and within protogalaxies under a very wide range of circumstances in the theory of galaxy and primordial star formation. In general, these circumstances can be divided into two broad categories, that of the dissipation of gravitationally-induced supersonic gas motions and that involving "positive energy" deposition in the gas. In the first category are included the shocks which result from the growth of cosmological density fluctuations and the gravitational collapse of density peaks. These include the accretion shocks of the Zel'dovich pancakes of a Hot Dark Matter (HDM) universe, the shocks which result when isolated density peaks in a so-called Cold Dark Matter (CDM) universe collapse or when primordial cloud-cloud collisions occur in a gravitational collapse within which there are smaller-scale density inhomogeneities, and the shocks which lead the expansion of a generic large-scale underdensity (i.e. "void") into the surrounding IGM. In the second category are the pregalactic and intergalactic blast waves in the explosive galaxy formation model. In all of these cases, radiative cooling behind these shocks is essential in order to dissipate enough gravitational or explosion energy to make gravitational instability and fragmentation possible in the shock-heated gas, a prerequisite for galaxy or star formation. As we describe below, a crucial factor in this postshock cooling is the nonequilibrium, gas-phase formation of H_2 molecules. We describe our calculations of this process in radiative shocks in primordial composition gas and the implications of these results for the gravitational instability behind such shocks, including an explanation for the origin of globular clusters. The work summarized here is described in more detail in references [1-7].

2. Hydrogen Molecules and Radiative Cooling

It is well known that a shock-heated gas composed only of atomic H and He is very inefficient at radiatively cooling below 10^4 K. Such a gas in collisional ionization equilibrium is mostly recombined at temperatures below 2×10^4 K. The resulting paucity of free electrons and the exponential cutoff in the temperature dependence of the H Lyα excitation rate combine to reduce the cooling rate below 10^4 K abruptly by many orders of magnitude. It is for this reason that such a gas cooling from a higher temperature is generally believed to relax to a thermally stable temperature just below 10^4 K. Without metals, such an atomic gas would not radiatively cool below 10^4 K (see the equilibrium curve in Figure 1). When a gas of H and He cools radiatively from a temperature well above 10^4 K, however, it cools faster than it can recombine. As a result, the recombination is out of equilibrium, and a greatly enhanced ionized fraction exists at temperatures below 10^4 K compared to the equilibrium value. This ionized fraction makes possible the gas-phase formation of H_2 molecules by the creation of the intermediaries H^- and H_2^+, as follows:

$$H + e \rightarrow H^- + \gamma \tag{1.1}$$

$$H + H^- \rightarrow H_2 + e \tag{1.2}$$

and

$$H + H^+ \rightarrow H_2^+ + \gamma \tag{1.3}$$

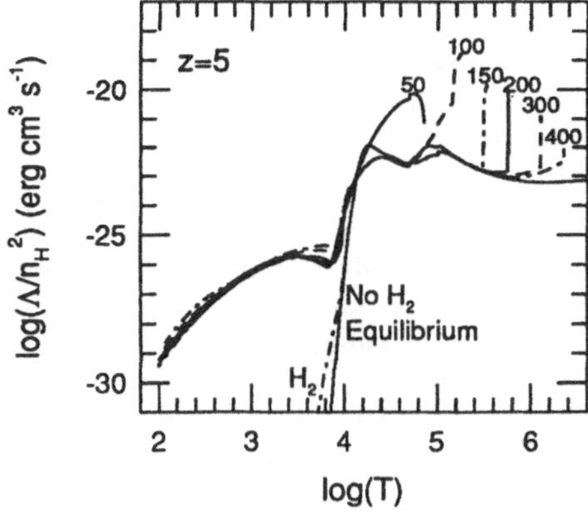

Figure 1. Total cooling rate coefficients ($\Lambda/n_H{}^2$). The solid line labeled by "No H_2" represents the equilibrium cooling calculated by using the equilibrium ionization fractions without H_2. The curve labelled "H_2" means equilibrium cooling including the equilibrium H_2 contribution. The numbers labelling the other nonequilibrium curves are shock velocities.

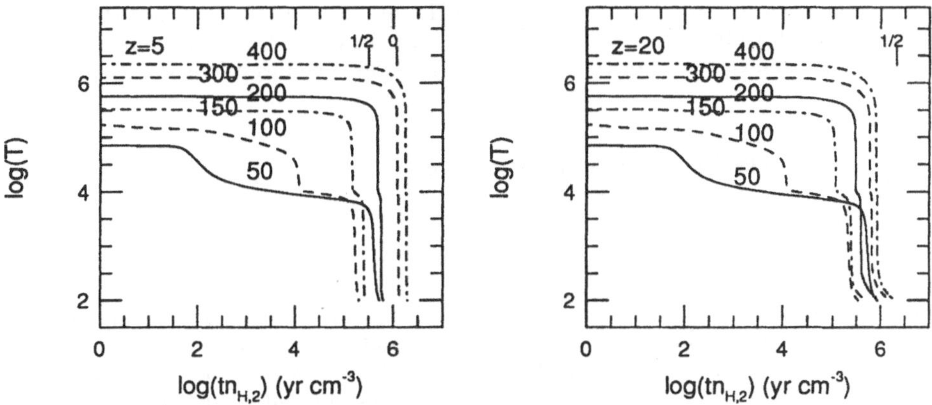

Figure 2 (a). (left) Temperature versus $tn_{H,2}$ $(years\ cm^{-3})$ for shocks in IGM at $z = 5$ with no external radiation field. The vertical markers labeled with 0 and 1/2 represent values of the age of the universe at $z = 5$, multiplied by $n_{H,2}$, with $q_0 = 0$ and 1/2, respectively, and $h = 1$. We have assumed $\Omega_b h^2 = 0.1$.

Figure 2 (b). (right) Same as (a), except $z = 20$. The markers for $q_0 = 0$ are outside the plot, to the right of the plot boundary [$\log(t_{age} n_{H,2}) = 7.06$].

$$H_2^+ + H \rightarrow H_2 + H^+ \tag{1.4}$$

The presence of the small concentration of H_2 molecules which results from reactions (1.1)–(1.4) then makes possible the further radiative cooling of the gas by H_2 rotational-vibrational line excitation to temperatures as low as 10^2 K.

The formation of H_2 molecules in a gas of primordial composition differs from that in the interstellar medium at recent times. Only gas phase reactions can occur in the primordial gas, since the dust grains which dominate the formation in the interstellar case are entirely absent. The simplest route for forming H_2 molecules in the gas phase, that of radiative association of two hydrogen atoms ($H + H \rightarrow H_2 + \gamma$), is unimportant, since H_2 has no dipole moment and the reaction rate is correspondingly very small. As a result, the formation can only occur by a more complicated route involving the additional step of forming certain trace ions as intermediaries.

The H_2^+ process above was first noted by [8], while the generally more important H^- process was pointed out by [9,10]. Since this early work, the formation of and radiative cooling by H_2 in contracting primordial gas clouds have been studied by a number of authors [e.g. 11–20]. In most of these studies, the initial condition of the gas was colder and more neutral than that of the shock-heated gas described here. As a result, the crucial new point made by the work described here, that a gas cooling from a hotter, ionized state cools radiatively faster than it can recombine, thereby greatly enhancing the ionized fraction and with it the efficiency of the H^- and H_2^+ processes, was either absent from or not applicable to this earlier work. Other work in which H_2 formation and cooling are discussed for the case of shock-heated, metal-free gas includes references [21–23].

3. Radiative Shocks in the IGM at High Redshift

3. 1. Shock Structure

We have numerically solved the hydrodynamical conservation equations, along with the rate equations for nonequilibrium ionization, recombination, molecule formation and dissociation and the equation of radiative transfer, in detail for steady-state, planar shocks of velocity in the range $50 \leq v_s \leq 400 km\ s^{-1}$ in gas of primordial composition (H and He only). The species e, H^o, H^+, He^o, He^+, He^{++}, H^-, H_2^+, H_2, and H_2^* (excited state of H_2) are included. We have taken self-consistent account of the effect of the diffuse postshock emission and a possible external radiation flux on the postshock flow and the preshock ionization levels. For shocks in the IGM, we take as the preshock hydrogen number density $n_{H,1} = 8 \times 10^{-6} \Omega_b h^2 (1+z)^3\ cm^{-3}$, where Ω_b is the IGM mass density in units of the critical value, $h = H_o/(100\ km\ s^{-1}\ Mpc^{-1})$, and z is the redshift. We take $\Omega_b h^2 = 0.1$ for the cases described here.

A selection of our results for the cases in which there is no external ionizing and dissociating radiation field (i.e. no additional radiation source besides the shocked gas itself) is shown in Figures 1–6. The radiative cooling curve in Figure 1 shows the substantial enhancement of the cooling rate at temperatures T below 10^4 K due to H_2 rotational-vibrational line cooling relative to the equilibrium cooling curve. The same curves show that the initial postshock cooling is also enhanced as a result of a lag between shock-heating and the

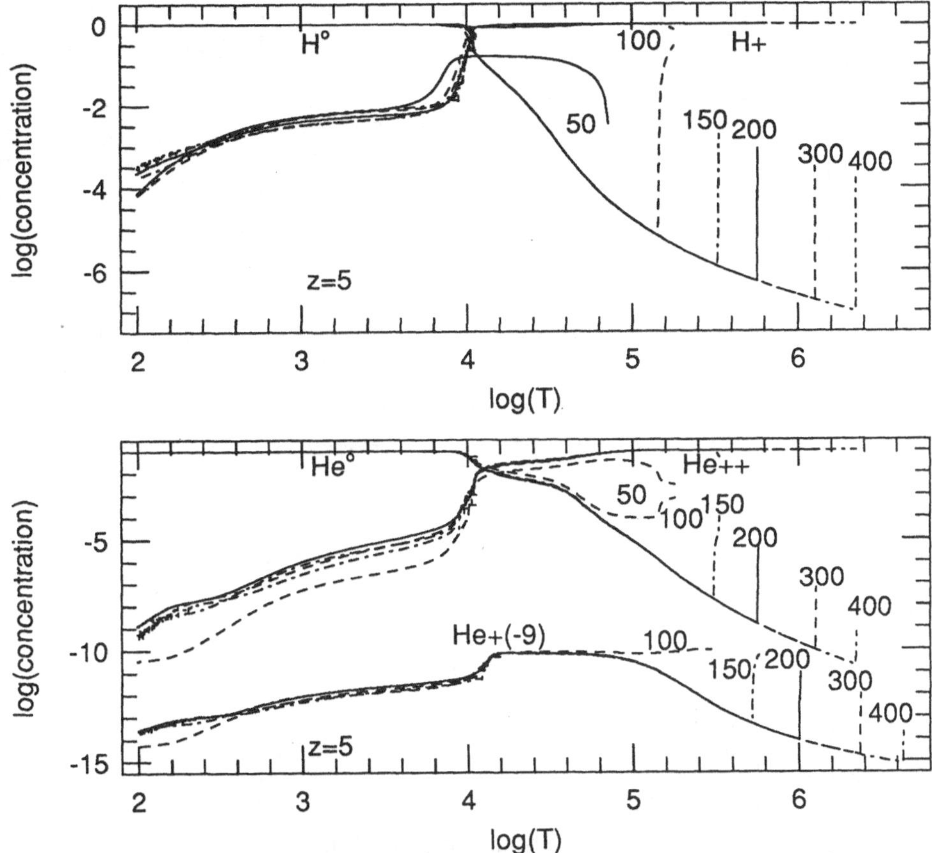

Figure 3 (a). (upper) Non-equilibrium relative ionization fractions, $y_i = n_i/n_H$, versus T, calculated for the same postshock flows as plotted in Figure 2 (a). Neutral hydrogen (solid lines) and ionized hydrogen fractions (dashed lines) are plotted.

Figure 3 (b). (lower) Same as (a), except He^0, He^+, and He^{++} fractions are plotted. The value of $\log(y_{He+})$ is offset by -9.

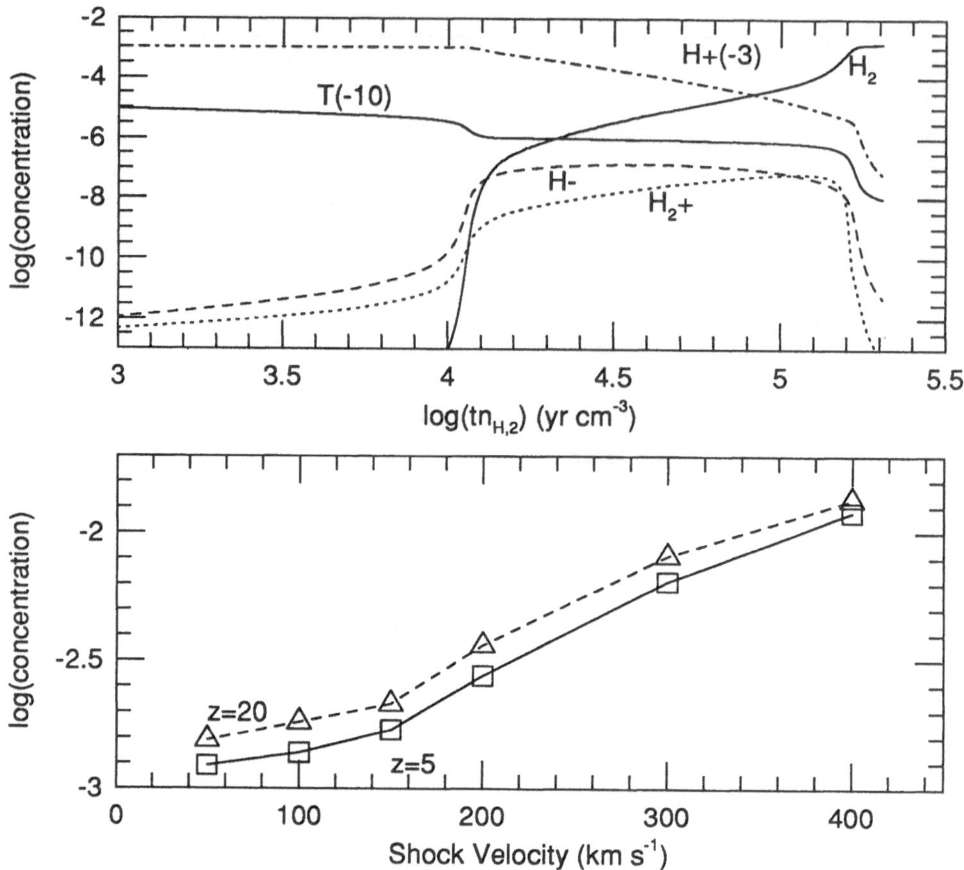

Figure 4. (upper) Temperature and the concentrations of H^+, H^-, H_2^+, and H_2 are plotted against $tn_{H,2}$ for 100 $km\ s^{-1}$ shock at $z = 5$ with no external radiation field. The value of $\log(T)$ is offset by -10, $\log(y_{H^+})$ by -3.

Figure 5. (lower) Final H_2 concentration at 100 K versus shock velocity for the same shock cases as plotted in Figure 2. The lines with boxes represent the shocks at $z = 5$, and the lines with triangles represent the shocks at $z = 20$.

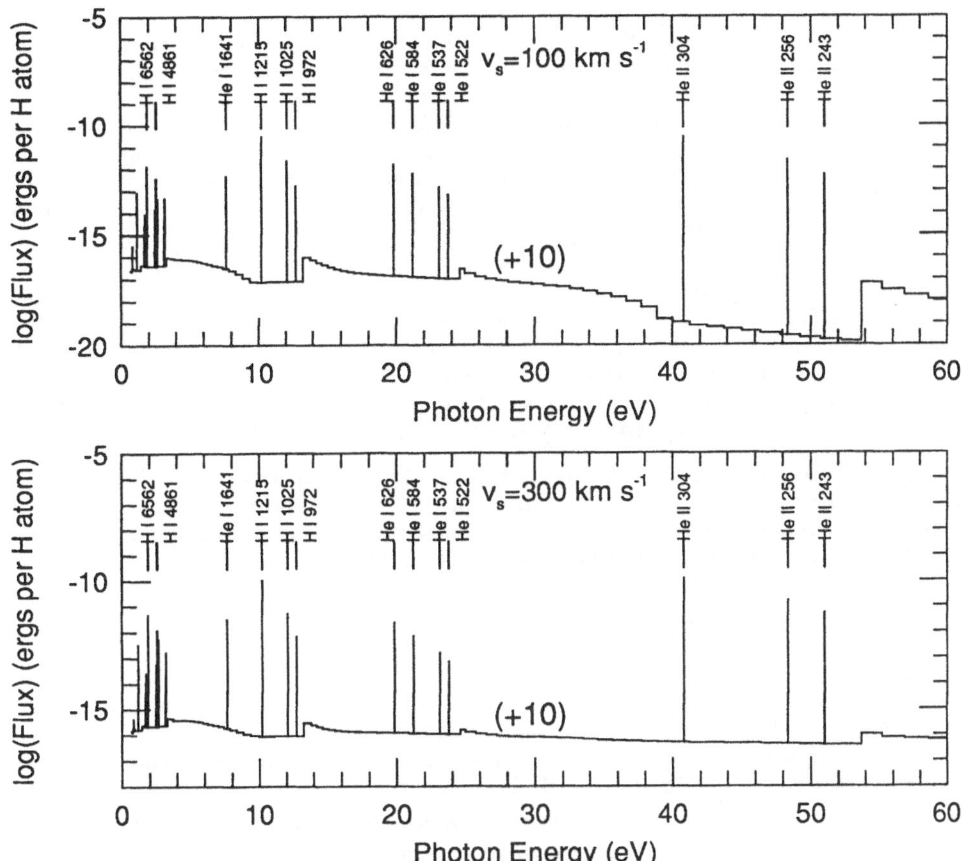

Figure 6 (a). (upper) Normalized radiation flux versus photon energy emergent from the same 100 $km\ s^{-1}$ shocks as presented in Figure 2 (a). The continuum plotted is $F_\nu/(n_{H,1}v_s)$, where $n_{H,1}$ = preshock H number density, $F_\nu = ergs\ cm^{-2}\ s^{-1}\ Hz^{-1}$, and the label "+10" means that the value plotted is the actual value multiplied by 10^{10}. The lines plotted are the quantity $F_l/(n_{H,1}v_s)$, where F_l = total flux integrated over the line profile ($ergs\ cm^{-2}\ s^{-1}$). The emitting atom or ion and the wavelength (in Å) of the strongest lines are labeled.

Figure 6 (b). (lower) Same as (a), except $v_s = 300\ km\ s^{-1}$.

nonequilibrium "ionizing-up" of the immediate postshock gas. Figures 2(a) and (b) for shocks at $z = 5$ and 20, respectively, show the time history of the postshock temperature T for each fluid element which passes through the shock, plotted against the product of time t since shock-heating and $n_{H,2}$, the postshock H density. There is a rough scaling law such that different values of n_H for the preshock gas give similar results when plotted against $tn_{H,2}$. The quantity $tn_{H,2}\ v_s$ gives the H column density between the shock and the fluid element at time t. The effect of the H_2 is to cool the gas rapidly once the temperature drops to 10^4 K, to final temperatures as low as 10^2 K. This is the result of the greatly enhanced nonequilibrium ionized fractions shown in Figures 3(a) and (b) for H and He relative to their values in ionization equilibrium at the same temperatures $T \lesssim 10^4\ K$. The elevated electron and proton concentrations enable enough H^- and H_2^+ to form so as to form H_2 faster than collisional dissociation can destroy it. The time and spatial variation of these quantities are shown together in Figure 4 for the case of $v_s = 100\ km\ s^{-1}$. The final values of the H_2 concentration $n(H_2)/n_H$ ($> 10^{-3}$) at $T = 10^2$ K are shown in Figure 5. The near-IR through extreme-UV radiation spectrum emergent from such shocks is plotted for two cases, $v_s = 100$ and 300 $km\ s^{-1}$, respectively, in Figures 6(a) and (b).

We have also considered the fate of these same shocks if there exists a strong external source of ionizing and dissociating radiation in the IGM at these high redshifts. We find that a sufficient external flux can inhibit H_2 formation and delay the cooling to $T < 10^4\ K$, thereby introducing a 10^4 K plateau phase in the postshock flow.

3. 2. Postshock Gravitational Instability

Since the cooling postshock flow is nearly isobaric, the extra cooling which occurs from $T \approx 10^4\ K$ to $\approx 10^2\ K$ due to H_2 cooling can have a dramatic effect on the gravitational stability of the postshock flow. A simple indication of the relative importance of gravity and gas pressure locally in the flow is given by the spherical Jeans mass for a static medium $M_J = (4/3)\pi\lambda_J^3\rho$, where the Jeans length $\lambda_J \propto T^{1/2}\rho^{-1/2} \propto T\ p^{-1/2}$ and, hence, $M_J \propto T^{3/2}\rho^{-1/2} \propto p^{3/2}\rho^{-2}$. If p is roughly a constant of the flow and $T = T_{final} = 10^4\ K$ (no H_2) or 10^2 K (H_2), then $M_J\ (H_2)/M_J(no\ H_2) \propto \left[T_{final}(H_2)/T_{final}(no\ H_2)\right]^2 = 10^{-4}$! In a realistic description of the gravitational stability of the postshock flow, however, the spherical Jeans criterion is not quite right. Let us consider one case, in particular, that of shocks from explosions.

In the case of the blast waves produced by explosions in the IGM, time-dependent hydrodynamics calculations [24] indicate that when the radiative cooling time drops below the dynamical evolution time for the shock, the description of the shock in terms of the steady state radiative shocks calculated here is valid for roughly one radiative cooling time. After this, the shock velocity decays on a time comparable to the radiative cooling time, and the gas that enters the shock is shock heated to progressively lower temperatures, from which radiative cooling is even more rapid than it was for the gas shocked at the onset of the radiative phase. The result is that, once the radiative phase is reached, virtually all of the gas which is subsequently shock-heated during this phase is rapidly converted to a cold, compressed layer at the temperature which corresponds to the endpoint of radiative cooling. Thus it is appropriate to describe the postshock flow in the radiative phase as a thin shell just behind the shock, *all* at this same final temperature. As such, the gravitational stability of the postshock flow for this case may be analyzed by application of perturbation calculations of the gravitational instability of a spherical thin shell bounded by a shock [25]. The results of such calculations, it turns out, are in qualitative agreement with a simple, approximate energy argument [26]. It is relatively straightforward to use this energy argument to see what effect the extra postshock radiative cooling which results from H_2 molecules has on the gravitational stability of the shocks generated by intergalactic explosions.

The energy per gram in a pillbox of radius λ cut out of a radiatively cooled shell of temperature T_{shell} of infinitesimal thickness, surface mass density Σ, radius R_{shell}, and expansion velocity V_{shell} is, for $\gamma = 5/3$, given by

$$E = E_{kinetic} + E_{thermal} + E_{gravitational}, \tag{3.1a}$$

or

$$E = (\lambda/R_{shell})^2 V_{shell}^2/4 + 3k_B T_{shell}/2\mu - 0.885\pi G\Sigma\lambda \tag{3.1b}$$

[24,26]. This is a quadratic equation in λ and implies that gravitational instability occurs only on scales such that $\lambda_{min} < \lambda < \lambda_{max}$, where λ_{min} and λ_{max} are the roots of the quadratic when $E = 0$. Of course, λ_{min} and λ_{max} must be real in order for gravitational instability to occur, which imposes the condition that

$$R_{shell}/V_{shell} > (0.885G\pi\Sigma)^{-1}(3k_B T_{shell}/2\mu)^{1/2}. \tag{3.2}$$

According to our discussion above, the appropriate value of T_{shell} to use is the value of the temperature after radiative cooling. If the inequality in equation (3.2) is satisfied, then we may ask what the effect on λ_{min} and λ_{max} is of reducing this value of T_{shell}. We can write $\lambda_{min}(\lambda_{max})$ according to

$$\lambda_{min}(\lambda_{max}) = (2a)^{-2}[-b \pm (b^2 - 4ac)^{1/2}], \tag{3.3}$$

where $a = (V_{shell}/R_{shell})^2/4$, $b = -0.885\pi G\Sigma$, and $c = 3k_b T_{shell}/2\mu$, and where the plus-sign refers to λ_{max}. The proper way to use equation (3.3) is to reduce T_{shell} while holding Σ and R_{shell}/V_{shell} fixed, as would be appropriate if the cooling end-point temperature is reduced but the overall cooling time is not changed. This is the case for a range of shock parameters considered in this paper. In that case, equation (3.3) clearly indicates that a reduction of T_{shell} *increases* λ_{max}, while *reducing* λ_{min}. According to inequality (3.2), the effect of our result that T_{shell} is substantially lower than previously thought is to make it easier for an explosion to satisfy inequality (3.2), for a given value of R_{shell}/V_{shell}, or, equivalently, a given value of the age of the explosion.

Whenever the condition for gravitational instability is well satisfied and the values of λ_{min} and λ_{max} are well separated, the wavelength with the maximum growth rate is of the order of λ_{min} (actually, $2\lambda_{min}$, according to [26]). In that case, λ_{min} gives a indication of the characteristic scale for the first gravitationally bound fragments. In this limit, λ_{max} is insensitive to T_{shell}, while λ_{min} can be estimated by assuming that the expansion kinetic energy term in equation (3.1) is small compared to the thermal energy term. In that case, we can define a critical λ, λ_c, such that $|E_{thermal}|$ equals $|E_{gravitational}|$, given by

$$\lambda_c = (3k_B T_{shell}/2\mu)(0.885\pi G\Sigma)^{-1}. \tag{3.4}$$

In order to estimate the effect of the additional cooling described here let us call our new value of this temperature $T_{shell,new}$ and the previously assumed value $T_{shell,old}$. The effect of the extra cooling found here is to leave Σ unchanged, but to change λ_c according to

$$\lambda_{c,new} = \lambda_{c,old}(T_{shell,new}/T_{shell,old}). \tag{3.5}$$

We note that this is exactly the same scaling with T_{shell} as for the spherical Jeans length. Similarly, if we define a characteristic mass scale M_c corresponding to λ_c, given by $\pi\lambda_c^2\Sigma$, then

$$M_{c,new} = M_{c,old}(T_{shell,new}/T_{shell,old})^2. \tag{3.6}$$

Once again, the scaling of M_c with T_{shell} is the same as that of M_J.

Since $T_{shell,new} \approx 10^2$ K while $T_{shell,old} \approx 10^4$ K, our results indicate that the characteristic minimum gravitational scale lengths for shocks produced by intergalactic explosions should be decreased by a factor of 10^2 from the values previously calculated, while the characteristic mass of the corresponding gravitationally bound fragment should be reduced by 10^4. If we assume that the value of Σ in an actual explosion-generated shell is within a factor of order unity of the mass per unit area overtaken by the shock in one radiative cooling time as measured at the onset of the radiative phase, then Σ corresponds roughly to the column density contained in one cooling length behind one of the steady state shocks calculated here. In that case, we can evaluate λ_c and M_c using our shock calculations to determine both T_{shell} and Σ. We find, for example, that for $v_s = 200$ km s^{-1} at $z = 5$, with $\Omega_b h^2 = 0.1$, that $\lambda_c \approx 4 \times 10^{20}$ cm and $M_c \approx 4 \times 10^4$ $M_\odot$. For the same shock velocity at $z = 20$, $\lambda_c \approx 6 \times 10^{20}$ cm, and $M_c \approx 6 \times 10^4$ $M_\odot$. We note that, unlike the Jeans length and mass, λ_c and M_c *are* affected by the presence of an external flux of ionizing radiation. Since the cooling time is increased, that is, when this radiation is present, Σ is also increased, even when T_{cold} is still 10^2 K.

4. Radiative Shocks Inside Protogalaxies and the Origin of Globular Clusters

The masses of globular clusters are confined to a remarkably narrow range, roughly 10^5 to 10^6 M$_\odot$. In contrast, stellar masses vary by three orders of magnitude and galactic masses by at least four orders of magnitude. Apart from globular clusters, there is little evidence for structure within the spheroidal components of galaxies (i.e., with masses from 1 to 10^5 M$_\odot$ or above 10^6 M$_\odot$). Moreover, the gross properties of globular clusters appear to be quite similar from one galaxy to another [27]. This suggests that globular clusters are the unique outcome of some process that operates in much the same way in different environments.

One possibility is that globular clusters formed during the collapse of protogalaxies [7]. The Jeans mass for a (spherical) cloud with a temperature T_c confined by an external pressure p_e is roughly $M_J \approx (kT_c/m_H)^2\, G^{-3/2}\, p_e^{-1/2}$. A natural value for T_c is 10^4 K, where the radiative cooling rate drops precipitously, and a natural value for p_e is $\rho_g\, v_g^2$, where ρ_g is the mean density within a protogalaxy and v_g is a typical collapse or virial velocity. The result is then $M_J \sim 10^6$ M$_\odot$, roughly the characteristic mass of globular clusters when some allowance is made for subsequent mass loss. Gunn [28] pointed out that these conditions would apply behind radiative shocks within a protogalaxy. This idea was also explored by McCrea [29], who noted that the necessary shocks would occur as large fragments collided in the buildup of protogalaxies. Fall and Rees [30] argued that a collapsing protogalaxy would be thermally unstable, giving rise to a two-phase medium with cool clouds at $T_c \approx 10^4$ K surrounded by hot diffuse gas at the virial temperature, $T_h \sim 10^6$ K. Apart from geometrical factors, these pictures are all manifestations of the same physical process; the cooling and compression of gas behind a radiative shock is just a thermal instability in one dimension. Therefore, the origin of globular clusters is hardly affected by the way or ways in which the two-phase medium is produced. One can show, using the Faber-Jackson and Tully-Fisher relations, that the Jeans mass given above has little variation from one galaxy to another [31].

The key issue regarding the origin of globular clusters is whether the clouds in a protogalaxy would remain at 10^4 K or cool to much lower temperatures. Fall and Rees pointed out that a necessary condition for the Jeans mass at 10^4 K to be "special" is that the cooling times within the clouds must be longer than their internal free-fall times [30]. Otherwise, the relevant value of M_{Jeans} is one at much lower temperatures. The main coolants below 10^4 K are molecular hydrogen and heavy elements produced by the first generations of stars. Fall and Rees argued that the H_2 would be dissociated by various sources of ultraviolet radiation within a protogalaxy, including the hot diffuse gas, massive young stars and an active galactic nucleus. In that event, when the metallicity was low, roughly $Z \lesssim 10^{-2}\ Z_\odot$, the clouds at 10^4 K would cool slowly enough to "imprint" a characteristic mass of order 10^6 $M_\odot$. However, the calculations by Fall and Rees were rather crude in that the gas was assumed to be in ionization equilibrium, leading to $n(H_2)/n_H \sim 10^{-5}$.

We have described above the importance of the nonequilibrium behavior of metal-free gas cooling radiatively from temperatures like those corresponding to the characteristic virial temperature of a protogalaxy. Our results demonstrate that the gas cools faster than it can recombine and, as a result, is able to form an H_2 concentration as high as 10^{-3} due to the greatly enhanced nonequilibrium ionized fraction at 10^4 K. In the absence of a strong external UV field, the gas does not "hang up" at 10^4 K in the sense discussed above. (A comment by Palla and Zinnecker [32] makes a related point about the nonequilibrium enhancement of H_2 cooling but fails to realize that this effect would actually eliminate the Fall and Rees suggestion for globular cluster formation.) Our studies of radiative shocks in the intergalactic medium (and [23]), however, indicate that a sufficiently intense UV field will dissociate the H_2 and delay the cooling below 10^4 K until a column density large enough to "self-shield" the H_2 accumulates behind the shock.

We have reconsidered this problem in detail in order to determine the conditions in a protogalaxy under which rapid H_2 formation and cooling are suppressed or delayed for a time long enough to imprint a characteristic mass of order 10^6 $M_\odot$ on the gas by gravitational instability [7]. We have applied the radiative shock code described earlier to calculate the thermal history of metal-free gas overtaken by steady-state radiative shocks of velocity 300 km s^{-1} such as would be produced by the typical gravitationally-induced motions within a protogalaxy, in a gas of density 0.1 to 1 cm^{-3} typical of that expected in the interstellar medium of the protogalaxy. We have considered the possibility that the gas is exposed to either radiation emitted by massive young stars or to the nonthermal continuum emitted by an active galactic nucleus (AGN). As we have already emphasized, the results should apply within factors of order unity to thermal instabilities of any geometry.

A selection of our results is shown in Figures 7 and 8, for a range of external flux levels for two illustrative external radiation spectra types, that of an AGN with a broken power-law energy flux going as $\nu^{-0.7}$ ($h\nu < 13.6\ eV$) and ν^{-1} ($h\nu \geq 13.6\ eV$) and that of a B0V star. It is convenient to express the incident flux level of these radiation fields in terms of the two dimensionless parameters ϕ, the ratio of the incident flux of ionizing photons to the incident flux $n_{H,1}v_s$ of H atoms crossing the shock, and Ψ, the ratio of the incident flux of H_2 dissociating photons at $h\nu \sim 12\ eV$ (i.e. in the Lyman-Werner bands) to $n_{H,1}v_s$. For the cases with an AGN-like radiation source (Figures 7a and 8a), the postshock T remains constant after dropping to $\approx 10^4$ K until the heating rate due to photoionization of He I by the deeply penetrating X-rays of the external flux is balanced by H_2 cooling. This He I photoionization is also important in generating secondary ionizations of H in the plateau at 10^4 K. The plateau ends in a vertical drop from $T \approx 10^4$ K to 10^2 K during which H_2 self-shielding in the Lyman-Werner bands allows the H_2 fraction to increase rapidly to $\approx 10^{-1.45}$. We have indicated on the plots the free-fall time $t_{ff} = [3\pi/(32G\rho)]^{1/2} \cong 1.4\times10^6 n_{H,1}^{-1/2}$ yrs for gas at the density and temperature of the plateau at 10^4 K as well as the more accurate gravitational instability growth time $t_G = [c_s/(2\pi G\rho_1 v_s)]^{1/2} = 6\times10^6 n_{H,1}^{-1/2}$ yrs based upon a thin sheet analysis for the gas in this plateau, where c_s is the sound speed of the plateau gas. For the cases shown in Figures 7b and 8b with early-type stellar incident radiation, which does not contain the X-ray photons of the AGN spectrum, the postshock gas can cool continuously below 10^4 K once it shields itself from H I ionizing photons, with a much smaller abundance of H_2 molecules than in the AGN-like flux case. In this case, it is primarily the photodissociating flux which determines the cooling time below 10^4 K.

Our results indicate that there is a critical threshold of external flux above which the time for the gas to cool below $\approx 10^4$ K after reaching this point exceeds the internal gravitational instability time for the gas, as required to explain globular cluster formation. For the AGN case, for $n_{H,1} = 0.1\ cm^{-3}$ and shocks located at distances out to 10 kpc from the AGN, the threshold flux is given by $\phi_{crit} \sim 30$, implying an AGN luminosity of $L_{AGN} \approx 2\times^{44}\ erg\ s^{-1}$. This is not unreasonable, since the relative rarity of quasars as luminous as $10^{46}-10^{47}\ erg\ s^{-1}$ ($\sim$1 per 50 galaxies at z = 2-2.5) nevertheless is consistent with every galaxy having an AGN with $L > 10^{46}\ erg\ s^{-1}$ for a few percent of its life. Moreover, the data do not preclude every galaxy at z=3 having a Seyfert-level nuclear luminosity $\gtrsim 10^{44}\ erg\ s^{-1}$. For the B0V case with $n_{H,1} = 0.1\ cm^{-3}$, $\phi_{crit} \sim 80$, with $\Psi_{crit} \sim 6\times10^3$. If such stars are distributed in a spherical, isothermal density profile out to a galactocentric radius of 10 kpc, then this implies a luminosity threshold of $L_* \approx 2\times10^{45}\ erg\ s^{-1}$. Similar values of the threshold L_* result if more massive O stars are considered instead of B0V stars. The star formation rate implied by an ordinary stellar origin for this UV field required to imprint the characteristic globular cluster mass is marginally consistent with constraints based on the abundances of heavy elements in the spheroidal components of galaxies. *In conclusion, we suggest that the formation of globular clusters is a threshold effect with the intensity of the UV/soft X-ray field in a protogalaxy as the controlling factor.*

This work was supported in part by Robert A. Welch Foundation Grant F-1115, Texas Advanced Research Program Grant 4132, and NASA NGT-50316. H. K. was partially supported by NSF Grant AST87-20285 and by the Minnesota Supercomputer Institute

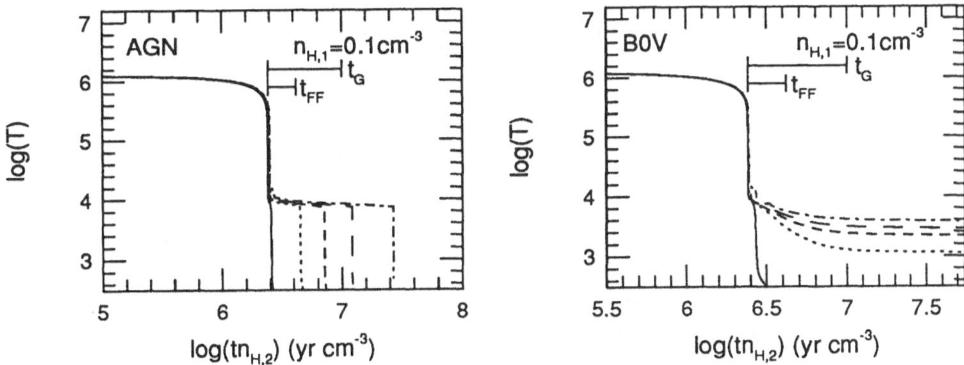

Figure 7 (a). (left) Postshock gas temperature vs. $tn_{H,2}(years\ cm^{-3})$ for cases with $n_{H,1} = 0.1\ cm^{-3}$, and with a AGN-like source. From the left to the right, the shocks with $\phi_{AGN} = 0$ (i.e. no external radiation field, solid line), 30 (dotted line), 60 (dashed line), 120 (long dashed line), and 240 (dot-dashed line) are plotted. Two pairs of vertical markers connected by a line represent the free-fall time (t_{FF}) and the characteristic gravitational time (t_G) of the gas when it reaches $10^4\ K$ multiplied by $n_{H,2}$ for the cases with $n_{H,1} = 0.1\ cm^{-3}$.

Figure 7 (b). (right) Same as (a), except that the external radiation sources are B0 V stars. From the bottom to the top, the shocks with $\phi_{B0V} = 8, 40, 80, 160$, and 400 are plotted.

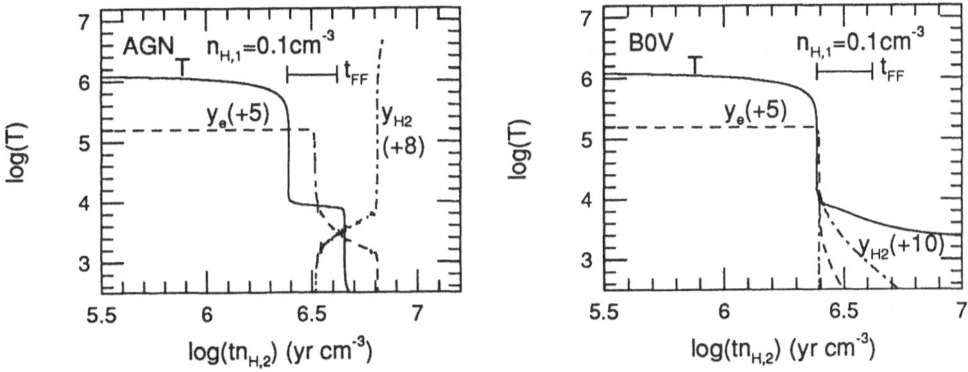

Figure 8 (a). (left) Temperature (solid line), the electron fraction (dashed line) and the abundance of H_2 (dot-dashed line) for the case with $\phi_{AGN} = 30$, and $n_{H,1} = 0.1\ cm^{-3}$. The value of $\log(y_e)$ is offset by 5 and $\log(y_{H_2})$ by 8. For example, the actual value of $\log(y_e)$ is the value plotted minus 5.

Figure 8 (b). (right) Same as (a), except $\phi_{B0V} = 80$. The value of $\log(y_{H_2})$ is offset by 10.

(MSI) during part of this work. Our numerical calculations were performed on The University of Texas Center for High Performance Computation Cray X-MP computer and on MSI's Cray 2.

References

1. Shapiro, P. R. 1986, in *Galaxy Distances and Deviation from Universal Expansion*, eds. B. F. Madore and R. B. Tully (Dordrecht: Reidel), pp. 203-214.

2. Shapiro, P. R. and Kang, H. 1987, in *IAU Symposium 117: Dark Matter in the Universe*, eds. G. R. Knapp and J. Kormendy (Dordrecht: Reidel), p. 365.

3. Shapiro, P. R. and Kang, H. 1987, *Ap. J.*, **318**, 32.

4. Shapiro, P. R. and Kang, H. 1987, *Rev. Mexicana Astron. Astrof.*, **14**, 58.

5. Shapiro, P. R. and Kang, H. 1987, in *High Redshift and Primeval Galaxies*, eds. J. Bergeron, D. Kunth, B.Rocca-Volmerange, and J. Tran Thanh Van (Paris: Editions Frontières), pp. 501-515.

6. Kang, H. and Shapiro, P. R. 1990, *Ap. J.*, in press.

7. Kang, H., Shapiro, P. R., Fall, S. M., and Rees, M. J. 1990, *Ap. J.*, in press.

8. Saslaw, W. C. and Zipoy, D. 1967, *Nature*, **216**, 976.

9. Peebles, P.J.E. and Dicke, R. H. 1968, *Ap. J.*, **154**, 891.

10. Hirasawa, T., Aiza, K., and Taketani, M. 1969, *Progr. Theor. Phys.*, **41**, 835.

11. Hirasawa, T. 1969, *Progr. Theor. Phys.*, **42**, 523.

12. Matsuda, T., Sato, H., and Takeda, H. 1969, *Progr. Theor. Phys.*, **42**, 219.

13. Yoneyama, T. 1972, *Pub. Astr. Soc. Japan*, **24**, 87.

14. Hutchins, J. B. 1976, *Ap. J.*, **205**, 103.

15. Silk, J. 1977, *Ap. J.*, **211**, 638.

16. Carlberg, R. G. 1981, *M.N.R.A.S.*, **197**, 1021.

17. Palla, F., Salpeter, E. E., and Staler, S. W. 1983, *Ap. J.*, **271**, 632.

18. Lepp, S. and Shull, J. M. 1984, *Ap. J.*, **280**, 465.

19. Izotov, Yu. I and Kolesnik, I. G. 1984, *Soviet Astr.*, **28**, No. 1, 15.

20. Murray, S. D. and Lin, D.N.C. 1989, *Ap. J.*, **339**, 933.

21. Struck-Marcell, C. 1982, *Ap. J.*, **259**, 116.

22. Struck-Marcell, C. 1982, *Ap. J.*, **259**, 127.

23. MacLow, M. M. and Shull, J. M. 1986, *Ap. J.*, **302**, 585.

24. Vishniac, E. T., Ostriker, J. P., and Bertschinger, E. 1985, *Ap. J.*, **291**, 399.

25. Vishniac, E. T. 1983, *Ap. J.*, **274**, 152.

26. Ostriker, J. P. and Cowie, L. L. 1981, *Ap. J. (Letters)*, **243**, L127.

27. Harris, W. F. 1987, *P.A.S.P.*, **99**, 1031.

28. Gunn, J. E. 1980, in *Globular Clusters*, eds. D. Hanes and B. Madore (Cambridge: Cambridge Uniersity Press), p. 301.

29. McCrea, W. H. 1982, in *Progress in Cosmology*, ed. A. W. Wolfendale (Dordrecht: Reidel), p. 239.

30. Fall, S. M. and Rees, M. J. 1985, *Ap. J.*, **298**, 18.

31. Fall, S. M. and Rees, M. J. 1988, in *I.A.U. Symposium 126: Globular Cluster Systems in Galaxies*, eds. J. E. Grindlay and A.G.D. Philip (Dordrecht: Reidel), p. 323.

32. Palla, F. and Zinnecker, H. in *The Harlow-Shapley Symposium on Globular Cluster Systems in Galaxies*, eds. J. E. Grindlay and A. G. David Philip (IAU), p. 697.

DISCUSSION

DI FAZIO: Paul, could you please comment on the energy-λ relation, from which we get the two λ_{min}, λ_{max} critical points? And also on the influence of the thin-sheet geometry on this sort of dispersion relation?

SHAPIRO: For the globular cluster formation problem, one can apply the thin sheet analysis to the postshock plateau layer at 10^4 K. In that case, there is a critical column density which must be accumulated in the plateau layer before gravitational instability occurs and a time t_G for the fastest growing mode corresponding roughly to 2 λ_{min} in the energy argument. The result for this thin sheet geometry is to make t_G somewhat larger than the free-fall time for a homogeneous sphere at the same density and to make the characteristic minimum fragment mass somewhat larger than the spherical Jeans mass M_J as well.

MATEO: Is there a way to interpret your conclusion that there is a minimum dissociative flux necessary to form globular clusters in terms of, say, a minimum mass that a galaxy could have and still have globulars? If not, is there some other observational consequence of this conclusion of yours?

SHAPIRO: The value of the characteristic mass imprinted gravitationally on the 10^4 K cooled gas does not itself vary much from one galaxy to another, as one can show using the Faber-Jackson and Tully-Fisher relations. However, the threshold luminosity requirements may vary with galaxy mass as the virial velocity and gas density vary, since the former sets the characteristic shock velocity while both effect the gas density at 10^4 K and the shocked H atom flux which enters the dimensionless parameters ϕ_{crit} and Ψ_{crit}. We are presently exploring this dependence. As for other observational consequences, there is one speculation worth mentioning if the radiation source is an AGN and if that emission is beamed; globular clusters would form preferentially "in the beam." Although the memory of this initial condition depends upon the dynamics, a nearly spherical galactic potential might permit cluster orbits to remain preferentially in certain quadrants even until the present.

CAYREL: How many frequencies have you carried on in your transfer equations?

SHAPIRO: We represent the diffuse radiation spectrum typically with 150 energy bins from 0.7 eV to 500 eV, while the energy bins for the external flux range to 12.4 keV. Resonance scattering effects are considered separately for the Lyman lines of H I, He I, and He II, as is the self-shielding of H_2 in the Lyman-Werner bands.

MOUSCHOVIAS: In a typical one of your runs, what is the largest H_2 density reached in the postshock H_2-cooling region, and what is the dominant ion (H^- or H_2^+)?

SHAPIRO: The largest H_2 concentrations typically range from 10^{-3} to a few times 10^{-2}. In general, the H^- process dominates.

The Intergalactic Medium: Initial and Boundary Conditions for Galaxy and Primeval Star Formation

Mark L. Giroux and Paul R. Shapiro
Department of Astronomy
The University of Texas at Austin
Austin, Texas 78712 USA

ABSTRACT. The formation of the first stars in gas of primordial composition depended upon the complex interplay of the hydrodynamics of gravitational collapse with nonequilibrium chemistry and its effect on the thermal evolution of the gas. The initial conditions and boundary values for the clouds which collapsed gravitationally to form galaxies and stars were determined by the thermal and ionization history of the general intergalactic medium. We describe new detailed numerical calculations of the evolution of the intergalactic medium in a post-recombination Friedmann universe, including a solution of the nonequilibrium rate equations for the ionization and recombination of H and He, the energy equation, and the equation of radiative transfer. The implications of this study for the characteristic mass scale and epoch of gravitational collapse for the average "first" star forming clouds will be discussed. The star formation rate and associated metallicity generation implied if the intergalactic medium was fully ionized by starlight by redshift $z > 4$ as suggested by recent interpretations of the Gunn-Peterson effect will also be discussed. Finally, we show that the relative strengths of metal lines from a quasar absorption-line Lyman limit system cloud at $z = 3$ photoionized by our intergalactic uv radiation background are compatible with the observations, at least for background sources of either AGN-type or star-forming galaxy spectra.

1. Introduction

Just as the state and evolution of the interstellar medium prescribe the formation of stars within our own galaxy, so the state and evolution of the intergalactic medium (IGM) at earlier cosmological epochs prescribe the formation of galaxies and primordial stars. The IGM determines the initial and boundary conditions for galaxy and primordial star formation and in turn is altered by feedback from the formation processes. The IGM, probed as it is by radiation from distant quasars, also provides crucial diagnostics of the history of galaxy and star formation in the universe. In what follows, we briefly describe some of our recent calculations of the thermal and ionization evolution of the IGM and their implications for galaxy and star formation. For a recent review of this subject, we refer the reader to reference [1].

R. Capuzzo-Dolcetta et al. (eds.), Physical Processes in Fragmentation and Star Formation, 71–79.

2. The Postrecombination IGM Prior to Reionization

The absence of a detectable H Ly α absorption trough in the spectra of high redshift quasars due to a smoothly distributed IGM (the so-called Gunn-Peterson effect) is an indication that the IGM, which recombined at redshift $z \sim 10^3$, when matter and radiation decoupled, must have been reionized at some epoch prior to the epoch corresponding to the known quasar redshifts (cf. reference [1] and references therein). At $z \sim 10^3$, however, just after the recombination epoch, the IGM must have been extremely uniform since no anisotropy in the cosmic microwave background (CMB) has yet been detected, other than the dipole anisotropy due to the Earth's motion relative to the CMB rest frame, down to levels below a part in 10^4 on any angular scale. In order for galaxies, stars, or any gravitationally bound clouds to have formed out of the IGM, therefore, it was first necessary for density fluctuations to grow on these mass scales and, if the growth was gravitational, this requires that the Jeans mass in the IGM not have exceeded those mass scales. Peebles and Dicke pointed out some time ago that the Jeans mass M_J in the IGM immediately following recombination in a baryon-dominated universe is $\sim 10^6\ M_\odot$, which they suggested could explain the origin of globular clusters [2]. In the so called Cold Dark Matter (CDM) model of galaxy formation [cf. 3,4], the first fluctuations on average to collapse out of the IGM are those typically of size somewhat less than the baryon Jeans length just after recombination with larger scales collapsing more recently. In fact, if the electron and radiation temperatures remain coupled by Compton scattering down to $z = 100$, then the average epoch for this smallest scale to complete conventional top hat collapse and thereby cease to be a part of the IGM is $1 + z \cong 13.7/b$, where b is the "biasing parameter" used to indicate the amplitude assumed for the initial fluctuation spectrum (e. g. Babul and Shapiro, private communication). For the standard CDM model, $b = 2.5$, in which case most of the baryons in the universe have not yet collapsed out of the background IGM until $z \lesssim 4.5$. Apparently, the evolution of M_J is an important ingredient in such models.

As part of our general study of the IGM, we have performed a new detailed calculation of the thermal and ionization evolution of a uniform IGM of primordial composition (H and He) in an expanding matter-dominated, Friedmann universe. Our calculation solves the rate equations for ionization and recombination, the equation of energy conservation, including the effects of cosmological expansion, radiative cooling, and cooling (or heating) by Compton scattering of CMB radiation, and the equation of radiative transfer, including the opacity of the IGM and its own diffuse emission. We present the results of this calculation in Figures 1 and 2 for the postrecombination IGM prior to reionization for the Baryon Jeans mass M_J and electron fraction n_e/n_H, for the case $\Omega_b h^2 = 0.1$, $q_0 = 1/2$, and $h = H_0/(100\ km\ s^{-1}\ Mpc^{-1}) = 1$, where Ω_b is the IGM mass density in units of the critical density, and we take $n(He) = 0.1\ n(H)$. Our initial conditions at $z = 800$ were $T = T_r = (2.7)(801)\ K$ and $n_e/n_H = 1.89 \times 10^{-3}$ [5]. The mass $M_J = 8.4 \times 10^5\ \Omega_b h^{-1}\ \Omega_{tot}^{-3/2}\ (T/T_r)^{3/2}\ M_\odot$, for a a CDM dominated universe where $\Omega_{tot} = 2q_0 = 1$ [3,6]. Our results indicate that M_J drops to half the constant value it would have if $T = T_r$ [i. e. $M_J(T_e = T_r) = 8.4 \times 10^5\ M_\odot$] when $z = 200$.

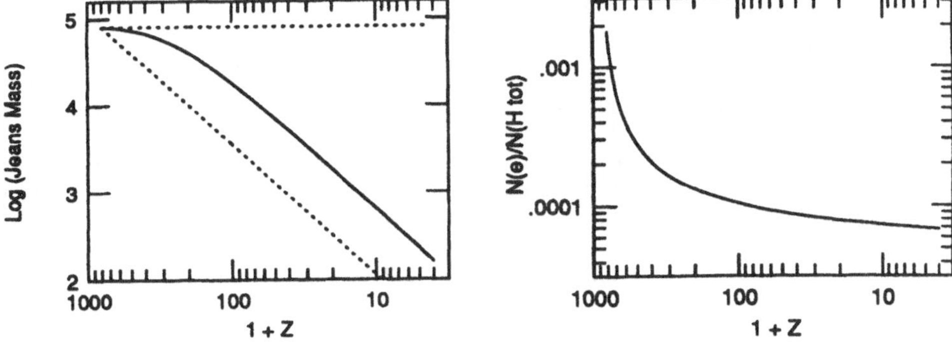

Figure 1. (left) $M_J/M_\odot$ for postrecombination IGM before reionization epoch. Dotted curve (upper) gives $M_J/M_\odot$ if $T = T_r \propto (1+z)$, while dotted curve (lower) assumes $T = T_r(z = 800)[(1+z)/801]^2$, the adiabatically cooling IGM.

Figure 2. (right) Electron fraction versus redshift for the same IGM as solid curve in Figure 2.

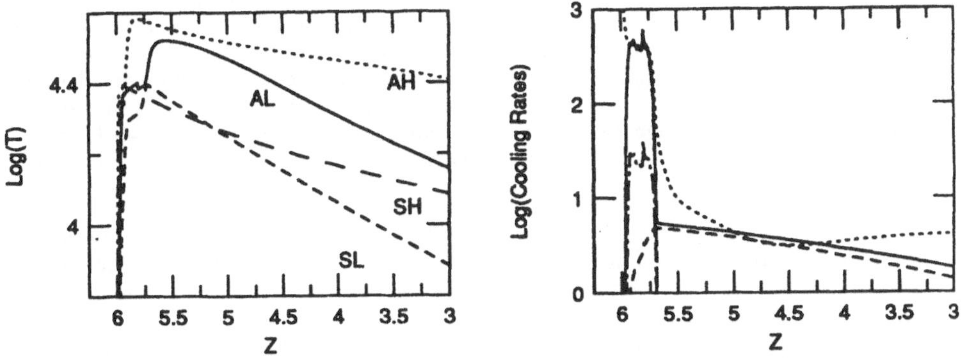

Figure 3. (left) Temperature of IGM versus redshift for $z_{turn-on} = 6$, $q_0 = 1/2$, $h = 0.75$, for case AL ($\zeta = 325$, $\Omega_b = 0.02$, AGN-like) (solid line), case AH ($\zeta = 1100$, $\Omega_b = 0.1$, AGN-like) (dotted line), case SL ($\zeta = 400$, $\Omega_b = 0.02$, Stellar) (short dash), case SH ($\zeta = 1200$, $\Omega_b = 0.1$, Stellar) (long dash). AGN-like sources have $\nu^{-1.5}$ energy spectra. Stellar sources have a Bruzual spectrum assuming a Salpeter IMF.

Figure 4. (right) Normalized cooling rates $\Lambda/\Lambda_{adiabatic}$ versus redshift z for case AL where $\Lambda_{adiabatic}$ is adiabatic cooling due to cosmological expansion. Solid line is the total radiative cooling and dotted is the total heating. In the ionizing up stage the cooling is dominated by line cooling (coincident with solid line) and collisional ionization cooling (dot-dashed). Afterwards, Compton cooling (short dash) is the dominant component.

3. The Reionization of the IGM

We have previously discussed the requirements for the reionization of the IGM in order to satisfy the Gunn-Peterson constraint, that $\tau_{GP} \leq 0.1$ at $z = 3.5$, where τ_{GP} is the H Ly α optical depth just to the blue of the Ly α emission line in the spectrum of a quasar at $z = z_{GP} = 3.5$ [1,7-11, see also 12-15]. For a photoionized IGM, it is useful to parameterize the required emissivity of uniformly distributed sources of ionizing photons in terms of the dimensionless ratio ζ defined as follows:

$$\zeta = 2n_x^0 N_{ph}(\geq 13.6eV)/(3H_0 n_H^0) \tag{3.1}$$

where n_x^0 is the number of sources per cm^3 (present comoving volume) assuming that $n_x(z)/n_x^0 = (1+z)^3$, $N_{ph}(\geq 13.6\ eV)$ is the luminosity per source in ionizing photons per second, and $n_H^0 = 8 \times 10^{-6}\Omega_b h^2\ cm^{-3}$ is the mean baryon density referred to the present [i. e. assuming $n_H(z)/n_H^0 = (1+z)^3$].

We have previously shown that assuming there is no high-z cutoff in the quasar distribution for $z > 3$, the required value of ζ, ζ_{GP}, exceeds the apparent quasar contribution at $z \sim 3$ by more than an order of magnitude unless the IGM density $\Omega_b < 10^{-2}$ [cf. 1, 11]. If the apparent decrease in the quasar number per comoving volume for $z \gtrsim 3$ is real, moreover, then the absence of any $\tau_{GP} \gtrsim 1$ in the spectrum of PC1158+4635 at $z = 4.73$, the highest redshift quasar known [16], makes the discrepancy between the quasar contribution and the required ζ_{GP} considerably greater.

An independent constraint on ζ at $z \gtrsim 4$ comes from the requirement that the Lyman α forest quasar absorption line clouds (QALC) be photoionized by the background UV radiation. Ionizing an IGM composed entirely of these discrete clouds, even if $\Omega_b = 0$ *between* the clouds, requires more than the observed quasar contribution even if there is *no* quasar cut-off for $z > 3$ [9]. An observed drop in the number of QALC's in a quasar's spectrum in the redshift range close to that of the quasar has also been used to infer an intensity for the metagalactic UV field at $z \sim 3$ which exceeds the known contribution from quasars [17].

In short, we are forced to conclude that some as yet undetected sources dominate the ionizing metagalactic radiation field at $z > 3$. We have calculated in detail the reionization of the uniform IGM by such high redshift sources for two limiting types of source spectra, that of an AGN (with energy flux $F_\nu \propto \nu^{-1.5}$ for $h\nu \geq 13.6\ eV$) and that of starlight from a galaxy undergoing star formation with a Salpeter IMF, as described by Bruzual [18]. Our results for two IGM densities, $\Omega_b h^2 = 0.05$ and 0.01, for $h = 0.75$, and a source turn-on epoch of $z_{turn-on} = 6$, including the important effect of the continuum opacity of the QALC's (of column density $10^{14} \leq N_{HI} \leq 10^{22}\ cm^{-2}$) due to both H I and He II in the clouds [11,14], are shown in Figures 3-9. Similar calculations are shown in [14].

In all of the cases shown in Figures 3-9, ζ values were chosen so that $\tau_{GP} = 0.1$ at $z = 3.5$ in accord with the observed upper limits. The quasar PC 1158+4635 at $z = 4.73$, on the other hand, has been observed with low spectral resolution to show a depression of the continuum to the blue side of its Ly α emission line[16]. This depression is higher than that expected from an extrapolation of the QALC H Ly α line opacity from its behavior at lower redshift. This implies that either the evolution of the QALC population with redshift changes for $z \gtrsim 4$ or else the excess depression is due to the previously undetected

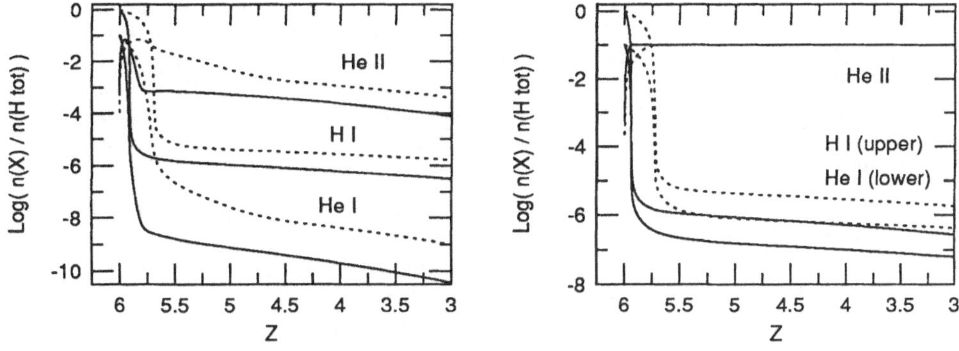

Figure 5. (left) Ionic Fractions n(X)/n(H) for cases AH(solid lines) and AL (dotted lines) as labelled.

Figure 6. (right) Same as Figure 6, except cases are SH(solid) and SL(dotted). The label "H I(upper)" refers to the upper dotted and solid lines in the lower half of the graph. The label "He I(lower)" refers analogously to the lower of these lines.

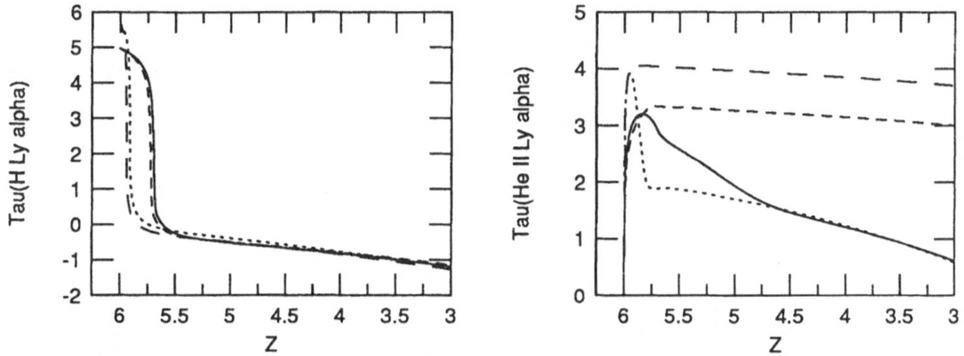

Figure 7. (left) τ_{GP}(H Lyman α) versus z for the same cases and corresponding line types as defined in Figure 4.

Figure 8. (right) τ_{GP}(He II Lyman α) versus z for the same cases and corresponding line types as defined in Figure 4.

Gunn-Peterson effect of a smoothly distributed IGM. If the former were the case and the depression is entirely accounted for by the effect of the QALC's, then our results indicate that ζ_{GP} should now be substantially *increased* in order to satisfy $\tau_{GP} < 0.1$, say, at $z = 4.73$. For example, if $\Omega_b h^2 = 0.05$, $\zeta_{GP} = 3100$ in this case, while for $\Omega_b h^2 = 0.01$, $\zeta_{GP} = 1040$. These values may still be underestimates, however, since adequate account must be taken of the high QALC opacity at $z \gtrsim 4$ implied by this interpretation of the continuum depression for PC 1158+4635.

As pointed out in [11], the emission of ionizing photons by stars implies the production of metallicity. Suppose the emissivity corresponding to such ζ_{GP} values as those above is supplied continuously until $z < 4.7$ by massive stars. Then, as described in [11], we can estimate that the metallicity generated by these stars corresponds to $\Omega_{metals} \gtrsim 7 \times 10^{-5} \zeta_3 (\Omega_b/0.1)$, or, since $\Omega_{gal,lum} \approx 10^{-2}$ for the luminous matter inside galaxies, the mass fraction in metals corresponds to $\Omega_{metals}/\Omega_{gal,lum} \gtrsim 7 \times 10^{-3} \zeta_3 (\Omega_b/0.1)$. For $h = 0.75$, $\Omega_b h^2 = 0.01$ and $\zeta_{GP} = 1040$, this means that more than $\sim 1/10$ of the Population I metal abundance was generated before $z \sim 4.5$. If, on the other hand, $\Omega_b h^2 = 0.05$, and $\zeta_{GP} = 3100$, then the implied metallicity by $z \sim 4.5$ exceeds the solar value!

Alternatively, if instead, we attribute half or more of the continuum depression of PC 1158+4635 to the Gunn-Peterson effect, which we estimate would imply $\tau_{GP} \gtrsim 0.7$, then an interesting constraint on the ionization source turn-on epoch results, as follows. Our results indicate that for epochs later than that at which the IGM ionization "break through" occurs, the curve of τ_{GP} versus z has a universal slope (e. g. see Figure 7). In particular, if $\tau_{GP} \lesssim 0.1$ at $z = 3.5$, then at $z = 4.73$, $\tau_{GP} \lesssim 0.3$ is predicted as long as the source turn-on epoch and "ionization breakthrough" are significantly earlier than $z = 4.73$. Hence, if $\tau_{GP}(PC1158 + 4635) \gtrsim 0.7$, then this is apparently too *large* a value of τ_{GP} if the sources turned on much before $z = 4.73$. *In short, if* $\tau_{GP} \sim 1$ *has been detected at* $z = 4.73$ *and yet* $\tau_{GP} \lesssim 0.1$ *at* $z = 3.5$, *then our results indicate that the ionization source turn-on redshift must be significantly less than 6 and relatively close to 4.73.* That being the case, we would further expect a significant increase of τ_{GP} with small increase of z beyond $z = 4.73$ for future high z quasar discoveries. We may finally be observing the epoch of reionization of the IGM.

4. AGN Versus Stellar Sources: Quasar Absorption Line Diagnostics

Steidel and Sargent [19] have made the important suggestion that the relative abundance of metal ions observed in Lyman limit system QALC's ($N_{HI} \sim 10^{17.5}$ cm^{-2}) at $z \sim 3$ provides a useful diagnostic of the radiation background at that epoch, since the QALC's are generally believed to be low-density, photoionized gas clouds. They have employed a photoionized nebula code in order to calculate the relative abundances expected if the metagalactic UV background has a spectrum like that of the sources in our two spectral cases, the AGN and stellar spectra. Motivated by the possibility that our detailed treatment of both cosmological effects and the continuum opacity of the IGM and of the QALC's in calculating the background spectrum might have resulted in a shape for the background radiation spectrum different from that of the intrinsic source spectra, we have reconsidered the ionic abundance ratios predicted for these photoionized Lyman limit clouds. We have used the most recent version of Ferland's photoionized nebula code CLOUDY [20], together with our own ionizing radiation spectra, as shown in Figure 9,

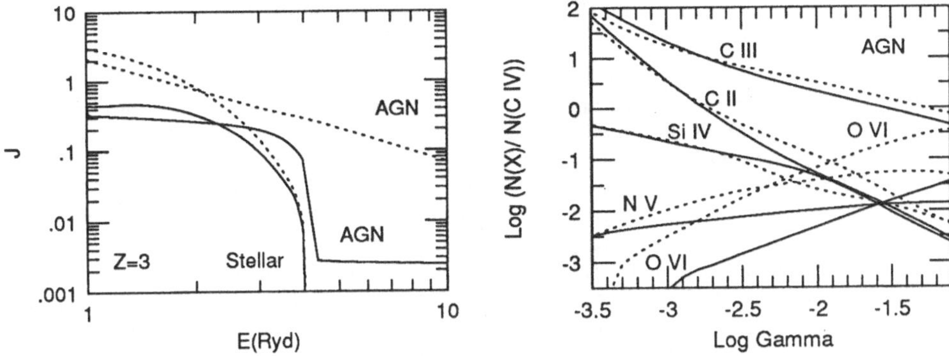

Figure 9. (left) Arbitrarily normalized mean intensity ($erg\ cm^{-2}\ s^{-1}\ Hz^{-1}\ ster^{-1}$) of intergalactic radiation field at $z = 3$ for cases AL and SL as attenuated by the IGM, including Ly α cloud opacity (solid curves), and as in the absence of attenuation (dotted).

Figure 10. (right) Predicted quasar absorption-line cloud column densities of the labelled species relative to that of C IV versus Γ for a cloud exposed to AGN-source radiation fields of Figure 9, attenuated (solid) and unattenuated (dotted).

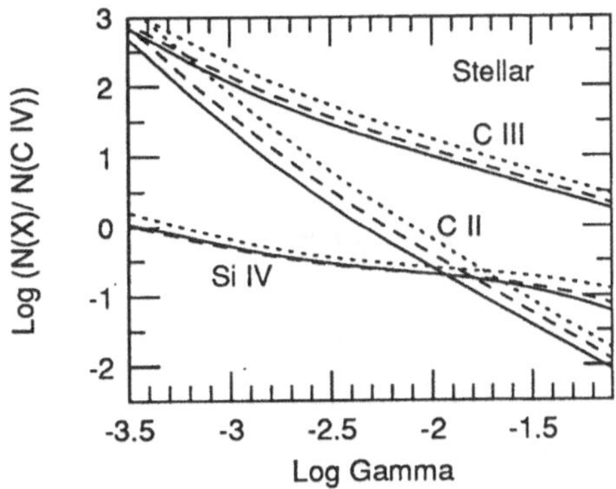

Figure 11. Same as Figure 10, except for the stellar radiation fields in Figure 9. The dashed curve gives results if spectrum is the intrinsic spectrum unattenuated and unmodified by cosmological expansion.

for clouds of H I column density $10^{17.5}$ cm^{-2} and metal abundance 10^{-2} times the solar value, like the clouds modelled by Steidel and Sargent.

Our results for each of our two assumed source spectra, the AGN and Bruzual stellar spectra, are shown in Figures 10 and 11. We show the differences which result both from taking proper account of the opacity of the IGM and of the effect of radiative transfer in a cosmologically expanding universe. According to Steidel and Sargent, the observations require $N(CII)/N(CIV) < 1$, $N(CIII)/N(CIV) < 10$, and $N(SiIV)/N(CIV) \lesssim 1$. We find that *both* kinds of source spectra are compatible with these simple observed abundances ratio requirements, contrary to the conclusion reached by Steidel and Sargent. They had concluded from their results that the shape of the background source spectra must be predominantly AGN-like, rather than stellar. While we do find a larger range of Γ values ($\Gamma = n_\gamma/n_H$, where n_H and n_γ are the number densities of H atoms inside the cloud and of ionizing photons in the the incident radiation field, respectively) allowed for the AGN case (i. e. $\log(\Gamma) \gtrsim -2.7$), we nevertheless find that a reasonable range of Γ values (i. e. $\log(\Gamma) \gtrsim -2$) also satisfies the observational constraint for the stellar case.

We note that the detection of or placement of upper limits on N V and O VI in the same clouds may provide a stronger discriminant between the two source spectra, since the stellar case does not have enough high energy radiation to generate these species. We also note, however, that, for these ions, the effect of properly including both the H I and the He II continuum opacities of the IGM and of the QALC's is quite large and should be taken into account.

We are grateful to Gary Ferland for supplying us with the latest version of his CLOUDY code. This work benefitted from the support of Robert A. Welch Foundation Grant F-1115, Texas Advanced Research Program Grant 4132, and NASA Training Grant NGT-50316.

References

1. Shapiro, P. R. 1989, in *Fourteenth Texas Symposium on Relativistic Astrophysics*, ed. E. J. Fenyves, *Ann. N. Y. Acad. Sci.*, **571**, pp. 128 – 150.

2. Peebles, P. J. E. and Dicke, R. H. 1968, *Ap. J.*, **154**, 891.

3. Blumenthal, G. R., Faber, S. M., Primack, J. R., Rees, M. J., 1984, *Nature*, **311**, 517.

4. White, S. D. M., Frenk, C. S., Davis, M., and Efstathiou, G., 1987, *Ap. J*, **313**, 505.

5. Jones, B. J. T. and Wyse, R. F. G. 1985, *Astron. and Astrophys.*, **149**, 144.

6. Peebles, P. J. E. 1984, *Ap. J.*, **277**, 470.

7. Shapiro, P. R. 1986, in *Galaxy Distances and Deviations from Universal Expansion*, eds. B. F. Madore and R. B. Tully (Dordrecht: Reidel), pp. 203 – 213.

8. Shapiro, P. R. 1986, *P. A. S. P.*, **98**, 1014.

9. Shapiro, P. R. and Giroux, M. L. 1987, *Ap. J. (Letters)*, **321**, *L*107.

10. Shapiro, P. R., Giroux, M. L., and Kang, H. 1987, in *High Redshift and Primeval Galaxies*, eds. J. Bergeron, D. Kunth, B. Rocca-Volmerange, J. Tran Thanh Van (Paris: Editions Frontiéres), pp.501 – 515.

11. Shapiro, P. R. and Giroux, M. L. 1989, in *The Epoch of Galaxy Formation*, eds. C. S. Frenk, R. S. Ellis, T. Shanks, A. F. Heavens, and J. A. Peacock (Boston: Kluwer Academic Publishers), pp. 153 – 161.

12. Donahue, M. and Shull, J. M. 1987, *Ap. J.*, **323**, L13.

13. Bond, J. R., Szalay, A. S., and Silk, J. 1988, *Ap. J.*, **324**, 627.

14. Miralda-Escude, J. and Ostriker, J. P. 1990, preprint.

15. Ikeuchi, S. and Ostriker, J. P. 1986, *Ap. J.*, **301**, 522.

16. Schneider, D. P., Schmidt, M., Gunn, J. E. 1989, *Astron. J.*, **98**, 1951.

17. Batjlik, S., Duncan, R. C., and Ostriker, J. P. 1988, *Ap. J.*, **327**, 570.

18. Bruzual, G. 1983, *Ap. J. Suppl.*, **53**, 497.

19. Steidel, C. C. and Sargent, W. L. W. 1989, *Ap. J.(Letters)*, **343**, *L*33.

20. Ferland, G. J. 1989, Ohio State University Astronomy Department Internal Report, 90 – 001.

SUPERSONIC TURBULENT FRAGMENTATION OF GIANT MOLECULAR CLOUDS

IGOR G. KOLESNIK and YAROSLAV YU. OHUL'CHANSKY
Main Astronomical Observatory
Ukrainian Academy of Sciences
252127 Kiev
USSR

ABSTRACT. The statistical description of compressible turbulence in application to GMC are carried out. The initial supersonic turbulent spectrum evolution on the stage of shock wave formation is considered. It is found that interecting shock fronts lead to the flat filamentary structures formation in the supersonic turbulent medium. The clumps size distribution and space filling factor are determined. It is found that these functions depend on a physics of clump's formation.

1. INTRODUCTION

For giant molecular clouds (GMC) two prominent properties are typical. GMCs consist of dense molecular gas clumps concentrating to the GMC centre and filling only a few persent of a whole its volume (Kahane et al.(1985)). Another typical feature of GMCs is related to the properties of clumps' motion. These clumps participate in chaotic motions with velocities v_t as a rule exceeding the sound velocity c_o at the temperature of molecular gas. This phenomenon is considered as a supersonic molecular cloud's turbulence.When velocity v_t exceeds about $0.3c_o$ a compressibility of the turbulent matter become very important. That stimulate clumps formation and fragmentation of molecular cloud. Thus in application to GMC it is necessary to develop the theory of turbulence and fragmentation under transsonic and supersonic random motions. These problems are considered in the given paper.

According to observations the velocity v_t is connected with turbulent length scale l by power relation $v_t \propto l^{\varkappa}$ (Larson (1981); Myers (1983); Solomon et al.,(1987)). According to Larson (1981) $\varkappa = 0.38$, but in the most cases $\varkappa \simeq 0.5$. The statistical dependence v_t on l is existed in a very wide range of length scales, from the smallest ones $l \simeq 0.01$ pc, that is close to the limit of spatial resolution, up to the largest ones $l \geq 10$ pc, that is comparable with the dimension of GMC. In some cases the statistical relations for density fluctuations in GMCs

R. Capuzzo-Dolcetta et al. (eds.), Physical Processes in Fragmentation and Star Formation, 81–86.

(Kleiner and Dickman (1984)) and for the size distribution of molecular clouds (Terebey et al. (1987)) have been determined.

The different aspects of molecular cloud's turbulence are discussed extensively now. But development of the statistical model for supersonic turbulence is collided with considerable difficulties. The results of Passot at al. (1988) that are related to the numerical simulations of turbulent medium are important from this point of view.

The hydrodynamic flow velocity field can be divided into the potential and vortical components. When transsonic or supersonic motions prevail the potential component is become more important that stimulates the shock wave's stochastic field development. Ohul'chansky (1988a) has described this process on the base of Burgers' equation treatment. In this paper we apply this approach for conditions of GMCs that permit the supersonic turbulence' spectrum evolution, the large density fluctuations development, and clumps formation to consider.

2. CLUMPS FORMATION IN MOLECULAR CLOUDS.

The GMCs are found to be maintained by turbulent pressure in virial equilibrium (Solomon et.al. (1987)) and their turbulent velocities don't exceed sound velocity in the warm interstellar gas having temperature T= 8000÷10000 K. This fact may be considered as a hint that the supersonic motions in molecular clouds have arisen from the incompressible subsonic turbulence with Kolmogorov's spectrum which take place in the warm interstellar gas. This possibility was argued by Kolesnik (1987; see this volume also). The existence of Kolmogorov's spectrum $E(k) \propto k^{-5/3}$, where k is the wave number , in interstellar neutral hydrogen is confirmed by observations (Kalberla and Stenholm (1983)).

It is known that in the limit of incompressible fluid having Kolmogorov's spectrum the vortex hierarchy with energy transfer from the largest energy-containing spatial scales to the smallest one being determined by viscous dissipation is existed. If medium temperature falls rapidly (Kolesnik (1987)), the compressibility effect becomes significant up to some length scale $l_o \simeq \alpha\, l_c$, where l_c is determined by condition $v_t(l_c) \simeq c_o$ in the presence of Kolmogorov's spectrum, and $\alpha \leq 1$ is a free parameter. In the range $l > l_o$ the potential component of velocity field begins to develope. Potential motions generate compression waves which are transformed into shock waves. These ones interact and the molecular cloud's clumpy structure formation are promoted. The scale l_o is defined as the minimal vortex' size up to which the clumpy structure will form. The typical temperature $T \simeq 10$ K of molecular gas corresponds to the sound velocity $c_o \simeq 0.2$ km/s. Taking into account that in the initial state turbulent velocity v_t depends on the length scale l according to the relation $v_t \simeq (1\mathrm{km/s}) \cdot (l/1\mathrm{pc})^{1/3}$, which is close to Larson (1981) one, for $c_o = 0.2$ km/s one obtaines $l_o \simeq 0.02$ pc. The existence of condensations with velocity dispersion $v_t \simeq c_o$ on such scales has been discovered by Falgarone and Puget (1988).

Thus in the range of scales $l_o \leq l \leq L_{max} \simeq 10$ pc the turbulent motions will have the supersonic velocities and their potential velocity components will determine the structural properties of turbulent medium.

Now let us consider the initial evolution of motions in the scales $l > l_o$ or for the wave numbers $k < k_o = 2\pi/l_o$. At supersonic velocities the inertial compression of the main bulk of gas predominates over the pressure expansion of compressed part of gas. Therefore at this stage for the main process we can use the uncollisional description by three-dimensional Burgers' equation

$$\partial\vec{v}/\partial t + (\vec{v}\nabla)\vec{v} = \mu\Delta\vec{v}, \tag{1}$$

where μ is the kinematic viscosity. For the k'th harmonics the inertial forces remain important during the time

$$t \leq t_*(k) \simeq \pi/2k \cdot v_o(k) \simeq 3 \cdot 10^5 \, l_{pc}^{2/3} \text{ yr.} \tag{2}$$

If $\mu \to 0$ the equation (1) has an implicit solution

$$\vec{v}(\vec{r},t) = \vec{v}_o(\vec{\xi}) \; ; \quad \vec{r} = \vec{\xi} + \vec{v}_o(\vec{\xi}) \cdot t \, , \tag{3}$$

which can be used for the description of initial evolution of turbulent spectrum into the compressible range (Ohul'chansky (1988a)).

The resulted spectrum at some moment t have some definite features. For the short wave length modes with $k > k_*(t)$, where $k_*(t)$ is determined from $t \simeq (\pi/2)k_* v_o(k_*)$, the spectrum is steeper than Kolmogorov's one. It has the spectral index in the range from −2 to −3 (Kaplan (1958); Passot and Pouquet(1987)). The long wave length modes with $k < k_*$ have not evolve yet, that is why the spectrum for $k < k_*$ differs from primordial Kolmogorov's one only slightly. Obviously, the time scale t_* derived from (2) determines the time interval during which the wave number k's harmonics are transformed into shock waves.

The further evolution of turbulent medium is following.The range of the shock wave forming harmonics is moved into the smaller wave numbers k gradually. The interacting shock wave's ansamble is formed. Because of isothermal conditions the magnitude of density fluctuations can rise to the large values. Most dense clumps are formed when shock waves collide in head-on. This promotes to the filamentary structures formation in molecular clouds. On the other hand the oblique shock waves interaction practically doesn't increase the density fluctuation. It leads only to the decreasing of the angle between interacting oblique shock fronts. This also provokes the plane dense structure formation. Therefore in the evolving supersonic turbulent medium the flat dense structures will be pronounced more and more by the stochastic shock fronts interactions. Such structures are revealed in observations (e.g.Bally et al.,(1987)) and in numerical simulations also (Passot and Pouquet (1987)).

3. THE STATISTICAL PROPERTIES OF CLUMPY CLOUDS.

In general case the statistics of density peaks is determined by the totality of all multy-point joint probabilities of physical quantity and its derivatives. Knowing $P(\nu,\eta,\zeta;\vec{r})$ – the one-point joint probability distribution of density $\nu=\rho/\rho_o$ and its first $\eta=\partial\nu/\partial\vec{r}$ and second $\zeta=\partial^2\nu/\partial\vec{r}^2$ derivatives the clump's size distribution can be find approximateely.

Let us consider the one-dimensional case for the illustration. Suppose that a density peak is situated in ε-vicinity of a point x. Then at this point $\eta=0$ and $\zeta<0$. The value of η in the limits of the ε-interval is changed from $-|\partial^2\nu/\partial x^2|\cdot\varepsilon/2$ to $+|\partial^2\nu/\partial x^2|\cdot\varepsilon/2$. So for the probability that the density peak of height ν is situated in ε-vicinity of the point x one can write

$$P_{\varepsilon}(x) = \int_{\zeta<0} d\zeta \int_{-|\zeta|\varepsilon/2}^{+|\zeta|\varepsilon/2} d\eta\, P(\nu,\eta,\zeta;x) \tag{4}$$

Hence the probability to have a density peak at the point x is equal

$$P_{pk}(x,\nu)=\lim_{\varepsilon\to 0} P_{\varepsilon}(x)/\varepsilon = \int_{\zeta<0} d\zeta\cdot P(\nu,\eta=0,\zeta)\cdot|\zeta| \tag{5}$$

In the multi-dimensional space in the similar way one obtaines

$$P_{pk}(\vec{r},\nu) = \int d\zeta\cdot|\det\zeta|\cdot P(\nu,\vec{\eta}=0,\zeta) \tag{6}$$

Here the integration is over all negatively determined values of matrix ζ.

Assuming that the space' distribution is equivalent to the ansamble's one it is possible to connect the density peak's differential distribution $dN(\nu,\zeta)/d\zeta$ with P. The value $dN/d\zeta$ is proportional to the under integral expression of (6). It can be connected with the clump's size distribution taking into account the relation $\zeta\simeq\nu/\lambda^2$, where λ is the clump's length scale. Then

$$dN(\nu,\lambda)/d\lambda = \frac{(dN/d\zeta)}{(d\lambda/d\zeta)} \propto \nu^{m+1}\cdot P(\nu,\eta=0,(\nu/\lambda^2))\cdot\lambda^{-(2m+3)} \tag{7}$$

Here m is the integer corresponding to the dimension of physical process in system. For supersonical turbulence m is equal to 1, but in the other processes (gravitational or thermal instabilties) the value of m may be equal to 2 or 3. Integrating (7) over all values of ν the size distribution of clumps can be obtained. Explicitly it can be done for the Gauss' probability distribution.

It is found that in the range of $\lambda \geq \sqrt{\langle\lambda^2\rangle}$, where $\langle\lambda^2\rangle \simeq \langle(\partial^2\nu/\partial x^2)^2\rangle^{-1/2}$, the probability distribution P weakly depends on λ. So for clumps in this range the relation (7) gives

$$(dN/d\lambda) \propto \lambda^{-2m-3}. \tag{8}$$

On the other hand for the clumps with $\lambda < \sqrt{\langle\lambda^2\rangle}$ after integration of (7) over variable ν/λ^2 it is obtained

$$(dN/d\lambda) \propto \lambda \tag{9}$$

Therefore, the size distribution has the maximum for clumps with length scales of the order $\sqrt{\langle\lambda^2\rangle}$. As it is seen from (8) the large scale clumps distribution is sensitive to the geometry of fragments. Received properties of the size distribution function can be used for conclusions what type of fragmentation processes is responsible for the clumps formation in the observing objects.

Now let us estimate the filling factor f that determines the relative volume of dense clumps in molecular cloud. The function f is determined by

$$f = \int d\nu \int d\zeta \cdot N(\nu,\zeta) \cdot V_{cl}(\nu,\zeta) \,, \tag{10}$$

where $V_{cl} \simeq (\nu/\zeta)^{m/2}$ is a clump's volume.

Except on dependence of integral parameters of physical system as a whole the statistical spatial properties of physical quantities are determined by the sequence of excited length scales l_1, l_2, ... only. If the edges of length scales are differed on more then one order of magnitude, so that $l_{min} \ll L$ (where l_{min} is the maximum scale in the group of the smallest ones and L is the minimal scale in the group of the largest ones), the fluctuation's spectra of physical quantities are likely to be self-similar ones in the spectral range $1/L \ll k \ll 1/l_{min}$. In the supersonic turbulent medium such spectra have the power shape $E \propto k^{-\alpha}$, where α is about 2 (Moiseev et al., (1976)). In this case from the scale properties of expression (10) it is followed

$$f \propto \delta \cdot (L/l_{min})^{m(\alpha-1)/2} \tag{11}$$

In the case of gaussian probability the available calculations (Bardeen et al.(1986);Ohul'chansky(1988b)) for the typical conditions of GMC lead to $\delta \simeq 10^{-3}-10^{-4}$. For $L/l_{min} \simeq 10^{3}$ and m=1 the relation (11) gives $f \simeq 10^{-2}$ that is corresponded to the observed in GMCs filling factor.

It should be noted that by the evaluation of the expression (11) the assumption of inertial spectrum ($L/l_{min} \gg 1$) has been used only and the specific nature of fluctuations has not been assumed. Therefore, if clumpy structure is the result of supersonical turbulence, then $l_{min} \simeq l_c$. If fragmentation is caused by other instabilities, then l_{min} is equal to the minimal scales of these ones.

REFERENCES.

Bally, J., Langer, W., Stark, A., Wilson, R.(1987) 'Filamentary structure in the Orion molecular cloud', Astrophys.J. 312, L45–L49.

Bardeen, J.M., Bond, J.R., Kaiser, N., Szalay, A.S.(1986) 'The statistics of peaks of gaussian fields', Astrophys.J. 304, 15–61.

Falgarone,E. and Perault,M.(1988) 'Structure at the 0.02 pc scale in molecular gas of low H_2 column density',Astron.and Astrophys. 205,L1–L4.

Kahane, C., Guilloteau, S., Lukas, K.A (1985) 'A multyline study of a typical GMC: S147/S153', Astron. and Astrophys. 146, 325-336.
Kalberla, P.M. and Stenholm, L.C. (1983) 'Evidenz fur turbulenz iminter-stellar medium', Mit.Astron.Gesel., 60, 397-401.
Kaplan, S.A. (1958) Interstellar Gasodynamics (in russian), Moscow.
Kleiner, S.C. and Dickman, R.L. (1984) 'Large-scale structure of the Taurus molecular complex.I', Astrophys.J. 286, 255-262.
Kolesnik, I.G. (1987) 'Formation of giant molecular cloudouds in superclouds and the origin of supersonic turbulence', Kinematika i Fizika Nebesnykh Tel 3, no.6, 47-56.
Larson, R.B. (1981) 'Turbulence and star formation in molecular clouds', Mon. Not. R. astr. Soc. 194, 809-826.
Moiseev, S.S., Tur, A.V., Yanovsky, V.V. (1976) 'The spectra and the ways of turbulence generation in compressible fluid', Sov. Zh. Exp. Teor. Fiz. 71, 1062-1069.
Myers, P.C. (1983) 'Dense cores in dark clouds.III.Subsonic Turbulence', Astrophys.J. 270, 105-118.
Ohul'chansky,Ya.Yu. (1988a) 'The evolution of supersonical turbulence in giant molecular clouds',Kinematika i Fizika Nebesnykh Tel 4,no.4,3-12.
Ohul'chansky, Ya.Yu. (1988b) 'Statistical characteristics of turbulence in giant molecular clouds',Inst.Theor.Phys. Preprint,Kiev,ITP-88-168E.
Passot,T. and Pouquet,A.(1987)'Numerical simulations of compressible homogeneous flows in the turbulent regime',Fluid Mech. 181, 441-463.
Passot,T., Pouquet,A., Woodward,P.(1988)'The plausibility of Kolmogorov-type spectra in molecular clouds' Astron. and Astrophys. 197,228-234.
Solomon, P.M., Rivolo, A.R., Barrett, J., Yahie, A. (1987) 'Mass, luminosity, and line width relations of galactic molecular clouds', Astrophys.J. 269, 531-539.
Terebey, S., Fich, M., Blitz, L., Henkel, C. (1987) 'The size spectrum of molecular clouds in the outer Galaxy', Astrophys.J. 308, 357-369

DISCUSSION

MOUSCHOVIAS: Does your calculation include the actual transmission of perturbations from the intercloud medium into a model cloud (across a density discontinuity), or do you assume a continuum of perturbation inside the cloud?

KOLESNIK: In considered model all spectrum of perturbation belongs to the cloud. It is corresponded to Kolmogorov's turbulent spectrum that was in the cloud before its rapid cooling.

ZINNECKER: In your picture of clump-clump collisions in GMCs you have not considered the effect of magnetic fields which will soften the collision and will make them less dissipative. How realistic is yours scenario of supersonic turbulent fragmentation if magnetic field are not taken into account?

KOLESNIK: It is a very complicated problem to distinguish the properties in the internal clumpy structure of GMC that are connected with the influence of magnetic fields only. I think that our results are concerned to the flat structures formation inside the GMC, and the size distribution function will not exchange drastically.

TURBULENCE IN THE FRAME OF THE EVOLUTION OF A SELF-GRAVITATING PROTOCLOUD

P. BATTINELLI[1], R. CAPUZZO-DOLCETTA[2], A. DI FAZIO[1]
V. A. URPIN[3], S. V. VERESHCHAGIN[4]

ABSTRACT. This paper studies the influence of turbulence generated by the supersonic motion of fragments formed during the collapse of a self-gravitating cloud. We investigate the dependence of the results upon the physical parameters, needed to describe turbulence. The results are given for three values of the gas cloud mass. The presence of turbulence with consequent dissipation is shown to raise appreciably the gas temperature. The results show that the turbulent velocity dispersion is sistematically smaller than the fragments' orbital velocity dispersion.

1. Introduction

The present work fits in the more general effort to construct a global model for the dynamical, thermodynamical and chemical evolution of a protogalactic cloud and of other, smaller size, self-gravitating protoclouds. The main problems to solve are, among others: *i)* when are the first stars born, and what is their characteristic mass spectrum; *ii)* in what environment (halo, protoglobular clusters, etc.) are they born and what is their resulting chemical composition; *iii)* what are the general driving mechanisms of present-day star formation (if different from primordial).

For what regards the aim of the present work, we note that to describe the gas temperature behaviour, the number density of the fragments, the mass spectrum of the formed fragments, and, as indirect consequences, the gas density and velocity dispersion distribution of the fragments, a description of turbulent motions (and their energetics) in the gas is necessary, as the mentioned quantities depend also on turbulence, often in a strong way. In particular, this work deals with the treatment of turbulence arising in the environment gas of a self-gravitating protocloud, in a specific scenario and given initial conditions. For example, the turbulence can be generated in the turbulent wakes that are left by fragments.

---

1) Osservatorio Astronomico di Roma, viale del Parco Mellini 84, I-00136 Roma, Italy;

2) Istituto Astronomico, Universitá La Sapienza, via Lancisi 29, I-00161 Roma, Italy;

3) Institute of Physics and Technology Ioffe of the USSR Acad. of Sciences, ul. Politechnich-eskaya 26, SU-194021 Leningrad, USSR;

4) Astronomical Council of the USSR Acad. of Sciences, ul. Pyatnitskaya 48, SU-109017 Moscow, USSR.

R. Capuzzo-Dolcetta et al. (eds.), Physical Processes in Fragmentation and Star Formation, 87–101.

These, in our scenario, have been formed through gravitational instability and orbit in the gaseous cloud under the action of the total gravitational field. The same scenario (but without turbulence) and the relative equations have been already presented in various forms in Di Fazio *et al.* (1980) and in Di Fazio (1986). The latter paper will be hereafter referred to as DF. According to this scenario, the protocloud undergoes the following evolution: the initially gaseous system is gravitationally unstable, starts collapsing and breaks into fragments; the chemical composition in the cloud changes continuously, due to changes of the state variables in the cloud. The changes in chemical composition affect the radiative properties of the gas (opacities, emission functions); the temperature of the gas strongly feels the effect of collapse and of radiative losses, and it strongly (non-linearly) affects the overall dynamics of the gas as well as the fragmentation rate and the instantaneous mass spectrum of fragments; the viscosity of the gas and the violent relaxation (collective processes) in the fragments' ensamble cause (respectively) the dissipation of the bulk kinetic energy of the gas in microscopic thermal motions and of the fragments' "fluid" in random orbital motions, thus causing a relaxation effect on the overall dynamics.

The present paper deals with the generalization of the above scenario, with the introduction of turbulence, generated in the turbulent wakes of fragments. We show that this turbulence can significantly affect the dynamics and in particular the thermal state of the system. Nevertheless, it is worthwhile to note that the procedure of calculation of the evolution of a self-gravitating cloud can be easily generalized also for other mechanisms of generation of turbulence.

2. The Main Equations

We consider a spherical gaseous cloud of initial mass (equal to the total mass) M_T, radius R_0, central temperature T_0 and chemical composition defined by the hydrogen and helium mass abundances X, and Y respectively. We assume that this cloud is gravitationally unstable both in view of general dynamics and of fragmentation. We then use the theory of gravitational instability for homogeneously contracting (expanding) media (e.g. Weinberg, 1972), and the consequently obtained analytical theory for the fragmentation rate mass spectrum and the number mass spectrum (mass function) (see DF). This way, we consider the instantaneous rate at which matter passes from gaseous to fragment state, for each fragment mass m following DF.
The fragmentation rate obtained from the mentioned theory,(i.e. the mass per unit time, per unit volume which passes from gas to fragments of mass in the interval $[m, m+\Delta m]$), is:

$$\left(\frac{d\rho}{dt}\right)_{Frag} = 3AG^{1/2}\rho^{3/2}\Big\{(f_J(m) - f_J(m+\Delta m) + -log_e\left[\frac{m^{1/3}(1+f_J(m))}{(m+\Delta m)^{1/3}(1+f_J(m+\Delta m))}\right]\Big\}, \qquad (1)$$

where $A = \pi^{3/2}/36$, G is the gravitational constant, $f_J(m) \equiv [(1-(M_J/m)^{2/3}]^{1/2}$ and $M_J = \pi/48(\pi\gamma K/m_H)^{3/2}(T/\mu)^{3/2}\rho^{-1/2}$ is the Jeans mass. The mass function is defined by:

$$dn(m,t) \equiv \phi(m,t)dm = AG^{1/2}\int_{t_0}^{t}\rho^{3/2}m^{-2}f_J(m)dt', \qquad (2)$$

$dn(m,t)$ is the number of fragments per unit volume, with mass in the infinitesimal interval $[m, m + dm]$, which exist at time t due to the process of gravitational instability alone and t_0 is the first instant at which $M_J \leq M$, (where M is the gas mass). It is then evident that to study in detail the spectrum of the fragments, and their dynamical evolution, it will be very convenient to choose a finite number – say $N+1$ – of points in some interesting range $[m_{min}, m_{max}]$. The fragments of class k have mass in the interval $[m_k, m_{k+1}]$ (where $k = 1, 2, ..., N$ and $m_{k+1} = m_k + \Delta m_k$).

All the equations regulating (in the non-turbulent case) *i)* the dynamics of fragments and gas, *ii)* the thermal balance and the radiation transport, *iii)* the time evolution of the chemical abundances, can be found in DF. In the latter work, the clouds are schematized as spheres of uniform density, using the Riemann-McLaurin-Jacobi (RMJ) formalism (see e.g. Chandrasekhar, 1969). Like in all the calculations of this kind (e.g. Gott and Thuan, 1976; Di Fazio and Palla, 1981; DF) the spatial dependence of the variables is lost, but an estimate of the time evolution of their space average is saved. This simplified model yields a good qualitative representation of the evolution of a self-gravitating cloud. In the present paper we adopt new terms regarding turbulence using the RMJ formalism.

Turbulence can exist in protoclouds due to different reasons. For example, it can be generated due to some kinds of instabilities (in particular, these can be hydrodynamical instabilities of the flow, that arise in the fragmenting protocloud). Turbulence can also be generated behind the shock fronts which are formed during fragmentation and during the violent bounce which takes place after a time in the range from 1.2 to about 3 free-fall times. Moreover, turbulence can arise in the wakes behind the fragments orbiting in the gas. It is not hard to estimate that, from the very beginning of fragmentation, the Reynolds number Re of moving fragments is very large, and thus the generation of turbulent wakes is very likely. The parameters of turbulence in the wakes can be estimated using standard formulas (e.g. Landau and Lifshits, 1971, p. 149÷155). According to these formulas, at a distance x behind the fragment of type k, the width a_k of the wake and the turbulent velocity v_{Tk} are of the order of magnitude:

$$a_k \simeq R_k(x/R_k)^{1/3}, \qquad v_{Tk} \simeq u_k(R_k/x)^{2/3};$$

where R_k is the characteristic scale of the k-fragment and u_k is its velocity relative to the gas. The Reynolds number in the wake decreases as $x^{-1/3}$, i.e. $Re(x) \simeq Re(0)(R_k/x)^{1/3}$, where $Re(0)$ is the Reynolds number just behind the fragment. For example, choosing typical values of the needed parameters for a protogalaxy: $u_k \approx 10^7$ cm/sec, $R_k \approx 10^{21}$ cm, and temperature $T \approx 10^4$ K, we get $Re(0) \approx 10^9$. In the case of a protocloud with $M_T \approx 10^6 M_\odot$, $R_k \approx 10^{20}$ cm and $T \approx 10^2$ K we obtain $Re(0) \approx 10^6$. At large distances from the moving fragments, when $Re \leq Re_{crit} \approx 10^3$ (Re_{crit} is the critical Reynolds number) the wake becomes laminar. Due to the large value of $Re(0)$, the length of turbulent wakes is extremely large. The length of the turbulent wake, x_{crit}, is: $x_{crit} \equiv R_k(Re(0)/Re_{crit})^3$ Taking the parameters of the mentioned two examples, we get $x_{crit} \approx 10^{18} R_k$ and $x_{crit} \approx 10^9 R_k$, respectively. Obviously then, the gas in the protocloud can be significantly turbulized, and thus the classical treatment of developed turbulence is suitable for our context. Let us introduce the resistivity force density exerted on fragments of class k by the gas. The resistivity force acting on one fragment has the classical form:

$$\vec{F}_k = -\alpha\rho\pi R_{Fk}^2 \, |\vec{w}_{Fk} - \vec{v}| \, (\vec{w}_{Fk} - \vec{v}), \tag{3}$$

where: ρ is the density of the gas in which the fragment moves, $\vec{v}$ is the hydrodynamical gas velocity, $\vec{w}_{Fk}$ is the speed of the fragment, and α is 0.5 times the so called coefficient of resistivity (e.g. Sedov, 1954). We use $\alpha \approx 1$ (for hard spheres moving supersonically in a gas with adiabatic index $\gamma = 5/3$, $0.881 \leq \alpha \leq 1.23$, see Landau and Lifshits, 1971). In our case, the velocity $\vec{w}_{Fk}$ of the generical fragment has an orderly component and a random one:

$$\vec{w}_{Fk} = \vec{v}_{Fk} + \vec{\sigma}_{Fk},$$

where $\vec{v}_{Fk}$, as described previously, is the radial velocity, ensemble-averaged of fragments at distance r from the centre, and $\vec{\sigma}_{Fk}$ is the velocity dispersion vector, i.e. is a vector with random orientation and with a modulus equal to the average velocity dispersion for fragments, $\sigma_{Fk} = \sqrt{2\epsilon_{Fk}/\rho_{Fk}}$, where ϵ_{Fk} is the internal energy density of the k-class fragments. Since in a given instant a chunk of k-fluid will contain fragments moving in all directions, and since our equation of motion is only radial, obviously we want the average value of the radial component of $\vec{F}_k$ with respect to the angle θ between the radial unit vector and the unit vector of the random component of $\vec{w}_{Fk}$, $\vec{\sigma}_{Fk}/\sigma_{Fk}$. Then, θ is defined by $\cos\theta \equiv \underline{\sigma}_{Fk} \cdot \underline{r}$ where the underlined quantities hereafter represent unit vectors. Let us also indicate as $\langle F_{kr} \rangle$ the average value of the radial component of $\vec{F}_k$, with respect to angle θ. This average quantity is:

$$\langle F_{kr} \rangle = -\Big\langle \alpha\rho\pi R^2_{Fk} \left|\vec{w}_{Fk} - \vec{v}\right| (\vec{w}_{Fk} - \vec{v}) \cdot \underline{r} \Big\rangle.$$

After the averaging, we get:

$$\langle F_{kr} \rangle = -\alpha\rho\pi R^2_{Fk} \left[\Delta v_{Fk} \langle B_k(\theta) \rangle + \sigma_{Fk} \langle \cos\theta B_k(\theta) \rangle \right],$$

where $\Delta\vec{v}_{Fk} \equiv \vec{v}_{Fk} - \vec{v}$,

$$\langle B_k(\theta) \rangle \equiv \frac{1}{6\Delta v_{Fk}\sigma_{Fk}} \left[(\Delta v_{Fk} + \sigma_{Fk})^3 - \left|\Delta v_{Fk} - \sigma_{Fk}\right|^3 \right],$$

and

$$\langle \cos\theta B_k(\theta) \rangle = \frac{1}{4\Delta v^2_{Fk}\sigma^2_{Fk}} \left\{ \frac{1}{5} \left[(\Delta v_{Fk} + \sigma_{Fk})^5 - \left|\Delta v_{Fk} - \sigma_{Fk}\right|^5 \right] + \right.$$

$$\left. - \frac{\Delta v^2_{Fk} + \sigma^2_{Fk}}{3} \left[(\Delta v_{Fk} + \sigma_{Fk})^3 - \left|\Delta v_{Fk} - \sigma_{Fk}\right|^3 \right] \right\}.$$

Then the density of the average radial resistivity force is given by: $f_k = n_{Fk} \langle F_{kr} \rangle$ where n_{Fk} is the k-fragment number density. The corresponding drag acceleration is $\ddot{R}_{Fk} = f_k/\rho_{Fk}$ to be added to the right hand side of the third of eq. (22) in DF. In the fourth of eq. (22) in DF we need to add a term which represents the rate of change of random kinetic energy of the fragment ensemble k, due to the work of the resistivity force $\vec{F}_k$, i.e. $(\dot{U}_{Fk})_{drag}$. The total work that the force $\vec{F}_k$ exerts on fragments of the species k per unit time is $\langle \vec{F}_k \cdot \vec{w}_k \rangle$. To change the energy of the regular motion, the needed power is $\langle\vec{F}_k\rangle \cdot \langle\vec{w}_k\rangle$. Consequently, the force $\vec{F}_k$ changes the kinetic energy of random motion by the quantity:

$$(\dot{U}_{Fk})_{drag} = N_{Fk}\left[\langle\vec{F}_k \cdot \vec{w}_k\rangle - \langle\vec{F}_k\rangle \cdot \langle\vec{w}_k\rangle\right] =$$

$$= N_{Fk}\alpha\pi R_{Fk}^2\rho\left[\Big\langle -|\Delta\vec{v}_{Fk} + \vec{\sigma}_{Fk}|\left(\vec{v}_{Fk} - \vec{v} + \vec{\sigma}_{Fk}\right) \cdot \left(\vec{v}_{Fk} + \vec{\sigma}_{Fk}\right)\Big\rangle + \right.$$

$$\left. -\langle\vec{F}_k\rangle \cdot \langle\vec{v}_{Fk} + \vec{\sigma}_{Fk}\rangle\right], \tag{4}$$

where N_{Fk} is the number of k-fragments. After evaluating average brackets, we obtain:

$$(\dot{U}_{Fk})_{drag} = N_{Fk}\alpha\rho\pi R_{Fk}^2\left[a_k\langle B_k(\theta)\rangle + b_k\langle\cos\theta B_k(\theta)\rangle - \langle F_{kr}\rangle\right],$$

where $a_k = vv_{Fk} - v_{Fk}^2 - \sigma_{Fk}^2$, $b_k = v\sigma_{Fk} - 2v_{Fk}\sigma_{Fk}$.

As mentioned above, we will conjecture that the gas in the protocloud may be turbulized due to different reasons. Consequently, to take into account the influence of turbulence on the dynamics of regular motion (or on the heat balance of the gas) is very complicated. In order to do that one uses very often some simplified model. Widely used in astrophysics (for example in the research on convection in the stars, disk accretion, etc.) is the so called "semi-empirical theory" by Heisenberg (1948) and von Weiszaecker (1948). According to this theory, the effect of turbulence on regular motion may be described by the renormalization of the viscosity coefficient. Such a simple model is very good for the explanation of a wide variety of different phenomena, as in the atmosphere, ocean, and laboratory conditions. To analize the processes taking place in protoclouds, we will use this model. In this fashion, the tension generated in the gas by regular motion, in the realm of small-scale turbulent pulsations, may be described by the "viscous" tensor Π_{ij}:

$$\Pi_{ij} = \rho\kappa\left(\frac{\partial v_i}{\partial x_j} + \frac{\partial v_j}{\partial x_i} - \frac{2}{3}\delta_{ij} div\ \vec{v}\right) \tag{5}$$

where the indexes i, $j \in \{1,2,3\}$ correspond to cartesian components, κ is the coefficient of turbulent viscosity $\kappa = v_T l_T$, where v_T and l_T are, respectively, the characteristic turbulent velocity and scale. We note that the "semi-empirical theory" by Heisenberg and von Weiszaecker was developed for non-compressible fluids, and for this reason, in the formulas they let $div\ \vec{v} = 0$; clearly, in our case, we should apply the entire formula (5). The necessary acceleration term due to (5), to be added to the rhs of the first of (17) in DF is

$$\ddot{R}_{vis} = \frac{\partial\Pi_{ij}}{\partial x_j}\frac{x_i}{r}$$

where we use the Einstein convention for indexes. Likewise, the contribution to the rhs of (17) due to the conservation of momentum arising from the drag force is given by

$$\ddot{R}_{drag} \equiv -\frac{M_F}{M}\sum_k \frac{f_k}{\rho_{Fk}},$$

where M, M_F are the total gas and fragment mass. The expression for the rate of transfer of energy from large-scale motions to small-scale ones and subsequenly transformed into heat, due to the dissipative effects, in the fashion of the theory by Heisenberg and von Weiszaecker has the form

$$\dot{U}_{diss} = \int \frac{\partial v_i}{\partial x_j} \Pi_{ij} dV.$$

This term is, of course, to be inserted in the gas energy balance equation, i.e. the second of eq. (17) of DF. In the second of eq. (17) of DF the cooling and opacity functions were those described in Di Fazio and Palla (1981). In the present work we use: *i)* the cooling function by Kaplan and Pikel'ner (1979) for a solar composition gas; *ii)* the Planck-mean opacities from Capuzzo-Dolcetta, Di Fazio and Palla (1985) for the hydrogen and helium components. The contribution from heavy elements is treated as in Di Fazio and Palla (1981).

We account for turbulent energy evolution through a suitable equation for the energy of turbulence. In doing this, we use the reasonable approximation (see Landau and Lifshits, 1971) of considering only the energy in the principal scale of turbulence. The needed equation is:

$$\dot{U}_T = -\sum_k (\dot{U}_{Fk})_{drag} - \frac{2^{3/2} U_T^{3/2}}{l_T M^{1/2}}.$$

From experiments on rigid bodies and atmospherical metereological observations of the disturbance fronts it is found $l_T \approx \beta L$, where L is the size of the moving object and β ranges in $[0.1, 0.5]$.

3. Numerical Models

Now we test in a straightforward way the influence of the initial conditions on the results. Our scheme should be able to describe the effects of turbulence in turbulized, self-gravitating gas clouds. The model contains all the physical input needed to give a good qualitative description of the evolution of such clouds. The numerical computations were performed through the fundamental 4th order Runge-Kutta method. Adequate time steps were adopted to control the maximum error in the calculations. The time step Δt was chosen as $\Delta t = min|a_i Y_i/\dot{Y}_i|$, where Y_i are the physical unknown quantities to be solved for in the integrations, and a_i are multipliers, $0.001 \leq a_i \leq 0.01$. For the internal energy, $U = \int \epsilon dV$, which, in our case, is the most strongly error-affected variable, we have $max|\Delta U/U| \approx 5 \cdot 10^{-3}$. First we will consider a model with initial radius $R_0 = 1000$ pc, mass $M_T = 10^7\ M_\odot$, central temperature $T_0 = 100$ K, and solar chemical composition. The chosen values of the parameters are typical of some giant molecular cloud. In what follows, we will call for brevity the above model the "reference model" (RM). It has the advantages of being physically interesting and not too much time-consuming.

In the RM we let the constant in the resistivity force (eq. 3) $\alpha = 1$ (see discussion in Section 2). We also let $\beta = 0.1$, β being the ratio between the main scale of turbulence and the radius of a fragment (as previously discussed, this value is typically obtained from experiments in very different conditions). Later, in this

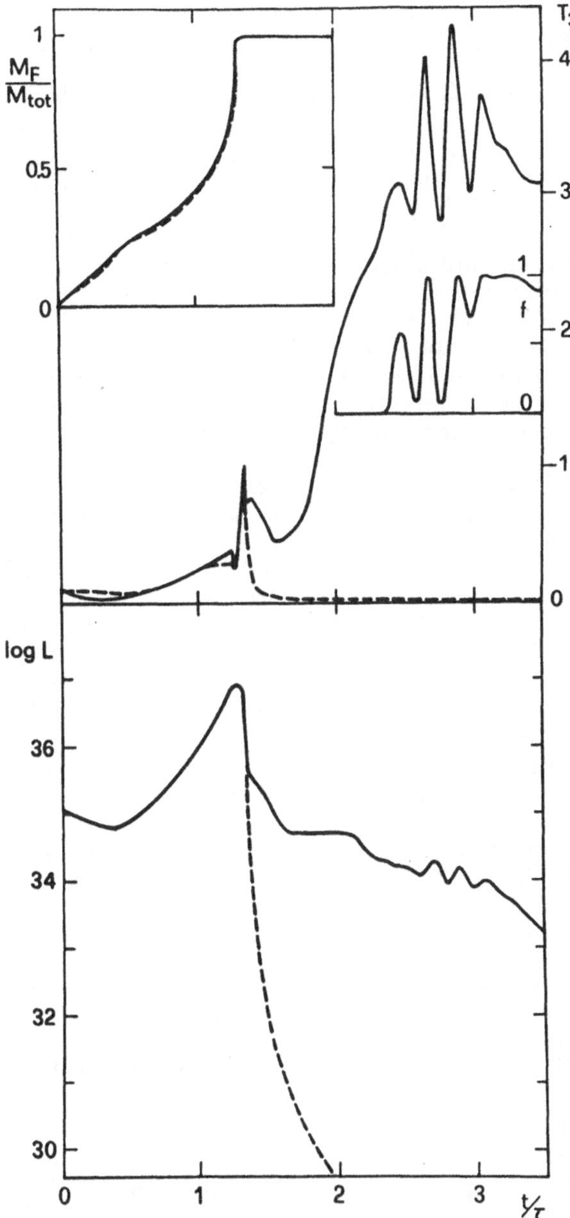

Figure 1. We display the mass fraction M_F/M_{TOT}, the temperature $T_3 \equiv T/10^3\ K$, the hydrogen ionization fraction f, the gas luminosity $L\ (erg/s)$, all four quantities vs. time. Solid lines refer to the turbulent case; broken lines to the non-turbulent case.

paper, we are going to study the influence of the parameters α and β on the results, and in order to do so we try the values: ($\alpha = 1$, $\beta = 0.1$); ($\alpha = 0.5$, $\beta = 0.1$); ($\alpha = 1$, $\beta = 0.5$).
For now, we integrate the RM, both with and without turbulence, and compare the results.

Figs. 1 and 2 compare the time evolution of the most relevant quantities for the RM with and without turbulence. In Fig. 1 we show the difference in the behaviour of the temperature of the gas between the turbulent and the non-turbulent case. This difference becomes significant after the first collapse, and more notably at $t/\tau \geq 1.3$, where τ is the initial free-fall time: $\tau = [R_0^3/(GM_T)]^{1/3}$. Up to this time, the number of fragments is not very high (as shown in the upper left hand side curves of M_F/M in Fig. 1), and they have not yet collapsed. Their radius R_{Fk} is still very large, and the rate of turbulent energy dissipation, which is inversely proportional to R_{Fk} is still small, and thus cannot significantly affect the heating. Up to 1.3τ, the gas radius decreases, the gas density increases, and the number of fragments also increases (since the fragmentation rate $\propto \rho^{3/2}$). The rapidly increasing number of fragments generates intensive turbulence, which results in an enhanced heating. In the non-turbulent case, such source of heating is absent, and the temperature decreases due to radiative losses and to the re-expansion of the gas. In the turbulent case, a re-heating of the very rarefied gas left-over after the first bounce is due to turbulent dissipation of energy. Qualitatively, this can be clarified in the following way. The onset of the regime of very rapid gas deplenishment, due to fragmentation, causes an abrupt decline of the cooling rate ($\lambda \propto \rho^2$), steeper than that of turbulent heating ($d\epsilon/dt \propto \rho$; where $d\epsilon/dt$ is the dissipation rate per unit volume). Thus, soon after the first bounce of the system ($t \gtrsim 1.5\tau$), in the energy balance equation the heating term due to dissipation overwhelmes the cooling term. This causes an increase of the temperature and hence a radius increase (see Fig. 1 and 2). The latter, in its turn, enhances the density decrease, creating a positive feedback. Around time $2\tau < t < 3\tau$ we note that, in the turbulent case, the temperature enters an oscillatory regime. These temperature oscillations cause an oscillation of the ionization fraction. At that time, the temperature reaches 3000 K, and the ionization begins to be relevant. Before that time, and after $t \approx 1.8\tau$, the b-b transitions dominate the cooling rate. The set-on of the ionization leads to an increase of the cooling rate due to f-b transitions, and the gas cools down. This continues up to when the ionization decreases, and the cooling becomes unimportant. After that, the gas heats again, and so on, until a time when, due to the gas expansion (see Fig 2c) and the consequent drop of the density, the characteristic relaxation time of ionization becomes very large. Thus, the ionization does not make it to follow the temperature oscillation, and the latter damps out.

In Fig 2c, we can see the time evolution of the radii of the gas and of the system of fragments R, R_F, both in the turbulent and non-turbulent cases. In the non-turbulent case, the gas is considerably colder, as the heating is absent, and it tends to concentrate to the center of the cloud. In the turbulent case, the dissipation of turbulent energy heats the gas, causing its expansion. The radius of the fragments' system in the turbulent case does not behave very differently from the non-turbulent case, as the evolution is mainly due to the adiabatic dynamics. Nevertheless, the value of the radius of the fragments' system in the turbulent case is a little smaller than that in the non-turbulent case (due to the loss of each fragment's orbital angular momentum caused by the drag force). The damped

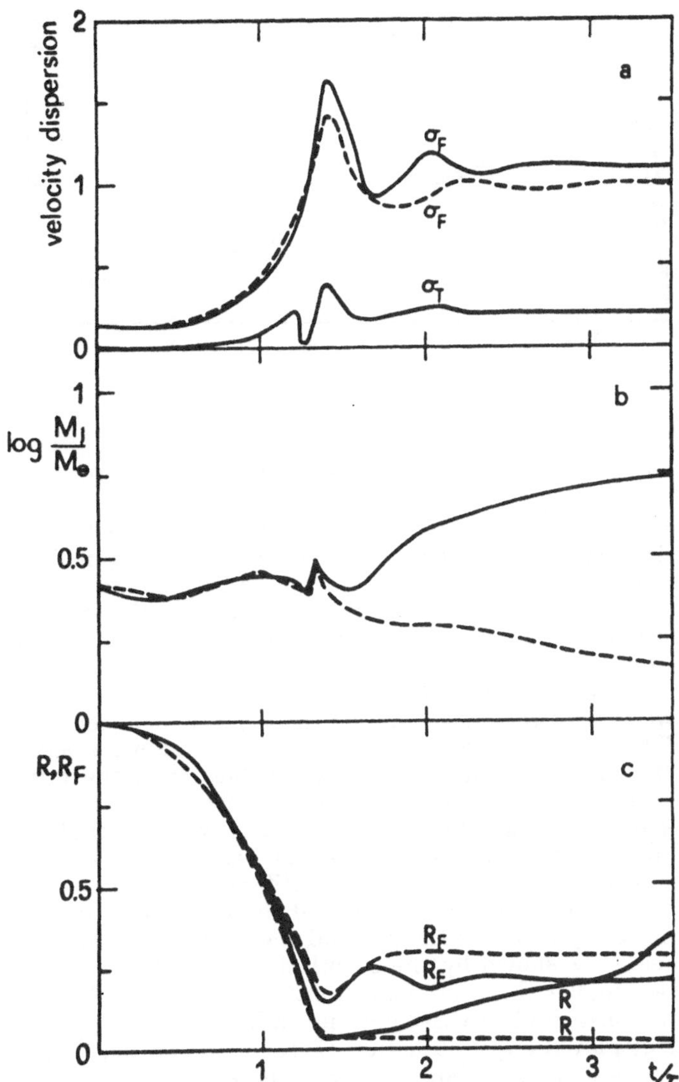

Figure 2. Panel a: the fragment velocity dispersions (σ_F) and the velocity dispersions of turbulence (σ_T) in units $10^6 cm/s$. Panel b: the Jeans mass in solar units. Panel c: fragment (R_F) and gas (R) spheres radii in units of their initial values. All quantities are plotted vs. time.
Solid lines refer to the turbulent case; broken lines to the non turbulent case.

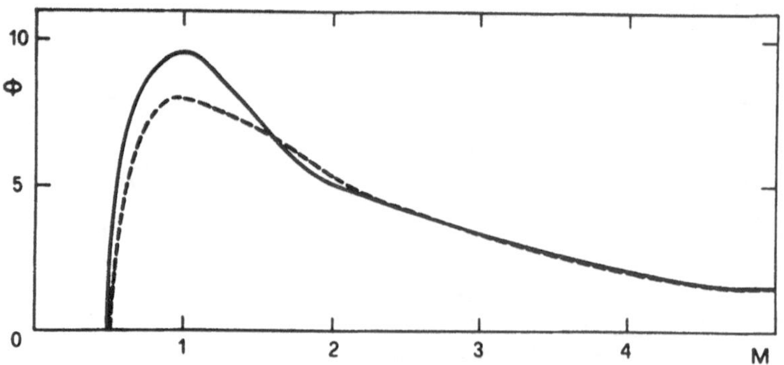

Figure 3. The mass function $\Phi = dN/dM$ in units $10^{-37}g^{-1}$ vs. mass (in $10^4 M_\odot$). Solid lines refers to model with turbulence; broken line without turbulence.

oscillations of the fragments' radius is simply typical of violent relaxating collisionless self-gravitating systems. The virialization of the fragments' system, i.e. the relaxation of the radius oscillations, is attained rather fast, at $t \approx 2\tau$. Since the gas radius increases, at $t > 3\tau$ it reaches the value of the fragments' system radius, and it continues to increase. But our model does not take into account the transport of internal energy due to turbulence in the cloud, and thus we consider the calculations as non-accurate some time after 3τ.

In Fig. 2a, we can see the typical adiabatic oscillations of the orbital velocity dispersion σ_F. The turbulent velocity dispersion σ_T, follows a similar damped behaviour, but it evolves at remarkably lower values than the orbital velocity dispersion . The relaxed ratio σ_F/σ_T is approximately 5.

For what regards the Jeans mass M_J (Fig. 2b), we note that the difference between the turbulent and the non-turbulent case becomes significant for $t > 1.3\tau$ (as well as for temperature and radius). This difference is caused by two factors: first of all due to the turbulent dissipation and the increase of temperature, and second, due to the re-expansion of the gas and the consequent decrease of the gas density. Nevertheless, as shown by the calculations, at $t > 1.3\tau$ the gas mass in the cloud becomes smaller than the Jeans mass. Hence, the difference in M_J in the turbulent and non-turbulent cases does not affect the fragmentation process.

In Fig. 3 we display the mass function of fragments obtained after the relaxation of the system's dynamics, both in the turbulent and in the non-turbulent case. Notwithstanding some differences at low masses, the mass functions in the two cases appear similar, and, in particular, they peak at $\approx 10^4 M_\odot$. For this reason, it is interesting to follow the evolution of such fragments, as the most abundant representatives of the first generation (we use the latter term due to the exhaustion of the fragmentation of the mother cloud). The newly formed fragments are unstable themselves, and start to evolve under the influence of their self-gravitation and, in their turn, start to fragment.

The initial conditions for the study of the evolution of such a sub-cloud are taken

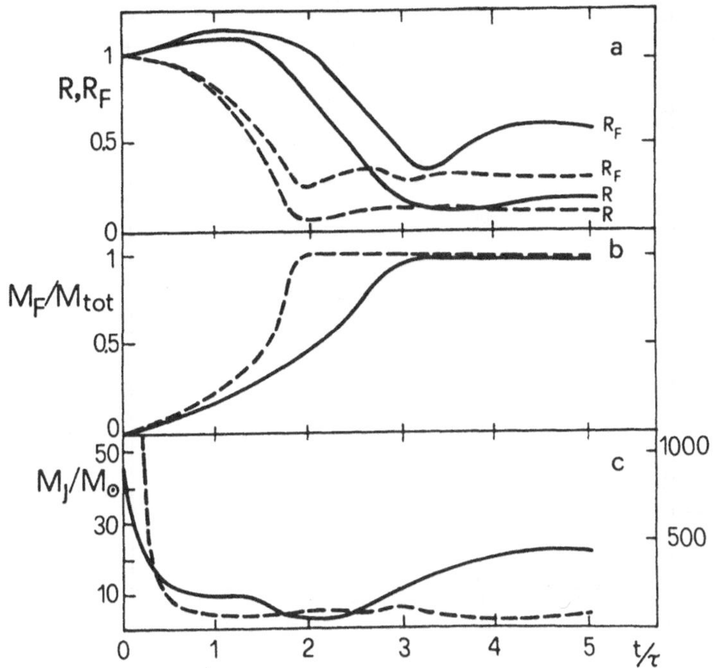

Figure 4. The time behaviours of the following quantities are displayed: R, R_F in units of their initial values (panel a); M_F/M_{tot} (panel c), the left ordinate refers to the $800 M_\odot$ model, the right to the $10^4 M_\odot$. Solid lines refer to the $10^4 M_\odot$ model; broken lines to the $800 M_\odot$ model.

from the ambient conditions at the time of formation of $10^4 M_\odot$ fragments. We then take: $M_T = 10^4 M_\odot$; $R_0 = 50$ pc; $T_0 = 150$ K.

The time evolution of the gas and fragment sistem's radii (R, R_F) can be seen in Fig. 4a. The 10^4 $M_\odot$ cloud has an initial slight expansion, due to the fact that its initial conditions (as given by the 10^7 $M_\odot$ mother-cloud evolution) are such that the internal energy is slightly larger than needed for equilibrium.

The strong cooling due to roto-vibrational transitions of H_2 molecule causes the internal energy of the gas to fall and to cause a collapse. The fragments' system also re-collapses, as the fragments lose orbital kinetic energy (this is due to the work of the resistivity force, and thus to the generation of turbulence, which increases its energy at the expense of the fragments' system energy in random orbital motions). The fragments' system reaches the minimum radius at about 3.2τ, while the gas at 3.6τ. We can see that the gas reaches soon a quasi-virial situation, while the fragments follow adiabatic oscillations damped only by violent relaxation. The temperature of the gas falls very fast (see Fig 5b), due to the cooling processes mentioned above, to a quasi-isothermal situation ($5 \div 10$ K). The orbital velocity dispersion σ_F of the fragments' system has the expected adiabatic behaviour, and (see Fig 5a) keeps well above ($5 \div 10$ times) the turbulent velocity dispersion, σ_T. As we have discussed above, the gas is very efficiently turbulized,

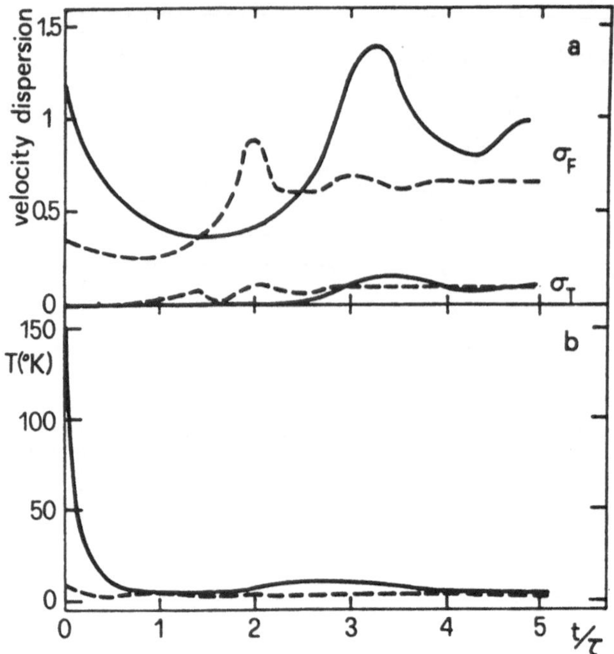

Figure 5. Fragment velocity dispersion and turbulent velocity dispersion in $10^5 cm/s$ (panel a) and temperature (panel b) vs. time. Solid lines refer to the $10^4 M_\odot$ model; broken lines to the $800 M_\odot$ model.

due to the very high Reynolds numbers typical of such clouds, but we can see that the velocity pulsation of non-regular motions (in the main scale of turbulence), which is given by σ_T, is never even close to σ_F.
The obtained mass function for this cloud – after the halt in the fragmentation process due to the gas exhaustion – peaks at about 130 $M_\odot$, with a slope (-2.49) just a bit steeper than the 10^7 $M_\odot$ cloud (-2.39). With the choice of initial conditions of this model, the fragmentation process is again very efficient ($M_F/M_T \approx 0.99$, see Fig. 4b). However, it is very interesting to note that this kind of cloud is very close to stability, as it can be seen from the initial expansion. Moreover, choosing a slightly higher value of the initial radius (which of course is realistic to happen to a considerable fraction of the first generation fragments of the 10^7 mothercloud) we obtain a substantial quasi-stability: we have, for $R_0 = 150$ pc, and $T_0 = 50$ K, about $\approx 90\%$ of fragments formed, in an environment of still dense gas, and a quasi-virial situation typical of many molecular and dust clouds. In this case, the fragmentation aborted not due to gas exhaustion, but due to the fact that the temperature (≈ 7 K) reached, together with the density (lower than in the case shown in this paper by about a factor 9) is sufficient to get the Jeans mass to outgrow the gas available mass. Moreover, in this case the obtained mass function has nearly the same slope of the shown case, but it peaks at ≈ 800 $M_\odot$. Both the discussed cases for the evolution of a 10^4 $M_\odot$ cloud are physically reasonable

choices. As the second, quasi-equilibrium choice is the more interesting of the two, due to the higher similarity with the known density, temperature and virial conditions in such molecular and/or dust clouds, we choose to follow the subsequent sub-fragmentation for a 800 $M_\odot$ cloud.
The initials conditions for the latter model are: $T_0 = 7$ K, $R_0 = 15$ pc. This model has a deeper collapse than 10^4 $M_\odot$ (as seen in Fig. 4a) as it is farther from equilibrium.
Fig. 5b shows that the evolution is practically isothermal. The third generation of fragments has a mass function peaking at $\approx$ 10 $M_\odot$ which is of the order of the masses of stars in OB associations. Note the difference, in the rate of fragmentation, between the 10^4 $M_\odot$ and the 800 $M_\odot$ models, the latter being considerably faster (Fig. 4b). A common feature for the above models is the initial rapid drop of the Jeans mass. It is easily seen from Fig 5a that, also in this case, the fragment's velocity dispersion is at a level about a factor $5 \div 10$ higher than σ_F. Consequently, if this kind of objects were observed, the line width would be dominated by clump orbital motions.

In table 1 we show the comparison with the reference model of the relevant variables in the two cases ($\alpha = 1$, $\beta = 0.5$) and ($\alpha = 0.5$, $\beta = 0.1$) at the virialization.

(α, β) :	$(1, 0.1)$	$(1, 0.5)$	$(0.5, 0.1)$
σ_F	11.1	26	15
σ_T	2.2	12	3.6
T	3300	1100	530

Table 1. We show the comparison with the reference model of some of the relevant variables in the two cases $(\alpha = 1, \beta = 0.5)$ and $(\alpha = 0.5, \beta = 0.1)$ at virialization. Velocity dispersions are in $10^5 cm/s$. T in K

Acknowledgements
This work has been accomplished thanks to Scientific Exchange Agreement between the Academy of Sciences of the USSR and Italian National Council of Researches (CNR), contract No. 2.16 between the Astronomical Council (Moscow) of the USSR Academy and the Astronomical Observatory of Rome. Many thanks to Dr. B. M. Shustov for offering computing facilities, ospitality and collaboration of the Moscow Institute. We thank Dr. A. Soloviev (Institute of Oceanology of the USSR Academy of Sciences) for a useful discussion on turbulent dissipation. Warm thanks are due to prof. R. Ruffini and to the International Center for Relativistic Astrophysics for a computing grant. We also thank G. Buonvino who drew the graphs.

References

Capuzzo-Dolcetta, R., Di Fazio, A., Palla, F.: 1985, *Astron. Astrophys.* **145**, 290
Chandrasekhar, S.: 1969, *Ellipsoidal Figures of Equilibrium* (New Haven, Conn.: Yale University Press)
Di Fazio, A.: 1986, *Astron. Astrophys.* **159**, 49
Di Fazio, A., Palla, F.: 1981, *Astrophys. Space Sci.* **76**, 391
Di Fazio, A., Vagnetti, F., Wilson, J.R.: 1980, *Astrophys. Space Sci.* **72**, 204
Gott, J.R. III, Thuan, T.X.: 1976, *Ap. J.* **204**, 649
Heisenberg, W.: 1948, *Zs. Phys.* **124**, 628
Kaplan, S.A., Pikel'ner, S.B.: 1979, *Fizika Mezhvezdnoj Sredi*, Nauka, Moscow; 1970, *The Interstellar Medium*, Harvard University Press, Cambridge MA.
Landau, L. P., Lifshits, E.M.: 1971, *Mecanique des Fluides*, MIR, Moscow
Sedov, N.: 1954, *Fizika Zhidkostej*, Nauka, Moscow
von Weiszaecker, C.F.: 1948, *Zs. Phys.* **124**, 614
Weinberg S.: 1972, *Gravitation and Cosmology: Principles and Applications of the General Theory of Relativity*, J. Wiley and Sons, N.Y., London, Sidney, Toronto

Discussion:

Bedogni: 1) Could the magnetic field reduce the turbulence scale length near the fragments? 2) What are the relative velocities of the interstellar fragments with respect to the surrounding medium?

Di Fazio: 1) This could happen in presence of a substantial ionization, and a magnetic field whose energy density $(B^2/8\pi)$ is not negligible with respect to the thermal and turbulent energy densities. 2) They are of the order of a fraction of the mean free collapse velocity, $(GM/R)^{1/2}$. In the physical conditions of the model, this corresponds to Mach numbers $M = |v_F - v|/c_s \epsilon$ [3,5], c_s being the sound velocity.

Scalo: I don't understand your use of the Heisenberg closure to represent the wake turbulence. It seems like the wake turbulence is assumed to dissipate by viscosity directly into heat. Aren' t the wake motions supersonic? If so, most of the dissipation would be in radiative shocks, and little heating should occur. I also question whether most of the drag dissipation goes into wake turbulence in your model, which involves supersonic drag, rather than being dissipated by radiation in the bow shock. Some 2-D numerical simulation of this problem were published by David Gilden in 1989.

Di Fazio: The renormalization of the viscous tensor, due to Heisenberg and von Weiszaecker deals only with the simplification of the problem, to avoid a very complex treatment of the spectrum of turbulence. A different thing is the dissipation of turbulence, which is known to take place through a non-dissipative cascade of scales up to the turbulent microscales for which the Reynolds number is of the order unity. Such microscales are the only ones that dissipate through viscosity into thermal motions, as well known also experimentally. For what regards further dissipations due to shocks, theory and experiments say that the dissipation of kinetic energy due to the formation of the bow shock is a small fraction of the energy lost in the wake (see eg. Sedov, 1954). Thus, the energy of the shock, which can be dissipated in its turn through radiative losses, is consequently not relevant, either.

Cayrel: In your scheme without turbulence the quantity of the gas left over in the cloud at the end of one fragmentation process was very small ($\approx$ 5%). Including turbulence is this fraction very much affected?

Di Fazio: The effect of turbulence on that fraction takes place through the dissipation of turbulent microscales into thermal energy of the gas. Due to the fact that the dissipation occurs mainly when a large part of the mother-cloud's mass has already been processed into fragments, said effect is not relevant, except when the initial conditions are very close to equilibrium. We have not completely explored the effect of quasi-equilibrium conditions on the mentioned fraction, nevertheless with the data at our disposal, we can estimate that, due to turbulence, the gas left increases to $\gtrsim$10%.

FRAGMENTATION AND SUPERSONIC TURBULENCE IN SELF-GRAVITATING GAS CLOUDS.

A. DI FAZIO (1), A. SOLOVIEV (2), V. A. URPIN (3), and S. V. VERESHCHAGIN (4).

(1) - Astronomical Observatory of Rome, Viale del Parco Mellini 84, I-00136 Rome, Italy.
(2) - Institute of Oceanology "Shirshov", Academy of Sciences, Moscow, USSR.
(3) - Physico-technological Institute "Ioffe" of the Academy of Sciences, Ul. PolitechnicheskaYa, D26, SU-194021 Moscow, USSR.
(4) - Astronomical Council of the Academy of Sciences, Ul. Pyatnitskaya, D. 48, 109017 Moscow, USSR.

ABSTRACT.

We test three different hypotheses to explain the Δv-L correlation between the velocity dispersion deduced from the line broadening in molecular clouds and the cloud sizes, namely: i) that the clouds are born essentially due to turbulence, with negligible evolutionary effects on the above spectrum, and that the relation is the universal specrum of turbulence, in the fashion of Kolmogorov; ii) that the clouds are born through some fragmentation process, and that the relation is nothing else than the virial correlation; iii) that both fragmentation and turbulence are present, and that the observed correlation is actually the superposition of the velocity dispersion of fragments' orbits and the velocity dispersion of turbulence. The virial slope in the LogΔv - Log L (0.5 - 0.62) is closer to that of the data (0.56) than that of Kolmogorov spectrum (0.33), although the quality of the data does not allow a net decision between the two extreme hypotheses. Nevertheless, the use of the model with turbulence and fragmentation shows that the obtained (at virialization) orbital velocity dispersion of fragments is always significatively larger than the turbulent velocity dispersion. If this model (see [10]) were correct, then the observed velocity dispersion would be mainly due to random orbits of sub-condensated globules or fragments, and <u>not</u> to turbulence. The latter seems to be the most plausible case.

R. Capuzzo-Dolcetta et al. (eds.), Physical Processes in Fragmentation and Star Formation, 103–115.

1. Introduction

A large part of the Galaxy is occupied by various gaseous structures, from superclouds with masses aroud 10^7 $M_\odot$ up to small dark clouds with protostellar masses. It is convenient to observe these objects in the radio and infrared bands, where they are transparent enough. In spite of much attention to these structures, the role of physical processes like gravitation, thermal instabilities or turbulence is far from complete understanding.

One of the keys for the investigation of physical processes inside gas clouds is a relation between the velocity dispersion deduced from Doppler broadening of spectral lines (Δv) and the clouds sizes (L) (see Larson, 1981, [1], Myers, 1983, [2], Sanders *et al.*, [3], 1985, Solomon *et al.* [4], 1987, Leung *et al.*, 1982 [5]). In this work we studied objects with diameters from 0.05 to 1000 pc. These values are typical of the scales of the density fluctuations inside gas clouds, globules, molecular clouds. Up to now it is known that this relation has a shape which can be very roughly approximated by a power law:

$$\Delta v \sim L^n \tag{1}$$

The parameter n is not well known, but it ranges approximately in the interval [0.3 , 0.6]. We underline that the question of the origin of the observational relation (1) is not settled, and for many reasons, the main one being the following: it is difficult to distinguish a virial correlation (i.e. with $n \gtrsim 0.5$) from, e.g., a turbulent one with n shown below (~ 0.33). In the frame of a turbulent origin of correlation (1), we need to use the idea of the universal turbulent spectrum, which was developed e. g. for the investigation of the velocity pulsations in the oceans of the Earth (see Monin and Ozmidov, 1981).

In this work, as a first approach, the hypothesis of the turbulent origin of correlation (1) was used in a first analysis of the observational data. Nevertheless, we also analized correlations of type (1) due to orbital motions of fragments, globules, and nodules inside gas clouds. We will show that the velocity dispersion of the fragments' (or other sub-condensations') motion is quite larger than that of turbulence, for gas clouds in wide interval of masses. This implies that if the observed clouds are sub-fragmented, or contain globules or any kind of sub-condensations that move in the clouds, the velocity dispersion in the observed lines would be mainly due to orbital motions inside the clouds themselves.

2. Observational relations between velocity dispersion and cloud sizes.

The observational correlations are shown in Fig.1, and in Table 1 is the information on how we used the data. For rendering these data comparable with each other, we have done the following:

1. We have averaged the data in [1] and [2] with the same method used in [3].
2. In [1] we have no diameters for the clouds, but only the maximum sizes. These sizes were reduced by a factor two.
3. The thermal motion of the gas particles was taken into account in [1] with σ_{Temp} = 0.32 Km/s. In [2] it was already done by the authors. In [3] this motion was very small and thus it has been neglected.

TABLE 1. Observational data

Authors	No. of objects	Objects	n	Observed molecules
[1]	57	density fluctu- ations, clouds	0.3	CO e.a.
[2]	43	density fluctu- ations, globules	0.3	NH, CO
[3]	80	clouds with sizes greater then 10 pc	0.6	CO

3. Universal turbulent spectrum or virial Δv-L correlation?

3.1 THE TURBULENT SPECTRUM HYPOTHESIS

The point scattering in Fig. 1 is large and it is difficult to determine exactly the slope of the correlation. Let us - as a first work-hypothesis - assume the following:

i) that the observed velocity dispersions are mainly due to turbulence (like in Dolotin and Fridman, 1989) and negligibly to other sources, such as e.g. orbits of sub-structures;

ii) that the clouds and sub-structures inside the clouds are born only due to turbulence in the Kolmogorov sense;

iii) that any evolutionary effects (such as mass-loss, collisions with disruption or with coalescence, accretion, etc.) were and are negligible. Under the above hypotheses, we need to use the universal turbulent spectrum. According to Kolmogorov [7] at sufficiently small scales the statistical structure of turbulence has a universal form and its scaling parameters depend only on turbulence energy dissipation and gas viscosity. In order to compare different observations it is necessary to use a method to bring observative spectra determined in different conditions to compatibility with each other. Usually, by "turbulent spectrum" we refer to the dependence of the kinetic energy of turbulence upon its scales. In this case, for the "compatibilization" of different spectra we can use the methods shown in [6]. In order to perform such an operation on the spectrums in the coordinate plane LogΔv - L the necessary transformations are developed and shown in [8].

Spectra shifted to the universal spectrum according to [8] are shown in Fig.2. Let us continue using our hypotheses i), ii), iii), and thus suppose that the so obtained total spectrum can be described by the shown continuous curve. We can divide the theoretical universal spectrum into three intervals with different slopes.

In the range of small scales we get a "*viscous*" interval with a slope n > 0.3. In this region there are small fluctuations of density inside the clouds. The steeply decreasing part of the curve is caused by the suppression of turbulence operated by the gas viscosity (here the mean free path of the gas particles is of the order of magnitude of the turbulent scale, and the turbulent pulsation is of the order of $(kT/m_H)^{1/2}$, so that $Re_\lambda \approx 1$, i.e. the Reynolds number of these irregularities in the flow is near unity).

In the intermediate - the "inertial"- interval, with Kolmogorov slope n=0.33, the energy transfers from large turbulent scales to small ones. This spectrum exists in scales smaller than the thickness of the galactic disk, where the turbulence is three-dimensional. The energy input to turbulence can be due to supersonic wakes, expanding HII regions, SN explosions and galactic rotation.

In the scales comparable with the thickness of the galactic disk, turbulence becomes "quasi-two-dimensional", and has the properties of two-dimensional turbulence. In particular, the slope of the spectrum is changed. In this case, the energy goes from small scales to large ones. The result is that small chaotic three-dimensional whirlpools disappear to form large two-dimensional structures. The reason of appearance of two-dimensional turbulence may be the internal gravitating Rossbi waves inside the rotating

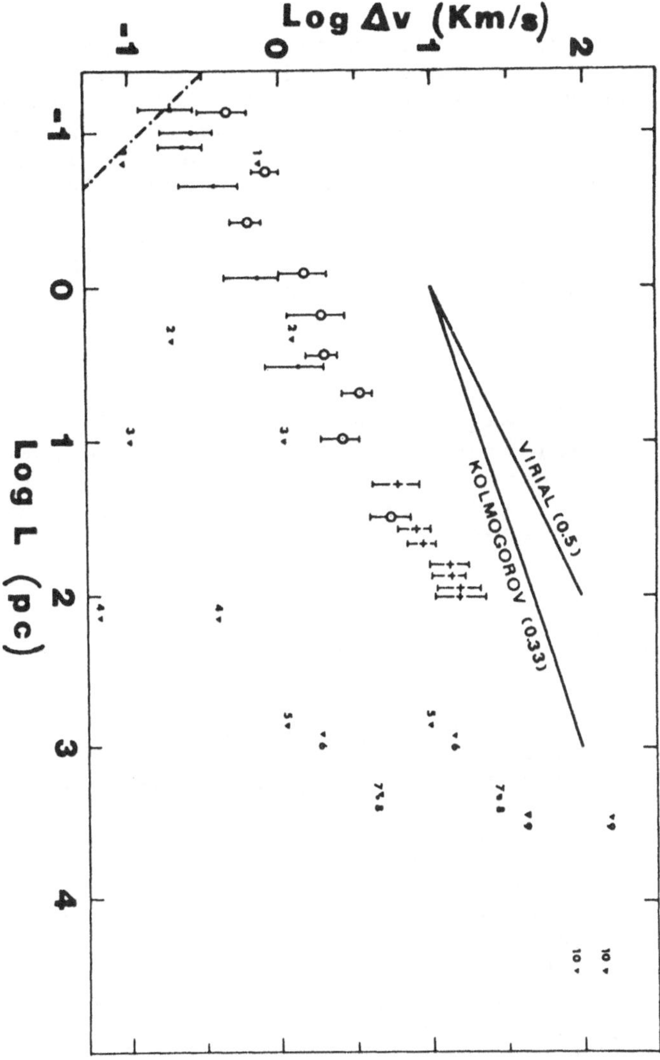

Fig. 1. Observational correlation between velocity dispersion and cloud sizes. Sources: [1] - points; [2] - circles; [3] - crosses. The two straight lines are the Kolmogorov (0.33) and the virial slope (0.5, considering $\rho \propto L^{-1}$; if $\rho \propto L^{-0.75}$, then the virial slope becomes 0.62).

The small triangles are the results of some numerical models: the numbers identify the model (see Sec-

./././

(cont. Fig. 1)

tion 3.3, Table 3, for the initial conditions of the models). Note that the triangles are in couples, with the same ascissa: the upper one represents the orbital velocity dispersion of fragments, σ_V, while the lower one is the turbulent velocity dispersion in the gas, σ_T. Arrows 1, 3, 5 represent the evolution of one $10^7 M_\odot$ mother protocloud, plus two smaller clouds which are its first two (out of three, see [10]) sub-fragmented generations. A least-square fit to the observative data yields a slope n = 0.56 and an intercept a = 0.034. The correlation coefficient is *r* = 0.98.

(cont. text)

cloud (see e. g. [9]).

Possibly, the increase of the slope at large scales can be caused by the violation of the Kolmogorov hypothesis due radiative losses: turbulent energy undergoes a cascade to small scales, then dissipates into heat, and finally is radiated away in the shockwave-compressed zones.

3.2 THE VIRIAL HYPOTHESIS

If, alternatively, we do not assume the hypotheses i) ÷ iii), we can reason in the following way: for the great majority of the molecular (as well as neutral hydrogen) clouds, when the clouds are observed in the radio bands, they show that there exists a more deeply granulated fine structure, made out of clumps, fragments, globules, and similar objects. These very often have radial velocity differences with the cloud's ambient gas. Let us then assume the following: *all* the clouds have such fine structure (as also suggested by the low Jeans mass values, see Larson, 1981, and Massi *et al.*, 1989), and the line widths that we observe are due to the inner random orbital motions. Then, we know immediately, from the virial condition, that $\Delta v \propto (M/L)^{1/2} \propto (\rho(L)L^2)^{1/2}$. Many authors (e.g. [1], [3], [14] and others) suggest that $\rho(L) \propto L^{-\alpha}$, with $0.75 < \alpha < 1$. Thus, we have $\Delta v \propto L^n$, with $0.5 < n < 0.62$. Since the data in Fig. 1 have a linear regression slope $n \sim 0.56$, the virial hypothesis for the correlation seems quite acceptable and simple.

3.3 A MODEL WITH TURBULENCE AND FRAGMENTS. A "MODIFIED" VIRIAL HYPOTHESIS.

If we now would like to compare the data with a third theoretical scenario, where turbulence and internal orbital motions are both present, we can use the models of evolution

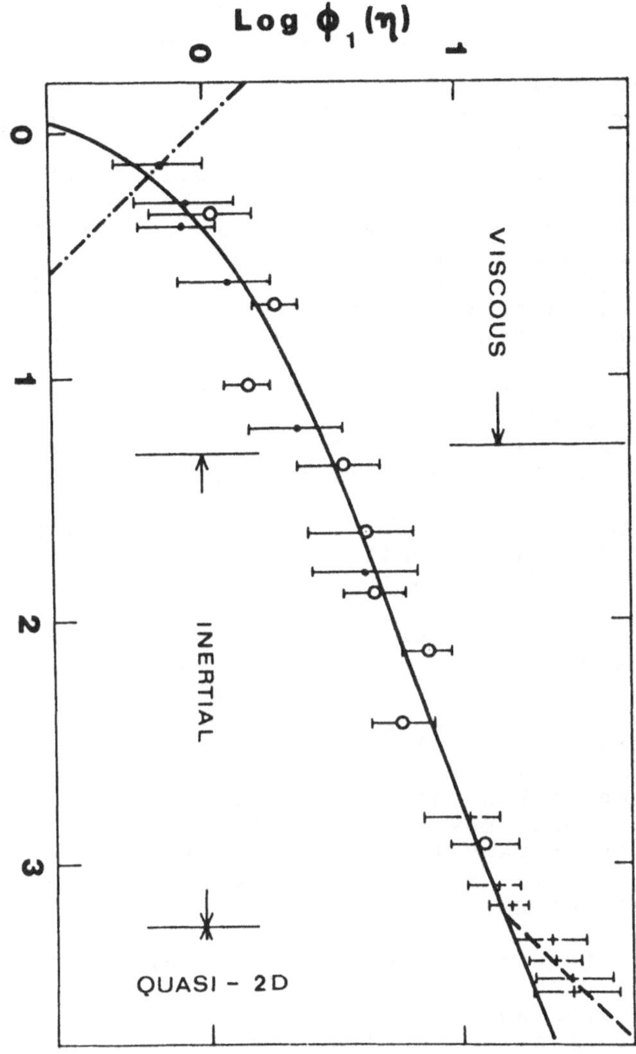

Fig.2.
Turbulent spectrum of the gas phase of the Galactic disk in dimensionless coordinates (Log(Φ)) and Log(η)). The dashed line shows the slope n=0.5, that here approximates the observational data at large scales. The boundaries of the intervals of the different turbulent energy dissipative mechanisms are shown.

of the gas clouds with fragmentation and turbulence (generated in supersonic wakes behind moving fragments), shown in this workshop [10]. We now shortly summarize and discuss some features of these models.

At the beginning, we have a total gaseous system with mass much greater than the Jeans mass. Initially, in the gas we have initial density perturbations with a "white noise" mass spectrum. A short time after the gas contraction begins, the fragmentation process starts. Part of the gas mass is transformed into fragments according to the fragmentation process due to gravitational instability proposed in [11]. When fragments move in the gas, turbulent wakes are formed behind them. Due to the work of the resistivity force, the fragments lose part of their energy. This energy is inputted with high efficiency into turbulence. The process of turbulent energy dissipation from large to small scales takes place. After the dissipation of the smallest scales (where $Re_\lambda \approx 1$) by gas viscosity, said energy heats the gas.

In this model, we do not make any particular assumptions about the relative magnitude of the turbulent part of the energy with respect to the orbital part (in random motions.

In the temperature region of calculated models (10 - 3500°K) the main source of the gas cooling is the de-excitation of the energy levels of molecular and atomic hydrogen. The excitation of said levels of molecular and atomic hydrogen is mainly due to collisions with free electrons and HI atoms.

We have not taken into account the heating the gas by shock waves around moving fragments and the subsequent cooling, but is is very likely that this process is not so important relatively to the heating due to other processes (compression of the gas due to collapse, dissipation through viscosity, etc.).

After contracting, bouncing, and following dynamical relaxation, the models achieve a virial stage.

It is important to stress that, in the model calculations presented in [10], we do not assume, but separately calculate the velocity dispersion σ_V of the fragments' orbital motion and the velocity disper ion of turbulence, σ_T (see [10]).

An example of application to gas clouds of the above described model is the following. In order to trace the evolution of super-molecular clouds in the galactic disk, we consider a model with super-cloud mass $10^7 M_\odot$, together with its sub-fragmentations (until we get stability, i.e. up to when the fragments that form are actually stars). Three generations of sub-fragments are obtained (we use the work by Battinelli et al., [10], this volume). Table 2 summarizes the evolution of the mentioned 10^7 $M_\odot$ mother protocloud, together with 2 of its 3 sub-fragmentations. The third subfragmentation data, not shown, can be found in [10]. These data refer to some sort of O-B type stars (the third generation's spectrum peaking at ~10 $M_\odot$). Further fragmentation does not take place, as these objects become very soon nuclear burners. Thus, the 800 $M_\odot$ cloud ends up in a cluster of OB stars, embedded, at the moment of their formation, in a dense, low temperature gas.

TABLE 2

The main parameters of the first two fragment generations in the evolution of a supercloud with M = 10^7 $M_\odot$.

M $M_\odot$	R pc	T °K	R pc	RF pc	T °K	NF	σ_F km/s	σ_T km/s
Initial condit.			Equilibrium stage					
supercloud								
10^7	1000	100	700	200	1000	100	10.0	2.0
first generation								
10^4	50	150	10	30	6	50	0.9	0.1
second generation								
800	15	7	2	5	6	10	0.7	0.09
third generation								
10	initially 7		STABLE (OB-type STAR)					

where:

M - the system's total mass
R, RF - radii of the gas cloud and fragments' system, respectively,
T - temperature of the gas,
NF - the number of formed fragments,
σ_F, σ_T - velocity dispersions of the fragments' orbital motions and of turbulence, respectively.

The briefly described evolution of a supercloud can be seen from Fig. 4 and 5 of [10]. The final velocity dispersions (both in orbital motions of the fragments and in turbulence of the gas) of the mother cloud and of the first 2 generations are plotted in Fig. 1 vs. their final sizes, and are marked by small arrows and by the model numbers 1, 3, 5 (in increasing order of mass). Two facts are evident: i) the orbital velocity dispersion (upper arrows for each model) is always several times above the velocity dispersion of turbulence, and the difference is from 3 to 4 times greater than the observative error bars for the objects in that zone of the graph; ii) The obtained LogΔv - Log L relation is not linear, and to be able to say about its slope we need to compute more models. Since a large volume of computations (slowed down by the dissipation of turbulence) is needed for this, we decided to calculate some representatively interesting?models. For this, we chose to follow:
i) (model 2) a gas cloud that can represent the protocloud of M17 (see [13]), with R_0 = 15 pc, T_0 = 20 K, M = 2700 $M_\odot$; ii) a cloud (model 4) cooler and more dilute than model 3, but with the same mass: R_0 = 150 pc, T_0 = 15 K, which is also a good representative of the first generation of fragments of the 10^7 $M_\odot$ protocloud; iii) a dwarf protogalaxy (model 7) with M = 10^9 M_0, R_0 = 10 Kpc, T_0 = 5000 K; iv) a protogalactic cloud (model 9) with M = 10^{11} $M_\odot$, R_0 = 30 Kpc, T_0 = 10^4 K; v) another protogalaxy (model 10) with the same initial mass and temperature just mentioned, but with a 100 Kpc initial radius. Models 6 and 8 correspond to the same data of the 10^7 $M_\odot$ cloud, except that they use different values of the turbulence parameters (α, β, see [10], this volume, respectively, the third and the second column of Table 1 of paper [10]). This was done in order to investigate what effect have our ([10]) turbulence parameters on the Δv - L relation. Looking at the arrows representing the models 5, 6, and 8, we can see that the relation is not sensibly affected by the variations of the resistivity coefficient$\alpha$and of β of paper [10]. Looking at the relation that is designed by the arrows, we must remark that at low sizes our models give some sort of plateau, while from ~ 100 pc on the relation is even steeper than the virial slope.

4. Conclusions

We can make the following conclusions from the shown observational data (the ΔV - L relation for the gas clouds) and from the results of our model for the radiative and dynamical evolution of gas clouds:

1) In the case we adopt the Kolmogorov spectrum interpretation for the data, we note that with the relation Δv - L for gas clumps, globules, molecular clouds we can compare the previously mentioned "universal turbulent spectrum" in a wide range of scales from 0.05 pc to 1kpc; the universal turbulent spectrum is similar to the data, but also a straight line of slope 0.5 (the virial relation Δv - L) seems to be quite a good fit;

2) the universal spectrum has three intervals. This means that the physical processes in different scales can be distinguished. Small scale turbulence disappears due to viscosity, in the intermediate scales we have a cascade transport of the energy, and at the large scales two-dimensional structures (whirlpools) can form. Nevertheless, the available data are too scattered, to make very clear and net distinctions of these intervals;

3) the least squares fit (see Fig.1) of the observed Δv - L correlation yields a slope - 0.56 - which is much closer to the virial one (0.5 - 0.62) than to the Kolmogorov one (0.33). This makes the virial hypothesis preferable, but, given the quality of the observations, we cannot make a strong statement against the turbulent-origin hypothesis for the correlation;

4) nevertheless: the velocity dispersion of the fragments' orbital motions is larger than the turbulent velocity dispersion for the model clouds of the examined size interval. This means that if in the clouds there were internal orbital motions of sub-condensations as well as turbulence, the orbiting clumps would account for the greatest part of the observed line width.

ACKNOLEWDGEMENTS

This work has been accomplished thanks to the Scientific Exchange Agreement between the Academy of Sciences of the USSR and the Italian National Council of Researches (CNR), contract No. 2.16 between the Astronomical Council (Moscow) of the USSR Academy and the Astronomical Observatory of Rome. Many thanks to Dr. B. M. Shustov for offering computing facilities, ospitality and collaboration of the Moscow Institute. Warm thanks are due to prof. R. Ruffini and to the International Center for Relativistic Astrophysics for a computing grant. We also thank G. Buonvino who drew the graphs.

REFERENCES

1. Larson, R.B.: 1981, Monthly Notices Roy. Astron. Soc., **194**, 809.
2. Myers, P.G.: 1983, Astron. J., **270**, 105.
3. Sanders, D.B., Scoville, N.Z., Solomon, P.M.: 1985, Astrophys. J., **289**, 373.
4. Solomon, P.M., Rivolo, A.R., Barret, J., Yahil, A.: 1987, Astrophys. J., **319**, 370.
5. Leung, C.M., Kutner, M.L., Mead, K.N.: 1982, Astrophys. J., **262**, 583.
6. Monin, A.S., Ozmidov, R.B.: 1981, "*Ocean turbulence*", Gidrometizdat, Leningrad.
7. Kolmogorov, A.N.: 1941, Dokl. Acad. Nauk USSR, **30**, 4, 299.
8. Vereshchagin, S.V., Soloviev, A.V.: 1990, Astron. J. USSR, in press.
9. Dolotin, V.V., Fridman, A.M.: 1989, J. Exp. Theor. Phys. USSR, in press.
10. Battinelli, P., Capuzzo Dolcetta, R., Di Fazio, A., Urpin, V. A., Vereshchagin, S.: 1990, this Workshop.
11. Di Fazio, A.: 1986, Astron. Astrophys., **159**, 49.
12. Di Fazio, A., Massi, M.: 1990, Astron. Astrophys., in preparation.
13. Massi, M., Churchwell, E., Felli, M.: 1989, Astron. Astrophys., in press.
14. Bhatt, H.C., Rowse, D.P., Williams, I. P.: 1984, Monthly Notices Roy. Astron. Soc. **209**, 69.

Discussion:

Cayrel: Could you specify to what phase was the viscosity ν of your transparencies applying to (gas, fragments' system)?

Vereshchagin: It was applying to the gas in: 1) monitoring the Reynolds number to decide in what regime is the so called coefficient of resistivity (the one that in our picture is responsible for creating turbulence, through the work of the resistivity force); 2) this second use of the viscosity was not on the transparencies, but we account also for the bulk dissipation in the gas (compression in non-reversible mode, etc.) and that is accounted for through the gas viscosity.

Perault: Could you describe where you make hydrodynamical calculations, and where you are using independent results,

for your multiscale model of fragmentation and turbulence?
Vereshchagin - Di Fazio: We solve together two density-homogeneous systems of hydrodynamical equations (the so called Riemann -MacLaurin - Jacobi formalism) for the turbulent gas and for the fragment system, formally divided in a multifluid group. The birth function of fragments (the "fragmentation rate" is used from Di Fazio, 1986, and it is calculated using the Bonnor-extended, time-dependent Jeans problem , and it is a function of temperature and gas density, plus, of course, it depends on the mass of the fragment to form). The assembling the needed dynamical equations (together with radiation transfer) and the equation of balance of the main scale of turbulence are our independent results. The fragmentation rate, on the other hand, is calculated without accounting for the presence of turbulence. This is the main simplification of this work, even though we have not heard, in the literature, of another analytical fragmentation theory which accounts for the turbolized status of the gas.

FRAGMENTATION AND COLLAPSE *IN* MAGNETIC MOLECULAR CLOUDS: NATURAL LENGTH SCALES AND PROTOSTELLAR MASSES

TELEMACHOS CH. MOUSCHOVIAS
University of Illinois at Urbana-Champaign
Departments of Physics and Astronomy
1011 West Springfield Avenue
Urbana, IL 61801, U. S. A.

ABSTRACT. Gravity is of course ultimately responsible for fragmentation and star formation. Magnetic forces dominate thermal-pressure and centrifugal forces over scales comparable to molecular cloud radii. Magnetic support of molecular clouds and the imperfect collisional coupling between charged and neutral particles introduce a natural "*Alfvén length scale*" ($\lambda_A = \pi v_A \tau_{ni}$) in the problem which together with a *thermal* ($\lambda_T = 1.09\, C_a \tau_{ff}$) and a *magnetic* ($\lambda_M = 0.62\, v_A \tau_{ff}$) "*Jeans length*" lead to the formation of fragments (or cores) in otherwise quiescent clouds and determine the sizes and masses of these fragments during the subsequent phases of contraction. (The quantity v_A is the Alfvén speed, τ_{ni} the mean neutral-ion collision time, C_a the adiabatic speed of sound, and τ_{ff} the free fall time scale.) Numerical calculations based on new adaptive-grid techniques follow the formation of fragments by ambipolar diffusion and their subsequent collapse up to an enhancement in central density above its initial equilibrium value by a factor $\simeq 10^6$ with excellent spatial resolution. The results confirm the existence and relevance of the three length scales and extend the analytical understanding of fragmentation and star formation derived from them. The relation $B_c \propto \rho_c^{\kappa}$ between the magnetic field strength and the gas density in cloud cores holds with $\kappa = 0.4 - 0.5$ even in the presence of ambipolar diffusion up to densities $\sim 10^9$ cm^{-3} for a wide variety of clouds. The value $\kappa \simeq 1/2$ is fairly typical. At the late stages of evolution, for example at a central density of about 3×10^8 cm^{-3}, a *typical* core is relatively uniform, contains 0.1 $M_\odot$ and a magnetic field $\simeq$ 3 mGauss, and is surrounded by a spatially rapidly decreasing, highly nonspherical (disk-like) density distribution. The amount of mass available for accretion onto the compact core is limited by magnetic forces, and is typically $\sim 1\ M_\odot$. These results are built into the detailed scenario for star formation described recently elsewhere.

1. INTRODUCTION

Complex as the structure of molecular clouds may be, particularly on small scales, those *individual* self-gravitating entities which have not yet given birth to stars are nearly isothermal ($T \simeq 10$ K) systems, typically characterized by a mean density $n \simeq 10^3$ cm^{-3} and a mass $\sim 10^2 - 10^4\ M_\odot$, and are typically threaded by a mean magnetic field 10 - 100 μGauss. At the one extreme, unbound molecular clumps of mass $\sim$ 1 - 10 $M_\odot$ are also observed (see Blitz 1987); they are thought to be transient structures that form behind interstellar shocks. At the other extreme, cloud complexes containing more than $10^6\ M_\odot$ and having mean densities $\lesssim$ 50 cm^{-3} often represent the extended neighborhoods of star

R. Capuzzo-Dolcetta et al. (eds.), Physical Processes in Fragmentation and Star Formation, 117–148.

forming molecular clouds, and they are shown to form by the Parker (1966) instability (Mouschovias 1974; Mouschovias, Shu, and Woodward 1974; Mouschovias 1978; Blitz and Shu 1980; see also the review by Mouschovias 1981). Of direct interest to star formation and to this meeting are the individual self-gravitating clouds. If it were not for the presence of the magnetic field they would all be collapsing on essentially the free fall time scale because the Bonnor-Ebert critical mass for a representative molecular cloud is only 5.8 $M_\odot$. In general, it is given by

$$M_{BE} = 1.2 \frac{C^4}{(G^3 P_{ext})^{1/2}}, \tag{1a}$$

$$= 5.8 \left(\frac{T}{10\ K}\right)^{3/2} \left(\frac{10^3\ cm^{-3}}{n}\right)^{1/2} \quad M_\odot . \tag{1b}$$

The quantity C [$= (k_B T/\mu m_H)^{1/2}$] in equation (1a) is the isothermal speed of sound in the cloud, P_{ext} is the fixed external pressure, and G the universal gravitational constant; k_B is the Boltzmann constant, and μ the mean mass per particle in units of the atomic hydrogen mass, m_H. In order to account for the cosmic abundance of helium, the value $\mu = 2.33$ was used in obtaining equation (1b).

Subsonic turbulence can be accommodated as an effective increase of T (by at most a factor of 2). Thus the combination of thermal and turbulent pressure forces can support at most a 16 $M_\odot$ molecular clump. Supersonic turbulence dissipates too rapidly to be of relevance for any significant length of time (see, also, discussion below).

Early observational evidence revealed a strong spatial correlation between young stars and relatively dense concentrations of interstellar matter (Baade 1944). This led to the hypothesis, which prevailed until the mid-1970s, that star formation was the result of the collapse (and fragmentation) of an interstellar cloud *as a whole*. However, neither observations (e.g., see review by Zuckerman and Palmer 1974), which could not detect velocities characteristic of collapse, nor theoretical calculations (see Mouschovias 1976a, b, 1978; Mouschovias and Spitzer 1976), which showed that even relatively weak magnetic fields could provide very effective support against gravity, lended any support to that popular notion. The predicted, relatively weak fields turned out to be consistent with those subsequently measured (see reviews by Heiles 1987, and Mouschovias 1981, 1987a). A new picture emerged in which formation of fragments (or cores) and their subsequent collapse was initiated in cloud interiors by ambipolar diffusion in otherwise magnetically supported clouds (Mouschovias 1976b, p. 156; 1977; 1978; 1979b). *These papers were the first to demonstrate, both qualitatively and quantitatively, that ambipolar diffusion* ***initiates*** *the quasistatic formation and, later, collapse of cores while the envelopes remain much better supported by magnetic forces* --this process of ***self-initiated collapse*** was subsequently referred to as "inside-out collapse" by some authors (e.g., see Shu, Adams, and Lizano 1987). The theoretical conclusion was reached that "*low-mass stars should preferentially form first, and perhaps only, in the core*[*s*] *of dense cloud*[*s*]" (Mouschovias 1978). Soon thereafter, observations confirming that prediction began to accumulate (see extensive discussion in Vrba, Coyne, and Tapia 1981). It should therefore be clear why the word "*in*" (as opposed to "*of*") is used in the title of this paper: *The collapse of a molecular cloud as a whole is not necessary for stars to form in its interior --nor is it, in fact, commonplace in nature.*

Observations also reveal that the topology of the magnetic field is too orderly (e.g., see Vrba *et al.* 1981; Moneti *et al.* 1984) and its pressure too large (compared to the thermal pressure) to permit the inference that chaotic motions (or turbulence) play a significant role in the overall support of molecular clouds (see review by Mouschovias 1987a, § 2.2.1). Interplay between theory and observations has led to more detailed calculations which represent improvements and refinements of the qualitatively novel theory of star formation. New observations were also undertaken. The updated, but still incomplete, theory is described in the papers by Mouschovias (1987a, b), which include a number of new predictions and a comparison with recent observations. Its key elements are shown in the schematic diagram on the following page. (A dashed line signifies a possible but unlikely evolutionary path. *Collapse* means indefinite contraction, but not necessarily free fall.) Since two recent summaries have been given from a different perspective (see Mouschovias 1989, § II; 1990), only my oral presentation at this symposium has dwelled on the theory of star formation. This written version of the paper focusses on the new results relating to the problem of fragmentation and star formation in magnetic clouds which I have summarized in my talk. More specifically, the following issues are addressed quantitatively.

1. Can fragmentation occur in a typical (model) molecular cloud? If so, how?
2. Once a self-gravitating fragment (or core) forms, what is its subsequent evolution?
3. How do the theoretical results compare with existing observations?
4. Does the theory have predictive power that can be tested by future observations?

2. CRITICAL MASSES FOR GRAVITATIONAL COLLAPSE: THE ROLE OF MAGNETIC BRAKING AND AMBIPOLAR DIFFUSION

The ratio of the magnitude of the gravitational potential energy (W_{gr}) and the rotational kinetic energy (W_{rot}) of a spherical blob of interstellar matter of density equal to the mean interstellar density ($\sim$ 1 cm^{-3}) and angular velocity equal to that of Galactic rotation in the solar neighborhood ($\Omega \sim 10^{-15}$ rad s^{-1}) is

$$\frac{|W_{gr}|}{2W_{rot}} = 2\pi \frac{G\rho}{\Omega^2}, \tag{2}$$

which is $\sim$ 1. *Therefore, formation of clouds, regardless of mass, that involves contraction perpendicular to the axis of rotation is forbidden if angular momentum is conserved.*

2.1. Magnetic Braking Time Scales

Magnetic braking transports angular momentum away from a cloud during its formation and contraction, thereby keeping the centrifugal forces small. It operates over a characteristic time *strictly smaller* than (see Mouschovias 1977, 1978, 1979a; Mouschovias and Paleologou 1980; or reviews by Mouschovias 1985, 1987b)

$$\tau_{\parallel} = \frac{\rho_{cl}}{\rho_{ext}} \frac{Z}{v_{A,ext}} \equiv \frac{\sigma_{m,cl}}{2\rho_{ext} v_{A,ext}} \equiv \left(\frac{\pi}{\rho_{ext}}\right)^{1/2} \frac{M}{\Phi_B} \equiv 0.4 \left(\frac{\rho_{cl}}{\rho_{ext}}\right)^{1/2} \frac{M}{M_{crit}} \tau_{ff}, \tag{3a}$$

where ρ_{cl}, R, and Z, are the density, equatorial and polar radius, respectively, of a cloud (or fragment) rotating about its (z-) axis of symmetry which is aligned with the magnetic field. The matter density and Alfvén speed in the external medium (or envelope) are denoted by ρ_{ext} and $v_{A,ext}$, respectively. The various forms of $\tau_{\parallel}$ given in equation (3a) are useful for different applications, and they involve the mass M of the cloud, the column (or surface) density of matter $\sigma_m \equiv M/\pi R^2$, the magnetic flux Φ_B threading the cloud, the free-fall time τ_{ff} (see eq. [8a] below), and the critical mass M_{crit} for gravitational collapse against the frozen-in magnetic flux Φ_B (see eq. [4a] below). Since the magnetic braking time scale for a perpendicular rotator has been found to be $\tau_{\perp} << \tau_{\parallel}$ (see Mouschovias and Paleologou 1979;

A SCENARIO FOR STAR FORMATION IN MAGNETIC CLOUDS

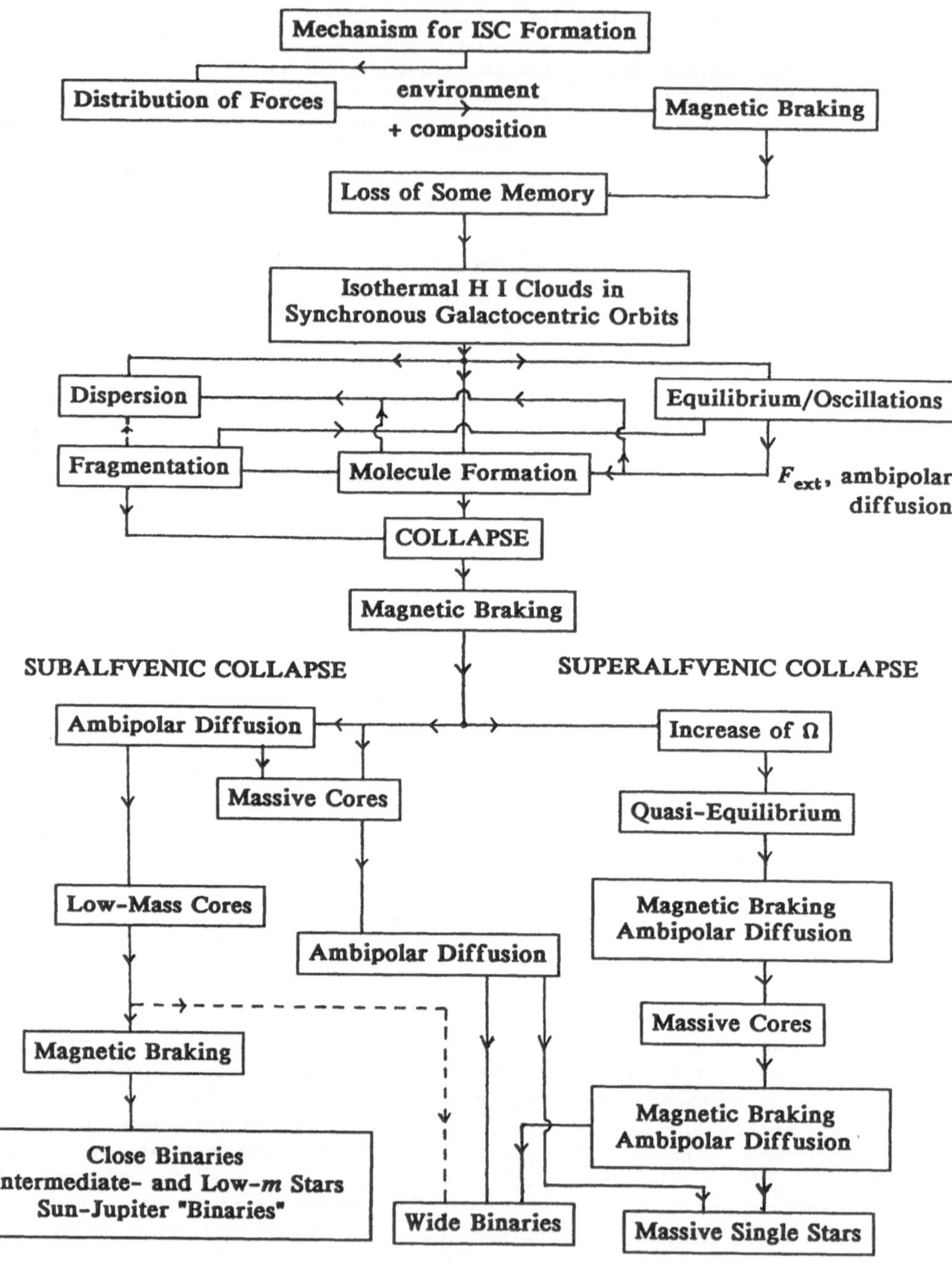

or summary in Mouschovias 1985, or 1987b, § 2.3.2), most clouds and fragments tend to become aligned rotators in time. For $\rho_{cl}/\rho_{ext} >> 1$, $\tau_\perp$ becomes

$$\tau_\perp = \frac{1}{2}\left(\frac{\rho_{cl}}{\rho_{ext}}\right)^{1/2} \frac{R}{v_A(R)} \equiv 2\left(\frac{\pi}{\rho_{cl}}\right)^{1/2} \frac{M}{\Phi_B} \equiv (2/3)^{1/2} \frac{M}{M_{crit}} \tau_{ff} , \qquad (3b)$$

where $v_A(R) = B(R)/(4\pi\rho_{ext})^{1/2}$ is the Alfvén speed *just outside* the cloud surface. It is indeed very clear from equations (3a) and (3b) that, in general, $\tau_\perp << \tau_\parallel$. *The expressions for* $\tau_\parallel$ *and* $\tau_\perp$ *can easily be recovered by simply calculating the time it takes for the torsional Alfvén waves generated by the rotation of a cloud (or core) to affect a moment of inertia in the surrounding medium equal to the moment of inertia of the cloud (or core)* --see Mouschovias and Paleologou 1979, 1980; or review by Mouschovias 1987b, in which alternative forms of $\tau_\perp$ and $\tau_\parallel$ are also given, including the effect of field lines fanning out away from an aligned rotator.[1]

Since supercritical cloud masses are rarely, if ever, observed, the last form of equation (3b) implies that $\tau_\perp \lesssim \tau_{ff}$; i.e., magnetic braking of a perpendicular rotator is always effective. Detailed collapse calculations have shown that the contraction time scale is typically $10\tau_{ff}$ (see summary in Mouschovias 1983). Hence, the last form of equation (3a) implies that magnetic braking of an aligned rotator is also very effective at least until a self-gravitating cloud (or core) has reached a density contrast with its background $\rho_{cl}/\rho_{ext} \simeq 10^3$. Dense cores, as opposed to clouds themselves, were predicted (see Mouschovias 1978), and are indeed observed, to rotate appreciably more rapidly than the surrounding envelopes (see Clark and Johnson 1978, 1981; Martin and Barrett 1978; Mattila, Winnberg, and Grasshof 1979; Young *et al.* 1981; Baudry *et al.* 1981; Schloerb and Snell 1984; Arquilla 1984; and Goldsmith and Sernyak 1984; see also review by Goldsmith and Arquilla 1985). Large as they may be, the observed angular velocities are at least one (and usually two or more) order(s) of magnitude *smaller* than those implied by conservation of angular momentum. *There is little doubt that magnetic braking has been at work from scales larger than* ~ 10 pc *to scales smaller than* ~ 10^{-2} pc.

2.2. Basic Physics of Fragmentation in Magnetic Clouds

The differential mass-to-flux ratio $[dm(\Phi_B)/d\Phi_B)]$ determines the relative magnitude of gravitational and magnetic forces, and affects crucially the evolution of a cloud (see Mouschovias 1976a, b). One can obtain reasonable theoretical estimates for this function, but no observational determination exists as yet. Observations such as those by Bregman *et al.* (1983) and by Schwarz *et al.* (1986) constitute an important first step toward the determination of $dm/d\Phi_B$.

---

[1]This result is unfortunately presented as new in a paper by Nakano 1989 which, in addition, follows step-by-step the formulation and solution of the magnetic braking problem by Mouschovias and Paleologou 1979 (MP79) but without reference. The equation of (rotational) motion of the cloud (or core) in the form $\partial^2\Omega/\partial\tau^2 \propto \partial\Omega/\partial\varsigma$ (see Nakano's eq. [18]), where t denotes time and ς a spatial coordinate, and the particular way of coupling it to the wave equation describing the propagation of torsional Alfvén waves in the external medium (or envelope) was first derived by MP79 (see their eq. [20]) for a perpendicular rotator and by MP80 for an aligned rotator (see their eq. [12b], which is identical with eq. [18] of Nakano 1989).

Unlike the case of a nonmagnetic isothermal sphere, for a magnetic isothermal cloud (or fragment, or core) to collapse *two* conditions must be satisfied. First, its mass-to-flux ratio must exceed a critical value given by (see Mouschovias and Spitzer 1976)

$$\left(\frac{M}{\Phi_B}\right)_{crit} = \frac{0.53}{3\pi}\left(\frac{5}{G}\right)^{1/2} = \frac{0.126}{G^{1/2}} \simeq \left(\frac{1}{63G}\right)^{1/2} . \tag{4a}$$

Second, the external pressure must also exceed a critical value given by

$$P_{ext} > P_{crit} = 1.89 \frac{C^8}{G^3M^2}\left[1 - \left(\frac{M_{crit}}{M}\right)^2\right]^{-3} , \quad \text{where} \quad M \geq M_{crit}. \tag{4b}$$

By solving equation (4a) for M_{crit} and substituting in the expression for P_{crit}, one obtains a single *sufficient* condition for the collapse of self-gravitating clouds supported by ***m***agnetic and ***t***hermal-pressure forces, namely,

$$M \geq M_{TM} = 1.38 \left\{1 - \left[\frac{0.126}{G^{1/2}(M/\Phi_{B,cl})}\right]^2\right\}^{-3/2} \frac{C^4}{(G^3P_{ext})^{1/2}} , \quad \text{for } [...] \leq 1; \tag{5a}$$

where $\Phi_{B,cl}$ and M are the total flux and mass of the cloud, respectively. Since the total mass-to-flux ratio $M/\Phi_{B,cl}$ of initially spherical clouds threaded by a uniform magnetic field is related to the central mass-to-flux ratio $(dm/d\Phi_B)_c$ by $M/\Phi_{B,cl} = (2/3)(dm/d\Phi_B)_c$ (see Mouschovias 1976a, eq. [44]), equation (5a) can equivalently be written in terms of the central mass-to-flux ratio as

$$M_{TM} = 1.38 \left\{1 - \left[\frac{0.19}{G^{1/2}(dm/d\Phi_B)_c}\right]^2\right\}^{-3/2} \frac{C^4}{(G^3P_{ext})^{1/2}} , \quad \text{for } [...] \leq 1. \tag{5b}$$

Recent numerical determination of critical masses by Tomisaka *et al.* (1988) is completely consistent with equations (5a) and (5b) and, in fact, the equation displayed in their Abstract is *identical* with equation (5b), except for the trivial fact that the number 0.19 in the brackets on the right-hand side of equation (5b) is replaced by 0.17, and the number 1.38 is replaced by $62/(4\pi)^{3/2}$ (= 1.39). Equation (5b) was the basis for the conclusion (see Mouschovias 1978, p. 218) that "*a distribution of flux strongly concentrated in a small cylindrical region threading the cloud core* [i.e., the quantity in brackets $\simeq$ 1 on the right-hand side of eq. (5b)] *can support significantly larger cloud masses than those given* [by M = 1 - 2 M_{crit}] *above. Conversely, a magnetic flux mainly threading a cloud's outer layers can allow a relatively low-mass core to collapse as long as it satisfies the Bonnor-Ebert condition*" --because, in this case, $(dm/d\Phi_B)_c$ >>> 1, hence the term in [...] $\simeq$ 0 on the right-hand side of eq. [5b] and, therefore, $M_{TM} \simeq 1.38\, C^4/(G^3P_{ext})^{1/2} \simeq M_{BE}$.

The effect of subsonic turbulence is easily included in equation (5a) or (5b), if one desires, as an increase of the isothermal sound speed C by at most 40%. However, since magnetic forces dominate thermal-pressure forces at typical molecular cloud sizes and densities, one is not usually concerned with the effect of subsonic (and, therefore, subAlfvénic) turbulence on the critical mass. Nevertheless, for individual fragments or cores deprived of magnetic support due to ambipolar diffusion, subsonic turbulence, if present, can increase the typical (i.e., at T = 10 K and n_n = 10^5 cm^{-3}) thermally supported mass from 0.58 to 1.6 $M_\odot$, as seen from equation (1).

An observationally useful expression is obtained by rewriting the necessary condition (4a) for collapse in terms of a critical mean column density of matter as

$$\langle\sigma_m\rangle_{crit} = \left(\frac{1}{63G}\right)^{1/2} \langle B\rangle = 2.4 \times 10^{-2} \frac{\langle B\rangle}{50\ \mu G} \quad g\ cm^{-2} . \tag{4c}$$

Other forms of this equation, for example in terms of a critical visual extinction, are given in Mouschovias (1987a). *It follows from equation (4c) that clouds, as opposed to cores, with supercritical masses should be very rare.*

These theoretical results and the observational fact that velocity fields characteristic of gravitational collapse are not observed suggest that very few, if any, clouds acquire a supercritical mass-to-flux ratio from the outset and begin to collapse as a whole. Instead, *clouds reach first a relatively quiescent state* (near hydrostatic equilibrium with magnetic and thermal-pressure forces balancing gravity), with possible leftover stable oscillations about the equilibrium state (Mouschovias 1975). It does *not* follow, however, that external triggers are necessary for star formation. As summarized in § 1 above, *self-initiated fragmentation (or core formation) due to ambipolar diffusion can lead to star formation in the cores of otherwise quiescent clouds* (Mouschovias 1976b, 1977, 1978, 1979b). A quantitative as well as physically transparent way to clarify this mode of star formation is as follows.

A form of equation (4a) especially suited for studies of fragmentation in magnetic clouds is obtained by rewriting it so as to express M_{crit} in terms of the cloud magnetic field (B) and the number density of protons (n_p):

$$M_{crit} = 5.0 \times 10^5 \frac{(B/3\ \mu G)^3}{(n_p/1\ cm^{-3})^2} \quad M_\odot , \tag{6}$$

(see Mouschovias and Spitzer 1976; Spitzer 1978; Mouschovias 1978). Even with the magnetic field still frozen in the matter, fragments with $M_{fr} \sim 150\ M_\odot$ can separate out in a cloud of mass $M \sim 10^4\ M_\odot$ at a density $n\ (= n_p/2) \sim 10^4\ cm^{-3}$. As ambipolar diffusion increases the mass-to-flux ratio in a fragment (or core) above the critical value for collapse, lower-mass fragments separate out gravitationally and begin to contract on their own. For example, even with a magnetic field as large as $\sim$ 100 μGauss in a core in which ambipolar diffusion is in progress (and, consequently, with the magnetic field only partially coupled to the neutrals), equation (6) implies that fragments with masses $M_{fr} \lesssim M_{crit} \simeq 3\ M_\odot$ can separate out at neutral densities $n_n \sim 4 \times 10^4\ cm^{-3}$. *A plethora of low-mass, low-spin fragments (or cores) should therefore be forming preferentially in (and perhaps only in) the cores of self-gravitating clouds* which (clouds) contract either subAlfvénically or not at all, while more massive fragments can form further out from a cloud's core (see also Mouschovias 1978 for the same, but somewhat less refined, prediction). This is consistent with observations of low-mass fragments in molecular clouds (see review by Myers 1985).

An important question must still be addressed: Does ambipolar diffusion progress rapidly enough for the above sequence of events to take place and lead to low-mass protostar formation in molecular cloud interiors?

2.3. Ambipolar Diffusion Time Scales

Ambipolar diffusion at *some* rate is unavoidable in the deep interior of a self-gravitating, magnetically supported cloud. The essence of ambipolar diffusion is to permit self-gravity to ***redistribute*** mass in the central flux tubes of a cloud and thereby induce fragmentation (Mouschovias 1977, 1978, 1979b; see also discussion associated with eq. [6] above), recently being referred to as core formation (Paleologou and Mouschovias 1983; Shu 1983; Scott 1984; Mouschovias, Peleologou, and Fiedler 1985; Myers and Goodman 1988), followed eventually by gravitational collapse of dense cores that have exceeded the critical mass-to-flux ratio specified by equation (4a) and the critical mass specified by equation (1). This picture is qualitatively different from Nakano and Tademaru's (1972) and from Mestel and Paris' (1979) notion that a cloud *as a whole* reduces its magnetic flux due to ambipolar diffusion, and is at the very heart of the new scenario for star formation we have originally proposed (Mouschovias 1977, 1978) and recently refined (Mouschovias 1987a, b). Its most direct, and perhaps most important, consequence is the one to which we referred above, namely, the formation of stars in the deep interior of otherwise quiescent, magnetically supported clouds.

Up to the stage that ambipolar diffusion increases the mass-to-flux ratio of the core above the critical value given by equation (4a), the contraction remains *quasistatic* and the time scale τ_{AD} for ambipolar diffusion (in an axisymmetric geometry) can be obtained analytically (see Mouschovias 1989, eq. [20]):

$$\tau_{AD} \equiv \frac{r}{|v_D|} = \frac{8}{1.4\pi^2}\frac{\tau_{ff}^2}{\tau_{ni}}\left(1 - \frac{8}{\pi^2}\frac{\tau_{ff}^2}{\tau_s^2}\right)^{-1}, \qquad \text{for} \quad \frac{8}{\pi^2}\frac{\tau_{ff}^2}{\tau_s^2} < 1\,, \tag{7}$$

where $v_D \equiv v_i - v_n$ is the relative ("drift") speed between ions and neutrals; τ_{ff}, τ_s, and τ_{ni} are the spherical free-fall time, the sound crossing time, and the neutral-ion collision time, respectively, and are given by

$$\tau_{ff} = \left(\frac{3\pi}{32G\rho_n}\right)^{1/2} = 1.07 \times 10^6 \left(\frac{10^3\ \mathrm{cm}^{-3}}{n_n}\right)^{1/2} \ \mathrm{yr}\,, \tag{8a}$$

$$\tau_s = \frac{r_c}{C}\,, \qquad \tau_{ni} = \frac{m_i + m_n}{m_i}\frac{1}{n_i\langle\sigma w\rangle_{in}}\,, \tag{8b, c}$$

where n_n and n_i are the number densities of neutrals and ions, respectively. All time scales refer to the core, which has radius r_c. The quantity $\langle\sigma w\rangle_{in}$ is the average collisional rate between ions of mass m_i and neutrals of mass m_n; it is equal to $1.69 \times 10^{-9}\ \mathrm{cm}^3\ \mathrm{s}^{-1}$ for HCO^+-H_2 collisions (see McDaniel and Mason 1973) --note that $m_i/m_n = 14.4$ for this particle pair. For *magnetically supported* clouds the term in brackets in equation (7) is essentially equal to unity and *the expression for the ambipolar diffusion time scale in the form* $\tau_{AD} \sim \tau_{ff}^2/\tau_{ni}$, *valid for quasistatic contraction, is a property of the general equations governing ambipolar diffusion and does not depend on geometry* (see Mouschovias 1987b, §§ 3.3 and 3.4). For example, the constant on the right-hand side of equation (7), which refers to an axisymmetric geometry, is $8/1.4\pi^2 = 0.58$, and the constant in cylindrical geometry is $2/1.4\pi = 0.45$ (Mouschovias 1979b; or review 1987a, eq. [12a]). In the cores of magnetically supported clouds, equation (7) can be put in the useful form

$$\tau_{AD} \simeq 2 \times 10^5 \left(\frac{n_i/n_{H_2}}{10^{-8}}\right) \ \mathrm{yr}. \tag{9}$$

This is a significantly short time scale for typical molecular clouds. [Note that the characteristic time $\tau_\Phi \equiv \Phi_B/|d\Phi_B/dt|$ for the reduction of the magnetic flux of a core is actually smaller than τ_{AD} by exactly a factor of 2; see eq. (29) below.]

After ambipolar diffusion has increased the mass-to-flux ratio of a quasistatically contracting core of a typical molecular cloud to the critical value given by equation (4a), accelerated contraction may ensue (if thermal pressure cannot support the core) and equation (7) may no longer describe the rate at which ambipolar diffusion progresses. We find instead that (see Mouschovias 1989, eq. [23])

$$\tau_{AD} = 0.27 \frac{\tau_{ff}^2}{\tau_{ni}} \left(\frac{\Phi_{B,crit}}{\Phi_B}\right)^2 , \qquad \text{for} \quad \Phi_B \le \Phi_{B,crit}; \tag{10a}$$

$$= 1.3 \times 10^3 \left(\frac{n_i/n_{H_2}}{10^{-10}}\right) \left(\frac{\Phi_{B,crit}}{\Phi_B}\right)^2 \quad \text{yr}, \tag{10b}$$

where Φ_B is the actual flux of the core at any stage past the onset of *dynamical* contraction, and $\Phi_{B,crit}$ is uniquely determined by the mass of the core from equation (4a) as $\Phi_{B,crit} = (63G)^{1/2}M$. The normalization of n_i/n_{H2} in equation (10b) refers to a neutral density ~ 10^9 cm^{-3}, above which n_i no longer increases with n_n (see Elmegreen 1979; Nakano 1979).

It is emphasized that equation (10) refers to a core, not to the entire cloud. The cloud as a whole has $M \simeq M_{crit}$ because it is magnetically supported. However, a core of mass ~ 1 $M_\odot$ has $M < M_{crit}$ or, equivalently, $\Phi_B > \Phi_{B,crit}$ for a relatively long time, i.e., until the quasistatic phase of contraction due to ambipolar diffusion manages to reduce the flux of this (fixed) mass so as to make the mass-to-flux ratio equal to its critical value. *If* this marks the onset of accelerated contraction of the core (see eq. [5b]), then equation (10a) shows clearly that this will also mark the stage at which τ_{AD} begins to increase above its quasistatic contraction value by the factor $(\Phi_{B,crit}/\Phi_B)^2$. It therefore becomes more difficult to lose flux during this late phase of accelerated contraction. New detailed collapse calculations which follow the evolution to densities ~ 10^9 cm^{-3} are summarized in §§ 5 and 6 below. The advanced stages of evolution beyond the density ~ 10^9 cm^{-3}, during which significant reduction in the electron density takes place and collisions no longer permit the dominant charge carriers to remain attached to the magnetic field, are only now beginning to be explored.

3. LOW- *VERSUS* HIGH-MASS STAR FORMATION

Although collapse at subAlfvénic speeds in cloud cores is the most common case (see Mouschovias 1978, Appendix B) and leads to the formation of low- and intermediate-mass stars and close binaries, Alfvénic or superAlfvénic collapse of a cloud or fragment (e.g., due to implosion by a shock) leads to a very different sequence of events and, most likely, to a different IMF; i.e., more high-mass stars and wide binary systems (see Mouschovias 1987a). One idea is that massive star formation is the result of the collapse of clouds with supercritical masses, $M > M_{crit}$ (e.g., see Mestel 1985; Lizano and Shu 1987). This idea is qualitatively different from our ideas based on the externally triggered superAlfvénic collapse of a cloud (or fragment) of mass $M \simeq M_{crit}$ in a cloud complex (see Mouschovias 1989, § 1.4). The difference and its consequences are summarized as follows.

Aside from the fact that supercritical cloud masses are rarely, if ever, observed, the contraction of a cloud with $M > M_{crit}$ leads to a rapid increase of the magnetic tension force (see Mouschovias 1978). It therefore remains to be demonstrated whether this kind of contraction can lead to a *qualitatively* different protostellar mass spectrum from that resulting from the self-initiated collapse due to ambipolar diffusion. On the contrary, during the superAlfvénic implosion of a clump with mass $M \simeq M_{crit}$ in a cloud complex, magnetic tension cannot stop or significantly slow down the collapse and, in addition, the torsional Alfvén waves remain trapped within the clump. Contraction with angular momentum nearly conserved implies that centrifugal forces will become progressively more important. Depending, however, on the precise density at which rapid collapse begins, they may not increase sufficiently to prevent the formation of very wide binaries, including rapidly rotating, massive single stars as members. For example, if angular momentum begins to be conserved above a density $\simeq 10^4$ cm^{-3}, the angular momentum left over in a fragment (due to earlier efficient magnetic braking) is exactly what is required to form a relatively wide (visual) binary system, with a period $\tau_b \simeq 100$ yr. This argument does not depend on mass. In fact, a simple model shows that binaries forming according to this scenario will have periods scaling with the density n_{cr}, above which angular momentum is nearly conserved, as $\tau_b/(100 \text{ yr}) \simeq [(10^4 \text{ cm}^{-3})/n_{cr}]^2$ (Mouschovias 1977). *Note that conservation of angular momentum during superAlfvénic collapse is due to trapping of the torsional Alfvén waves inside the cloud, not due to rapid ambipolar diffusion.* This is an important qualitative difference from our early work, and has significant implications for massive star formation.

If Alfvénic or superAlfvénic implosion begins at $n < 10^3$ cm^{-3}, centrifugal forces will lead to a quasi-equilibrium configuration by balancing gravity perpendicular to the axis of rotation. Further contraction takes place only as rapidly as magnetic braking can remove angular momentum to the envelope. Since the fragment still has most of its magnetic flux trapped inside at this stage and since it was *magnetically supported before implosion* (a key difference from the supercritical-mass idea of Mestel and of Lizano and Shu), *magnetic forces will also contribute significantly to this quasi-equilibrium, and both magnetic braking and ambipolar diffusion are expected to occur simultaneously, over comparable time scales.* Although this complex configuration has not been studied in any satisfactory fashion, it seems likely that stars which will form in such fragments will be characterized by both rapid rotation *and* strong magnetic fields. Since these are signatures of massive stars, it is tempting to suggest, at least tentatively, that this is how such stars form (see Mouschovias 1987a, 1989, for a strengthening of this suggestion and for a scenario of high-mass star formation even in subAlfvénically contracting but very massive clouds).

Although calculations do exist, simplifying assumptions concerning the geometry do not presently allow definitive conclusions on the issue of whether a protostar (low-mass or massive) becomes opaque while possessing a relatively weak magnetic field which will be amplified by a dynamo process, or whether it does so while having too much magnetic flux which will be dissipated during this or subsequent stages of contraction. We believe, nevertheless, that it is during these relatively late stages of protostar formation that its magnetic field will detach from the background. This view is based on calculations (see Mouschovias 1976a, b; 1978) which show that conditions conducive to magnetic detachment do not arise during the earlier, relatively diffuse stages of fragment contraction and protostar formation. Once magnetic detachment takes place, further loss of angular momentum becomes very inefficient (compared to earlier stages) and occurs through a magnetic wind.

The issue still remains whether these analytical considerations concerning low- and high-mass star formation, which are based on the calculations of Mouschovias (1976a, b) and on the implied new scenario for star formation (see Mouschovias 1977), will be confirmed by more realistic numerical calculations modeling the formation and collapse of fragments in molecular clouds. Before we describe the results of our recent numerical experiments, we gain new insight into the process of ambipolar diffusion and its consequences relating to the sizes, masses, and magnetic fluxes of fragments by obtaining and comparing *three natural length scales unavoidably present in molecular clouds.*

4. PROTOSTELLAR FRAGMENTS: SIZES, MASSES, AND MAGNETIC FLUXES

4.1. Analytical Considerations in Three Dimensions

The magnetic field unavoidably introduces an "*Alfvén length scale*" in molecular clouds (see Mouschovias 1987a, § 2.2.6[b]; Mouschovias, Morton, and Ciolek 1990), namely,

$$\boxed{\lambda_A \equiv \lambda_{A,\mathrm{cutoff}} = \pi v_A \tau_{ni}\,,} \tag{11}$$

below which ambipolar diffusion prevents Alfvén waves from propagating in the neutrals. The quantity v_A is the Alfvén speed in the neutrals and is given by

$$v_A = \frac{B}{(4\pi\rho_n)^{1/2}} = 1.36 \left(\frac{B}{30\ \mu\mathrm{G}}\right)\left(\frac{10^3\ \mathrm{cm^{-3}}}{n_n}\right)^{1/2} \quad \mathrm{km\ s^{-1}}\,. \tag{12}$$

The quantity τ_{ni} is the mean (momentum exchange) collision time of a neutral particle in a sea of ions, and is given by equation (8c) above, which may be written as

$$\tau_{ni} = 6.68 \times 10^{3+2k} \left(\frac{3 \times 10^{-3}\ \mathrm{cm^{-3}}}{K}\right)\left(\frac{10^3\ \mathrm{cm^{-3}}}{n_n}\right)^{k} \quad \mathrm{yr}\,. \tag{13}$$

where in equation (13) we adopted collisional rates and particle masses appropriate to HCO^+-H_2 collisions (see discussion following eq. [8c]) and we used the n_i - n_n relation

$$n_i = 3 \times 10^{-3} \left(\frac{K}{3 \times 10^{-3}\ \mathrm{cm^{-3}}}\right)\left(\frac{n_n}{10^5\ \mathrm{cm^{-3}}}\right)^{k} \quad \mathrm{cm^{-3}}\,, \tag{14}$$

which is valid in the density interval $10^3 \lesssim n_n \lesssim 10^9$ cm^{-3}. Observations and theoretical considerations suggest that $1/3 \leq k < 2/3$ (see Oppenheimer and Dalgarno 1974; Elmegreen 1979; Nakano 1979; Langer 1985; Falgarone and Pérault 1987), with 1/3 - 1/2 being the most likely range. Uncertainties in the cosmic-ray ionization rate and in the amount of metal depletion and charge neutralization onto grains combine for an uncertainty in the value of K perhaps by as much as an order of magnitude. (For this reason, the numerical solutions summarized in §§ 5 and 6 below treat k and K as free parameters and examine the dependence of the solution on their values within the observationally allowed range.)

As explained in Mouschovias 1987a [see § 2.2.6(b)], the physical meaning of the length scale λ_A is the following. If a disturbance in the magnetic field has wavelength $\lambda < \lambda_A$, it *diffuses* (or decays) before collisions between neutrals and ions have had time to transmit to the neutrals the magnetic force associated with the disturbance. In different terms, the ambipolar diffusion time τ_{AD} inside this length scale is shorter than the Alfvén crossing time τ_A and, therefore, the neutrals are essentially unaffected by magnetic forces. However, over length scales $> \lambda_A$, magnetic forces are transmitted to the neutrals very effectively by collisions with ions (or charged grains)[2] because $\tau_{AD} > \tau_A$. The characteristic length λ_A is therefore expected to set the length scale and, at a typical density (~ 10^4 cm^{-3}; see

---

[2]The effect of charged grains in coupling the magnetic field to the neutral matter is to replace τ_{ni} by the harmonic mean $\tau_{ni}\tau_{ng}/(\tau_{ni} + \tau_{ng})$ in the expression for λ_A, where τ_{ng} is the mean collision time between neutral particles and charged grains. This effect is accounted for in a calculation by Mouschovias, Ciolek, and Morton (1990).

Mouschovias 1977, and contrast with Nakano and Tademaru's 1972 density of ~ 10^9 cm^{-3}) characterizing the onset of ambipolar diffusion at a significant rate, the mass scale (~ 1 $M_\odot$) for molecular cloud cores and protostars. This does not mean that the size of a core is exactly equal to λ_A; it depends on the relative magnitude of λ_A and the other two natural length scales of the system (see below).

The interplay of gravitational and thermal-pressure forces introduces a second length scale in the problem which we take to be the radius of an isothermal sphere on the verge of collapse. From the work of Bonnor (1956) and Ebert (1955; 1957), this length scale, which we denote by λ_T, is $\lambda_T \equiv R_{BE} = 0.41\ GM/C^2$. In order to make its meaning more transparent and completely analogous to that of λ_A, we express λ_T in terms of the adiabatic speed of sound C_a (= $\gamma^{1/2}C$, where γ is the ratio of specific heats) and the spherical free fall time τ_{ff} (see eq. [8a]) to find that

$$\boxed{\lambda_T = 1.09\ C_a\tau_{ff}\ ,} \tag{15}$$

where the free-fall time refers to the *mean* density $\bar{\rho}_n$ of the critical isothermal sphere, and C_a is given by

$$C_a \equiv \gamma^{1/2}C = 0.24\ (T/10\ \mathrm{K})^{1/2} \quad \mathrm{km\ s^{-1}}\ . \tag{16}$$

We have taken the ratio of specific heats $\gamma = 5/3$ because neither vibrational nor rotational internal degrees of freedom of molecular hydrogen can be collisionally excited at the low temperatures present in molecular clouds which have not yet given birth to stars. The scale λ_T is smaller than the traditional Jeans radius at a density $\bar{\rho}_n$, namely, $R_J \equiv \lambda_J/2 = C_a(\pi/4G\bar{\rho}_n)^{1/2}$, by a factor of 1.5. It is more reliable than R_J as a measure of the mass on the verge of collapse at the density $\bar{\rho}_n$ because it acounts for the development of a central concentration, which makes it easier for gravitational forces to overwhelm the thermal-pressure forces; hence the fact that $\lambda_T \equiv R_{BE} < R_J$. The physical meaning of λ_T, analogous to that of λ_A, is that in order for thermal-pressure forces to prevent a self-gravitating object (cloud or core) from collapsing, the size (radius) of the object must be small enough ($< \lambda_T$) so that a fluid element at the center can communicate with another fluid element at the surface via a sound wave within one free-fall time. If the object's size is greater than λ_T, the deep interior will collapse before the outermost envelope has had time to learn about it and readjust its structure in response; i.e., gravity suppresses sound waves with period $\tau \gtrsim \tau_{ff}$, for it overwhelms the thermal pressure on length scales $> \lambda_T$.

The interplay between magnetic and gravitational forces introduces yet a third length scale in the problem. In a fashion analogous to the derivation of λ_T in terms of C_a and τ_{ff}, the critical mass-to-flux ratio given by equation (4a) can equivalently be written as a critical magnetic length scale (radius) λ_M in terms of the Alfvén speed in the neutrals (see eq. [12]) and the free-fall time as

$$\boxed{\lambda_M \equiv R_{crit} = 0.62\ v_A\tau_{ff}\ .} \tag{17}$$

A region of size (radius) $R < \lambda_M$ is magnetically supported, while a region of size $R > \lambda_M$ contains enough mass at its magnetic flux for gravitational forces to dominate the magnetic forces and, in the absence of other expansive forces such as thermal pressure, lead to collapse. In other words, Alfvén waves with period $\tau \gtrsim \tau_{ff}$ are suppressed by gravity, for it overwhelms the magnetic forces on scales $> \lambda_M$. In the presence of thermal pressure, the condition $R > \lambda_M$ is only necessary but not sufficient for collapse with frozen-in magnetic flux. The "external" thermal pressure (in the case of a core this is the pressure of the envelope) must also exceed the value given by equation (4b) for collapse to set in. The condition $R > \lambda_M$ is sufficient for collapse if thermal-pressure forces are negligible compared to magnetic forces. Since the critical mass-to-flux ratio used to derive equation (17) was obtained from exact equilibrium states on the verge of collapse calculated by Mouschovias (1976a, b), which have already developed a central concentration, the length scale λ_M is more reliable than a magnetic Jeans length which could be obtained from a linear stability analysis performed on a usually uniform (nonequilibrium) state or from the virial theorem. A simple but approximate way to make the condition $R > \lambda_M$ sufficient for collapse without reference to equation (4b) is to replace the Alfvén speed in equation (17) with the magnetosonic speed, $v_{ms} = (v_A{}^2 + C_a{}^2)^{1/2}$. (The effect of subsonic turbulence and long-wavelength hydromagnetic waves can be included in a similar fashion by adding the squares of their characteristic speeds under the square root on the right-hand side of the expression for v_{ms}.)

We now calculate typical values for λ_A, λ_T, and λ_M in molecular clouds, at the representative mean density $n_n = 10^3$ cm^{-3}, mean magnetic field strength $B = 30$ μGauss, and temperature $T = 10$ K, and then use the relative magnitude and the dependence of these characteristic lengths on neutral density to gain a unique insight into the formation, evolution, and self-initiated collapse of molecular cloud cores due to ambipolar diffusion.

We find from the expressions for λ_A, λ_T, and λ_M (see eqs. [11], [15], and [17], respectively) that

$$\lambda_A = 0.29 \left(\frac{B}{30\ \mu\mathrm{G}}\right) \left(\frac{1}{10^{1-2k}}\right) \left(\frac{10^3\ \mathrm{cm}^{-3}}{n_n}\right)^{0.5+k} \left(\frac{3 \times 10^{-3}\ \mathrm{cm}^{-3}}{K}\right) \quad \mathrm{pc}\ , \qquad (18a)$$

$$\lambda_T = 0.29 \left(\frac{T}{10\ \mathrm{K}}\right)^{1/2} \left(\frac{10^3\ \mathrm{cm}^{-3}}{n_n}\right)^{1/2} \quad \mathrm{pc}\ , \qquad (18b)$$

$$\lambda_M = 0.91 \left(\frac{B}{30\ \mu\mathrm{G}}\right) \left(\frac{10^3\ \mathrm{cm}^{-3}}{n_n}\right) \quad \mathrm{pc}\ , \qquad (18c)$$

where we have used the expressions for the Alfvén speed (eq. [12]), the neutral-ion collision time (eq. [13]), the sound speed (eq. [16]), and the free-fall time (eq. [8a]). Evidently, typical molecular cloud interiors have $\lambda_M > \lambda_T \simeq \lambda_A$ (at $n_n \simeq 10^3$ cm^{-3} and $k = 1/2$). *Since gravitational forces dominate thermal-pressure forces on scales* $\gtrsim \lambda_T$ *and magnetic forces are rendered ineffective on scales* $\lesssim \lambda_A$ *by ambipolar diffusion, it follows that gravitationally driven ambipolar diffusion operates effectively on scales* $\simeq \lambda_A \simeq \lambda_T$ *and causes infall velocities which are a significant fraction of the sound speed. Core formation should consequently be commonplace in magnetically supported molecular clouds, as is indeed revealed by observations* (e.g., see review by Myers 1985).

The inequality $\lambda_M > \lambda_T$ is just another statement of the fact that, before cores form, typical molecular clouds are primarily magnetically supported. The inequality $\lambda_M > \lambda_A$ essentially means that $\tau_{ff} > \tau_{ni}$, which implies (see eq. [7]) that $\tau_{AD} > \tau_{ff}$; i.e., *magnetic fields regulate the contraction of the neutrals through ambipolar diffusion and keep it quasistatic at least during the initial phases of core formation.* In quantitative terms, the ambipolar diffusion time scale in *magnetically supported, axisymmetric clouds* (i.e., during the quasistatic phase of contraction) is given by (see Mouschovias 1989)

$$\frac{\tau_{AD}}{\tau_{ff}} = \frac{8}{1.4\pi^2} \frac{\tau_{ff}}{\tau_{ni}} = 0.58 \frac{\tau_{ff}}{\tau_{ni}}\ . \qquad (19)$$

The proportionality constant on the right-hand side is only slightly different in different geometries; e.g., it is equal to $2/1.4\pi = 0.45$ in cylindrical geometry (see Mouschovias 1987a, eq. [12]). Equation (19) may be written in terms of the characteristic lengths, by using equations (11), (15), and (17), as

$$\frac{\tau_{AD}}{\tau_{ff}} = 2.93 \frac{\lambda_M}{\lambda_A}, \tag{20a}$$

$$= 1.29 \frac{v_A}{C} \frac{\lambda_T}{\lambda_A} = 1.82 \sqrt{\alpha} \frac{\lambda_T}{\lambda_A}, \tag{20b}$$

where the definitions of the Alfvén speed v_A and the adiabatic speed of sound C_a have also been used, and the pressure ratio α is defined by

$$\alpha = \frac{B^2}{8\pi\rho_n C^2} = 26.0 \left(\frac{B}{30\ \mu G}\right)^2 \left(\frac{10^3\ cm^{-3}}{n_n}\right)\left(\frac{10\ K}{T}\right). \tag{21}$$

As explained in § 2.3, equation (20) breaks down when the right-hand side approaches unity (from above); beyond that stage of a core's evolution, equation (10) for τ_{AD}, capable of describing the dynamical phase of contraction, takes over.

The relative magnitude of the three length scales λ_A, λ_T, and λ_M and their dependence on ρ_n reveal, in fact determine, how molecular cloud cores can evolve. It follows from their definitions (eqs. [11], [15], and [17]) that

$$\lambda_A \propto v_A \tau_{ni} \propto \frac{B}{\rho_n^{1/2}} \frac{1}{\rho_n^{k}} \propto \frac{1}{\rho_n^{0.5+k-\kappa}}, \tag{22a}$$

$$\lambda_T \propto C_a \tau_{ff} \propto \left(\frac{T}{\rho_n}\right)^{1/2}, \tag{22b}$$

$$\lambda_M \propto v_A \tau_{ff} \propto \frac{B}{\rho_n} \propto \frac{1}{\rho_n^{1-\kappa}}, \tag{22c}$$

where equations (12), (8c), (14), and (8a) have been used, and a relation $B \propto \rho_n^{\kappa}$ has been assumed to hold *in the core* (see cautions in Mouschovias 1978, §§ III[A] and III[C]), with κ to be determined from the solution of the collapse problem in the presence of ambipolar diffusion. The variation of the relative magnitudes of the three length scales with density is now obtained from the relations (22a) - (22c):

$$\frac{\lambda_T}{\lambda_A} \propto \frac{1}{\rho_n^{\kappa-k}}, \tag{23a}$$

$$\frac{\lambda_M}{\lambda_A} \propto \frac{1}{\rho_n^{0.5-k}}, \tag{23b}$$

$$\frac{\lambda_T}{\lambda_M} \propto \frac{1}{v_A} \propto \rho_n^{0.5-\kappa}. \tag{23c}$$

It is clear from the relations (23a) - (23c) that the only way in which the initial relative magnitudes of the three length scales ($\lambda_M > \lambda_T \simeq \lambda_A$) will be preserved during the formation and contraction of a cloud's core is for the condition $k = \kappa = 1/2$ to be satisfied at all times. If this turns out to be the case, then the contraction of the core and the plasma drift, although rapid in that the velocities can become sonic (but $\lesssim v_A$), will remain quasistatic up to densities $n_n \sim 10^9$ cm^{-3}, at which n_i no longer increases with n_n; i.e., $k = 0$. More generally, under the diverse physical conditions found in different molecular clouds, it is likely the case that $k \neq \kappa$ and that either both quantities simultaneously or one at a time are different from the value 1/2. This means that the initial physical conditions in the parent molecular cloud will be erased from a core's memory at some stage during its evolution.[3] For example, although λ_T/λ_M is (initially) less than unity in magnetically supported molecular clouds (see eqs. [18b] and [18c]), ambipolar diffusion *in a core* may decrease the local Alfvén speed (i.e., $\kappa < 1/2$), with the consequence that (see eq. [23c]) λ_M may become smaller than λ_T for a core. This means that thermal-pressure forces now become more important than magnetic forces on this (relatively small) scale. Similarly, if $k < \kappa$, then λ_A can exceed λ_T as the density of the core increases (see eq. [23a]), and if in addition $k < 1/2$, then λ_A can exceed λ_M as well at some stage (see eq. [23b]). This means that virtually unhindered collapse will set in inside λ_A which will also be accompanied by very effective ambipolar diffusion, in contrast to the claim/assumption that the onset of collapse *necessarily* marks the onset of magnetic flux trapping in the core (see review by Mouschovias 1987a, § 2.2.5). In summary, the precise values of the exponents κ and k in the respective relations $B_c \propto \rho_c^{\kappa}$ and $n_i \propto n_n^{k}$ are crucial for determining the evolution of molecular cloud cores and for permitting a simple analytical understanding of that evolution.

Since in the density range 10^3 - 10^4 cm^{-3}, at which ambipolar diffusion sets in at a rate rapid enough to begin to affect a cloud's evolution, we have that $\lambda_T \simeq \lambda_A$ ($\simeq$ 0.1 - 0.3 pc $< \lambda_M$), the "initial" core mass ($\simeq 1\ M_\odot$) cannot be prevented by magnetic or thermal-pressure forces from eventually collapsing and forming a protostar, despite the fact that λ_T and λ_A (and the mass inside these length scales) evaluated at the *instantaneous* density of the core decrease upon contraction (see eqs. [22b] and [22a]). In other words, a runaway compact core of mass significantly less than the original core mass is expected to form, but the original core mass represents a reservoir of matter (with a spatially rapidly decreasing density, and infall speeds comparable to the sound speed) which will eventually accrete onto the compact core and thus determine the protostellar mass. *Because of effective magnetic support, matter significantly farther than the original λ_A from the core's center can accrete only extremely slowly and, therefore, it is not available for increasing the protostellar mass by any appreciable amount.*[4] At the later stages of contraction, accretion may be further limited by the presence of angular momentum. Moreover, at a distance greater than $2\lambda_A$ ($\simeq 2\lambda_T$) from the center of a forming core, gravitationally driven ambipolar diffusion can lead to the formation of another, independent core much more rapidly than it would take for this material to be accreted by the first core. Each core can form either a binary or a single star, depending on its leftover angular momentum. ***This, we suggest, is the origin of stellar clusters in otherwise quiescent molecular clouds.***

Although the condition $\lambda_T \gtrsim \lambda_M \gtrsim \lambda_A$ is not met by a typical molecular cloud as a whole (i.e., molecular clouds are not thermally supported), it can be met by an evolving *core* in an otherwise magnetically supported cloud as follows. As ambipolar diffusion during the

---

[3]The initial typical (i.e., suggested by observations) physical conditions, however, remain important in the surrounding envelope and *prevent any significant amount of mass originally outside the region $r \simeq \lambda_A \simeq \lambda_T$ from accreting on the evolving core.*

[4]We thus see that the premise of the argument, that an effectively infinite reservoir of mass surrounds and can accrete onto each molecular cloud core and that, therefore, the onset of a stellar wind is necessary to cut off the accretion and determine the stellar mass (see Shu *et al.* 1987; Lizano 1990, in this volume), is invalid. It is based on the *assumption* that there is no characteristic length in the problem.

quasistatic phase of contraction redistributes mass in the central flux tubes of a cloud, *a stage is reached at which* λ_M *decreases sufficiently to become comparable to* λ_T *in the core* because isothermal contraction maintains C_a = *const* while v_A most likely decreases (i.e., $\kappa < 1/2$), however slowly, due to ambipolar diffusion and flattening along field lines (see eq. [23c]). *If at this stage* λ_A *is significantly smaller than* λ_T *and* λ_M, *the mass inside* $r \simeq \lambda_M \simeq \lambda_T$ *collapses with little additional loss of magnetic flux. If, however, at this stage* λ_A *is comparable to (or slightly greater than)* λ_T *and* λ_M, *then rapid collapse will ensue inside* λ_A *accompanied by rapid flux loss as well.*

4.2. Two-Dimensional, Cylindrically Symmetric Geometry: A Good Testing Ground

The crucial difference among one-dimensional (rectilinear), two-dimensional (cylindrically symmetric), and three-dimensional (axially symmetric) geometries (hereinafter referred to as 1D, 2D, and 3D, respectively) as far as the collapse of molecular clouds is concerned, whether self-initiated due to ambipolar diffusion or externally triggered by an increase of the surrounding pressure, lies in the number of characteristic lengths that can *intrinsically* exist in each geometry. In 1D, the length scales λ_M and λ_T do not exist, unless $T = 0$ in which case $\lambda_T = 0$. This is so because there is no critical mass for collapse against a finite magnetic flux frozen in the matter or against a finite temperature. Thermal-pressure forces and magnetic forces independently forbid indefinite contraction, although rapid and significant contraction may take place before either kind of force halts the collapse. The length scale λ_A, however, is always present and, consequently, ambipolar diffusion can initiate rapid and, depending on the collisional coupling between the neutrals and the plasma, even dynamic but not indefinite contraction (for finite T), as clearly demonstrated by the calculations of Paleologou and Mouschovias (1983), and Mouschovias, Paleologou, and Fiedler (1985). Those 1D results showed quantitatively how in *magnetically supported* clouds ambipolar diffusion alone can lead to the formation of cores with densities 10^4 - 10^8 cm^{-3} within diffuse envelopes of densities 10^2 - 10^3 cm^{-3}. Shu's (1983) 1D quasistatic calculations used model clouds in which thermal pressure forces were important from the outset, and the evolution due to ambipolar diffusion led to an enhancement of the central density typically by only a factor 2 - 10 over its initial value. The ultimate (but not necessarily realizable) fate of 1D cores is for gravity to be balanced by thermal pressure and the magnetic flux to slowly diffuse out so that the field strength in the core is the same as that in the envelope.

In 2D geometry, a critical mass (M_C) per unit length along the symmetry axis exists, above which thermal pressure cannot support a cloud against its self-gravity; it is given by

$$M_C = \frac{2C^2}{G} = 16.5 \left(\frac{T}{10\ \mathrm{K}}\right) \quad M_\odot\ \mathrm{pc}^{-1} . \tag{24}$$

This means that, in addition to the ever-present length scale λ_A, the characteristic length (radius) $\lambda_{T,2D}$ also enters the problem. It is obtained from M_C in the same fashion that λ_T was obtained from the Bonnor-Ebert critical mass in 3D (see eq. [15]). We find that

$$\boxed{\lambda_{T,2D} \equiv R_{C,2D} = 1.24\, C_a \tau_{ff,2D} ,} \tag{25}$$

where the free fall time in cylindrical geometry is

$$\tau_{ff,2D} = (4G\rho_n)^{-1/2} . \tag{26}$$

By comparing equations (25) and (15), we find that $\lambda_{T,2D} = 1.05\, \lambda_T$; i.e., for practical purposes, the characteristic length introduced by thermal pressure in self-gravitating clouds in 2D is the same as that introduced in 3D. However, the length scale λ_M is absent in 2D geometry; i.e., a *frozen-in* magnetic flux can always support any mass per unit length (see Mouschovias and Morton 1990a, § IIIc[iii]), although the final equilibrium state may be characterized by an extremely high degree of central concentration. The consequence of the absence of a critical mass-to-flux ratio for the evolution of clouds in 2D due to ambipolar diffusion is that it is easier for magnetic forces to impose quasistatic contraction in this geometry than in 3D geometry. Hence, if dynamical contraction is found in 2D for a set of physical parameters, it is more likely that similar values of those parameters will lead to dynamical contraction in 3D. The converse, however, is not true; i.e., if quasistatic contraction is found in 2D, it may still be the case that the same set of physical parameters may lead to dynamical contraction in 3D. As long as this and other corollary cautions are kept in mind, the very weakness of collapse calculations in 2D is also their strength, in that the absence of one of the three characteristic lengths makes the physics associated with the other two more transparent and permits a systematic building of our understanding of the role of ambipolar diffusion in the evolution of molecular clouds and star formation. This, after all, is how a theory is developed. Calculations which have rushed to include higher dimensions have, as we shall see below, either arrived at incorrect conclusions or missed the essence of their own incomplete results.

4.3. Reduction of the Flux-to-Mass Ratio

4.3.1. The Effect of Geometry. One of the corollary cautions mentioned in the last paragraph concerns the reliability of the precise factor by which the flux-to-mass ratio of a core is reduced due to ambipolar diffusion in 2D, as opposed to 3D, geometry. The most meaningful comparison is made by asking what this factor will be in the two geometries if the model clouds start from "identical" initial, magnetically supported *equilibrium* states (i.e., from initial states characterized by the same values of the three dimensionless free parameters ν_{ff}, α_c, and k, appearing in the differential equations describing the evolution of the system, and by the same $dm(\Phi_B)/d\Phi_B$, with $\lambda_M >> \lambda_A$, λ_T in 3D) and they are then allowed to evolve up to the same *enhancement* of the central density $\rho_{n,c}$, e.g., from 10^3 to 10^9 cm^{-3}.[5] (The free parameter ν_{ff} is essentially the ratio τ_{ff}/τ_{ni} in the core of the initial equilibrium state. Hence, it follows from eq. [19] that, during the quasistatic phase of contraction, $\tau_{AD}/\tau_{ff} \simeq \nu_{ff}$.) The magnetic flux of a ~ 1 $M_\odot$ blob of interstellar matter at the mean density and magnetic field ($\simeq 3\ \mu$G) of the interstellar medium is estimated to exceed that of magnetic stars by 2 - 5 orders of magnitude, depending on the assumed geometry (thin cylindrical magnetic flux tube or spherical blob, respectively). This is the "*magnetic*

---

[5]It is emphasized here that the 3D initial equilibrium state is an oblate cloud with thermal-pressure forces (plus wave and subsonic-turbulent pressures, if one desires) balancing gravity along field lines, and that its mass-to-flux ratio in the core is no closer to the critical value $(63G)^{-1/2}$ given by equation (4a) than it is in a more uniform state. Equation (4a) already includes the effects of flattening along fields lines and of the presence of thermal pressure, the latter through the accompanying condition (4b). This point is missed by some recent papers, which then proceed to rediscover the Mouschovias and Spitzer (1976; see also § 2.2 above) result concerning the role of thermal pressure in the determination of a sufficient condition for collapse. Assigning a new name or symbol to a previously determined physical quantity hardly justifies its presentation as a new result.

flux problem" of star formation. The most likely discrepancy is probably 2 - 3 orders of magnitude. However, the critical mass-to-flux ratio given by equation (4a) is smaller than its value in magnetic stars by only 1 - 2 orders of magnitude. Hence, a protostellar fragment (or core) with near critical mass-to-flux ratio must have typically lost 1 - 2 orders of magnitude of its original flux. Therefore, the central flux tube of the initial equilibrium states of any calculation must be given a mass-to-flux ratio which is $\lesssim$ 0.1 of the critical value before ambipolar diffusion is turned on; otherwise the calculation cannot pretend to address the magnetic flux problem of star formation, useful as it may be for other purposes.

In 2D, the entire enhancement of $\rho_{n,c}$ is due to motions perpendicular to the field lines. In 3D, part of the density enhancement is due to motions along the field lines. We estimate that approximately 10^3 of the 10^6 enhancement in density is due to contraction parallel to the field lines and that, with κ in the reasonable range 0.4 - 0.5, the reduction of the flux-to-mass ratio of a protostellar fragment due to ambipolar diffusion in 3D prior to the stage $n_n \sim 10^9$ cm^{-3} will be by half as many orders of magnitude (i.e., typically 1 - 2) as it is in the 2D geometry. ***The important point, however, to be emphasized is that ambipolar diffusion during the early quasistatic phase of contraction in 3D will reduce the flux-to-mass ratio of a core by whatever factor necessary to bring it near its critical value given by equation (4a).*** At a later stage, after collapse ensues in a thermally and magnetically supercritical core, as discussed in § 4.1 above and from a different point of view in Mouschovias (1989), the additional reduction of the flux-to-mass ratio depends on whether $\lambda_A \gtrsim \lambda_M \simeq \lambda_T$ or $\lambda_A << \lambda_M \simeq \lambda_T$ at the onset of collapse. In the former case ambipolar diffusion will continue to be very effective in reducing the flux of the core, whereas in the latter case collapse will occur with flux trapping. It is clear that the magnetic flux problem of star formation consists of two distinct questions:

1. How does a cloud core of mass M_c ($\sim 1\ M_\odot$) reduce its magnetic flux from that implied by the general interstellar conditions to the critical value $\Phi_{B,crit} = (63G)^{1/2}M_c$ (see eq. [4a]), i.e., by a factor 10^1 - 10^2?
2. How does the critical flux $\Phi_{B,crit}$ at the mass M_c get further reduced (by an additional factor 10^1 - 10^2) to agree with observed fluxes of stars with mass M_c?

4.3.2. Lingering Controversies. Nakano (1979; 1984, and references therein) finds through quasistatic calculations that the flux-to-mass ratio of an axisymmetric core is reduced by an insignificant amount (typically by a factor < 2) due to ambipolar diffusion. Lizano (1990, in this volume) also finds a similar result. Can one conclude that in nature the typical reduction of a core's flux achieved by ambipolar diffusion is only a factor of 2?

First, it should be pointed out that the assumed *initial* states in Nakano's (1979) calculations have a mass-to-flux ratio within about 20% of the critical value given by equation (4a) (see Nakano 1979, beginning of § 5). Hence, by construction, these calculations cannot address the first part of the magnetic flux problem of star formation, namely, how the flux of the cloud managed to drop near its critical value; the problem is swept under the rug by the assumed initial conditions. Only a small reduction of the magnetic flux leads to dynamical contraction even if the magnetic field were to be frozen in the matter. The point is that ambipolar diffusion, operating at a stage earlier than the assumed initial states (which have a central density $\gtrsim 10^7$ cm^{-3}), *can* reduce the core's flux-to-mass ratio very substantially and, therefore, this substantial reduction must also be credited to ambipolar diffusion, not just the small reduction found for the restricted stage of evolution followed by these calculations. The time scale for flux loss during the early, quasistatic phases of contraction is $\tau_\Phi = \tau_{AD}/2 = 0.29\ \tau_{ff}^2/\tau_{ni} = 10^5\ (x/10^{-8})$ yr, as we have seen in § 2.3 above. (The quantity x denotes the degree of ionization.) Hence, the time required for the magnetic flux of a core to decrease by 2 - 3 orders of magnitude is only (4.6-6.9) $\times$ $10^5(x/10^{-8})$ yr.

Second, since the quasistatic approach cannot follow the evolution past the onset of dynamical contraction, these calculations stop or become unreliable after the central density typically increases by a factor $\simeq 10^2$; i.e., the evolution cannot be followed long enough to address the second part of the magnetic flux problem, as expressed by question #2, § 4.3.1.

The magnetic flux of Lizano's initial states is also very close to (and even below) the critical value given by equation (4a), and the (quasistatic) evolution due to ambipolar diffusion is pursued to a central density enhancement $\lesssim 10$. Hence, for the same reasons given above in the case of Nakano's work, this calculation cannot address the magnetic flux problem either. However, it contains a conceptual difference from that of Nakano. It studies clouds in which thermal pressure, enhanced by a simulated subsonic turbulent pressure, plays a significant role in supporting the cloud against its self-gravity. The relative effect of turbulence is then gradually reduced in a predetermined (and arbitrary) fashion as the cloud contracts either due to this reduction itself or due to ambipolar diffusion. Not surprisingly, when the contribution of the thermal pressure to the support of a cloud dominates that of the turbulent pressure, the cloud reaches an equilibrium or collapses depending on whether its mass is, respectively, smaller than or greater than $\simeq M_{BE}$ (see our discussion of critical masses in § 2.2). Lizano's initial equilibrium states are more realistic than those of Nakano, in that their central densities are 10^4 - 10^5 cm^{-3}, but these densities are still too high and, in these models, ambipolar diffusion has played absolutely no role in achieving them. Self-gravity, and the consequent flattening along field lines, is responsible for them. Since these densities (*put into the model from the outset* as part of the initial equilibrium states with a frozen-in magnetic flux) are comparable, and even exceed, the observed densities of NH_3 cores (e.g., see Myers and Benson 1983), and since the calculation ends after ambipolar diffusion has led to an increase of the central density above its initial equilibrium value typically by less than a factor $\simeq 10$, it is hardly legitimate to claim that the calculation has studied or explained the formation of cores by ambipolar diffusion.

Partly because the core is not resolved by the stationary grid at the end of the calculation, the existence of a characteristic size and mass for the core has been emphatically missed. The conclusion was also emphasized that nothing seems to distinguish one isodensity contour from another at the end of the calculation and, therefore, the mass of the entire model cloud will eventually be accreted by the core, unless a stellar wind cuts off this accretion. As we discussed at length in § 4.1 above, this argument (which is a conjecture, not a result of the calculation) ignores the ability of magnetic forces to severly limit accretion from a volume of radius significantly larger than the initial (i.e., typical of quiescent molecular clouds) $\lambda_A \simeq \lambda_T = 0.1$ - 0.3 pc (see eqs. [18a] and [18b]). Although the input parameters chosen for a typical model cloud used in these calculations may allow the entire cloud to eventually fall in, the star that would form would have a mass equal to 6.7 $M_\odot$. This is hardly an unlimited reservoir of matter that needs to be prevented from falling in by invoking the onset of a stellar wind. The issue is not whether stellar winds can reverse any leftover accretion present when they turn on; they most likely can. The issue is (1) whether *they* determine the masses of protostellar fragments, and (2) whether the need for such a mechanism, present at a very advanced stage of the star formation process, has been demonstrated by eliminating all other likely mechanisms that can operate at earlier stages. The quantitative arguments presented in § 4.1 and our numerical calculations conclude that magnetic support of envelopes, the formation of disks about each protostellar core, and the observed as well as theoretically expected tendency of protostars to form in clusters, rather than in isolation, all independently and collectively limit the amount of matter that can be accreted by each protostellar core: *Each core, because of ambipolar diffusion, can at best empty a region containing one to a few Bonnor-Ebert masses (evaluated at typical densities ~ 10^4 cm^{-3} of* ***quiescent*** *molecular clouds at which ambipolar diffusion can also progress* *over a relevantly short time), i.e., 1 - 10* $M_\odot$.

Our calculations (in 1D, 2D, and 3D; see below) reliably follow the evolution, including the dynamical phase of contraction in 3D, up to an enhancement of the central density by a factor $\simeq 10^6$. First, ambipolar diffusion reduces the flux of a core by whatever factor necessary to bring it down to its critical value. Second, the dynamical contraction that eventually follows (in 3D) never actually evolves into a free fall; the maximum infall acceleration for typical clouds is only $(0.1 - 0.5)g$, where g is the acceleration of gravity, and occurs in the immediate neighborhood of a compact, high-density core. The flux of a core keeps decreasing but not as rapidly as it did during the quasistatic contraction phase (see § 6 below).

We show for the first time through collapse calculations that the least the Alfvén length scale λ_A does is to set the scale for the size of molecular cloud cores, in agreement with the analytical considerations of Mouschovias (1987a) and with the results of § 4.1 above. Its magnitude relative to the thermal length scale in 2D (see eq. [25]) or relative to the thermal and magnetic length scales in 3D (see eqs. [15] and [17]) determines the very nature of the evolution of a core, including the issues of whether the evolution is dynamic or quasistatic and whether a core can continue to lose magnetic flux at a significant rate even during the dynamical phase of contraction.

5. CYLINDRICAL COLLAPSE DUE TO AMBIPOLAR DIFFUSION

We have considered model clouds initially in exact equilibrium states in the absence of ambipolar diffusion, with magnetic and thermal-pressure forces balancing gravity everywhere within a cloud. An external medium of fixed thermal and magnetic pressure bounds each cloud. *Thus any evolution at all is entirely the result of ambipolar diffusion.* Such initial states, as in the studies of Mouschovias (1979), Paleologou and Mouschovias (1983), and Mouschovias, Paleologou, and Fiedler (1985), demonstrate clearly the onset of ambipolar diffusion in cloud cores, in fact the very formation of such cores, the enchancement of their densities (by up to a factor $\simeq 10^6$), and their *self-initiated collapse* that leads to star formation, as originally suggested by Mouschovias (1977) (see also the early review by Mouschovias 1978).

The non-ideal MHD differential equations describing the evolution of a model cloud away from its initial equilibrium state due to ambipolar diffusion contain *three dimensionless free parameters* (ν_{ff}, α_c, and k), which are judiciously chosen to refer to the *core* of the initial state (see Mouschovias and Morton 1990a, § IIId for details). The parameter α_c is defined by equation (21), while k is the exponent in the n_i - n_n relation (eq. [14]). The parameter ν_{ff} is defined by

$$\nu_{ff} \equiv \frac{1}{\pi^{1/2}} \frac{\tau_{ff,c}}{\tau_{ni,c}} = 8.31 \left[\frac{K}{3 \times 10^{-3}\ cm^{-3}}\right] \left(\frac{10^5\ cm^{-3}}{n_{n,c}}\right)^{0.5-k} . \tag{27}$$

The boundary conditions introduce *one* additional free parameter in the case of initial states with $\alpha = const$, and *two* additional parameters in the case $\alpha \neq const$. *The initial conditions introduce no new dimensionless free parameters in the problem.* We therefore expect the evolution of the cores, but not necessarily of the envelopes, of all model clouds to be insensitive to the initial conditions and to become more so as time progresses and the cores further separate (in density) from their envelopes. The numerical results confirm this expectation.

Observations and theoretical considerations suggest that $1/3 \leq k < 2/3$ for $10^3 \lesssim n_n \lesssim 10^9$ cm^{-3} (see discussion following eq. [14]). We studied the full range 1/3 - 2/3, but we put more emphasis on the range 1/3 - 1/2 in presenting the results. For the range of values of k specified above, we studied representative model clouds in the parameter space $0.5 \leq \nu_{ff} \leq 25$ and $1.20 \leq \alpha_c \leq \infty$, which is sufficient to reveal all the important features of the solution associated with the formation and collapse of cores in massive molecular clouds. Values of ν_{ff} smaller (larger) than 8.31 correspond to fewer (more) ions at a given neutral density than implied by the value of $K = 3 \times 10^{-3}$ cm^{-3} in equation (14). A typical initial equilibrium state has temperature $T = 10$ K, central and surface density of neutrals $n_{n,c} = 3 \times 10^3$ cm^{-3} and $n_{n,S} = 3 \times 10^2$ cm^{-3}, respectively, corresponding to an external thermal pressure $P_{ext} = n_{n,S} k_B T = 3 \times 10^3\ k_B$ (where k_B is the Boltzmann constant), an external magnetic field $B_{ext} = 4\ \mu$Gauss, and a central magnetic field $B_c = 65.3\ \mu$Gauss. The mean density of this state is $\langle n_n \rangle = 1.31 \times 10^3$ cm^{-3}, and the *mean* magnetic field is $\langle B \rangle = 27.7\ \mu$Gauss, which is closer than the central field B_c to the value that OH Zeeman observations with a single-dish telescope would reveal. With these values of the relevant dimensional physical quantities, equation (21) yields $\alpha_c = 41.0$ as a typical value. Values of α_c smaller (larger) than 41.0 correspond to larger (smaller) temperatures (or effective tempertures, if one desires to include the effect of subsonic turbulence) than 10 K. To complete the specification of the typical initial equilibrium state we take $k = 1/2$.

The initial states represent clouds in exact equilibrium under flux freezing, with some degree of central concentration (an inevitable consequence of self-gravity) and with deep interiors that are primarily magnetically supported (initially), while near the surface the thermal pressure is comparable (and in a number of cases even exceeds) the magnetic pressure.

It follows that the above typical model cloud has mass per unit length $M_S = 398\ M_\odot$ pc^{-1} and radius $R = 1.29$ pc. Hence, a cylinder with height equal to the cloud diameter $2R$ contains a mass $M_{cl} = 1.03 \times 10^3\ M_\odot$. This is approximately the mass of a three-dimensional cloud which would result by gravitational breakup of the cylindrer along its axis (parallel to the field lines) and which would form a disk-like cloud with thermal pressure balancing gravity along the field lines. In this paper we discuss only those features of the typical model cloud which relate directly to the role of the characteristic lengths λ_A and λ_T in the formation and evolution of the core. A much more complete discussion, including a comprehensive parameter study, is given in Mouschovias and Morton (1990b). Concerning the parameter study, it will suffice for the purposes of this paper to mention only that the late stages of evolution of the core (but not necessarily the envelope) were found to be rather insensitive to the two free parameters introduced by the boundary conditions as long as the parameter α_c is sufficiently larger than unity to make the cloud thermally supercritical (see eq. [24]) but magnetically supported, as suggested by observations.

Figure 1a shows the density of neutrals, n_n, normalized to its initial central value $n_{c0} \equiv n_{n,c0} = 3 \times 10^3$ cm^{-3}, as a function or radius r, normalized to the initial cloud radius $R_0 = 1.29$ pc, at seven different times t_j ($j = 0, 1, ..., 6$) chosen so as to have an enhancement of the central density by the factor 10^j ($j = 0, 1, ..., 6$) compared to its initial value. The formation of a high-density core outside which the density decreases rapidly with r until it reaches its (fixed) surface value $n_S \equiv n_{n,S} = 3 \times 10^2$ cm^{-3}, seven orders of magnitude lower than the core density at the end of a run, is demonstrated dramatically. The positions of the characteristic lengths λ_A and λ_T at each time are marked on the corresponding density curve by a "*star*" and an *open circle*, respectively. The magnetic field, normalized to its initial central value $B_{c0} = 65.3\ \mu$Gauss, as a function or radius r, normalized again to the initial cloud radius, is shown in Figure 1b at the same times as those chosen for Figure 1a. The positions of λ_A and λ_T are marked as in the case of the density curves. It is clear from both figures that *the characteristic length* λ_A *and its relation to* λ_T *set the scale for the size of the core, the more so the later the stage of evolution.*

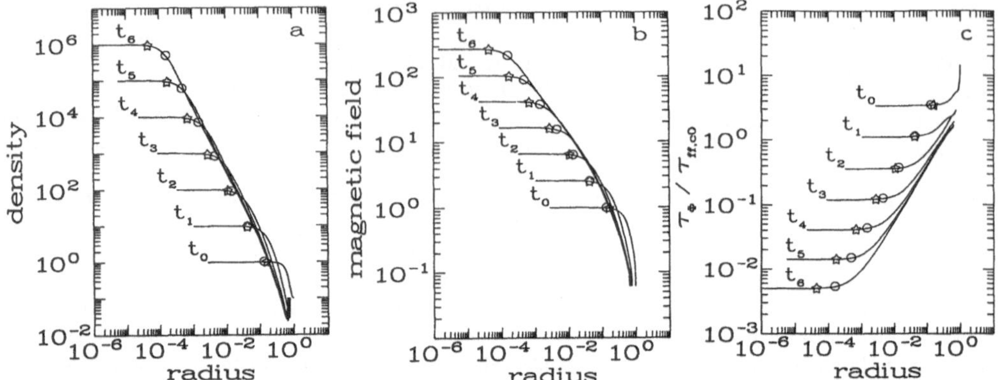

Figure 1. (**a,** *left*) *The density, normalized to its initial central value, as a function of distance r from the axis of symmetry, normalized to the initial cloud radius* R_0, *at seven different times* t_0, t_1, ... t_6. In units of 10^6 yr these times are 0, 1.669, 2.182, 2.352, 2.409, 2.429, and 2.435. The two characteristic lengths λ_A and $\lambda_{T,2D}$ (see text) are marked on each curve by a *star* and an *open circle*, respectively. As the contraction progresses, they define the instantaneous size of the core better and better. (**b,** *center*) *The magnetic field, normalized to its initial central value, as a function of radial distance at different times.* The normalization of *r*, the times t_0 ... t_6, and the display of the length scales λ_A and $\lambda_{T,2D}$ are as in Fig. 1a. (**c,** *right*) *The flux loss time scale, normalized to the initial central free fall time, as a function of r.* The normalization of *r* the times t_0 ... t_6, and the display of the length scales λ_A and $\lambda_{T,2D}$ are also as in Fig. 1a.

Since initially $\lambda_A \simeq \lambda_T$, the analytical arguments presented in §§ 4.1 and 4.2 imply that the contraction of the core should verge on being dynamic at first but, after λ_A drops below λ_T (at a density enhancement $\simeq$ 10), it should turn quasistatic. This is exactly what is revealed by the acceleration in the core: It executes one oscillation [from about $-0.2|g|$ to $+0.1|g|$; a negative (positive) sign denotes accelerated (decelerated) infall] and it practically vanishes above a central density enhancement of about 30.

The mass (per unit length along the symmetry axis) inside λ_A is initially slightly greater than the critical mass M_C for collapse against thermal pressure but, as time progresses, it decreases as $\rho_{n,c}\lambda_A^2 \propto \rho_{n,c}^{-0.2}$ beyond the stage at which the central density enhancement is 10 and the contraction turns quasistatic. Thus it falls by an order of magnitude by the time the calculation is stopped, when the central neutral density has increase by six orders of magnitude in all, to $n_{n,c} = 3 \times 10^9$ cm^{-3}, and the isothermality assumption as well as the n_i - n_n relation (14) break down. The critical mass is, of course, independent of time and equal to $4.1 \times 10^{-2} M_S$, where M_S is the total mass (per unit length) of the cloud. Hence, its position λ_T is by definition the radius inside which the mass is exactly equal to M_C.

The conclusion that the length scale λ_A and its relation to λ_T set the scale for the size of the core while the envelope remains relatively well supported by magnetic forces can be tested for consistency with other results of the calculation by examining the variation of the characteristic time for loss of magnetic flux by any given mass shell as a function of radius at any given time. This is given by

$$\tau_\Phi \equiv \frac{\Phi_B}{|d\Phi_B/dt|} = \frac{\Phi_B}{|\mathbf{v}_D \cdot \nabla\Phi_B|}, \tag{28}$$

where $d/dt = \partial/\partial t + (\mathbf{v}_n \cdot \nabla)$ is the time derivative comoving with the neutrals; the fact that the time derivative of Φ_B comoving with the ions vanishes was used to obtain the last part of equation (28). The quantity $\mathbf{v}_D \equiv \mathbf{v}_i - \mathbf{v}_n$ is the drift velocity of the plasma relative to the neutrals. The time scale τ_Φ, normalized to the initial central free-fall time $\tau_{ff,c0} = (4G\rho_{n,c0})^{-1/2} = 5.67 \times 10^5$ yr, is shown as a function of r on a log-log plot in Figure 1c at the same seven times chosen for Figure 1a. The radial coordinate r is normalized to R_0 as in previous figures. It is clear that τ_Φ is constant in space in the core (in the innermost few percent of the mass) but increases dramatically with r in the envelope, especially during the later stages of contraction, revealing the much better coupling of the magnetic flux and the neutral particles outside the core. In the present cylindrical geometry, this implies a very effective support of the envelope. (It does so in axisymmetric geometry as well in the common case in which the cloud *as a whole* is magnetically supported, but not in the rare case in which the mass of the cloud is both magnetically and thermally supercritical.)

In the core, the time scale for flux loss τ_Φ defined by equation (28) can be shown by expanding Φ_B about $r = 0$ to be given by

$$\tau_{\Phi,c} = \frac{r}{2|v_D|} = \frac{1}{2}\tau_{AD} = \frac{1}{2}\left|\frac{\partial v_D}{\partial r}\right|^{-1}. \tag{29}$$

It would therefore be physically more meaningful to define τ_Φ, rather than $\tau_{AD} \equiv r/v_D$, as *the* ambipolar diffusion time scale. We do *not* do so here only to avoid confusing a reader used to the traditional definition, but we suggest that the change be made in the future.

The detailed solution shows that the motion of the neutrals in the core remains subsonic ($|v_n| \lesssim 0.5C$) but that it becomes supersonic outside. Nevertheless, the motion of the neutrals remains quasistatic even in the envelope, except very near the cloud surface, because of the effective magnetic support. *Hence, a comparison of speeds, as opposed to accelerations, is not a good way to determine the onset of dynamical contraction and, certainly,* $v_n \simeq C$ *does not necessarily mark the onset of dynamical contraction (collapse) in magnetically supported clouds.* The contraction speed remains subAlfvénic everywhere in the cloud.

Once the quasistatic phase of contraction is established, the flux loss time scale at the center, $\tau_{\Phi,c}$, decreases almost as $\tau_{ff,c}^2/\tau_{ni,c} \propto n_{i,c}/n_{n,c} \propto n_{n,c}^{-(1-k)} \propto n_{n,c}^{-1/2}$, in good but not excellent agreement with the analytical expression $\tau_{\Phi,c} = 0.22\, \tau_{ff}^2/\tau_{ni}$ with $k = 1/2$ (see § 2.3). The flux loss time scale $\tau_{\Phi,c}$ remains greater than $\tau_{ff,c}$ by a factor 3 - 5 during the quasistatic phase of this model, while the density e-folding time $\tau_{\rho,c}$ is greater than $\tau_{ff,c}$ by a factor 2 - 2.8. At the end of the calculation, $\tau_{\rho,c} = 4.3 \times 10^{-2}\, \tau_{ni,c0} = 1.66 \times 10^3$ yr; i.e., the density is increasing very rapidly despite the fact that the contraction is quasistatic. The small departure of $\tau_{\Phi,c}$ from the variation $\tau_{ff}^2/\tau_{ni} \propto n_{n,c}^{-(1-k)}$ is due to the fact that the pressure force in the core becomes relatively more important in time and by the end of the run it is nearly 50% of the magnetic force. Under such circumstances, the full equation (7) should be used (with the constant on the right-hand side appropriate for cylindrical geometry being $2/1.4\pi$), which accounts for the presence of thermal pressure during quasistatic contraction. Although initially the term in parentheses is essentially equal to unity (i.e., the effect of thermal pressure on ambipolar diffusion is negligibe), at the end of the calculation this term becomes equal to 0.75, which means that thermal pressure lengthens $\tau_{AD,c}$ (and, therefore, $\tau_{\Phi,c}$) by the factor 1.3. This is exactly the factor by which $\tau_{\Phi,c}$ exceeds the value it would have at $n_c/n_{c0} = 10^6$ if its slope were exactly equal to -0.5 during the quasistatic phase of contraction, i.e., above $n_c/n_{c0} \simeq 30$.

The time required for the central density to increase by a factor 10^6 is slightly less than 2.44×10^6 yr. This represents $4.30\tau_{ff,c0}$, or $63.2\tau_{ni,c0}$, but only $0.64\tau_{AD,c0} = 1.28\tau_{\Phi,c0}$. The ordering of time scales $\tau_{\Phi,c0} > \tau_{ff,c0} > \tau_{ni,c0}$ and the fact that the cloud is initially in equilibrium mean that ambipolar diffusion is relatively slow and therefore controls the cloud's evolution, at least until a time $t \simeq \tau_{\Phi,c0}$ has elapsed. *It follows that the value of the free parameter* $\nu_{ff} = 8.31$ *is almost large enough for the solution to approach the limiting behavior in which the evolution of a model cloud is primarily controlled by the initial ambipolar diffusion time scale.* We have found that if $\nu_{ff} = 25$ it takes approximately 2.8 times as long for the central density to increase by a factor 10^6, but this longer time is still only $\simeq 1.2\tau_{\Phi,c0}$.

The flux-to-mass ratio inside a cylindrical shell of matter of fixed height equal to its *initial* diameter and which contains exactly 1 $M_\odot$ at all times decreases in time in a fashion almost indistinguishable from that of the central flux-to-mass ratio. In this 2D geometry, then, ambipolar diffusion in this representative model cloud reduces the flux of a 1 $M_\odot$ core to 2.7×10^{-3} its initial value by the time the density increases to just above 10^9 cm^{-3}. Because the ion density does not increase with neutral density above $n_n \sim 10^9$ cm^{-3}, if this calculation were pursued further in time the flux-to-mass ratio would continue to decrease rapidly. As we have already cautioned, however, in 3D axisymmetric geometry only half as many orders of magnitude of flux are expected to be lost by the same density enhancement achieved in the present 2D geometry (see discussion in § 4.3.1 above, and § 6 below) --not just a factor ≤ 2 obtained by Nakano (1979) and by Lizano (1990), the origin of which we explained in § 4.3.2.

Finally, we note that at the time at which we stop the calculation the grid resolves the innermost $\simeq 10^{-6}$ of the original radial extent of the cloud. With a high-resolution grid the danger of overresolving the core, in which the physical quantities are flat as functions of space, arises. One would then be dealing with noise about the true values of the physical quantities. This noise would decay or grow, depending on the stability of the numerical scheme. Although our implicit time integrator is stable, we take great care to avoid overresolution. We have found empirically that the best compromise between excellent resolution and no numerical oscillations in the core is achieved when, at the end of a run, the size of the first mass cell is smaller than the smallest characteristic length of the problem (λ_A, or λ_T, or the size of the flat-density core) by a factor 10^1 - 10^2.

6. AXISYMMETRIC COLLAPSE DUE TO AMBIPOLAR DIFFUSION: ORIGIN OF COMPACT CORES, STRONG FIELDS, AND ACCRETION DISKS

We consider an isothermal cloud (or fragment) of mass $M = 45.5$ $M_\odot$, temperature $T = 10$ K, initially uniform density $n_{n,u} = 300$ cm^{-3}, threaded by a uniform magnetic field $B_u = 30$ μGauss along the (z-) symmetry axis, and with a "canonical" relation between the ion and neutral densities, namely, $n_i = K\,(n_n/10^5 \text{ cm}^{-3})^k$, where $K = 3 \times 10^{-3}$ cm^{-3} and $k = 1/2$ (as found by Elmegreen 1979, and Nakano 1979). Its initial radius and half-height are $R_0 = Z_0 = 0.75$ pc. This cloud is embedded in a massive cloud complex, the effects of which are simulated by proper boundary conditions. The cloud is allowed to contract to an equilibrium state with the magnetic field frozen in the matter. The central density becomes $n_{n,c0} \simeq 10^3$ cm^{-3}, while the central field increases only to $B_{c0} = 30.6$ μGauss. If ambipolar diffusion were neglected, the model cloud would remain in its equilibrium state indefinitely. The onset of ambipolar diffusion in the core, however, causes slow contraction at first and more rapid contraction later, after the central mass-to-flux ratio increases sufficiently. The general differential equations describing the evolution of the cloud are given in Paleologou and Mouschovias (1983, § IIa; or Mouschovias 1987b, § 3.3). They contain three dimensionless free parameters characterizing the core at equilibrium, namely, the ratio of the magnetic and thermal pressures $\alpha_c = B_{c0}^2/8\pi\rho_{n,c0}C^2$ (= 26.4 for the above model), the ratio of ambipolar diffusion and free-fall time scales $\nu_{ff} = \tau_{AD}/\tau_{ff}$ (= 9.2), and the exponent k (= 0.5) in the n_i-n_n relation. A detailed description of the solution together with a study of its dependence on the free parameters is given eslewhere (see Fiedler and Mouschovias 1990). Here we present only a few features of the solution relating to the formation and evolution of high-density cores, the B_c-ρ_c relation, and the increase of the central mass-to-flux ratio. Because of the relatively large value of ν_{ff}, the evolution is controlled by ambipolar diffusion for a long time (see Mouschovias and Morton 1990b for a physical discussion).

Since the early evolution of the model cloud is physically slow and since it takes place essentially over the analytically determined flux loss time scale $\tau_\Phi = \tau_{AD}/2 = 0.3\ \tau_{ff}^2/\tau_{ni} = 10^5\ (x/10^{-8})$ yr, it becomes unnecessary to follow *numerically* the earlier phases of contraction. Simply, ambipolar diffusion reduces the central flux-to-mass ratio by whatever factor necessary to make it equal to its assumed initial equilibrium value which, in any case, represents a magnetically subcritical but thermally supercritical case in accordance with observations. A parameter study then determines the dependence, if any, of the solution on initial conditions.

Figure 2a exhibits isodensity contours (*dashed lines*) and field lines (*solid curves*) in the innermost 10% of the original extent of the cloud after the cloud has evolved to a central density $n_c = 10^6 n_{c0} = 3 \times 10^8$ cm^{-3} in a time $\simeq 1.6 \times 10^7$ yr, most of which was consumed by ambipolar diffusion in achieving an increase of the central density by a factor of 20 (see Fig. 2b). Both the r and z coordinates are normalized to the original radius of the cloud (R_0 = 0.75 pc). The long dashes represent isodensity contours with an enhancement over the initial density by a factor 10^5, 10^4, 10^3, 10^2, 10^1, and 10^0, while the short dashes are drawn at the levels 3×10^5, 3×10^4, etc. The flux tubes shown contained initially a total mass 0.013, 0.051, 0.11, 0.20, 0.36, 0.45, 0.62, and 0.81 $M_\odot$. By the time t_6, to which this figure refers, ambipolar diffusion has increased the total mass in each of these flux tubes to 0.12, 0.45, 0.90, 1.4, 2.0, 2.7, 3.4, and 4.2 $M_\odot$, respectively. The tenfold increase of the mass-to-flux ratio of the innermost flux tube occurred despite the fact that the cloud's initial mass-to-flux ratio was within about a factor of two of its critical value for collapse.

Even in the absence of angular momentum, the formation of a disk due to the anisotropy of the magnetic force is quite evident. Only matter within the original λ_A can accrete onto the compact core. The rest is well supported by magnetic forces. Despite the fact that this particular model cloud had an initial central mass-to-flux ratio near critical, the flux tube which had an initial radius in the equatorial plane equal to λ_A has contracted by only a few percent and has increased its mass (due to ambipolar diffusion) by less than a factor of 2.

The density, normalized to its initial central value, is shown in Figure 2b as a function of (cylindrical) radial distance r in the equatorial plane (*solid curves*) and as a function of distance z along the axis of symmetry (*dashed curves*) at seven different times, chosen so as to have an increase of the central density by a factor of 10 from one output time to the next. The formation of a nearly uniform, high-denisty core that runs away from the surrounding envelope is evident. At the last output time, the mass inside the flat core is only 0.1 $M_\odot$. The size of the core is set by the relation between the three characteristic lengths present in the problem, namely, λ_A (marked on the curves by a *star*), λ_M (*open circles*), and λ_T (*filled circles*). Above a density enhancement of 10^2, thermal-pressure forces become comparable to magnetic forces in the high-density core, but not in the envelope, where magnetic forces dominate the thermal-pressure forces. Since this model cloud was chosen to have a mass-to-flux ratio near critical, a substantial fraction of it is falling in, with the density decreasing as $z^{-2.7}$ along the axis of symmetry and as r^{-2} in the equatorial plane in the region immediately outside the high-density core. The maximum infall velocity at time

t_6 occurs at $r \simeq 2 \times 10^{-2} R_0$ and is equal to 1.4*C*. Although supersonic, this velocity is only 19% of the local magnetosonic speed. In other words, despite the supercritical state achieved in the central flux tubes of the cloud, magnetic forces in the envelope and a combination of magnetic and thermal-pressure forces in the core prevent free fall from setting in even at central densities $\simeq 3 \times 10^8$ cm^{-3}. The inward acceleration of fluid elements in the high-density core at t_6 is only about 13% of the local acceleration of gravity. Farther out, however, near the region of maximum infall velocity, the acceleration reaches $\simeq 0.5g$; i.e., the evolution is dynamic.

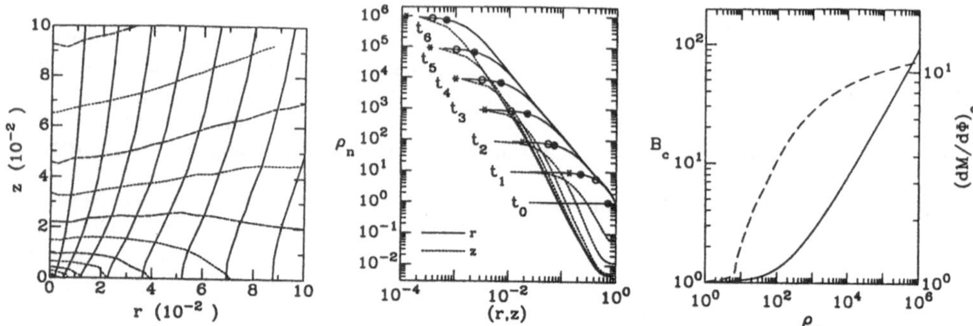

Figure 2. (**a**, *left*) *Isodentity contours (dashed lines) and magnetic field lines (solid curves) in the innermost 10% of a model cloud at time* t_6, *at which the central density has increased by a factor of* 10^6. The levels of the isodensity contours and the total mass inside each flux tube are given in the text. (**b**, *center*) *The density as a function of radial distance r in the equatorial plane (solid curves) and distance z from the equatorial plane along the axis of symmetry (dashed curves) at seven different times* $t_0, t_1, \dots t_6$. In units of 10^6 yr these times are 0, 10.5, 15.6, 16.37, 16.51, 16.55, and 16.56. The three characteristic lengths λ_A, λ_T, and λ_M (see text) are marked on each curve by a *star*, *filled circle*, and *open circle*, respectively. (**c**, *right*) *The central magnetic field (solid curve) and mass-to-flux ratio (dashed curve, with scale shown on the right side of the frame) as functions of the central density*. All quantities are normalized to their initial central values.

The central magnetic field B_c, normalized to its initial value in the uniform state (= 30 μG), as a function of the central density ρ_c, also normalized to its initial value (= 300 cm^{-3}), is shown in Figure 2c as a *solid curve*. After the initial flattening of the cloud along field lines (which does not appreciably increase the field strength), the field B_c increases as $\rho_c^{0.48}$ although ambipolar diffusion is in progress. In other words, the relation obtained earlier for self-gravitating clouds with *frozen-in* fields, namely,

$$\frac{B_c}{B_0} = \left(\frac{\rho_c}{\rho_0}\right)^{1/2}, \qquad n_0 \simeq 137 \left(\frac{B_0^3}{M}\right)^{1/2}, \qquad (30a, b)$$

(Mouschovias 1976b) can be used as a reasonable approximation even in the presence of ambipolar diffusion at least up to densities ~ 10^9 cm^{-3} although the physics enforcing this relation is quite different in the two situations. The quantity ρ_0 in equation (30a) is the gas density at which gravitational forces become comparable to magnetic forces (at field strength

B_0) and significant contraction perpendicular to field lines begins. Converted to a number density of *protons* (n_0), it is approximately given by equation (30b), where B_0 is measured in *microgauss* and the cloud's mass M in *solar masses* (see brief review by Mouschovias 1985). Equation (30b) yields a density range $n_0 \simeq 1$ - 320 cm^{-3} for $B_0 = 3$ μGauss and for M in the corresponding range 5×10^5 - 5 $M_\odot$. The exponent κ in the $B_c \propto \rho_c^{\kappa}$ relation is essentially zero below the density n_0 and becomes nearly equal to 1/2 above n_0. (For the spatial variation of κ within a cloud and its physical explanation, see Mouschovias 1978, § IIIa2.) Figure 2c shows that the central field is enchanced by a factor 10^2, so that its value becomes 3 mG at a density $\simeq 3 \times 10^8$ cm^{-3}. It is emphasized, however, that the region of strong field involves a contracting core mass $\lesssim 0.1$ $M_\odot$. The strong fields measured in H_2O masers (e.g., see Moran 1990) are consistent with the predictions of our calculations.

Figure 2c also exhibits the central mass-to-flux ratio (*dashed curve*), normalized to its initial value, as a function of the normalized central density. The scale is read on the right side of the frame. Despite the fact that this model cloud was initially close to the critical state for collapse, ambipolar diffusion has increased the central mass-to-flux ratio by slightly more than a factor of 10 by the time t_6, at which the central density has increased by a factor 10^6. A detailed parameter study shows that a larger increase of the central mass-to-flux ratio is achieved for other observationally reasonable sets of parameters. On the basis of the single model presented here, however, one must conclude that the issue of whether ambipolar diffusion resolves the magnetic flux problem of star formation remains unsettled by numerical calculations, despite the insight gained from our analytical results (see § 2.3). Two extensive parameter studies of axisymmetric collapse due to ambipolar diffusion address this question directly (see Fiedler and Mouschovias 1990; Morton and Mouschovias 1990).

7. CONCLUSION

A dozen years ago we suggested that, while magnetic braking prevents centrifugal forces from interfering, ambipolar diffusion sets in at an appreciable rate to have unavoidable and observable consequences at relatively low densities ($n_n \sim 10^4$ cm^{-3}, as opposed to the previously thought $\sim 10^{10}$ cm^{-3}) in the interiors of otherwise quiescent, self-gravitating, interstellar molecular clouds. Ambipolar diffusion does not lead to a loss of magnetic flux by a cloud as a whole. It simply *redistributes* mass in the central flux tubes of a cloud (see review by Mouschovias 1978). This amounts to *self-initiated* fragmentation (or core formation) and, later, collapse of cores while the envelopes remain relatively well supported by magnetic forces. The process of core formation and contraction is initially quasistatic, controlled by the flux loss time scale $\tau_\Phi = \tau_{AD}/2 = 0.3\ \tau_{ff}^2/\tau_{ni}$, where τ_{ff} is the free fall time scale and τ_{ni} the mean collision time of a neutral particle with ions. After the mass-to-flux ratio of a core increases above the critical value $(M/\Phi_B)_{crit} = (63G)^{-1/2}$, it begins to contract dynamically provided also that it is thermally supercritical (see conditions [4a] and [4b] or, equivalently, [5a] or [5b]). Once dynamical contraction begins, the flux loss time scale increases and, as long as the magnetic forces dominate the thermal-pressure (or turbulent) forces, it is given by $\tau_\Phi = \tau_{AD}/2 = 0.15\ (\tau_{ff}^2/\tau_{ni})(\Phi_{B,crit}/\Phi_B)^2$, where $\Phi_{B,crit} = (63G)^{1/2}M$ is the critical flux below which a core of mass M may begin to contract dynamically. In typical cores of self-gravitating clouds, the magnetic field scales with the gas density as $B_c \propto \rho_c^{\kappa}$ with $\kappa \simeq 1/2$ even while ambipolar diffusion is in progress.

At a typical density $n_n \sim 4 \times 10^4$ cm^{-3}, a plethora of low-mass ($\lesssim 3$ $M_\odot$), low-spin fragments have been predicted to form preferentially in (and perhaps only in) the deep interiors of self-gravitating clouds. Observations have confirmed this prediction. By contrast to the subAlfvénic contraction involved in low-mass star formation, the formation of massive stars most likely involves externally triggered (e.g., implosion by a shock) superAlfvénic collapse, with consequent trapping of magnetic flux as well as angular momentum. A quasi-equilibrium is reached with both centrifugal and magnetic forces balancing gravity. The subsequent evolution is simultaneously controlled by magnetic braking and ambipolar diffusion, and the massive protostar is likely to be characterized by both rapid rotation and strong magnetic fields. In clouds with primordial chemical composition, protostars with $M \gtrsim 10\ M_{\odot}$ are favored (see Mouschovias 1989).

We find that there are three natural length scales in molecular clouds. The "*Alfvén length scale*" λ_A is always introduced by ambipolar diffusion and is comparable with the "*thermal Jeans length*" λ_T but smaller than the "*magnetic Jeans length*" λ_M. Inside λ_A the neutral particles are not affected by magnetic forces, while beyond λ_A the neutrals are well coupled to, and therefore very effectively supported by, the magnetic field. This, together with the fact that gravitational forces overwhelm the thermal-pressure forces on scales $\gtrsim \lambda_T$, implies that the fragment that will form by this gravitationally driven ambipolar diffusion will have available to it for eventual accretion not an unlimited reservoir of matter (which would have been the case in the absence of the magnetic field or if $\lambda_A >> \lambda_T$) but only the mass originally in a volume of radius comparable to $\lambda_A \simeq \lambda_T$, which is $\sim 1\ M_{\odot}$. Axisymmetric collapse calculations provide evidence, but not yet proof, that this is indeed the case. Moreover, at a distance greater than $2\lambda_A$ ($\simeq 2\lambda_T$) from the center of a forming core, gravitationally driven ambipolar diffusion can lead to the formation of another, independent core much more rapidly than it would take for this material to be accreted by the first core. Each core can form either a binary or a single star depending on its leftover angular momentum. We suggest that this is the origin of stellar clusters in otherwise quiescent molecular clouds. We predict that compact cores ($n_n > 10^8\ cm^{-3}$, $M \lesssim 0.1\ M_{\odot}$), strong magnetic fields (> 1 mGauss), and accretion disks go hand-in-hand.

It has been claimed on the basis of quasistatic calculations that ambipolar diffusion can decrease a core's flux-to-mass ratio by less than a factor of two. The initial states of these calculations, by construction, are very close to (and sometimes beyond) the critical state for gravitational collapse even with a magnetic field frozen in the matter. In reality, it is ambipolar diffusion that must have been responsible for achieving a state so close to collapse against magnetic forces. During the quasistatic phase of evolution, ambipolar diffusion *can* reduce the flux-to-mass ratio of a core by whatever factor necessary to achieve the critical state because the flux loss time scale *is* very nearly the evolutionary time scale. In order for the magnetic flux problem of star formation to be resolved, however, the critical flux $\Phi_{B,crit}$ at a core mass $M_c \simeq M_{star}$ must be further reduced by a factor 10^1 - 10^2. It seems that ambipolar diffusion can achieve a reduction by a factor of 10 by the time the density has increased to $\simeq 10^{10}\ cm^{-3}$. Thus it may still resolve the problem by a density $\sim 10^{12}\ cm^{-3}$. The jury is still out on the resolution of the magnetic flux problem by ambipolar diffusion.

Acknowledgements. This work was supported in part by the National Science Foundation. All computations and graphics were performed at the National Center for Supercomputing Applications, at the University of Illinois, Urbana-Champaign.

8. REFERENCES

Abt, H. A. 1983, *Ann. Rev. Astr. Ap.*, **21**, 343.

Arquilla, R. 1984, Ph. D. thesis, University of Massachusetts.

Baade, W. 1944, *Ap. J.*, **100**, 137.

Baudry, A., Cernicharo, J., Perault, M., Noe, J., and Despois, D. 1981, *Astr. Ap.*, **194**, 101.

Blitz, L. 1987, in *Physical Processes in Interstellar Clouds*, eds. G. E. Morfill and M. Scholer (Dordrecht: Reidel), p. 35.

Blitz, L., and Shu, F. H. 1980, *Ap. J.*, **238**, 148.
Bonnor, W. B. 1956, *M. N. R. A. S.*, **116**, 351.
Bregman, J. D., Troland, T. H., Forster, J. R., Schwarz, U. J., Goss, W. M., and Heiles, C. 1983, *Astr. Ap.*, **118**, 157.
Clark, F. O., and Johnson, D. R. 1978, *Ap. J.*, **220**, 550.
____. 1981, *Ap. J.*, **247**, 104.
Ebert, R. 1955, *Zs. f. Ap.*, **37**, 217.
____. 1957, *Zs. f. Ap.*, **42**, 263.
Elmegreen, B. G. 1979, *Ap. J.*, **232**, 729.
Falgarone, E., and Pérault, M. 1987, in *Physical Processes in Interstellar Clouds*, eds. G. E. Morfill and M. Scholer (Dordrecht: Reidel), p. 59.
Fiedler, R. A., and Mouschovias, T. Ch. 1990, *Ap. J.*, *to be submitted.*
Field, G. B. 1965, *Ap. J.*, **142**, 531.
Goldsmith, P. F., and Arquilla, R. 1985, in *Protostars & Planets II*, eds. D. C. Black, and M. S. Matthews (Tucson: Univ. of Arizona Press), p. 137.
Goldsmith, P. F., and Sernyak, M. I., Jr. 1984, *Ap. J.*, **283**, 140.
Heiles, C. 1987, in *Physical Processes in Interstellar Clouds*, eds. G. E. Morfill and M. Scholer (Dordrecht: Reidel), p. 429.
Goldsmith, P. F., and Strom, K. M. 1987, *Ap. J.*, **321**, 855.
Jeans, J. H. 1928, *Astronomy and Cosmogony* (Cambridge: Cambridge Univ. Press).
Larger, W., D. 1985, in *Protostars and Planets II*, eds. D. C. Black, and M. S. Matthews (Tucson: Univ. of Arizona Press), p. 650.
Lizano, S. 1990, *in this volume.*
Lizano, S., and Shu, F. H. 1987, in *Physical Processes in Interstellar Clouds*, eds. G. E. Morfill and M. Scholer (Dordrecht: Reidel), p. 173.
Martin, R. N., and Barrett, A. H. 1978, *Ap. J. Suppl.*, **36**, 1.
Mattila, K., Winnberg, A., and Grasshof, M. 1979, *Astr. Ap.*, **78**, 275.
McDaniel, E. W., and Mason, E. A. 1973, in *The Mobility and Diffusion of Ions in Gases* (New York: Wiley).
Mestel, L. 1985, in *Protostars and Planets II*, eds. D. C. Black, and M. S. Matthews (Tucson: Univ. of Arizona Press), p. 81.
Mestel, L., and Paris, R. B. 1979, *M. N. R. A. S.*, **187**, 337.
Mestel, L., and Spitzer, L., Jr. 1956, *M. N. R. A. S.*, **116**, 503.
Moneti, A., Pipher, J. L., Helfer, H. L., McMillan, R. S., and Perry, M. L. 1984, *Ap. J.*, **282**, 508.
Moran, J. 1990, in *Galactic and Intergalactic Magnetic Fields*, eds. R. Beck, P. P. Kronberg, and R. Wielebinski (Dordrecht: Reidel), *in press.*
Mouschovias, T. Ch. 1974, *Ap. J.*, **192**, 37.
____. 1975, *Ph.D. Thesis*, Univ. of California at Berkeley.
____. 1976a, *Ap. J.*, **206**, 753.
____. 1976b, *Ap. J.*, **207**, 141.
____. 1977, *Ap. J.*, **211**, 147.
____. 1978, in *Protostars and Planets*, ed. T. Gehrels (Tucson: U. of Ariz. Press), p. 209.
____. 1979a, *Ap. J.*, **228**, 159.
____. 1979b, *Ap. J.*, **228**, 475.
____. 1981, in *Fundamental Problems in the Theory of Stellar Evolution*, eds. D. Sugimoto, D. Q. Lamb, and D. N. Schramm (Dordrecht: Reidel), p. 27.
____. 1983, in *Solar and Stellar Magnetic Fields: Origins and Coronal Effects*, ed. J. O. Stenflo (Dordrecht: Reidel), p. 479.
____. 1985, *Astr. Ap.*, **142**, 41.

____ . 1987a, in *Physical Processes in Interstellar Clouds*, eds. G. E. Morfill and M. Scholer (Dordrecht: Reidel), p. 453.
____ . 1987b, in *Physical Processes in Interstellar Clouds*, eds. G. E. Morfill and M. Scholer (Dordrecht: Reidel), p. 491.
____ . 1989, in *The Physics and Chemistry of Interstellar Molecular Clouds*, eds. G. Winnewisser and J. T. Armstrong (Berlin: Springer-Verlag), p. 297.
____ . 1990, in *Galactic and Intergalactic Magnetic Fields*, eds. R. Beck, P. P. Kronberg, and R. Wielebinski (Dordrecht: Reidel), *in press.*
Mouschovias, T. Ch., Ciolek, G., and Morton, S. A. 1990, *in preparation.*
Mouschovias, T. Ch., and Morton, S. A. 1985a, *Ap. J.*, **298**, 190.
____ . 1985b, *Ap. J.*, **298**, 205.
____ . 1990a, *Ap. J.*, *to be submitted.*
____ . 1990b, *Ap. J.*, *to be submitted.*
Mouschovias, T. Ch., Morton, S. A., and Ciolek, G. 1990, *in preparation.*
Mouschovias, T. Ch., and Paleologou, E. V. 1979, *Ap. J.*, **230**, 204.
____ . 1980, *Ap. J.*, **237**, 877.
____ . 1986, *Ap. J.*, **308**, 781.
Mouschovias, T. Ch., Paleologou, E. V., and Fiedler, R. A. 1985, *Ap. J.*, **291**, 772.
Mouschovias, T. Ch., and Spitzer, L., Jr. 1976, *Ap. J.*, **210**, 326.
Mouschovias, T. Ch., Shu, F. H., and Woodward, P. R. 1974, *Astr. Ap.*, **33**, 73.
Myers, P. C. 1985, in *Protostars & Planets II*, eds. D. C. Black, and M. S. Matthews (Tucson: Univ. of Arizona Press), p. 81.
Myers, P. C., and Benson, P. J. 1983, *Ap. J.*, **266**, 309.
Myers, P. C., and Goodman, A. A. 1988, *Ap. J.*, **329**, 392.
Nakano, T. 1979, *Publ. Astr. Soc. Japan*, **31**, 697
____ . 1984, *Fundam. Cosmic Phys.*, **9**, 139.
____ . 1989, *M. N. R. A. S.*, **241**, 495.
Nakano, T., and Tademaru, T. 1972, *Ap. J.*, **173**, 87.
Nakano, T., and Umebayashi, T. 1986, *M. N. R. A. S.*, **218**, 663.
Oppenheimer, M., and Dalgarno, A. 1974, *Ap. J.*, **192**, 29.
Paleologou, E. V., and Mouschovias, T. Ch. 1983, *Ap. J.*, **275**, 838.
Parker, E. N. 1966, *Ap. J.*, **145**, 811.
Schloerb, F. P., and Snell, R. L. 1984, *Ap. J.*, **283**, 129.
Schwarz, U. J., Troland, T. H., Albinson, J. S., Bregman, J. D., Goss, W. M., and Heiles, C. 1986, *Ap. J.*, **301**, 320.
Scott, E. H. 1984, *Ap. J.*, **278**, 396.
Shu, F. H. 1983, *Ap. J.*, **273**, 202.
Shu, F. H., Adams, F. C., and Lizano, S. 1987, *Ann. Rev. Astr. Ap.*, **25**, 23.
Spitzer, L., Jr. 1962, *Physics of Fully Ionized Gases*, 2nd ed. (New York: Interscience).
____ . 1978, *Physical Processes in the Interstellar Medium* (New York: Wiley-Interscience).
Tomisaka, K., Ikeuchi, S., and Nakamura, T. 1988, *Ap. J.*, **335**, 239.
Vrba, F. J., Coyne, G. V., and Tapia, S. 1981, *Ap. J.*, **243**, 489.
Young, J. S., Langer, W. D., Goldsmith, P. F., and Wilson, R. W. 1981, *Ap. J. Lett.*, **251**, 81.
Zuckerman, B., and Palmer, P. 1974, *Ann. Rev. Astr. Ap.*, **12**, 279.

DISCUSSION

Cayrel: The mass interval for star formation associated with your magnetically controlled mechanism is 0.1 to 20 $M_\odot$. How is this range modified in a zero-metal environment?

Mouschovias: In primordial protoglobular cluster clouds, massive star formation ($\gtrsim$ 10 $M_\odot$) is favored. The reason is that these clouds are expected to be more massive than typical present-day clouds; certainly they must have been more massive than observed cluster masses (~ 10^5 $M_\odot$). The mass inside the length scale λ_A, which cannot be prevented by magnetic forces from collapsing, varies (at a given density) with the cloud mass as $M_{cl}^{3/4}$ (see Mouschovias 1989, § IIc). The degree of ionization x also has some effect. We did not recalculate x; we took the number from the literature. The two effects combine to give the protostellar masses I quoted.

Tscharnuter: Have you included Ohmic dissipation in your calculation?

Mouschovias: No. Our collapse calculations in both the cylindrically symmetric and in the axisymmetric case are pursued to densities no higher than 3×10^9 cm^{-3}. (Beyond those densities, both the isothermality assumption and the relation between the ion density and the neutral density used in the calculations break down.) For these relatively low densities, Ohmic dissipation is not important. In Maxwell's equation $\partial B/\partial t$ = ..., the Ohmic term is smaller than the electron advection term by a factor ~ 10^{10} for typical molecular cloud cores (see Mouschovias 1987b, eq. [75a]).

Zinnecker: Despite trying hard, I still have not fully understood the discrepancy between your results and those of Nakano regarding the critical density at which ambipolar diffusion becomes important. Could you, please, explain the situation once more?

Mouschovias: A short question but no short answer! Let me try in a different way. This story started with the paper by Nakano and Tademaru (1972), which concluded that ambipolar diffusion is not important below densities ~ 10^9 cm^{-3}. They studied the loss of magnetic flux by a cloud *as a whole*. This, as I have discussed, is conceptually incorrect. Ambipolar diffusion leads to a *redistribution* of mass only in the central flux tubes of a cloud, not to a loss of flux by the whole cloud (see § 2.3 in my paper). The precise origin of the large density they obtain is the notion/requirement they imposed, namely, that ambipolar diffusion becomes important only if it progresses over a time scale comparable to the free fall time. This requirement is directly responsible for their ~ 10^9 cm^{-3}. However, interstellar clouds are far from free-falling; the free fall time scale is an irrelevant time scale for most of the lifetime of a molecular cloud (see Mouschovias 1977). In that paper I posed the question of what the time scale of gravitationally driven ambipolar diffusion is in the cores of self-gravitating clouds --it is too long in the low-density envelopes. I found that, at a density ~ 10^4 cm^{-3}, τ_{AD} is significantly short ($\lesssim 10^7$ yr) to affect the cloud's evolution. In fact, τ_{AD} *is* the evolutionary time scale. This is how fragments (or cores) form. I caution, however, that there is no magical density at which ambipolar diffusion suddenly sets in. It is always present in self-gravitating molecular clouds. However, at densities significantly smaller than 10^4 cm^{-3}, τ_{AD} is too long for most purposes (see eq. [9]).

A second source of misundertanding is the fact that one does not often distinguish between (1) the onset of ambipolar diffusion at some rate to be of significance for the evolution of a cloud and (2) the resolution of the magnetic flux problem of star formation. The density ~ 10^4 cm^{-3} which I obtained is for purpose #1. The magnetic flux problem is not resolved by this density. On the latter issue, Nakano's view is that ambipolar diffusion cannot resolve the magnetic flux problem, period; and that Ohmic dissipation does so above 10^{12} cm^{-3}. He claims that it reduces the flux of a protostar by an insignificant amount (see discussion in § 4.3.2 in my paper). Our calculations show that that is an artifact of his initial conditions (his model clouds are too close to collapsing even with flux freezing). We have not by any means proven that ambipolar diffusion can resolve the magnetic flux problem below densities ~ 10^{12} cm^{-3}, but we have not finished our investigations yet. It looks promising for the moment, but as I said in my talk, the jury is still out.

Palla: I have two questions related to the effect of the magnetic field in the formation of primordial stars. (1) The minimum mass you quoted of $\simeq 10\ M_{\odot}$ was calculated by assuming some initial value of the magnetic field. What is this value, and what do you consider to be its origin? (2) What is the explanation for the high ionization rate you quoted, given the fact that the relevant ions, H^- and H_2^+, are expected to have a very small abundance?

Mouschovias: (1) I have assumed that, at any given density, the field is of comparable strength in primordial clouds as it is in present-day clouds of the same density. Nobody knows whether this is so. It is an *assumption*. The origin of the field is presumably the dynamo mechanism. This, as we know, requires a weak seed field, which is then amplified. Does nature do it that way? Your guess is as good as mine. The uncertainties that exist in the input parameters is the reason for which I have avoided working on primordial star formation. I nevertheless thought it was interesting, given the unsolved problem of metal contamination of globular cluster stars, that a simple assumption on the strength of the magnetic field leads to the conclusion that massive stars are the first to form in primordial clouds and thus, perhaps, contaminate the parent cloud before low-mass stars form (see Mouschovias 1989). (2) I have taken the degree of ionization from the literature (see my answer to Dr. Cayrel's question). I have done nothing original and nothing out of the ordinary on this. However, if the ion density is smaller than previously thought, the neutral-ion collision time increases and, therefore, λ_A and the selected masses increase (see eq. [11]).

Boss: You have been very conservative in making comparisons between observations and the predictions of your magnetic field theories. Let me aid you in your caution by noting that there is an alternative explanation for the velocity field (ascribed to retrograde rotation) in some dark cloud cores, namely, nonmagnetic collapse onto a binary or multiple protostellar system.

Mouschovias: Yes, I am aware of your paper on the subject. Our prediction (Mouschovias and Paleologou 1979) was that retrograde rotation of a fragment in a cloud or of a star in a stellar system is a natural consequence of the magnetic braking of a perpendicular rotator. The infall velocity field which you describe is not, of course, true retrograde rotation but might *appear* as such. I agree. The difference is that magnetic braking of a perpendicular rotator can predict or explain the *real* phenomenon of retrograde rotation, not something that may just look like it.

Mateo: What is the implication for your model of star formation of the (still controversial) observation that halo stars occur less frequently in binaries than disk stars do?

Mouschovias: It is my impression that 70% of all stars are in binary or multiple systems. I am not aware of any observation indicating that single stars are more abundant in nature than multiple ones. From a dynamical point of view, it is easier to form a binary (or multiple) stellar system than a single star because the orbital motion of a binary can and does store typically three orders of magnitude more angular momentum than the spin of its individual components. Hence, formation of these stars is not as demanding on magnetic braking. If single stars were more abundant than multiple systems, it would imply, at least in the context of our work, that magnetic braking was too efficient in removing angular momentum from a collapsing fragment. Although we cannot exclude that possibility, it would surprise me if that turned out to be the case.

2. STAR FORMATION IN DIFFERENT ASTROPHYSICAL ENVIRONMENTS.

J. SCALO - Perception of interstellar structure: facing complexity.

A. LIOURE, J.P. CHIÈZE - Interstellar gas cycling powered by star formation.

I.G. KOLESNIK - Star complex formation in differentially rotating superclouds.

E.A. LADA - Global star formation in the L1630 molecular cloud.

H. ZINNECKER - Observations of fragmentation.

E. KRÜGEL - Extrasolar planetary material.

C.J. CLARKE - Star formation and dissipation in galactic discs.

E. KRÜGEL - Star formation in star burst galaxies.

Yu. I. IZOTOV, V.A.LIPOVETSKII, N.G. GUSEVA, A. Yu. KNIAZEV, J.A. STEPANIAN - The new brilliant blue compact dwarf galaxy: a test on the early stages of galaxy evolution.

H.W. YORKE, R. KUNZE - Star formation in elliptical galaxies.

B. ROCCA–VOLMERANGE - Star formation in distant galaxies.

L. ANGELETTI, P. GIANNONE - The early evolution of spheroidal star systems.

PERCEPTION OF INTERSTELLAR STRUCTURE: FACING COMPLEXITY

John Scalo
Astronomy Dept.
University of Texas
Austin, TX 78712
U.S.A.

ABSTRACT. This paper challenges some orthodox notions concerning the structure and evolution of star-forming regions, proposing that they arise largely by a dual process in which conceptual models are fashioned after categories which are in great part reflections of observational limitations, and the models are projected onto interpretations of data, an example of hypostatization of categories. Several examples are discussed. The need for internal support of molecular clouds is questioned. It is suggested that the inverse density-size relation often claimed for clouds and accounted for by several theoretical models is an artifact caused by limited dynamic range in column density detectability, selection bias, distance uncertainties, and internal density gradients, and is contradicted by several unbiased surveys. Limited spatial dynamic range (ratio of image size to resolution) in maps of column density structure results in a "Mr. Magoo effect" which tends to accommodate quasi-static evolutionary concepts. Column density structures mapped with a large spatial and column density dynamic range are dominated by irregular, connected, and nested forms on all scales. Contour shapes of both atomic and molecular clouds exhibit self-similar irregularity with a common fractal dimension over a large range in scale. These features and a technique for the quantification of complex structure are illustrated with a densely-sampled column density image of the Taurus region constructed from IRAS data. A comparison with 553 high-accuracy polarization vectors in the region is also given.

The question is not how the phenomena must be turned, twisted, narrowed, crippled so as to be explicable, at all costs, upon principles that we have once and for all resolved not to go beyond. The question is: To what point must we enlarge our thought so that it shall be in proportion to the phenomena?

—Schelling, *Philosophie der Mythologie*

Theory like mist on eyeglasses. Obscure facts.

—Charlie Chan, *Charlie Chan in Egypt*

You can observe a lot just by watching.

—Lawrence (Yogi) Berra

R. Capuzzo-Dolcetta et al. (eds.), Physical Processes in Fragmentation and Star Formation, 151–177.

1. Introduction

Our conceptions and theories of interstellar structure and its relation to star formation are shaped in large part by our impressions of observational data. Each of the various techniques used to map interstellar structure are, however, rather severely limited in different, often non-overlapping, ways involving resolution, density of mapping, and temperature, column density and volume density sensitivity. Furthermore, published spectral line maps of star-forming regions usually consist of less than a thousand or so pixels of spatial information, enough to show the presence of some substructure but insufficient to reveal detail which may be crucial for understanding the actual physical processes involved. This situation renders us susceptible to a "Mr. Magoo" syndrome, in which phenomena are misinterpreted because of perceptual deficiencies, in this case near-sightedness.

Combined with our natural tendency to classify, these effects have resulted in a sort of consensus view of the interstellar medium as consisting of some smooth intercloud "phase" in which are imbedded "clouds" of various types: dark clouds, diffuse clouds, globular filaments, low-mass cores, high-mass cores, giant molecular clouds, cirrus clouds, superclouds. Although it is recognized that there is some overlap and duplication among these categories, it is generally tacitly assumed that the categories do refer to physically and spatially distinct entities. This assumption allows a further abstraction of the data into models for cloud evolution and star formation, without reference to the original data on which the categories were based. The acceptance of particular models in turn influences our impressions of the observations. The tendency to verify conceptual models by observing selected features and suppressing awareness of other features is an example of the general human tendency to *hypostatize* categories, what A. N. Whitehead termed "the fallacy of misplaced concreteness." This tendency occurs in fields as diverse as theories of music (Serafine 1988), evolution of painting styles (Peckham 1965), and everyday experience (e.g. Bohm 1980).

For example, tabulations of the mean properties of "diffuse clouds" identified by optical absorption lines, reddening, and HI emission (e.g. Spitzer 1968a,b) led to models in which the cool atomic interstellar medium is distributed as hundreds of thousands of similar "standard clouds" which are independent entities except for occasional collisions. These conceptions are still prevalent, as in theories involving mass exchange processes between phases of the ISM, and the growth of massive clouds by the coalescence of small diffuse clouds. Such models persist in spite of the fact that high-resolution HI maps of nearby galaxies (e.g. Wright *et al.* 1972, Rots and Shane 1975, Allen and Goss 1979, Unwin 1980a,b, van Albada 1980, Bosma *et al.* 1981, Ball 1986) consistently reveal a nested distribution of highly irregular and connected structures, as well as holes (see Brinks and Bajaja 1986), covering scales from kiloparsecs down to the resolution limit: the HI is not distributed as a sea of small discrete clouds.[1]

Similarly, summaries of results of Galactic CO surveys have led to the conception of the molecular material in galaxies distributed as thousands of discrete "giant molecular clouds" with sizes around 100 pc, even though the "clouds" identified in these surveys exhibit a large range in sizes (e.g. see Figure 2 below based on data from Solomon *et al.* 1987) limited at small sizes by resolution, and their derived properties depend on a number of selection factors (cf. Casoli *et al.* 1984, Myers *et al.* 1986). Furthermore,

[1]These same maps are also strong evidence against a 3-phase interstellar medium in which a hot component occupies a significant fraction of the volume, another model whose persistence in light of a number of contradictions (most are summarized in Heiles 1987b), some long-standing, is another example of the type of hypostatization being discussed here.

densely sampled maps of nearby molecular clouds (discussed below) show these regions to possess extremely irregular and nested forms at all resolvable scales, and to be spatially related to the larger HI structures (themselves containing irregular substructure) in which they are imbedded. Yet most theoretical models assume discrete standard "GMCs" whose internal structure consists of a large number of "dense cores" of similar properties. Unlike the case with HI, discussions of recent high-resolution CO mapping of nearby galaxies (e.g. Lo *et al.* 1987, Vogel *et al.* 1987, Lada *et al.* 1988) tend to reinforce the idea that the (molecular) interstellar medium can be reduced to a swarm of typical "GMCs" due to the facts that the attainable linear resolutions (about 25 pc to 300 pc in the papers listed above) are similar to the accepted typical dimensions of GMC-category clouds (a primary motivation for the observations was the detection of such clouds), and that the spatial dynamic range is usually only an order of magnitude. Thus statements like "most of the molecular gas resides in GMCs" is operationally correct, but may completely misrepresent the spatial structure of that gas. For example, the preliminary complete CO map of the central region of M51 by Adler *et al.* (1989) shows irregular nested forms covering a large range of scales down to the resolution limit.

These examples illustrate the strong reductionist tendency to conceive the gas in galaxies in terms of a small number of "building blocks" whose properties are based on categories abstracted from data which is limited by selection. This process tends to be self-justifying. For example, even though the estimated mean star formation efficiency among clouds of various masses (e.g. Myers *et al.* 1986) does not demand long cloud lifetimes in order to account for the Galactic star formation rate, the idea that clouds require internal support persists, in part because a coalescence model which *assumes* the existence of separate randomly distributed long-lived clouds requires $\gtrsim 10^8$ yr to build massive clouds from small ones. Since the only *direct* evidence on cloud lifetimes gives a lower limit of about 10^7 yr (Bash *et al.* 1977, Leisewitz *et al.* 1989), and since no one has yet shown how to support a massive irregular structure for at least 10 dynamical timescales, it would seem natural to consider rejecting the coalescence model and accept the impermanence of clouds. This rejection may require that small molecular clouds are destroyed and reformed in interarm regions (e.g. Lo *et al.* 1987); we are now so accustomed to thinking of long-lived clouds that this somehow seems more problematic than supporting a cloud for ten dynamical timescales! Besides, as discussed below, the complex appearance of all "clouds" observed with large spatial dynamic range makes it difficult to imagine that they are quasi-equilibrium structures.[2]

An illustration of the subjectivity of even operational definitions of clouds and cloud categories is the case of "dense cores" (see the summaries by Myers 1987, 1990). Zhou *et al.* (1989) have found that analysis of CS lines toward dense cores reveal physical properties and conditions greatly different than previously inferred from NH_3 observations; in particular the masses are much larger, densities higher, and the linewidths are supersonic. That the linewidths may be supersonic at the scale of "cores" is also indicated by VLA H_2CO observations of clumps associated with H II regions (Martin-Pintado *et al.* 1985) and in much smaller condensations such as the 0.03-0.04pc C_3H_2 clumps in the Heiles Cloud 2 region of Taurus (Guelin and Cernicharo 1988). Besides showing how structure depends on method of observation, the available high resolution observations suggest that clumpiness is present at scales considerably smaller than the

---

[2] Space precludes a discussion of the relation of supersonic linewidths to the idea that clouds must be supported, except for the reminders that highly supersonic linewidths have long been known for cool HI "clouds," that an approximate (or even exact!) balance of energies in the sense of the virial theorem does not necessarily imply a quasi-static equilibrium, and that it is not so clear how to calculate the appropriate gravitational energy for a fractally irregular (sec. 6 below) cloud.

canonical NH_3 cores (e.g. Mezger *et al.* 1988, Falgarone and Perault 1987). Considering that a large part of currently popular theoretical ideas about star formation are based on the properties of NH_3 cores, and their assumed lack of internal substructure, this example suggests that we pause to consider the degree to which our speculations and theories are products of observational limitations and categorization.[3]

The purpose of the present paper is to point out that observations of cloud structure which possess a fairly large column density dynamic range and spatial dynamic range (size of map divided by resolution) bear little resemblance to orthodox concepts and models, the observed structure being dominated by irregularity, connectedness and continuity at all scales. The construction of a densely-sampled column density map of the Taurus region with large column density dynamic range is presented and used to illustrate these features. Methods for quantifying the apparently complex structure are outlined. The general conclusion is that models based on conceptions of distinct clouds and quasi-equilibrium assumptions are inadequate and must be supplanted by more dynamical approaches, especially high resolution numerical simulations of the hydrodynamic equations, in order to allow for the full range of dynamical phenomena associated with the nonlinear advection term in the momentum equation.

2. The Density-Size Relation

An instructive example of the danger of basing models on observations without consideration of selection effects involves column densities. Extinction maps derived from star counts can generally only be constructed for regions in which the visual extinction $A_V \gtrsim 0.5$ mag, while above $A_V \sim 5\text{-}7$ mag there are too few background stars to establish anything but a lower limit. Mapping with ^{13}CO usually requires $A_V \gtrsim 0.5$ for a detectable line, and saturates when $A_V \gtrsim 4$. Furthermore, the LTE approximation is only valid for number densities $\gtrsim 10^3$ cm^{-3} and fractionation may enhance the ^{13}CO line when $A_V \lesssim 2\text{-}3$. Except for fractionation, similar restrictions must apply to column densities derived from ^{12}CO. In general, then, the commonly used column density indicators have a dynamic range of about a factor of ten.

It is commonly thought that volume densities and sizes of clouds are related by $n \sim r^{-1}$, so that cloud column densities are approximately constant over at least three orders of magnitude in size (e.g. Larson 1981, Leung *et al.* 1982, Myers 1983, Solomon *et al.* 1987, Falgarone and Perault 1987). However the data on which this correlation is based contain many selection effects in addition to the dynamic range limitations just discussed: Some surveys are biased in the sense that objects are selected on the basis of previous identification using a particular technique; some surveys sample clouds at different distances with the same angular resolution, so the linear resolution is variable, which means that internal density gradients can masquerade as an inverse density-size relation (cf. Cernicharo *et al.* 1985); distance uncertainties can make a given correlation appear to be of the form $n \sim r^{-1}$, as recently shown for a sample of clouds by Leisawitz (1989), or can even create a correlation where there is none.

---

[3] This is not meant to imply that categorization of "cloud types" is not a useful tool for summarizing observational data, conveying the diversity of cloud properties in compact form, and suggesting the relative importance of physical processes (e.g. Myers 1990). The concern here is that models based on cloud categories will be regarded as representations of the actual interstellar medium rather than of abstractions. Prior to the advent of molecular line studies, this tendency was sometimes referred to as the "standard cloud syndrome;" see sec. 7 below for more discussion.

Figure 1 shows the n-r data compiled by Myers and Goodman (1988) from several sources. (The data from Leung *et al.* 1982 can actually be shifted in column density and size by an uncertain amount because they used a very small $^{13}CO/H_2$ conversion factor but compensated by adjusting the size to account for an assumed internal density profile.) Lines of constant column density corresponding *approximately* to the dynamic range of the observations are also shown. The fact that these limits bracket most of the observed points, along with the other selection effects listed above, is cause for concern. I have also indicated the rough positions of some additional observations (e.g. Mezger *et al.* 1988), following Falgarone and Puget (1986), which are limited in different ways than the data points, to emphasize the wide range of column densities available to interstellar structures.

In some cases the claim of constant column density is not substantiated by the very data which led to the claim. Figure 2 shows data for clouds identified in a ^{12}CO survey

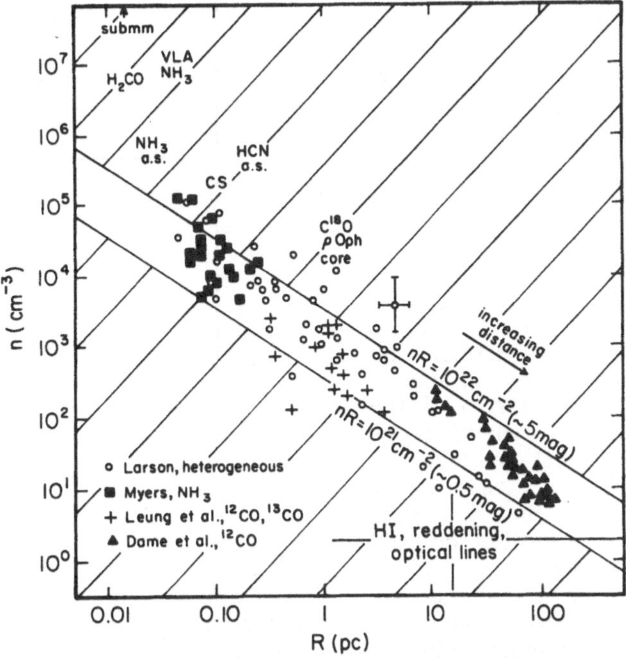

Figure 1. Relation between volume density and size for selected interstellar cloud structures. The data from Larson (1981), Leung *et al.* (1982), Myers (1983) and Dame *et al.* (1986) are those compiled by Myers and Goodman (1988). The Leung *et al.* points can be shifted by an uncertain amount in n and r because of the small adopted $^{13}CO/H_2$ ratio and their definition of cloud size. Other labels refer to clouds observed in different species and/or with different techniques, after Falgarone and Puget (1986); "a.s." denotes aperture synthesis observations. The two solid diagonal lines represent the approximate column density limitations of ^{12}CO, ^{13}CO, and extinction observations for the data shown.

by Solomon *et al.* (1987). If the assumption of virial equilibrium is accepted on the basis of their arguments, then the column density of an individual cloud is proportional to the square of the linewidth divided by the cloud size. Fig. 2a shows column densities calculated in this way for all the clouds in their catalogue. The spread of a factor of 5-10 is either a real dispersion in column density, a reflection of large departures from virial equilibrium, or both. The virial assumption was justified in Solomon *et al.* by the near-constancy of the conversion factor $N(H_2)/I_{CO}$; however, when this quantity is plotted as a function of size (Fig. 2b) a clear trend, roughly varying as $R^{-0.5}$, is apparent. (The trend of course appears much weaker if plotted against mass.) But if we abandon the virial assumption and assume instead that the ^{12}CO luminosity L_{CO} measures the cloud mass, the column density of a cloud should be proportional to L_{CO} divided by the cloud area, which is shown as a function of cloud size in Fig. 2c. One sees a scatter of a factor of 5-10, and, if anything, a trend for column density to increase with size. Since the spread is large in both 2a and 2c, and might be larger if the observations had a larger dynamic range in column density, I must conclude that there is little basis for the assertion by Solomon *et al.* (1987) that their data support the idea of constant column density for molecular clouds.

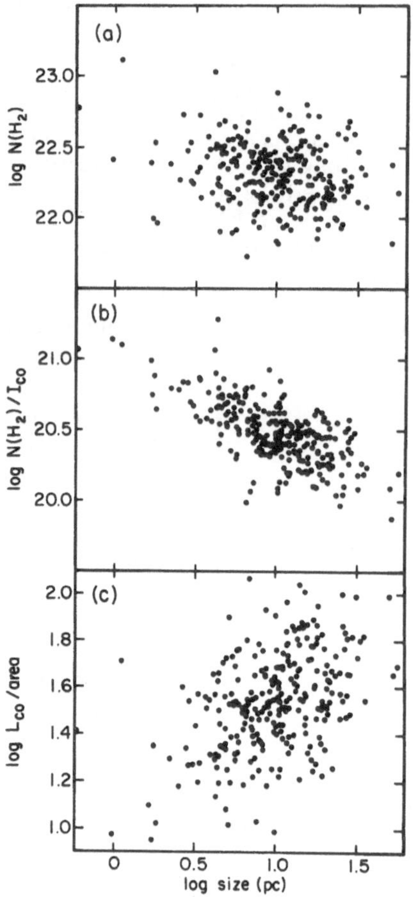

Figure 2. Cloud properties from the ^{12}CO survey of Solomon *et al.* (1987). (a) Column densities, computed assuming virial equilibrium, as a function of cloud size; (b) conversion factor $N(H_2)/I_{CO}$ which should be a constant independent of cloud size if the virial assumption is valid; (c) column densities computed assuming that the ^{12}CO luminosity measures the cloud mass.

Actually the situation concerning the purported correlations is even more perplexing. Densely sampled unbiased ^{13}CO surveys of individual star-forming complexes by Carr (1987, Cep OB3) and Loren (1989a,b, ρOph region) find that column density *increases* with cloud size, with $N \propto r^{0.5}$ and $r^{0.8}$, respectively, with large scatter. These results cannot be dismissed as an anomaly that only occurs within individual cloud complexes: The same sort of result was found by Casoli *et al.* (1984) in their unbiased ^{13}CO survey of Orion and Perseus arm clouds, and by Leisewitz (1989) in his survey of clouds associated with young clusters. The result is difficult to reconcile with the fact that examination of just about any well-sampled extinction or ^{13}CO column density map shows that column density *decreases* with increasing size of contour for spatially separate contours. Projection might account for this latter correlation, but only if the clouds are hierarchically structured, and then with difficulty.

Whatever the resolution of these seeming contradictions, it seems clear that there is good reason to discard the notion of constant column density. Unfortunately, the constant column density result lent itself to a variety of theoretical interpretations (e.g. Henriksen and Turner1984, Silk 1985, Chieze 1987, Fleck 1988, Maloney 1988, Elmegreen 1989, Bigliari and Diamond 1988, McKee 1989). The discussion given here suggests that these models are in need of serious revision or abandonment. It should also be noted that several authors have suggested that the apparent linewidth-size correlation may also be contaminated by selection effects (e.g. Scalo 1987, Blitz 1988).

3. Effects of Spatial Dynamic Range

Until recently, most studies of the column density structure of star-forming regions were based on images containing no more than ~ 10^3 pixels. Typically these maps reveal several roughly spherical or moderately elongated condensations embedded in a smooth lower density background. Most models for cloud evolution and star formation are based, often implicitly, on conceptual abstractions of such images. However,the fact that these maps always have similar generic appearance even though they correspond to linear scales which differ by orders of magnitude in some cases (see, for example, Fig. 1 in Scalo 1988, Fig. 2 in Falgarone and Phillips 1989) suggests that the observed structure is a very crude representation of a much richer spatial organization. Indeed, the available column density images containing on the order of 10^4 pixels or more (a list is given in Scalo 1988) do not resemble any of the quasi-equilibrium models but instead suggest a much more dynamic situation in which the dominant characteristics are irregularity and variety on all resolvable scales. Oort (1974) pointed out that "even the denser dust clouds often have shapes which show clearly that they are not equilibrium structures." Instead of the "raisin pudding" or "meatball in sauce" models on which most theories for interstellar cloud evolution are based, these structures remind one more of egg drop soup.

A good example is the extensive ^{13}CO mapping of the Orion A region by Bally *et al.* (1987), which has a spatial dynamic range of about 200 and about 10^5 pixels of integrated column density information (as well as velocity channel maps). These images exhibit an enormous variety of structure, with filaments, bubbles, cavities and more indescribable configurations covering the entire accessible range of size scales. Bally *et al.* (1988) conclude that "Molecular clouds do not have a 'relaxed' structure which might be expected of an object whose components have survived many dynamic time-scales... Clouds cannot be modeled as a 'gas' consisting of individual particles, in this case cloud cores, whose motion is responding to the overall gravitational potential of the entire cloud." Most theorists do not consider this complexity as a constraint on simple models because it may only represent an example of the effects of outflows, ionization, heating, stellar winds and explosions associated with the many massive young stars in Orion; certainly these processes must play some role in shaping at least part of the observed structures. However similar indications of complex structure can be seen in other less densely sampled maps of less active cloud complexes (e.g. Rossano 1978, Falgarone and Perault 1988). Falgarone (1990) remarks that the latter study finds that column density enhancements at any background column density between 5×10^{20} cm^{-2} and 10^{22} cm^{-2} "are not islands in a sea of low density gas but present a clear connectedness in space and velocity."

At the other extreme, IRAS images of low column density "cirrus" clouds (see Weiland *et al.* 1986, Bazell and Desert 1988, Deul and Burton 1989), which are supposedly devoid of any star formation activity, also show surprisingly complex and varied structures. Since their average properties are very similar to the "diffuse clouds" observed in HI emission, optical absorption lines, and reddening, the high degree of irregularity makes it difficult to accept that diffuse clouds are confined by external thermal pressure, as assumed in many theoretical models. It may be tempting to dismiss the complexity of cirrus clouds as irrelevant for models of star-forming clouds because they are not self-gravitating.[4] In the next section an example is given of complex structure in a self-gravitating cloud complex without massive star formation.

4. A Densely Sampled Column Density Map of Taurus

In order to construct a densely sampled column density map of a cloud complex which is both self-gravitating and not (yet?) stirred up much by star formation, a column density image of the Taurus region has been constructed from IRAS data (Scalo and Houlahan 1990). The primary drawback to using the IRAS data for this purpose is that it contains no velocity information, and the possible importance of projection effects must be kept in mind (see Discussion). For molecular spectral line work on selected regions of the Taurus complex see Duvert *et al.* (1986), Cernicharo and Guelin (1987), Heyer, Vrba *et al.* (1987), Nercessian *et al.* (1988), and references therein. The large-scale structure was studied by Kleiner and Dickman (1984, 1985) in ^{13}CO, and on a larger scale by Wouterlout and Habing (1985) in OH and Ungerechts and Thaddeus (1987) in ^{12}CO. A survey of earlier studies is given in Scalo (1985).

Destriped 60 μm and 100 μm IRAS images of a 9° x 9° area (1° = 2.4 pc at a distance of 140 pc) centered on $\alpha(1950) = 4^h30^m00^s$, $\delta(1950) = 26°00'00''$ in the core of the Taurus complex were obtained from IPAC. Each image contained 540^2 pixels of size 1 arcmin (=0.04 pc). The effective resolution is estimated at 2-3 arcmin. Subtraction of zodiacal light and galactic emission was performed by fitting two-dimensional polynomials to a number of low-intensity spots on each image. For each pixel the dust temperature was taken as the color temperature derived from the observed 60 μm/100 μm flux ratio, assuming a λ^{-n} wavelength dependence of the far infrared emissivity. The 100 μm optical depth could then be derived from the Planck function using the observed 100 μm intensity. Assuming that the warm dust fraction is a constant, as suggested by the work of Langer *et al.* (1989) and Snell *et al.* (1989) and by the extensive comparisons mentioned below, the 100 μm optical depth is proportional to the total column density of gas. It was found that the resulting *relative* column density map (the absolute scale of the column densities is irrelevant for the present discussion) was virtually independent of the choice of emissivity law for n = 1 and 2, and also was not sensitive to different choices of background subtractions, except for the smallest optical depths.

The estimated noise at 100 μm is about 0.2 to 1 MJy/sr. For comparison, after subtraction of background (5-9 MJy/sr), the 100 μm intensities were >5 MJy/sr over most of the 100 μm image (>100 MJy/sr in the brightest spots), and were only as low as 1-3 MJy/sr in the darkest "holes." At 60 μm the relative background subtraction is larger, but the noise estimate is only about half as large as at 100 μm. It therefore appears that only the very smallest column densities may be affected by noise.

The resulting range of 100 μm optical depth τ_{100} is 1×10^{-5} to 4.4×10^{-3} (in the core of the L1495 cloud) for a λ^{-1} emissivity law. Comparison with available studies of

---

[4] Stacy *et al.* (1989) have recently discovered several "dense cores" within diffuse clouds, one of which contains an infrared point source, at least in projection.

extinction and ^{13}CO gives $A_V \approx 2000\ \tau_{100}$ for this emissivity law, so the column density range corresponds to $A_V \approx 0.02$-0.05 mag (roughly at the noise level) to $A_V \approx 10$ mag. This range includes both conditions in which the gas is mostly atomic *and* in which it is mostly molecular, eliminating the artificial separation between (HI, reddening) and (CO, extinction) studies.

Figure 3a shows the Taurus column density image. The dots are T Tauri stars; the plus signs are the IRAS 60 μm point sources which are not known T Tauri stars and which satisfy the color criteria for embedded sources given by Emerson (1987). Figure 3b serves partly as a finding chart, showing the approximate locations of Barnard and Lynds dark clouds. Also shown are 553 polarization vectors taken from Hsu (1984; 261 vectors), Moneti *et al.* (1984; 171 vectors mostly in the eastern portion of the region), and Heyer, Vrba *et al.* (1987; 121 vectors along the filaments associated with B217 and B18). Polarization vectors with estimated uncertainties in position angle greater than 5 degrees have been omitted.

The validity of the derived column density structure was checked by comparison of various higher-column density ($0.5 \lesssim A_V \lesssim 5$) subregions of the map with gray scale representations of extinction maps by Cernicharo and Bachillar (1984) for the dark clouds Heiles Cloud 2 (L1534 region), L1495, L1506, L1529, and L1539, and with ^{13}CO maps by Heyer, Vrba *et al.* (1987) for Heiles Cloud 2, B216-217-218, and B18(=L1529 region). The agreement between these maps is for the most part very good, and in fact the pixel-to-pixel noise level appears significantly smaller in the IRAS structure, especially compared to the extinction maps. An example of the ability of the 100μm column density map to faithfully portray the actual column density structure is the fact that the ring of enhanced ^{12}CO line widths in B18 found by Murphy and Myers (1985) and confirmed by Heyer, Snell *et al.* (1987) in ^{13}CO line widths, is only hinted at in the ^{13}CO or A_V column density maps, but can be clearly seen in the τ_{100} map as the "loop" at $\alpha \approx 4^h29^m$, $\delta \approx 24°$ 30'. One disagreement appeared to occur in the core of the L1495 cloud, but it turns out that the ^{13}CO and extinction are saturated there; the C^{18}O map of this region by Duvert *et al.* (1986) is in good agreement with the τ_{100} structure, showing that the IRAS data can be used to probe column densities as large as $A_V \approx 10$ mag, even when there is no internal heat source. Much of the lower-column density structure can be seen by careful inspection of POSS plates (cf. Herbig 1977). These comparisons, along with independent comparisons of τ_{100} with ^{13}CO column densities in the cloud B5 by Langer *et al.* (1989) and in Heiles Cloud 2 and B18 by Snell *et al.* (1989) and of τ_{60} with A_V and $\tau(^{13}\mathrm{CO})$ in ρOph by Jarrett *et al.* (1989), demonstrates the ability of IRAS to probe the relative column density structure over a range of at least a factor of 100 in column density. The only major exceptions occur around the locations of embedded IRAS point sources, where the column densities come out very small. As discussed by Langer *et al.* (1989) and Snell *et al.* (1989), this effect is due to temperature gradients along the lines of sight to the point sources, which cause an overestimate of the appropriate mean temperature and an underestimate of the optical depth. These stellar heating regions can be easily recognized as small dark circular disks in the column density image. Although the effect is minor for Taurus, it should be much more serious in regions with massive star formation.

Although the reproduction does not do justice to the detail of the original (e.g. many condensations within the B210-212-215 filament and the L1495 cloud are resolved), the overall visual impression of Fig. 3a is a welter of irregular and interconnected forms. Of particular note is the fact that Heiles Cloud 2 and L1495 are not independent entities but are the major condensations within a large ($\gtrsim$ 20 pc) elongated structure; the connecting region apparently has clumps with appreciable column densities. Kleiner and Dickman (1984) suggested from their ^{13}CO observations that this structure is a "bar," and demonstrated the existence of a velocity gradient along the bar. Inspection of images

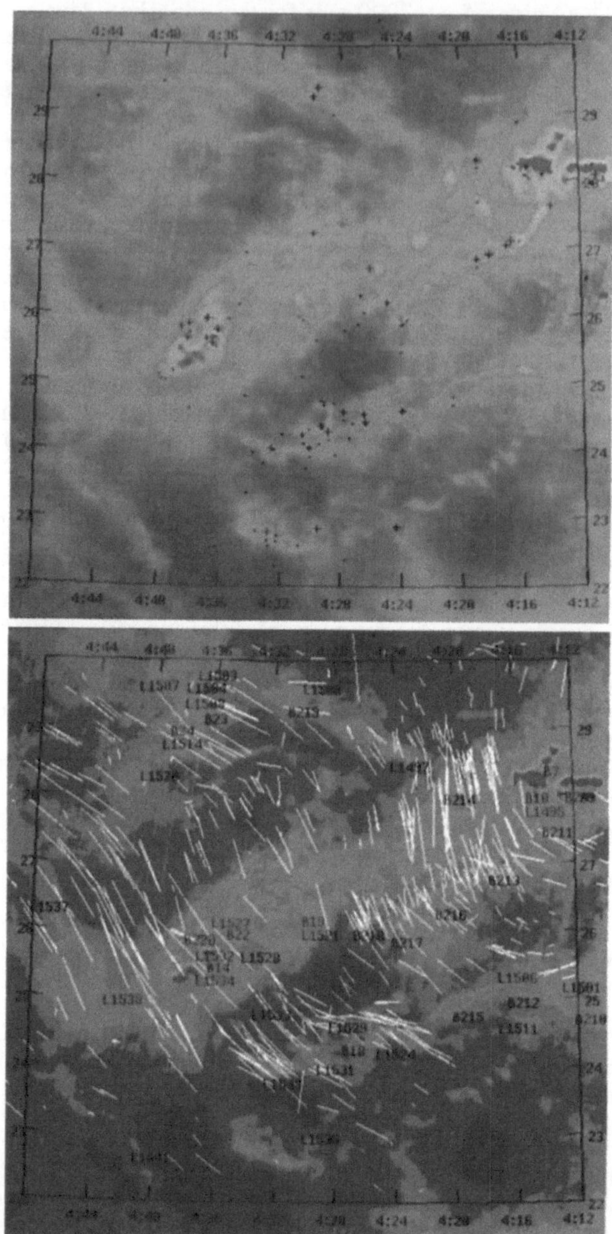

Figure 3. (a) Taurus 100μm column density image constructed from IRAS data. Dots are T Tauri stars, plus signs are IRAS 60μm point sources which are not known T Tauri stars and which satisfy the color criteria for embedded sources given by Emerson (1987). (b) Finding chart for Taurus column density image. Polarization vectors are from Hsu (1984), Moneti *et al.* (1984), and Heyer, Vrba *et al.* (1987), and only include vectors for which the position angle uncertainty is less than five degrees.

constructed using smaller numbers of grey levels and different contrasts suggests that the large elongated structure is connected to the approximately parallel filaments to the south. In fact the large-scale structure of the entire image is remarkably similar to the sketch of the $A_V \approx 1.5$ mag contour by Herbig (1977), which might be described as a "tendril." However when viewed in full detail, one might describe the structure as "cirrus-like." The long irregular filamentary-type structures to the south appear like chains of "smoke puffs." The curved chain associated with L1536 at its eastern end can actually be traced over four degrees.

The main conclusion to be drawn from this image is that, when viewed with large dynamic range in spatial scale and column density, one sees complex, irregular, interconnected structure on all scales. This structure does not resemble the ideas of quasi-static evolution of virialized "clouds" or "clumps" popular in current models, but instead suggests a more dynamically active organizational process. The degree to which dynamic range in spatial scale and column density affect one's visual impression and associated conceptual modelling can be seen by comparing Figure 3a with Figure 4, which shows Kleiner and Dickman's (1984) 15 arcmin resolution contour map of ^{13}CO emission in Taurus, constructed with 1200 pixels (already a large number compared to most millimeter spectral line maps). Although irregularity is becoming apparent for the largest contours, the general impression is of a parent cloud containing a number of "blobs" or "cores" of similar size. Figure 3a, with about 50 times more resolution elements, reveals much more variety and irregularity. In fact the irregularity and continuity of structure makes it difficult to clearly identify any separate entities which correspond to discrete "clouds," although of course regions with various density contrasts and forms can be operationally distinguished. The two highest-column density regions in Taurus, the cores of L1495 and Heiles Cloud 2, might seem like the best candidates for "clouds" in the conventional sense because of their somewhat more regular appearance. However it must be remembered that these "cores" are only imaged with $\sim 10^2$ resolution elements in the present map, and yet already exhibit substructure, so that there is good reason to suspect that, if we could zoom into these cores with the same spatial dynamic range as in the large scale map, they would exhibit a similar level of complexity. For example, Cernicharo and Guelin (1987) interpret their higher-resolution molecular line observations of Heiles Cloud 2 in terms of two bent filaments containing clumps like TMC1, which is itself known to contain substructure.

Lack of space precludes a full discussion of the magnetic field structure. There is a generally ordered large-scale component oriented roughly perpendicular to the material connecting Heiles Cloud 2 and L1495; the field around the B216-217 filament shows a strong degree of local perpendicular orientation, as shown by Heyer, Vrba *et al.* (1987). Closer inspection shows that there are many positions at which the field direction deviates significantly from the mean large scale direction, and regions of enhanced column density whose orientation is neither roughly parallel nor perpendicular to the large scale or local polarization direction The polarization directions in the B18 "filament" (around α=4^{h}28^m, δ=24°) are not parallel or perpendicular to the filament itself, but *do* appear approximately aligned with the longer axes of the "smoke puff" condensations within the filament, which are themselves roughly parallel.

The dispersion in polarization direction for the combined Taurus data is 41°, which is *larger* than the 33° dispersion found in the Orion L1641 cloud using 219 stars by Vrba *et al.* (1988). Although some of this dispersion comes from larger-scale changes in direction of the "ordered" component rather than local variations, this result still brings into doubt the conclusion of Vrba *et al.* that the magnetic field plays a more dominant role in Taurus than in L1641, a conclusion that was based on a comparison of dispersions using only Heyer, Vrba *et al.*'s (1987) data covering a small area of Taurus compared to the L1641 data. More generally, the assumption that an ordered or aligned field implies "magnetic dominance" is questionable, a point recognized by Vrba *et al.* Available Zeeman measurements of the line of sight component of the magnetic field in Taurus suggests a dynamically weak field; Crutcher (1988) lists results toward five high-column density positions in Taurus between 0 and 8 μG, with "solid upper limits" between 7 and 16 μG. Similarly weak line of sight field components have been measured at several positions in Taurus by Heiles (private communication; see Heiles 1987a). Of course one cannot rule out the possibility that the field is almost entirely in the plane of the sky.

The highly elongated structures, some of which like B217 *are* oriented roughly perpendicular to the field, seem clearly filamentary, *not* flattened structures viewed edge on, as would be expected in all models for gravitational contraction against a magnetic field. Note that several papers refer to these elongated structures as "disks" and misquote Zeeman results for Taurus as indicating a strong field, (e.g. Tomisaka *et al.* 1988). Heyer, Vrba *et al.* also discuss B216-217-218 in terms of a rotating disk model despite the fact that the tendency for velocity gradients to occur along *filaments* is well-known from HI observations (see Heiles 1974 for a review).

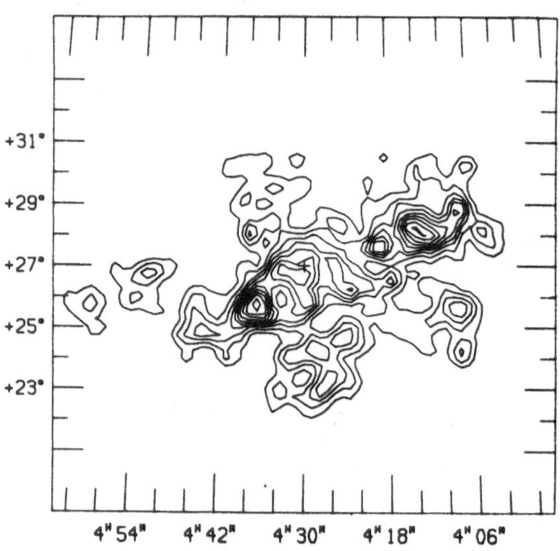

Figure 4. Contours of equal ^{13}CO emission in Taurus from Kleiner and Dickman (1984), derived from a 1200-pixel image.

The only thing that seems clear is that existing models explain little about the relation between the observed field morphology and the column density structrue in Taurus. How do bars, filaments, and "stringy"-looking structures form across the magnetic field if the ordered component of the field is dynamically important? How can we understand how the very irregular (in fact fractal—see sec. 6 below) matter distribution could be coupled to a field exhibiting such little apparent evidence for tangling? This perhaps suggests that either the energy spectrum of the tangled field has most of its power at the large scales, as occurs in incompressible MHD turbulence (Pouquet 1988, private communication), or that MHD instabilities and reconnection, which have been ignored in nearly all studies (the only exception I know of is Norman and Heyvaerts 1985), are capable of dissipating the field locally while preserving the ordered appearance of the field on large scales.

I cannot resist noting my subjective impression that the magnetic field structure has a helical component that appears to wrap around elongated structures like the B217-217-218 filament and the "bar" containing Heiles Cloud 2 and L1495. Heiles (1987a) has pointed out the possibility of a helical interpretation of the Zeeman effect direction reversal in Orion and Bally (1989, private communication) notes the same impression in comparing the Zeeman results with the optical polarization survey of Vrba *et al.* (1988). Mathewson (1968) interpreted the local (<500 pc) large-scale polarization in terms of a helical model. Moneti *et al.* (1984) point out that the mean field direction in Taurus is consistent with this model. I am unaware of any specific theoretical model which predicts a helical magnetic field.

The distribution of T Tauri stars and embedded IRAS point sources is clustered. Both types of objects are found in Heiles Cloud 2, the B18 chain, and the eastern end of the L1536 chain to the far south. Several embedded sources and one T Tauri star are found in the B213 portion of the filament extending southeast from L1495, but the B210-212-215 filament is nearly devoid of young objects. Particularly interesting is the fact that several T Tauri stars and embedded sources are associated with regions of relatively low column density ($A_V \lesssim$ 0.2-0.5mag) far enough from high column density regions that they probably could not have wandered there in 10^7 yr if the objects were born with a velocity dispersion of 0.1 km s^{-1}. Examples can be seen in the L1536 chain to the far south and the B219-L1500 gas to the north. Evidently star formation does *not* occur solely in regions of large column density, although of course we expect that the gas very close to the stellar source has a large density. This is also shown by the IRAS point source in the "cirrus cloud core" found by Stacy *et al.*(1990) and, most dramatically, by the finding of T Tauri stars far from any clouds (de la Reza *et al.* 1989 and references therein).

In closing this section, it is worth pointing out that Taurus actually exhibits the properties of most categories of interstellar "clouds:" diffuse clouds, cirrus clouds, dark clouds, dark cloud cores, globular filaments, and Bok globules are found in different subregions. Furthermore, the Taurus complex is itself a subregion of a much larger entity, seen in HI (McGee and Murray 1961) and reddening (e.g. Lucke 1978, Perry and Johnston 1982), which includes the Perseus complex and the Per OB2 association (see also Wouterlout and Habing 1985, Ungerechts and Thaddeus 1987). Lucke's (1978) Figure 8 and 9 indicate that the Taurus-Perseus complex is imbedded in a huge "supercloud" with size in excess of a kiloparsec. Thus in the case of Taurus, and presumably in other regions, observations sensitive to a large range of column densities and spatial scales reveals the continuity and coexistence of the various operationally-defined cloud categories.

5. Aspects of Complexity: Hierarchical Structure

While the visual impression of a densely sampled map of a star-forming region can be quite informative, it is of obvious interest to develop quantitative descriptors of structure which can be used to directly compare the observed structure with future numerical hydrodynamic simulations of large spatial dynamic range. In the past, most empirical studies have concentrated on estimating total or average properties for an entire region and cataloguing and searching for correlations between the properties of operationally defined clouds within the mapped region, but not on characterizing the spatial structure itself.

Two-point second order spatial statistics such as the power spectrum, correlation function, and structure function have been evaluated for some regions (see Dickman 1985, Perault *et al.* 1985), but they involve such a severe compression and smearing of the spatial *relational* information, and are so affected by structures whose sizes are a significant fraction of the image size (Houlahan and Scalo 1989), that they cannot provide an adequate characterization of the complex column density structures being discussed here. Even very high-order structure functions, which are used to characterize intermittency in incompressible turbulence, smear out most of the relational information in the data. Most proposed measures of complexity (see Lindgren and Nordahl 1988 and references given there) are actually measures of randomness or information, or of the degree to which a string of symbols can be compressed, and only have applicability to one-dimensional structures.

One of the characteristic features of complex systems is hierarchical structure, which is apparent in comparisons of maps of interstellar structures at different resolutions (see Scalo 1985, Falgarone and Perault 1987) and has figured prominently in many older theoretical discussions of fragmentation, as reviewed in Scalo (1988). The recognition and description of a hierarchical spatial structure is a problem which has apparently not been discussed in the literature. For interstellar structures which can only be viewed as two-dimensional projections, the difficulties are magnified by the fact that projection will make a random three-dimensional distribution of density enhancements with a variety of scales appear somewhat hierarchical, while even a strictly hierarchical three-dimensional structure will appear more randomized due to the effects of projection.

With these considerations in mind, a new method of image analysis, called "structure tree analysis" (Houlahan and Scalo 1990), was designed to recognize and characterize complex structure, especially hierarchical structures, in a manner well-suited for comparison of observations with theory. In addition, the technique automatically produces a catalogue of operationally defined clouds and their properties, and can be used to calculate the fractal dimension of boundary irregularities and estimate the topological genus.

Briefly, the procedure consists of successively thresholding the image at increasing grey levels (e.g., column densities) and identifying "clouds" as areas of connected pixels at each grey level, retaining information on the lineage of each cloud to larger "parent" clouds which were identified at smaller grey level thresholds. If the image is viewed as analogous to a mountain range, with height corresponding to column density, then the clouds are those parts of the plane, at a given height (grey level), that intersect the mountain range. A "path" is a sequence of clouds that preserves connectivity (or lineage). Paths can be illustrated by plotting the position of the centroid of each cloud in the sequence against the corresponding intensities. A "structure tree" is the set of all paths found in a given image. The tree construction is schematized in Figure 5.

Because the structure tree is basically a "skeleton image" or "primal sketch" of the observed structure, there is little loss of spatial relationship information in its construction; we have simply reduced the task of describing the complete structure to the more tractable problem of describing a tree. The usefulness of this method for the problem of identifying and characterizing hierarchical structure can be seen by noting that a randomly distributed collection of clouds with a range of scales will produce a tree with a very large branching ratio, while a strictly self-similar hierarchical arrangement yields a "fractal" tree.

Figure 6 (top row) shows the structure tree for a projection of a simulated 3-dimensional hierarchical cloud model with a branching factor of 3 and three levels above the "root". The left panel is the top view of the tree (i.e., the projection onto the spatial

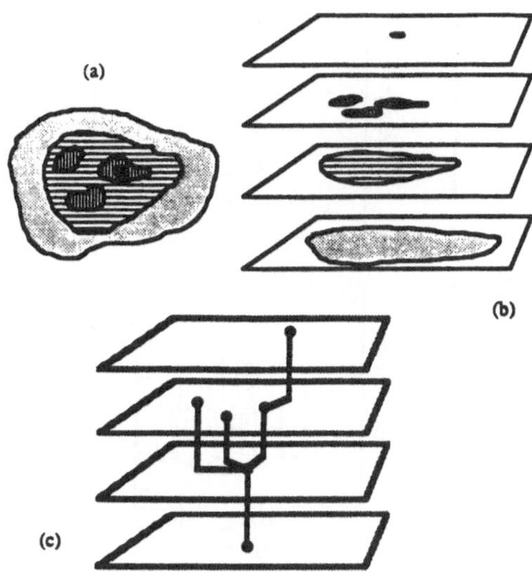

Figure 5. Schematic illustration of structure tree construction for an image. (a) Original cloud. (b) Result of thresholding to partition the image into "clouds." (c) Structure tree connecting "cloud" centers of mass according to lineage.

axes) while the other two panels are two different side views of the tree. The bottom row shows the tree for a randomized version of the hierarchical model, in which the cloud positions have been assigned randomly while the cloud sizes and densities are identical to those in the hierarchical model.

In order to actually use the structure trees, it is necessary to find descriptors which are sensitive to the various aspects of the tree structure. These aspects can be classified into two categories depending on whether or not they are invariant to "rubber-sheet" distortions applied to the tree. For example, branching nodes (branching sites) remain invariant whether the tree is stretched or expanded in either intensity or space—or any other transformations that preserve lineage. On the other hand, the density contrast and scale reduction encountered in going from one level to the next (which are measures of how a hierarchical system distributes its mass among its children and its levels) are not invariant to rubber-sheet distortions.

Therefore it is to be expected that branching nodes should play a key role in any tree descriptor designed to be sensitive to features such as the number of levels and the degree of fragmentation present in any hierarchical structure. For example, the average number of branching sites encountered in following a path from tip down to the root, and

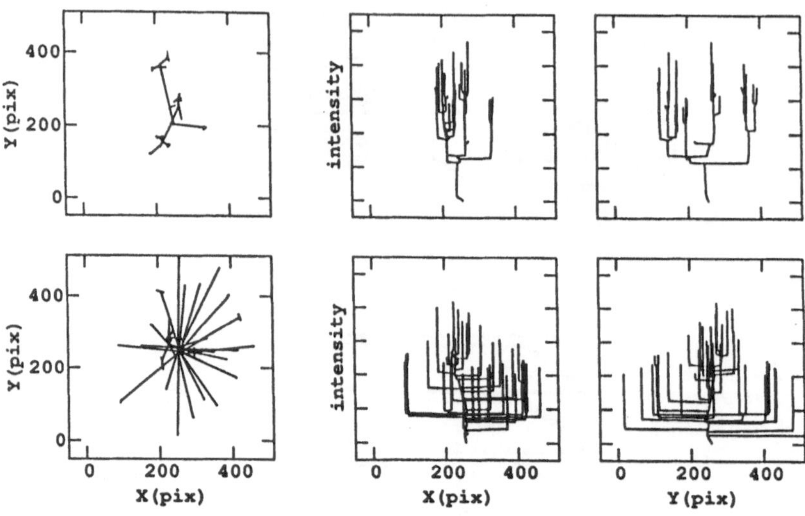

Figure 6. Top and two side views of structure trees constructed for projections of a 3-dimensional simulation of hierarchically nested clouds (top) and its randomized counterpart (bottom).

the progeny ratio, which is the ratio of progenies for successive branching nodes (a node's progeny is the total number of branching nodes on the branch between it and the tree tips), were descriptors designed to respond to invariants like fragmentation and lineage. Examples of quantities that were to measure non-invariants like the scale reduction and density contrast were the average density contrast and separation between the branches and their parent branching nodes. Descriptors of irregularity, like the dispersion in children separations at each branching node, were also used.

There is of course no guarantee that all of the descriptors from the invariant class are independent of those that are non-invariant, or that an individual descriptor may be able to itself to say whether an image is generally hierarchical or random. For these reasons, the descriptors were applied to an ensemble of trees obtained for 120 hierarchical and randomized projected simulated images with a variety of assigned parameters like image type (hierarchical/random), the total number of levels, the scale reduction factor per level, and the branching factor, and regressed against the known values of the underlying parameters. The resulting linear combinations for each parameter were found to be able to estimate that parameter's value for any of the individual models in the ensemble. The success with the simulated structures led us to attempt an application to the Taurus 100 μm column density map. There is no assurance, of course, that the Taurus structure is reasonably close enough to one of the types of simulated structures which were constructed assuming spherical independent clouds so that the derived parameters are meaningful. For this reason we feel that the major importance of the structure trees will be comparisons of observations with future numerical hydrodynamic calculations and quantitatively comparing the structures of regions with different levels of star formation activity.

Figure 7 shows two versions of the structure tree for the Taurus column density map; each version shows two side views and the top view of the tree. The 100μm optical depths cover the range 1 x 10^{-5} to 2 x 10^{-3}. The top tree was constructed using the "raw" column density map, and is extremely crowded, especially at small optical depths. The density of branches by itself poses no problem (except for the eye trying to study the tree); the problem is that most of these branches are due to very small column density fluctuations, on the order of 1 to 10 percent of their immediate surroundings. These fluctuations are not due to noise in the IRAS intensities, since most of them have appreciable areas (>25 armin^2). They form a sea of small bumps and ripples all over the column density surface, being especially plentiful on the large areas with low average column densities. Whether they represent fluctuations in the Taurus complex or in the background (which was assumed to be smooth), such a population of "clouds" was not incorporated in the simulated structures which were used to derive the linear combination

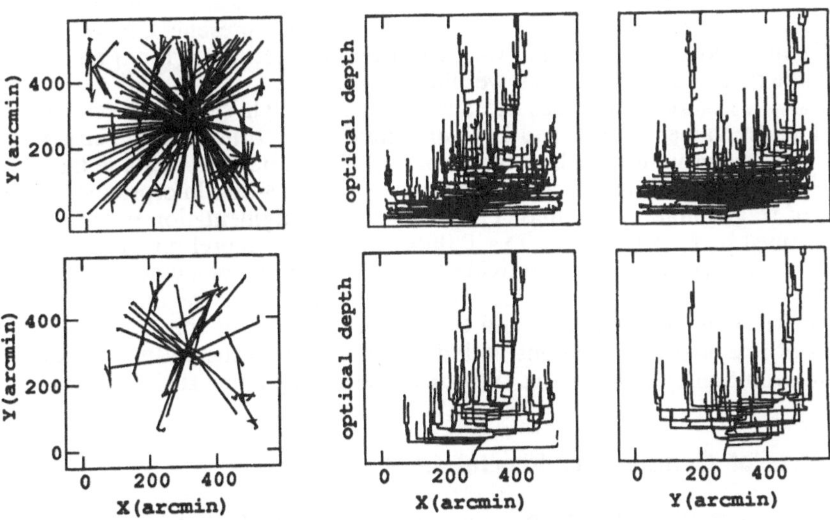

Figure 7. Top and two side views of structure tree constructed from the Taurus 100μm column density image. Top row shows the tree for the complete image, bottom rows shows the "pruned" tree obtained by rejecting column density fluctuations less than about 10 percent.

of tree statistics that we are using to estimate parameters for Taurus, and so they were removed from the tree by rejecting all branches with heights less than 1×10^{-4} in τ_{100}; the corresponding rejected fractional fluctuations were mostly less than 5 percent, although the actual fractional rejection criterion depends on column density.

The "pruned" structure tree is shown below the raw tree; the removal of the "sea of grass" reveals the underlying structure. The one-to-one correspondence with features in the original image (Fig. 3) can be seen. (A useful place to begin is to notice that the tallest tree tips in both projections of the tree are clumps within the L1495 cloud.) Some hierarchical structure can be seen even in these projections.

What do the tree statistics tell us about the prevalence of hierarchical or random spatial structure, and what parameters describe the hierarchical component in Taurus? The preliminary results are as follows. The hierarchical indicator has a value intermediate between the hierarchical and randomized simulations, suggesting that Taurus has a column density structure which is a mixture of both components, or that Taurus cannot be represented by the systems of nested or random clouds which comprised the simulations. Assuming that the former is true, the parameters of the hierarchical component were estimated using the linear regressions of statistics which reliably estimated the parameters of the simulations. This yielded an average branching ratio, or number of children per parent, denoted η, of about 7, an average change in scale or "shrinkage factor," per level of hierarchy, denoted θ, of about 0.09, and an average column density contrast D between child and parent clouds of about 1.7. The average number of levels of hierarchical structure per path through the tree (from root to every terminal tree tip) was 2.8; this estimate does not include the "root," and refers to structure with linear scale between about 5-7 pc (sizes of largest clouds above the root) and about 0.2 pc (size of smallest parents). A physically significant quantity is the average number of hierarchical levels per unit of spatial dynamical range, which is $2.8/(6/0.2) \approx 0.1$, which is an independent and consistent estimate of the shrinkage factor θ. If Taurus represents a mixture of lthe two types of simulated structures, then we expect that the estimates for η and D are upper limits, while those for θ and L are lower limits.

As an illustration of how this tree analysis can yield physically interesting quantities, I will adopt the values of η, θ and D given above and assume that the structure is self-similar. The mean volume filling factor of children in parent clouds is then $\varepsilon = \eta\theta^3 \approx 5.1 \times 10^{-3}$. The mean volume density contrast is $\delta = D/\theta \lesssim 20$. Interpreting the hierarchy in terms of fragmentation, the mass efficiency of the fragmentation process per level of hierarchy must be $f = \eta\theta^3\delta = \eta\theta^2 D \approx 0.096$. While only a preliminary result, it should be clear that this type of estimate provides a direct constraint on numerical hydrodynamic models for cloud fragmentation (values of η, θ, δ for numerical fragmentation calculations are summarized in Scalo 1985), and that an application of this approach, including measures of irregularity, to regions with different levels of star formation could provide important evolutionary information. We are currently beginning such a study.

6. Aspects of Complexity: Fractal Contours

A striking feature of contour maps of interstellar structures observed with large spatial dynamic range is the irregularity and convolution of the contours at all scales. An obvious question is whether this irregularity is self-similar, or "fractal."

The fractal dimension of a collection of two-dimensional objects can be determined by plotting log (perimeter) as a function of log (area). (Note that the fractal dimension here refers to the irregularity of the column density contours, not the hierarchical internal structure, as explained in Scalo 1988; the fractal dimension of the hierarchical structure is not very informative.) For planar objects with smooth shapes, $P \propto A^{1/2}$. But for objects with irregularity at all scales, $P \propto A^{D/2}$, where D, the fractal dimension ($1 \le D \le 2$), characterizes the irregularity of the boundary. As the boundaries get more and more complex, they eventually fill the plane, and $P \propto A$, while $D \rightarrow 2$. As part of its compression of an image, the structure tree automatically computes and stores the perimeter and area of every cloud (set of connected pixels) at every grey level, where the perimeter is the number of non-cloud pixels bordering a cloud. The Taurus complex is of course a 3-dimensional object, so we are only estimating the dimension of its 2-dimensional projection. It is known that 2-dimensional slices of a 3-dimensional fractal with $D>2$ are fractals with fractal dimension D-1. However little is known about the relation for projected fractals, although it is sometimes assumed that the same relation holds (a discussion is given in Sreenivasan and Meneveau 1986), and we will tentatively do the same.

Figure 8 shows the logP-logA relation for all the clouds in the IRAS τ_{100} map. The cloud properties were defined as averages of values at the two branching nodes defining the lowest and highest column density in a given cloud; plots using different choices (e.g. values at branching nodes) gave the same result. Note that clouds of all column densities, including regions which, if mapped in isolation or with a more limited range of sensitivity might be called "cirrus", "dark clouds," and "dark cloud cores," are included in Fig. 8. The presence of the "ripples" with very small density contrasts causes the increased scatter at the smallest sizes. The relation is fit well by a single power law of slope 0.7, corresponding to a 2-dimensional fractal dimension of 1.4, over a factor of about 50 in size (taken as $A^{1/2}$). (The turnover at large sizes is due to contours which extend outside the image boundaries, so the areas are systematically underestimated.) A similar result was found independently by Bazell and Desert (1988) who obtained a fractal dimension of

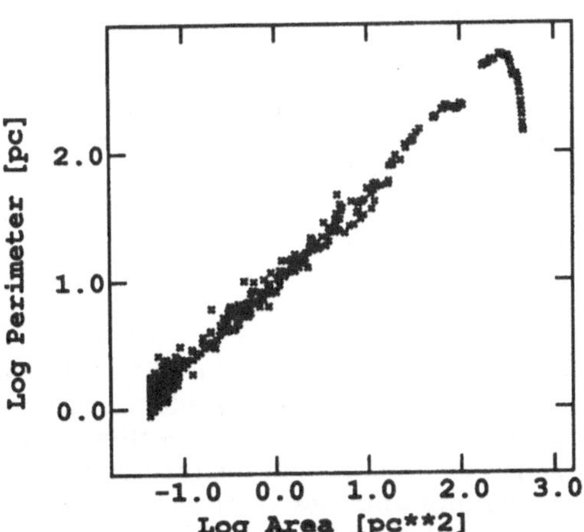

Figure 8. Log (perimeter) vs. log (area) for all sets of connected pixels ("clouds") at all column densities for the Taurus 100μm column density image.

1.26 for the 100 μm intensity contours for three regions of infrared cirrus (low column density) over a factor of about 30 in size, although this was only for a narrow range of intensities (2-8 MJy/sr) compared to the Taurus 100 μm image intensities (1-100 MJy/sr after background subtraction). Beech (1987) determined the fractal dimension for 24 Lynds clouds of opacity class 5 and 6, many of which are in the Taurus region, from cloud outlines traced from photographic plates. Beech found D = 1.4 over a factor of about 10 in size. Robbins (1987, unpublished) estimated a fractal dimension of 1.46 for the ^{12}CO contours in the map of the Orion-Monoceros complex presented by Maddalena *et al.* (1986) over a factor of 10 in size. Falgarone and Phillips (1989, quoted in Falgarone 1990) combined the large scale ^{12}CO ($J = 1 - 0$) map of Taurus with high-resolution ^{12}CO ($J = 2 - 1$) and ($J = 3 - 2$) observations they made in a small subregion and found a projected dimension of 1.34 over a factor of nearly 10^3 in scale (again taken as $A^{1/2}$). Wokker (quoted in Falgarone 1990) obtained D = 1.4 for contours of HI 21 cm emission.

The result that $D \approx 1.4$ over such a large range of size and column density has some interesting and important implications for both theoretical and empirical studies of cloud evolution and star formation.

1. As pointed out by Falgarone (1990), the similarity of the dimensions measured over such a large range of column densities indicates that there is no fundamental difference between the dominant physics controlling the shapes of molecular and atomic clouds. This suggests that the separation of HI and H_2 clouds into distinct conceptual categories is largely artificial, and illustrates how theoretical speculations may have mistakenly generalized an operational distinction based on detection method to a distinction in spatial distribution, origin, and evolution. Another indication that the common theoretical separation of these two types of clouds is partly an artifact comes from comparisons of the spatial distributions of HI and ^{13}CO, which suggest that they trace out the same kpc-scale structures and that "giant molecular clouds" are the denser condensations within HI "superclouds" (Elmegreen and Elmegreen 1987 and references therein, Grabelsky *et al.* 1987, Jacq *et al.* 1988). Of course there *are* physical differences, such as relative importance of self-gravity and the response of sub-Jeans mass clouds to mild disturbances investigated by Tohline *et al.* (1987); the similar fractal dimensions suggest that the dominant physics controlling cloud shapes is not related to these properties.

2. The dimension $D \approx 1.4$ is the same as that found for the turbulent-nonturbulent interface of incompressible fluids (using 2-dimensional slices or one-dimensional cuts), essentially independent of the type of flow (e.g. boundary layer, jet, wake), as shown by Sreenivasan and Meneveau (1986). This is also very close to the fractal dimension of the boundaries of terrestrial clouds and rain areas found by Lovejoy (1982) and of hail clouds by Rys and Waldvogel (1986), covering nearly four orders of magnitude in size. Thus, while the result does not yet provide any constraint on the specific types of physical processes at work, it does at least suggest a possible connection with turbulence. The connection must be generic, since interstellar cloud "turbulence" is expected to be considerably different than ordinary incompressible turbulence (compressibility, shock dissipation, presence of magnetic fields, presence of stellar energy sources). The major feature in common would seem to be the nonlinear advection term in the momentum equation, whose power to generate, amplify and distort fluctuations is ignored in all discussions based on linear stability analyses or the virial theorem, and is largely suppressed by existing numerical simulations because of lack of spatial dynamic range.

3. Observational estimates of cloud masses and densities invariably assume a simple geometric form. The most common procedure is to measure the average column density N in some region within which a cloud is detected by some method (e.g. extinction, ^{13}CO) and then estimate the density by $\rho \propto N/r$ (r = size of region) and the mass by m $\propto$

Nr^2. This estimate is inappropriate for a fractal for which the mass scales with region size as $m \propto r^{D_3}$, where D_3 is the dimension of the three-dimensional fractal.

4. Fractal contour structure also raises several theoretical questions. For example, what does it mean for some of these irregular structures to appear approximately "virialized?" Is it reasonable to imagine any sort of quasi-hydrostatic equilibrium for such clouds? Shouldn't such irregularity imply a highly disordered magnetic field? Two possible resolutions for the latter question were suggested in sec. 4 above.

7. Conclusion: Deconstructing the Cloud Model

The orthodox view of the interstellar medium as consisting of two basic components—independent regions of high density, called "clouds," which occasionally interact through collisions, and a lower-density smooth intercloud medium—traces back at least as far as Ambartsumian's (1944) and Chandrasekhar and Munch's (1952) statistical studies of Milky Way brightness fluctuations. This picture was reinforced for a new generation of astronomers by Clark's (1965) "raisin-pudding" model based on his HI observations and Scheffler's (1967) estimate of mean cloud properties based on reddening data. Most influential were the tabulations of mean cloud properties in Spitzer's review (1968a) and book (1968b). Theoretical models for the interstellar medium concentrated on accounting for the properties of "standard clouds" or, in a few papers, on the mass spectrum of clouds. Virtually no attention was paid to the spatial distribution or form of the interstellar "clouds" even though the predominance of irregularity and continuity could be seen in many published studies, such as the detailed illustration of the dark cloud distribution over much of the sky given by Lynds (1962) and the densely sampled HI emission maps presented by Verschuur (1974a,b). Minkowski (1955) warned that "... it becomes at once obvious that the schematic picture of separate clouds of 10 parsecs diameter has little resemblance to reality. An entirely chaotic mass of dark clouds of all possible shapes and sizes appears projected on the background of stars and faint emission nebulosity. ... Huge complexes of clouds covering many square degrees are seen broken up into numerous irregular details. ..." Chandrasekhar and Munch (1952) had emphasized that the amplitudes of density fluctuations were small and that the spatial distribution appeared quite continuous, although they still proceeded to derive the average properties that a random ensemble of independent, identical spherical clouds would require to match the angular correlation function. Donn (1958) argued that the distribution of matter is much more continuous and coupled than the cloud model would suggest, and a cautious critical discussion of the cloud model was given by Heiles (1974).

Despite these doubts, and the contrary evidence from photographic plates and densely sampled maps, the cloud model continued to thrive, no doubt in part because it accommodated simple visual imagery and analytical calculations. Perhaps not surprisingly, the flood of data from molecular line studies beginning in the 1970s only led to the introduction of more cloud "types" rather than any substantial revision or abandonment of the cloud model, which had become a "collective representation" (Barfield 1965) whose convenience made it impervious to displacement. More recent suggestions, motivated by numerical hydrodynamic simulations, that the concept of well-defined and separate "clouds" which occasionally interact may be fundamentally flawed (Lattanzio *et al.* 1985, Hunter *et al.* 1986) have gone essentially unnoticed. As the interpretation and even design of observations come to contain projections of the models, we face the danger that, in the context of Myers' (1987) cartoon of astronomer types studying the process of falling leaves, we see trees only as the mental toy in Myers' Figure 4. The situation described by Heiles (1974, p. 14) remains true today: "Authors, when carefully describing the observational situation in review articles such as this one, have generally been careful to point out many of these uncertainties [in the cloud model]. But many other authors, both theoretical and observational, go right ahead and ignore them. After enough repetition, the standard assumptions have come to be regarded as being observational fact."

In the present paper I have suggested that spatial complexity, in the form of irregularity, connectedness, lack of preferred scales, and nested forms, is not mere detail, but a ubiquitous and essential feature of interstellar structure, and that most current theoretical models are based on a categorization of "clouds" which completely ignores this spatial information. The conept of distinct and separate "clouds" finds little counterpart in densely-sampled column density images of cool interstellar gas. Structures which give the appearance of well-defined clouds at a given resolution simply "dissolve" into connected complexity at higher resolution and column density dynamic range.

It is important to realize that the only other complex fluid system with which we have any experience is incompressible turbulence, whose intractability in terms of analytical conceptual models is well-known. In this field the details of the spatial velocity field, such as fractal structure, high-order structure functions and long-lived coherent structures, have played a crucial role in understanding the dynamics. Progress in this area has only come with the advent of numerical simulations with large spatial dynamic range (e.g. McWilliams 1984, Brachet *et al.* 1986, Benzi *et al.* 1987, Santangelo *et al.* 1989), and we can expect a similar situation in the study of interstellar structure and star formation. For example, the recent compressible turbulence simulations of Passot *et al.* (1988) demonstrate the ability of the nonlinear advection terms to generate complex filamentary structures, even in the absence of self-gravity and magnetic fields. As mentioned earlier, all existing linear stability analyses, virial theorem arguments, and quasi-equilibirum theories ignore or suppress the potentially dominating effects of this nonlinearity. It may be that facing nonlinear complexity requires abandoning our almost total reliance on conceptual, analytical approaches in favor of high-resolution simulations. For most astronomers this will be a hard pill to swallow.

I want to thank Padraig Houlahan for his collaboration in much of the work presented here, to Mark Heyer, Bob McMillan, and J.-C. Hsu for providing data in convenient form, to Neal Evans for his encouragement and open-minded attitude, and to Betty Friedrich for her patient preparation of several drafts of this paper. This work was supported by grants JPL 958003, NSF AST8619738, and a grant from Texas Advanced Research Projects.

References

Adler, D. S., Lo, K. Y., and Allen, R. J. 1989, paper presented at the Second Wyoming Conference on *The Interstellar Medium in External Galaxies*, Grnd Teton, July 3-7, 1989.

Allen, R. J. and Goss, W. M. 1977, *Astr. Ap. Suppl.*, **36**, 135.

Ambartsumian, V. A. 1944, *Dokl. Akad. Nauk. SSSR*, **44**, 244.

Ball, R. 1986, *Ap. J.*, **307**, 453.

Bally, J., Langer, W. D., Stark, A. A., and Wilson, R. W. 1987, *Ap. J. (Letters)*, **312**, L45.

Bally, J., Stark, A. A., Wilson, R. W., and Langer, W. D. 1990, in *IAU Symposium No. 120*, ed. G. Tenorio-Tagle, in press.

Barfield, O. 1965, *Saving the Appearances* (Middletown: Wesleyan Univ. Press).

Bash, F. N., Green, E., and Peters, W. L., III 1977, *Ap. J.*, **217**, 464.
Bazell, D. and Désert, F. X. 1988, *Ap. J.*, **333**, 353.
Beech, M. 1987, *Ap. Sp. Sci.*, **133**, 193.
Benzi, R., Patarnello, S., and Santangelo, P. 1987, *Europhys. Lett.*, **3**, 811.
Bigliari, H. and Diamond, P. H. 1988, *Phys. Rev. Lett.*, **61**, 1716.
Blitz, L. 1988, in *Millimeter and Submillimetre Astronomy*, ed. R. D. Wolstencroft and W. B. Burton, in press.
Bohm, D. 1980, *Wholeness and the Implicate Order* (London: Ark Paperbacks).
Bosma, A., Goss, W. M., and Allen, R. J. 1981, *Astr. Ap.* **93**, 106.
Brachet, M. E., Meneguzzi, M., and Sulem, P. L. 1986, *Phys. Rev. Lett.*, **57**, 683.
Brinks, E. and Bajaja, E. 1986, *Astr. Ap.*, **169**, 14.
Carr, J. S. 1987, *Ap. J.*, **323**, 170.
Casoli, F., Combes, F., and Gerin, M. 1984, *Astr. Ap.*, **133**, 99.
Cernicharo, J. and Bachillar, R. 1984, *Astr. Ap. Suppl.*, **58**, 327.
Cernicharo, J. and Guelin, M. 1987, *Astr. Ap.*, **176**, 299.
Cernicharo, J., Bachiller, R., and Duvert, G. 1985, *Astr. Ap.*, **149**, 273.
Chandrasekhar, S. and Munch, G. 1952, *Ap. J.*, **115**, 103.
Chièze, J. P. 1987, *Astr. Ap.*, **171**, 225.
Clark, B. G. 1965, *Ap. J.*, **142**, 1398.
Crutcher, R. M. 1988, in *Molecular Clouds in the Milky Way and External Galaxies*, ed. R. L. Dickman, R. L. Snell, and J. S. Young (New York: Springer-Verlag), p. 105.
Dame, T. M., Elmegreen, B. G., Cohen, R. S., and Thaddeus, P. 1986, *Ap. J.*, **305**, 892.
de la Reza, R., Torres, C.A.O., Quast, G., Castillo, B. V., and Vieira, G. L. 1989, *Ap. J. (Letters)*, **343**, L61.
Deiss, B. M. and Kegel, W. H. 1986, *Astr. Ap.*, **161**, 23.
Deuel, E. R. and Burton, W. B. 1989, preprint.
Dickman, R. L. 1985, in *Protostars and Planets II*, ed. D. C. Black and M. S. Mathews (Tucson: Univ. of Arizona Press), p. 150.
Donn, B. 1958, in *Proc. Third Symposium on Cosmical Gas Dynamics*, *Rev. Mod. Phys.*, **30**, 940.
Duvert, G., Cernicharo, J., and Baudry, A. 1986, *Astr. Ap.*, **164**, 349.
Elmegreen, B. G. 1989, *Ap. J.*, **338**, 178.
Elmegreen, B. G. and Elmegreen, D. M. 1987, *Ap. J.*, **320**, 182.
Emerson, J. 1987, in *Star Forming Regions*, ed. M. Peimbert and J. Jugaku (Dordrecht: Reidel), p. 19.
Falgarone, E. 1990, in *Structure and Dynamics of the Interstellar Medium*, IAU Colloquium No. 120, ed. A. Webster, in press.
Falgarone, E. and Perault, M. 1987, in *Physical Processes in Interstellar Clouds*, ed. G. E. Morfill and M. Scholer (Dordrecht: Reidel), p. 39.
_______. 1988, *Astr. Ap.*, **205**, L1.
Falgarone, E. and Phillips, T. G. 1989, in *Submillimetre and Millimeter Wave Astronomy*, ed. R. D. Wolstencroft and W. B. Burton, in press.
Falgarone, E. and Puget, J. L. 1986, *Astr. Ap.*, **162**, 235.
Fleck, R. C. 1988, *Ap. J.*, **328**, 299.
Grabelsky, D. A., Cohen, R. S., Bronfman, L., Thaddeus, P., and May, J. 1987, *Ap. J.*, **315**, 122.
Guelin, M. and Cernicharo, J. 1988, in *Molecular Clouds in the Milky Way and External Galaxies*, ed. R. L. Dickman, R. L. Snell, and J. S. Young (New York: Springer-Verlag), p. 81.

Heiles, C. 1974, in *IAU Symposium No. 60, Galactic Radio Astronomy*, ed. F. J. Kerr and S. C. Simonson III (Dordrecht: Reidel), p. 13.
Heiles, C. 1987a, in *Physical Processes in Interstellar Clouds*, ed. G. E. Morfill and M. Scholer (Dordrecht: Rediel), p. 429.
Heiles, C. 1987b, *Ap. J.*, **315**, 555.
Henriksen, R. N. and Turner, B. E. 1985, *Ap. J.*, **287**, 200.
Herbig, G. H. 1977, *Ap. J.*, **214**, 747.
Herbig, G. H. and Rao, N. K. 1972, *Ap. J.*, **174**, 401.
Heyer, M. H., Snell, R. L., Goldsmith, P. F., and Meyers, P. C. 1987, *Ap. J.*, **321**, 370.
Heyer, M. H., Vrba, F. J., Snell, R. L., Schloerb, F. P., Strom, S. E., Goldsmith, R. F., and Strom, K. M. 1987, *Ap. J.*, **312**, 855.
Houlahan, P. and Scalo, J. M. 1990, *Ap. J. Suppl.*, in press.
Hsu, J.-C. 1984, unpublished Ph.D. dissertation, University of Texas at Austin.
Hunter, J. H., Sandford, M. T., Whitaker, R. W., and Klein, R. I. 1986, *Ap. J.*, **305**, 309.
Jacq, T., Baudry, A. and Walmsley, C. M. 1988, *Astr. Ap.*, **207**, 145.
Jarrett, T. H., Dickman, R. L., and Herbst, W. 1989, *Ap. J.*, in press.
Jones, B. F. and Herbig, G. H. 1979, *Astr. J.*, **84**, 1872.
Kleiner, S. C. and Dickman, R. L. 1984, *Ap. J.*, **286**, 255.
______. 1985, *Ap. J.*, **295**, 466.
Lada, C. J., Margulis, M., Sofue, Y., Nakai, N., and Handa, T. 1988, *Ap. J.*, **328**, 143.
Langer, W. D., Wilson, R. W., Goldsmith, P. F., and Beichman, C. A. 1989, *Ap. J.*, **337**, 739.
Larson, R. B. 1981, *M.N.R.A.S.*, **194**, 809.
Lattanzio, J. C., Monaghan, J. J., Pongracie, H., and Schwarz, M. P. 1985, *M.N.R.A.S.*, **215**, 125.
Leisewitz, D. 1989, *Ap. J.*, submitted.
Leisewitz, D., Bash, F. N., and Thaddeus, P. 1989, *Ap. J. Suppl.*, **70**, 731.
Leung, C. M., Kutner, M. L., and Mead, K. N. 1982, *Ap. J.*, **262**, 583.
Lindgren, K. and Nordahl, M. G. 1988, *Complex Systems*, **2**, 409.
Lo, K. Y., Ball, R., Masson, C. R., Phillips, T. G., Scott, S., and Woody, D. P. 1987, *Ap. J. (Letters)*, **317**, L63.
Loren, R. B. 1989a, *Ap. J.*, in press.
______. 1989a, *Ap. J.*, in press.
Lucke, P. B. 1978, *Astr. Ap.*, **64**, 367.
Maddalena, R. J., Morris, M, Johnston, K. J., and Henkel, C. 1986, *Ap. J.*, **303**, 375.
Maloney, P. 1988, *Ap. J.*, **334**, 761.
Martin-Pintado, J., Wilson, T. L., Johnston, K. J., and Henkel, C. 1985, *Ap. J.*, **299**, 386.
Massi, M., Churchwell, E., and Felli, M. 1988, *Astr. Ap.*, **194**, 116.
Mathewson, D. S. 1968, *Ap. J. (Letters)*, **153**, L43.
McGee, R. X. and Murray, J. D. 1961, *Aust. J. Phys.*, **14**, 260.
McKee, C. F. 1989, *Ap. J.*, **345**, 782.
McWilliams, J. C. 1984, *J. Fluid Mech.*, **146**, 21.
Mezger, P. G., Chini, R., Kreysa, E., Wink, J. E., and Salter, C. J. 1988, *Astr. Ap.*, **191**, 44.
Minkowski, R. 1955, in *Gas Dynamics of Cosmic Clouds*, IAU Symp. No. 2 (Amsterdam: North Holland Publishing Company), p. 6.

Moneti, A., Pipher, J. L., Helfer, H. L., McMillan, R. S., and Perry, M. L. 1984, *Ap. J.*, **282**, 508.

Murphy, P. C. and Myers, P. C. 1985, *Ap. J.*, **298**, 818.

Myers, P. C. 1983, *Ap. J.*, **270**, 105.

_______. 1988, in *Galactic and Extragalactic Star Formation*, ed. R. Pudritz and M. Fich, p. 331.

Myers, P. C. 1990, in *Molecular Astrophysics*, ed. T. W. Hartquist (Cambridge Univ. Press), in press.

Myers, P. C. and Goodman, A. A. 1988, *Ap. J.*, **329**, 392.

Myers, P. C., Dame, T. M., Thaddeus, P., Cohen, R. S., Silverberg, R. F., Dwek, E., and Hauser, M. G. 1986, *Ap. J.*, **301**, 398.

Nercessian, E., Castets, A., Cernicharo, J., and Benayoun, J. J. 1988, *Astr. Ap.*, **189**, 207.

Oort, J. 1974, in *IAU Symposium 60, Galactic Radio Astronomy*, ed. F. J. Kerr and S. C. Simonson III (Dordrecht: Reidel), p. 43.

Passot, T., Pouquet, A., and Woodward, P. 1988, *Astr. Ap.*, **197**, 228.

Peckham, M. 1967, *Man's Rage for Chaos: Biology, Behavior, and the Arts* (New York: Schocken Books).

Perry, C. L. and Johnston, L. 1982, *Ap. J. Suppl.*, **50**, 451.

Rossano, G. S. 1978, *Astr. J.*, **83**, 241.

Rots, A. H. and Shane, W. W. 1975, *Astr. Ap.*, **45**, 25.

Rozyczka, M. Tscharnuter, W. M., and Yorke, H. W. 1980, *Astr. Ap.*, **81**, 347.

Rydgren, A. E., Schmelz, J. T., Zak, D. S., and Vrba, F. J. 1984, Publ. U. S. Naval Obs., Vol. XXV, Part I.

Santangelo, P., Benzi, R., and Legras, B. 1989, *Phys. Fluids A.*, **1**, 1027.

Scalo, J. M. 1985, in *Protostars and Planets II*, ed. D. C. Black and M. S. Mathews (Tucson: Univ. of Arizona Press), p. 201.

_______. 1988, in *Molecular Clouds in the Milky Way and External Galaxies*, ed. R. L. Dickman, R. L. Snell, and J. S. Young (New York: Springer-Verlag), p. 201.

Scheffler, H. 1967, *Zeit. f. Ap.*, **65**, 60.

Serafine, M. L. 1988, *Music as Cognition* (New York: Columbia University Press).

Silk, J. 1985, *Ap. J. (Letters)*, **292**, L71.

Solomon, P. M., Rivolo, A. R., Barrett, J., and Yahil, A. 1987, *Ap. J.*, **319**, 730.

Spitzer, L. 1968, *Physical Processes in the Interstellar Medium*

Spitzer, L. 1968a, in *Stars and Stellar Systems VII: Nebulae and Interstellar Matter*, ed. B. M. Middlehurst and L. H. Aller (Chicago: Univ. Chicago Press), p. 1.

______. 1968b, *Diffuse Matter in Space*, Interscience Publishers.

Sreenivasan, K. R. and Meneveau, C. 1986, *J. Fluid Mech.*, **173**, 386.

Stacy, J. G., Myers, P. C., and de Vries, H. W. 1989, in *The Physics and Chemistry of Interstellar Molecular Clouds*, ed. G. Winnewisser and J. T. Armstrong (New York: Springer-Verlag), in press.

Tohline, J. E., Bodenheimer, P. H., and Christodoulou, D. M. 1987, *Ap. J.*, **322**, 787.

Tomisaka, K., Ikeuchi, S., and Nakamura, T. 1988, *Ap. J.*, **335**, 239.

Ungerechts, H. and Thaddeus, P. 1987, *Ap. J. Suppl.* **63**, 645.

Unwin, S. C. 1980a, *M.N.R.A.S.*, **190**, 551.

______. 1980b, *M.N.R.A.S.*, **192**, 253.

van Albada, G. D. 1980, *Astr. Ap.*, **90**, 123.

Verschuur, G. L. 1974a, *Ap. J. Suppl.*, **27**, 65.

_______. 1974b, *Ap. J. Suppl.*, **27**, 283.

Vogel, S. N., Boulanger, F., and Ball, R. 1987, *Ap. J. (Letters)*, **321**, L145.

Vrba, F. J., Strom, S. E., and Strom, K. M. 1988, *Astr. J.*, **96**, 680.

Weiland, J. L., Blitz, L. Dwek, E., Hauser, M. G., Magnani, L., and Rickard, L. J. 1986, *Ap. J. (Letters)*, **306**, L101.
Wouterloot, J.G.A. and Habing, H. J. 1985, *Astr. Ap. Suppl.*, **60**, 43.
Wright, M.C.H., Warner, P. J., and Baldwin, J. E. 1972, *M.N.R.A.S.*, **155**, 337.
Zhou, S., Wu, Y., Evans, N. J., Fuller, G. A., and Myers, P. C. 1989, *Ap. J.*, **346**, 168.

Krügel: To derive column densities for a cold cloud from IRAS seems problematic. You may miss a lot of cold matter. τ_{100} will also be very sensitive to the color temperature between 60 and 100 μm for a cold cloud.

Answer: We were also very concerned about this when we began the study, but our comparisons of 100 μm column density structures with those derived from ^{13}CO, $C^{18}O$, and extinction, along with similar comparisons by other groups (see text), indicate that the fraction of warm-to-cold dust must be nearly constant throughout the region. The 60 μm/100μm color temperature is very uniform over the entire Taurus region, except, of course, in local spots around embedded sources. This uniformity is unexpected. Perhaps the fractal forms of the density contours makes the effective UV optical depth much smaller than would be estimated on the basis of a "smooth" cloud model.

Zinnecker: One disadvantage of using the IRAS data for your purpose is that it lacks velocity information; thus you suffer from projection effects. I should think that the structure tree approach could be applied to velocity-resolved $C^{18}O$ J = 2–1 maps (data cubes) of giant molecular clouds. Such data would be free from projection effects.

Answer: I agree that projection effects are a major problem and that channel maps can help alleviate the problem. However, even in channel maps there is always the chance that two volumes of gas at the same velocity can be at very different positions along the line of sight. More seriously, the channel maps could artificially destroy the "lineage" or "containment" relations between the structures on various scales, a property which I feel is essential. For example, "child" and "parent" clouds might not be identified as such just because they differ appreciably in radial velocity. Also, it is not so obvious how to apply the structure tree approach to velocity data, since contours of constant velocity will in general not be closed curves. Notice that the parameters we derived for Taurus from the structure trees were based on descriptors which had been trained on simulations which *were* projections of 3-D structures.

Shapiro: With regard to the genus as a measure of the structure in the clouds you have studied, do the data favor a structure which is sponge-like or meatball like? That is, are the empty regions connected, or are they surrounded on all sides by filled regions?

Answer: In my talk I discussed some preliminary results for the behavior of topological genus as a function of column density threshold, following the success of this approach in studies of extragalactic large scale structure by Gott and co-workers. However, I can't see how to overcome the effects of projection. As a rather extreme example, a three-dimensional arrangement of randomly-oriented long filaments contains only one "hole"—the space between the filaments; but a two-dimensional projection would give the appearance of a net, full of polyhedral holes. My subjective visual impression of the Taurus region, including available line data, is that it's more like a sponge than meatballs or Swiss cheese.

Larson: What do structure analyses like these really teach us about the physics of molecular clouds? Isn't there a danger of seeing just what you look for, e.g. correlations, filaments, or hierarchies?

Answer: The structure analyses themselves are meant to provide an objective and relatively compact means for characterizing the structure of star forming regions which can be compared with high-resolution numerical simulations when they become available. I expect that such simulations will predict different structures for different initial conditions and physical processes included, giving us a direct method for establishing the validity of theories. Such structure indicators have not been used previously to characterize observed structure, and models have been free to essentially ignore much of the available spatial information because there was no way to quantify it. As I try to explain in my paper, I think such spatial information is at least as fundamental a constraint as the mean properties of ill-defined cloud categories. As a specific example, I think the fact that contour shapes of both atomic and molecular clouds of all sizes are extremely irregular and self-similar and don't at all resemble what one would expect from any quasi-equilibrium model tells us a lot about the physics—that the evolution is probably much more dynamical than was thought.

As far as the danger of seeing what you look for, I would say that the danger lies in looking at images with a small number of pixels, so that you can always see the nicely regular and separate "clouds" that theorists thrive on. When you look at the structure with lots of pixels, in detail, and you see what looks like a mess, you learn that your models were far too simple or way off-base or, more likely, both. To see this, consider that your eyes provide you with about 10^9 pixels of constantly changing information. Imagine that our eyes only had a hundred or a thousand pixels, so that we could only see some "blobs," some moving and some not, a few different sizes and shapes, and so on. I'm sure we could all come up with some pretty amusing "blob theories" for our everyday experience, but think how little they would have to do with the actual phenomena. The situation with interstellar structure is not so different. In the text I referred to this as the "Mr. Magoo syndrome."

INTERSTELLAR GAS CYCLING POWERED BY STAR FORMATION

Alain LIOURE and Jean-Pierre CHIEZE
Commissariat à l'Energie Atomique
Centre d'Etudes de Bruyères-le-Châtel
Service P.T.N.
B.P. 12
91680 Bruyères-le-Châtel FRANCE

Abstract : Star formation in molecular clouds is a low efficiency process. The mean conversion efficiency of gas into stars is about $1-10\%$, but lower values are often derived. Star formation is followed by a rapid dispersion of the unprocessed material from the active region. As a result, a large mass flux from the cold molecular phase of the interstellar medium towards its more diffuse phases is produced. When stationary state is assumed, this imprints a cycle of the interstellar matter, powered by star formation and feeded by the disruption of molecular clouds. We suggest that the flow is dominated by condensation processes which in sequence bring the gas to the highest densities found in protostellar clumps : cooling processes and related instabilities drive heated gas to HI state followed by mechanical processes as collisions of HI clouds or gravitation inside individual molecular clouds. Simplifying assumptions upon the condensation rates of each process allow for the calculation of the mass distribution over these various phases. The model presented here is a dynamical version of the standard two phase model, and predicts a nonequilibrium -though stationary- continuous distribution of lukewarm gas.

I - Introduction

Although the estimated amount of mass involved in star formation is small ($3-10M_{\odot}yr^{-1}$, [25]), young stars are very effective in destroying molecular clouds. In fact, a strong interaction occurs between newly formed stars and the neighbouring interstellar medium, through sellar winds, molecular outflows and ionizing OB stars, and the gas is heated torwards high temperatures. As a result, the *efficiency*, defined as the ratio of the mass of stars formed in a cloud to the total mass of this cloud, is low, a typical value being around 5% [26]. The disruption of molecular material induces a mass flux of interstellar gas from high densities and low temperatures to a warm and more diffuse state. The mass carried by this flow can be simply calculated, given an observed star formation rate (R_{SF}) and a typical value of the efficiency (ϵ): $\phi_0 = (1-\epsilon)R_{SF}/\epsilon \sim R_{SF}/\epsilon \sim 60-200M_{\odot}yr^{-1}$. This considerable amount of low density gas is injected in the diffuse phase of the interstellar medium, and should condense back, if one assumes a stationary state. Various physical processes dominate the gas evolution through this cycle, each

R. Capuzzo-Dolcetta et al. (eds.), Physical Processes in Fragmentation and Star Formation, 179–186.

at a different stage. We suggest the following evolutionary track, which will be described with more details in the followings : ... $\rightarrow$ formation of stars $\rightarrow$ molecular cloud disruption by ionizing stars $\rightarrow$ cooling and condensation of warm gas $\rightarrow$ formation of HI clouds by thermal instability $\rightarrow$ formation of molecular clouds by coalescence of HI clouds $\rightarrow$ formation of dense cores by a limited gravitational collapse $\rightarrow$ formation of stars $\rightarrow$ ... . This cycle, which drains the constant mass flux ϕ_0, is powered by star formation which is only a small leakage from this mass flow. We thus attribute the general name of *cascade* to this model, stressing the unavoidable evolution torwards high densities and low temperatures.

The aim of the model is to quantitatively account for the mass distribution of intestellar gas throughout the whole density spectrum. We choose as a main coordinate the function $g(n_H,t) = \partial M/\partial Log n_H$, defined as the mass contained in an elementary logarithmic interval of density. Moreover, we define $\phi(n_H,t)$ as the mass flux evolving from density n_H to $n_H + dn_H$, depending generally on density and time. In this frame, we suppose that a condensation rate $\omega(n_H,t) = \frac{1}{n_H}\frac{d\,n_H}{dt}$ can be connected with each physical process. With these definitions, we have: $\phi(n_H,t) = \omega(n_H,t)\; g(n_H,t)$. The density distribution function g obeys the continuity equation:

$$\frac{\partial g}{\partial t} = -\frac{\partial \phi}{\partial Log\; n_H} = -\frac{\partial\,(\omega g)}{\partial Log\; n_H} \tag{1}$$

We will be interested only in the *stationary* solution. This means that the mass flux depends neither on time nor on density, and is thus a constant throughout the whole density spectrum. This crucial parameter can be thus attributed an intrinsic value from observations, independant of the model itself. Hereafter, we will write $\phi(n_H,t) = \phi_0 = const.$ As a consequence of these assumptions,the mass distribution is simply related to the condensation rate by the expression:

$$g(n_H) = \frac{\phi_0}{\omega(n_H)} \tag{2}$$

The problem is now to identify the relevant physical processes governing each phase and to evaluate the corresponding condensation rate.

II - Thermal Cascade

Previous works [24],[26] show that a great amount of interstellar matter can be heated and ionized by a single O star through the 'champagne phase' phenomenon. Moreover, a careful analysis of I.R.A.S. data [14] as well as theoretical arguments [15], show that a massive star can ionize the gas far away its own location. Because the thermal time scale is far shorter than the dynamical one, we argue that the condensation rate is dominated by thermal processes at this stage. Following Spitzer [23], we will assume that the evolution of the heated gas is isobaric, or, at least, that the process of recovering pressure equilibrium with the ambient medium is quick. Hence, we suppose that the heated gas enters the cycle at a temperature of about 8000K - 10000K, with a density convenient to assure global pressure balance. Kulkarni and Heiles [12] adopt a mean interstellar pressure of

4000 $K cm^{-3}$, not far from the value theoretically derived earlier by Mac Kee and Ostriker [16]: 3800 $K cm^{-3}$.

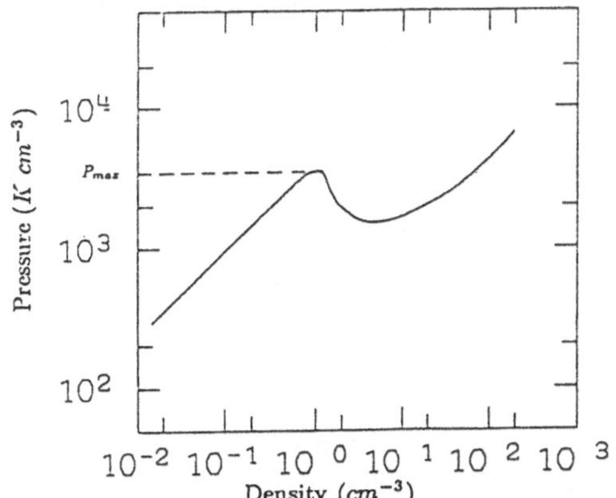

Figure 1 : Equilibrium curve for the global cooling function adopted; constant heating rate ($\Gamma = 4 10^{-26} erg\ s^{-1}$), and cooling function of Shull and Woods (1985). We have labeled as P_{max} the pressure corresponding to the low density thermal instability threshold.

The precise knowledge of the physical processes which govern the thermal evolution of the gas is still a challenge. We adopted the cooling function of Shull and Woods [21] and the heating rate recently derived by d'Hendecourt and Léger [8] due to Polycyclic Aromatic Hydrocarbons (P.A.H.). The presence of these molecules is supported by some infrared lines and this mechanism is at present the only one able to maintain an observed warm phase [12]. Here, we have supposed a constant heating rate per H atom, $\Gamma = 4\ 10^{-26} ergs\ s^{-1}\ H^{-1}$, according to these authors. Thus, the net heating rate per unit volume is equal to: $B(n_H, T) = \Gamma n_H - \Lambda(T) {n_H}^2\ ergscm^{-3}s^{-1}$. We have plotted on fig. 1 the thermal equilibrium curve of the interstellar gas obtained with these assumptions. The classical S shape curve delineates the frontier between net gas heating (lower region) and cooling (upper region).

We are interested in the isobaric evolution of thermally unbalanced gas. The gas evolution is completely described by the condensation rate (the opposite of the cooling rate) which is just proportional to the value of the heat exchange rate:

$$\omega_{th} \equiv dlog\ n_H/dt = -\frac{2}{5}\frac{B}{P_0} \tag{3}$$

where we have assumed a 5/3 adiabatic exponent. For a given interstellar pressure, the condensation rate depends only upon the density n_H. The condensation rate is displayed on fig. 2 for different pressures.

We assume that the gas always lies in the cooling region and that the mean interstellar pressure is always greater than P_{max} (see fig. 1), as it is suggested by the figures given above. In that case, the bulk of the gas heated by the young stars cools down to its HI equilibrium temperature.

However, one can easily show that the isobaric evolutionary path is partly unstable in the sense that small density perturbations will grow exponentially. A simple linear analysis shows that the perturbation develops if the quantity $\omega_F = \frac{2}{5P}(B_T T - B_{n_H} n_H)$ is positive, the subscripts meaning derivation relatively to the corresponding variable. This criterion has the same analytical expression as the classical Field one, except that it is written in a very general out of equilibrium situation (though at constant pressure). We have displayed on fig. 3 the values of ω_F, indicating the density range of instability, and on fig. 4 the evolution with time and density of the ratio of two initially slightly different densities. This can be interpreted as follows: the bulk of the mass cools down but small perturbations can be enhanced and grow into isolated cloudlets, with a characteristic size

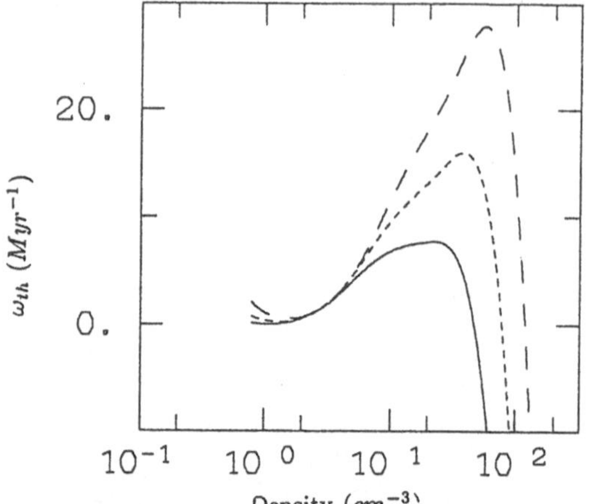

Figure 2 : The condensation rate ω_{th} calculated at constant pressure against the density. The plain curve corresponds to $\tilde{P} = 3800Kcm^{-3}$, the short dashed one to $\tilde{P} = 4500Kcm^{-3}$ and the long dashed one to $\tilde{P} = 6000Kcm^{-3}$.

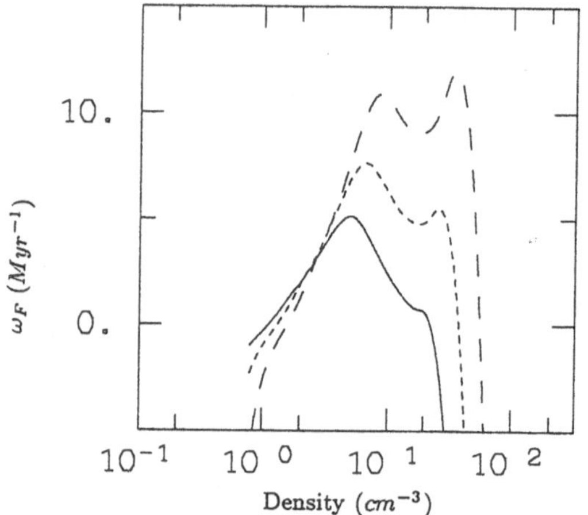

Figure 3 : The stability criterion (ω_F, see text) of the cooling phase at constant pressure. The plain curve corresponds to $\tilde{P} = 3800Kcm^{-3}$, the short dashed one to $\tilde{P} = 4500Kcm^{-3}$ and the long dashed one to $\tilde{P} = 6000Kcm^{-3}$.

related with thermal instability. Due to this cooling instability, we suggest that the gas is already fragmented in small cloudlets when it enters the thermally stable HI phase, where the continuity of the condensation flow is carried out by the collisional processes discussed in the next section.

We have examined the thermal evolution for interstellar pressures in the range $3800Kcm^{-3} \leq \tilde{P}_0 \leq 6000Kcm^{-3}$. The expected mass distribution of warm neutral gas is obtained from eq.(2). The major contribution to the total mass arises naturally from the regions where the condensation rate is low. Accordingly, most of the warm gas lies in the vicinity of the low density bump of the thermal equilibrium curve. The mass distribution we obtain is represented in fig. 5. For the reference interstellar pressure we adopt, $\tilde{P}_0 = 3800Kcm^{-3}$, a mass $M_w = 810^8 M_\odot$ of warm gas is concentrated in the density range $0.8cm^{-3} \leq n_H \leq 1.6cm^{-3}$, while a smaller fraction, $M_{lw} = 3.510^7 M_\odot$ of lukewarm gas ($T_{lw} \leq 2000K$) is distributed at higher densities. This distribution differs from the predictions of the two-phase model, in that the bulk of the warm gas is *distributed* in a finite temperature range $3000K \leq T_w \leq 4000K$, and that there exists a minor component at lower temperatures. In other words, this is a dynamical version of the Field model, in which the gas is always in a transient state. The characteristic cooling time to reach thermal equilibrium is about $t_{cool} \sim 20Myr$.

III-Collisional cascade

Since the HI gas in clouds is in thermal equilibrium, it can only evolve through mechanical interactions dominated by collisions, [13],[2],[7],[9]. Chièze and Lazareff [2] self-consistently derive a stationary *cloud mass spectrum*, taking into account thermal evaporation, erosion by supernovae remnants, fragmentation by grazing collisions and coalescence. The essential feature is that altogether

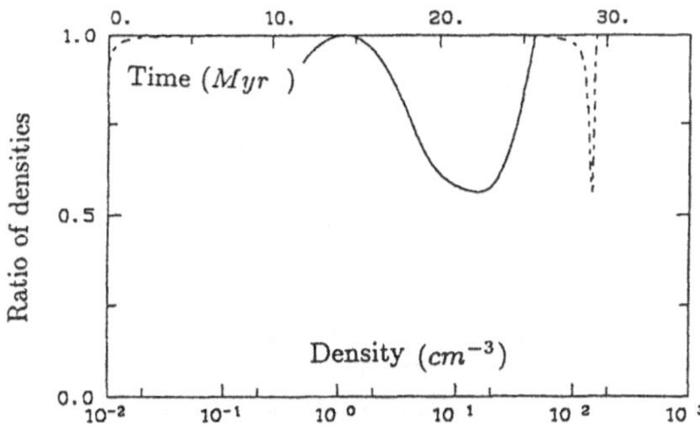

Figure 4 : Evolution with time (dashed curve) and density (plain curve) of the ratio between densities of a background medium and a small perturbation in pressure equilibrium.

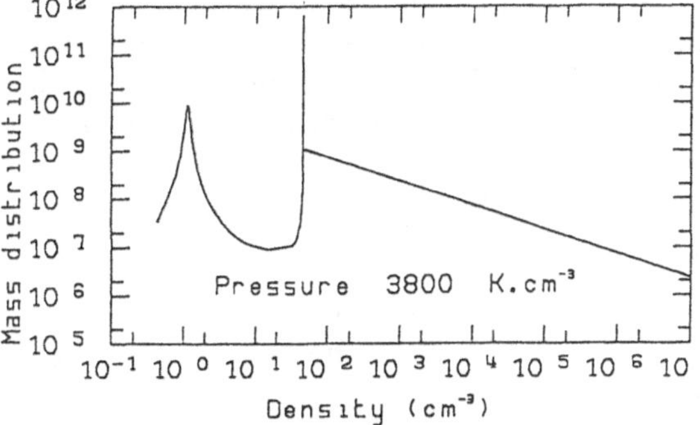

Figure 5 : The mass distribution throughout the whole spectrum of densities. The low density part addresses the warm and lukewarm phase ($8000K \geq T \geq 100K$), whereas the molecular phase is concerned with the high density part.

these processes drive a net mass flux carrying the mass distributed among low mass clouds into much massive clouds. We argue that the high mass cut-off of this cloud mass spectrum can be identified with the elementary molecular clouds which constitute the building blocks of the much more massive cloud complexes. Hydrostatic self-gravitating cloud models [5] and the normal mode analysis of their gravitational stability, including thermal processes [3] show that the limiting mass of individual clouds -given a $\tilde{P}_0 = 3800\ K\ cm^{-3}$ ambient pressure- is of the order of $100\ M_\odot$ when turbulent support is neglected. This sets up the natural cut-off of the collisional cloud mass spectrum, and the starting point of the gravitational cascade to be discussed in the next section. Furthermore, these clouds have a fully developped molecular core ($n_H \sim 10^3\ cm^{-3}$, $A_v \sim 1\ mag$) surrounded by an extended envelope, mostly atomic in the outer parts. The mean cloud density is $n_H \sim 75\ cm^{-3}$. We suggest that these clouds make the transition from HI clouds to individual molecular clouds.

With a high mass cut-off of $3.10^2\ M_\odot$, and a mean HI density $< n_{HI} >= .3 cm^{-3}$, one obtain a mass flux $\phi_{coll} = 1.3\ 10^7\ M_\odot Myr^{-1}$ in good agreement with the mass flux adopted in this paper. Moreover, the total HI mass of these colliding clouds roughly fits the observation, most of it being locked in the massive ones: $M_{HI} \sim 2\ 10^9 M_\odot$.

IV - Gravitational cascade

We focus in this section on the possible role of self-gravitation in the formation of dense condensations in molecular clouds and eventually of stars. Some features are in favour of this hypothesis, even if it is well known that the naive picture of free-falling molecular material grossly overestimates the observed star formation rate. In fact, massive giant molecular complexes are fragmented into numerous clouds of a few $100 M_\odot$, called for this reason *building blocks* of these complexes [5]. A far higher density contrast is found between a molecular cloud as a whole and its dense cores than between a giant complex as a whole and the building blocks, suggesting the presence of a very efficient compression process. Moreover, the mere existence of dense cores shows that gravity overcomes at least locally the stabilizing mechanism, for which turbulence is often addressed. However, a proper theory of turbulence is still to be made. The only available works are concerned with large scale structures [1] or dissipation by cloud collisions [20],[9] or energy injection in molecular clouds by stellar winds [22]. The few studies of clumps are still not reliable, in that different authors draw different conclusions as far as correlation lenghts are concerned: 0.1 pc [11], 0.3 pc [18] or no correlation at all [19]. One more argument is given by the observation of chemical composition of building blocks, especially as far as atomic C and CO are concerned. They are conspicuously out of equilibrium and it can be shown [4] that a dynamical mixing inside molecular clouds with a characteristis time close to the local Jeans time could fairly explain this feature. Thus, we propose that cloud material is in a state of stretched gravitational collapse, that is either the condensation rate is lowered by various effects such as turbulence, or collapse is incomplete, with a fraction z of matter returning to its initial state (see also Chièze, this volume).

Hence, we choose in the followings to scale the condensation rate with the usual local Jeans rate: $\omega_J = (4\pi G\rho)^{1/2}$, and use an *effective* rate $\omega = z\ \omega_J$. Thus, with the mass flux ϕ_0 of the cascade expressed in $M_\odot Myr^{-1}$, the mass of molecular gas in $M_\odot$ and the particle density, n_H, in cm^{-3}, the mass distribution reads,

$$\frac{dM}{dLog\ n_H} = 25\ \frac{\phi_0}{z}\frac{1}{\sqrt{n_H}} \tag{4}$$

In these expressions, the factor $0 < z \leq 1$ is a measure of the efficiency of the gravitational cascade. The total mass of molecular gas in the Galaxy is given by integrating the mass spectrum (eq.(2))

$$M(H_2) = 50\ \frac{\phi_0}{z}\frac{1}{\sqrt{n_{H_l}}}$$

where the density cutoff $n_{H_l} \sim 100\ cm^{-3}$ is the typical density of the individual clouds in which most of the molecular gas is locked. Equating ϕ_0 to the destruction/formation rate of molecular clouds due to star formation, we find with $R_{SF} = 3\ 10^6\ M_\odot Myr^{-1}$ and $r = 0.1$:

$$M(H_2) = 1.5\ 10^9\ \left(\frac{z}{0.1}\right)^{-1}\left(\frac{n_{Hl}}{100\ cm^{-3}}\right)^{-1/2} M_\odot \tag{5}$$

A further mechanism we have not addressed is the possible role of magnetic fields. We simply omit them at present, remaining aware that they could fairly well slow gravitational collapse by an order of magnitude (see Mouschovias, this volume).

V - Discussion and Conclusion

This model gives a comprehensive global vision of the interstellar medium. Many attempts of this kind have already been published: Mac Kee and Ostriker

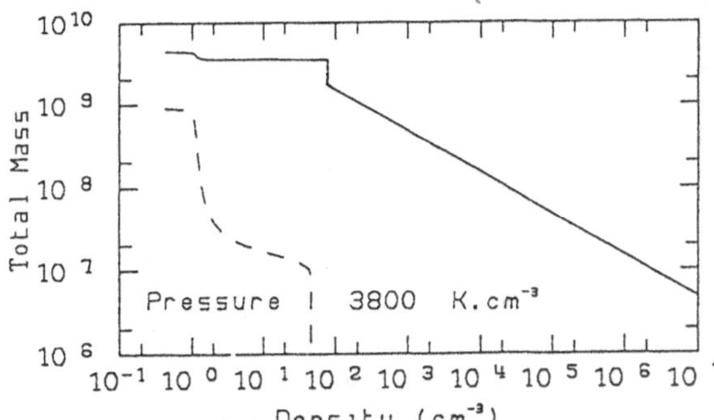

Figure 6 : The integrated mass of the Galaxy. This curve is obtained by simple integration of curve on fig. 5 over density. The pressure is the standard mean interstellar pressure : $\tilde{P} = 3800Kcm^3$, roughly corresponding to P_{max} (see fig. 1). The integration is performed from right (high densities) to left (low densities). The dashed line gives the detail of the integration for the warm phase. The discontinuity near $n_H = 100cm^{-3}$ takes into account the mass locked in HI clouds.

[16] (MO) and Ikeuchi et al. [10] (IK) to quote but the main ones. MO do not take into account the molecular phase at all and the warm phase is a serious shortcoming of their model in that its mass is probably well underestimated. IK have elaborated a global model including six different phases and complex mechanisms of mass exchange as well as those already considered by MO. IK search the stationary solution of a set of non linear differential equations which naturally leads to different regimes: a quiescent one where the various phases are assigned a precise mass, and an oscillatory one, where one can see the periodic occurence of violent phenomenon, as starbursts.

Conversely, we try to account for a *distribution* of mass among the whole density spectrum, disregarding the discontinuous description with few phases. This model has been built making as few assumptions as possible (stationary state, pressure equilibrium) and emphasizing some conspicuous observational facts (the prominent role of molecular clouds of $1 - 3pc$ and $100 - 300M_{\odot}$, the importance of gravity inside these clouds, the presence of mass at various temperatures, the power of star formation in ionizing molecular gas). The results are displayed on fig.6 for the reference pressure adopted: $\tilde{P} = 3800Kcm^{-3}$.

In the molecular phase, most of the mass is contained in the building blocks, the total mass being very close to the observed one ($\sim 210^9 M_{\odot}$) provided that the slowed down collapse is correct. A more accurate theory of turbulence within molecular clouds is needed. However, the role of magnetic fields could be important too in stretching free-fall collapse. A strong observational test, whould concern the mass distribution of clumps inside the building blocks. We predict a $-\frac{1}{2}$ slope in a $log(M) - log(n_H)$ diagram, and an obervational confirmation would be in favour of gravity as a dominant process.

As far as the warm phase is concerned, it is accounted for by the simple fact that the cooling rate is low in the temperature range 3000K - 5000K. Its mass content strongly depends on the ambient mean pressure following the law: the lower the pressure, the more mass. For the standard ($3800Kcm^{-3}$) pressure, we obtain $M_w \sim 10^9 M_{\odot}$, close to the estimated value with observational data [12]. This rises the usual question: is the warm gas in a stable or transient (cooling) state? This model favours the latter.

The main shortcoming of the model discussed here is that it is not self-regulated. The star formation rate is a parameter as well as the mean pressure of the interstellar medium. Moreover, we do not take into account the role of supernovae which have been shown to play a crucial role in the dynamics of interstellar medium flux could be calculated consistantly with the star formation rate as well as the mean interstellar pressure. Attempts of that kind have been made by Parravano ([17], see also this volume). He demonstrates that the mean interstellar pressure should be maintained close to P_{max} by some self regulating mechanism, and gives an observational test, fairly well confirmed for a large sample of galaxies of different types. From these result, we can infer that the mean interstellar pressure is close to P_{max}, depending on the cooling function adopted. We have shown previously that this case was in favour of a large mass of warm gas. As far as the star formation rate is concerned, its value is often parametrized with other better known quantities, as the total mass content. However, due to the current lack of precise knowledge about physical mechanisms for star formation, we chose to rely on observed values of the star formation rate. We have overlooked the role of supernovae remnants as sources of energy and mass flux among phases; though they will have minor influence on the mass distribution, as the total mass of hot gas is very small, they are of prominent importance to infer the respective filling factors. The question to be addressed is whether the Galaxy is filled with hot gas, as in MO, or mainly with warm gas, as proposed earlier by Field et al. [6].

References

[1] Chièze, J.P.: 1987, *Astron. Ap.* **171**, 287

[2] Chièze, J.P., Lazareff, B.: 1980, *Astron. Ap.* **91**, 290

[3] Chièze, J.P., Pineau des Forêts, G.: 1987, *Astron. Ap.* **183**, 98

[4] Chièze, J.P., Pineau des Forêts, G.: 1989, *Astron. Ap.* in press

[5] Falgarone, E., Puget, J.L.: 1985, *Astron. Ap.* **142**, 157

[6] Field, G.B., Goldsmith, D.W., Habing, H.J.: 1965, *Ap. J. (Letters)* **155**, L149

[7] Hausman, M.A.: 1982, *Ap. J.* **261**, 532

[8] d'Hendecourt, L.B., Léger, A.: 1987, *Astron. Ap.* **180**, L9

[9] Henriksen, R.N., Turner, B.E.: 1984, *Ap. J.* **287**, 200

[10] Ikeuchi, S., Asao, H., Yukata, D.T.: 1984, *M. N. R. A. S.* **207**, 909 (IK)

[11] Kleiner, C., Dickman, R.C.: 1985, *Ap. J.* **295**, 466

[12] Kulkarni, S.R., Heiles, C.: 1987, *Interstellar Processes*, edited by Hollenbach and Thronson (Reidel Publishing Company), page 87

[13] Kwan, J.: 1979, *Ap. J.* **229**, 567

[14] Leisawitz, D., Hauser,M.G.: 1988, *Ap. J.* **332**, 954

[15] Mac Kee, C.F.: 1986, *Astrophysics and Space Science* **118**, 383

[16] Mac Kee, C.F., Ostriker, J.: 1977, *Ap. J.* **218**, 148 (MO)

[17] Parravano, A.: 1988, *Astron. Ap.* **205**, 71

[18] Pérault, M., Falgarone,E., Puget, J.L.: 1986, *Astron. Ap.* **157**, 139

[19] Scalo, J.M.: 1984, *Ap. J.* **277**, 556

[20] Scalo, J.M., Pumphrey, W.A.: 1982, *Ap. J. (Letters)* **258**, L26

[21] Shull, J.M., Woods,D.T.: 1985, *Ap. J.* **288**, 50

[22] Silk, J.: 1985, *Ap. J. (Letters)* **292**, L71

[23] Spitzer, L.: 1978, *Phys. Processes in the Inters. Medium*, Wiley, New York

[24] Tenorio-Tagle, G.: 1979, *Astron. Ap.* **79**, 287

[25] Tinsley, B.M.: 1980, *Fund. of Cosmic Phys.* **5**, 287

[26] Whitworth, A.: 1979, *M. N. R. A. S.* **186**, 59

STAR COMPLEX FORMATION IN DIFFERENTIALLY ROTATING SUPERCLOUDS

IGOR G. KOLESNIK
Main Astronomical Observatory
Ukrainian Academy of Sciences
252127 Kiev
USSR

ABSTRACT. Conditions of giant molecular cloud's complexes formation in central parts of rotating superclouds are investigated. It is shown that only in the most massive superclouds having masses $M \geq 10^7 M_\odot$ the central density may become high enough for the rapid gas cooling and therefore for the cold core formation. The formulas for core parameters estimations are obtained. The forming cores have masses $(1-4) \cdot 10^6 M_\odot$ and equatorial radii about 300-500 pc that is very close to the star complexes parameters. So we can think that their origin is the result of action of proposed mechanism. In the rapidly cooling core the initially existing subsonic motions will become supersonic ones. This process can explane the nature of giant molecular cloud's supersonic turbulence having the velocities up to 6-8 $km \cdot s^{-1}$ that is typical for the warm interstellar matter from wich the superclouds are formed.

1. PARAMETERS OF STAR COMPLEXES AND SUPERCLOUDS. THE PRINCIPAL IDEAS OF THEIR FORMATION.

There are two types of connected large-scale structures in galaxes: superclouds and star complexes. Superclouds can be seen in neutral hydrogen or molecular gas distributions. They have masses of order $10^7 M_\odot$ and sizes up to 2-2.5 kpc (Elmegreen and Elmegreen (1983)). Star complexes contain stellar population with dispersion of ages up to the 10^8 years, OB associations, and giant molecular clouds. They have masses near $10^6 M_\odot$ and sizes 500-600 pc (Efremov (1989)).

Superclouds and star complexes have similar properties both in different galaxes and in different places of the same galaxy. This is possible only if their nature is determined by the common physical processes and general mechanisms.

Elmegreen (1987) has argued that gravitational instabilities in sheared magnetic galaxy disks produce the structures with supercloud parameters. As was shown by Kolesnik (1987) if in the central part of such

R. Capuzzo-Dolcetta et al. (eds.), Physical Processes in Fragmentation and Star Formation, 187–192.

structure the density exceeds some critical value–the level density n_{LD}, the gas cools and the dense cold core is formed. Then if it is surrounded by the HI envelope with column density N_{env} greater then $N_1 \simeq 5\cdot 10^{20}$ cm^{-2} the gas in the core is transformed into the molecular form (Kolesnik(1977); Arshutkin and Kolesnik(1984)).In this way the giant molecular clouds can be formed and the star foming regions can be develope.

The main idea of proposed mechanism is based on the well known fact that thermal properties of interstellar matter strongly depends on gas density. For densities lower then n_{LD} the gas temperature is approximately $T_1 \simeq 10^4$K but when the density rises to values greater then n_{LD} the temperature rapidly decreases to the level $T_0 \simeq 100$K (Kaplan and Pikel'ner (1979)). As a result the interstellar matter brakes up into the cold cloudletts which are in pressure equilibrium with the tenuous warm intercloud gas. Inside the selfgravitating supercloud when in the central regions the density exceeds the critical level n_{LD} the gas cools and contracts into the dense low temperature core. In the frame of spherical models it was shown that only inside the superclouds with masses of order $10^7 M_\odot$ the necessary conditions for the dense core formation can be obtained (Kolesnik(1987)). The core has parameters that are very close to the giant molecular cloud's ones: the masses of order (1-3)$\cdot 10^5$M and radii 30-50 pc.Therefore the conditions

$$n > n_{LD} \quad \text{and} \quad N_{env} \geq 5\cdot 10^{20} cm^{-2}$$

can be considered as necessary and sufficient ones for the supercloud's molecular core formation.

Here the case of rotating two-dimentional superclouds will be considered where the giant molecular cloud complexes formation can be expected.

2. THE ROTATING SUPERCLOUD INTERNAL STRUCTURE MODEL

Supercloud formation take place in the two phase interstellar gas. In this case practically the whole considered volume is filled by the warm gas having temperature T_1 equal about 10^4 K. So we can describe supercloud's structure using isothermal models. The value of the level density depends on the physical properties, of radiation fields, and chemical composition of interstellar matter where the supercloud is formed. For the gas of galaxy disks we can take $n_{LD} \simeq 0.7$-0.9 cm^{-3} (Kaplan and Pikel'ner (1979)).

Let us consider the structure of the supercloud having the mass M_* which is surrounded by the gas with pressure P_{ext}. The density $\rho(s,z)$ and velocity v(s) distributions inside the supercloud can be described using the analytical solution received by Schmitz (1983) for isothermal gas with differential rotation

$$\rho(s,z)=\frac{8c_1^2}{\pi G}\cdot\frac{A^2+s^2}{(A^2+2s^2+4z^2)^2}\,,\quad v(s)=c_1\sqrt{2}\,\frac{s}{\sqrt{A^2+s^2}}\,. \tag{1}$$

Here c_1 – is an isothermal sound velocity in the gas having temperature T_1 and molecular weight μ_1, A – is a free parameter that can be determined through the central density ρ_c and the total mass M_*.

Using this solution for the supercloud's modelthe next relations can be obtained
for the mass

$$M_*=8.93\cdot 10^7\ \tilde{c}_1^4\ \tilde{P}_{ext}^{-1/2} w_*\ M_\odot\ , \tag{2}$$

for the total angular momentum

$$J_*=1.07\cdot 10^{69}\ \tilde{c}_1^7/\tilde{P}_{ext}\,\delta_* j_*\ g\cdot cm^2\cdot s^{-1}, \tag{3}$$

for the central density

$$n_c=(0.22/\delta_*)\ \tilde{P}_{ext}/\ \tilde{c}_1^2\ cm^{-3}, \tag{4}$$

and for the equatorial radius

$$s_*=\ 0.7\ (\tilde{c}_1^2/\ \tilde{P}_{ext}^{\ 1/2})\left[\sqrt{8\delta_*+1}\ -4\delta_*+1\right]^{1/2} kpc \tag{5}$$

In these relations $\tilde{c}_1=c_1/7km\cdot s^{-1}$, $\tilde{P}_{ext}=P_{ext}/2.5\cdot 10^{-13}dyn\cdot cm^{-2}$; $w(\delta)$ and $j(\delta)$ are undimensional functions of the density contrast $\delta=\rho/\rho_c$ presented in the table 1; δ_* – is a density contrast between the surfase and the center of the model.

Table 1

δ	w	j	δ	w	j
0.99	$2.982\cdot 10^{-4}$	$3.489\cdot 10^{-7}$	0.20	0.2712	0.1814
0.95	$3.530\cdot 10^{-3}$	$2.482\cdot 10^{-5}$	0.10	0.3365	0.5400
0.90	$1.010\cdot 10^{-2}$	$1.522\cdot 10^{-4}$	0.06	0.3660	1.061
0.80	$2.913\cdot 10^{-2}$	$1.009\cdot 10^{-3}$	0.02	0.3983	3.872
0.70	$5.460\cdot 10^{-2}$	$3.317\cdot 10^{-3}$	0.008	0.4089	10.21
0.60	$8.588\cdot 10^{-2}$	$8.316\cdot 10^{-3}$	0.004	0.4126	20.97
0.50	$1.229\cdot 10^{-1}$	$1.834\cdot 10^{-2}$	0.001	0.4155	85.83
0.40	$1.657\cdot 10^{-1}$	$3.827\cdot 10^{-2}$	0.0008	0.4156	107.5
0.30	$2.148\cdot 10^{-1}$	$8.008\cdot 10^{-2}$	0.0006	0.4153	143.8

The function $w(\delta)$ has a maximum w_m=0.419 at δ=0.0008.This indicates the existence of the upper mass limit for equilibrium superclouds for given corresponding external pressure P_{ext}. The relations (2)-(5) allow to find masses to the case of dense cores formation .For the typical conditions they are equal $(0.5-2)\cdot 10^7 M_\odot$. The lower mass limit is determined by the condition that the central density became equal to the critical level n_{LD} , the upper one corresponds to the Jeans mass.

3. DENSE CORE' PARAMETERS.

Evidently near the centre of superclouds density changes very slowly and almost equal to the level density. So the relation

$$(n_c - n_{LD})/n_c = 1-\delta_{LD} = \varepsilon \leq 0.1-0.2 \qquad (6)$$

is a small parameter. This permits the linearization of relations (1). After that for the core parameters we have:

the equatorial radius

$$s_1 = 1.16 \ \varepsilon^{0.5} \ \tilde{c}_1^{2} \ \tilde{P}_{ext}^{-1/2} \ \delta_*^{1/2} \ \mathrm{kpc}, \qquad (7)$$

the cooled mass

$$M_1 = 1.4\cdot 10^7 \varepsilon^{1.5} \ \tilde{c}_1^{3} \ n_c^{-1/2} \ M_\odot \ , \qquad (8)$$

the angular momentum

$$J_1 = \ 10^{67} \ \varepsilon^{2.5} \ \tilde{c}_1^{5} \ n_c^{-1} \ g\cdot cm^2\cdot s^{-1}. \qquad (9)$$

Therefore for the typical range of input parameters:

$M_1 = (5-10)\cdot 10^5 M_\odot$, $s_1 = 300-500$ pc, $J_1 = (0.5-2)\cdot 10^{65}$ $g\cdot cm^2\cdot s^{-1}$.

As it is seen they are very close to the parameters of the star complexes.

The numerical models calculated by Kiguchi et al.(1987) corresponds to the case of more rapid rotation. At the large distances from the center these models give a flat rotation with velocity $2c_1$ in contrast to $c_1\sqrt{2}$ in Schmitz solution (1). As a result numerical models give a higher mass concentration to the centre that leads to the more massive and flatten cold cores. Our calculations carried out with the approximate algorithm described by Hachisu and Eriguchi (1985) show that the forming cores have masses $M_1 = (1-4)\cdot 10^6 M_\odot$,axes ratio $(z_1/s_1) = 0.4-0.5$, and equatorial radii in accordance with (7) 300–500 pc. These results are also in good agreement with parameters of star complexes.

4. SUPERSONIC TURBULENCE DEVELOPMENT. GIANT MOLECULAR COMPLEXES FORMATION.

Warm interstellar gas is turbutent. The turbulent velosities are subsonic. Shear galactic motions, the large-scale explosive processes generate irregular motions up to the length scale $l_0 \sim 100$ pc. Thus the lower limit of turbulence decay time scale can be estimated by relation $\tau_t \geq (l_0/c_1) \sim 10^7$ years. In the internal part of a supercloud where density is over n_{LD} the rapid cooling in the time scale $\tau \leq 10^6$ years takes plase (Spitzer (1978)). So in the forming dense core we have an unusual situation with $\tau_t \gg \tau_c$ and therefore the initially subsonic motions convert in a supersonic turbulence.

Turbulent pressure becomes the only force that maintains the core in quasiequilibrium state. It is surrounded by extended envelope having the column density greater then $N_1 = 5 \cdot 10^{20} cm^{-2}$ that is critical for molecules formation. The description of subsequent core evolution is a very complicated problem. But it is most plausible that a cooling and cotracting core convertes into the complex of semispherical star forming giant molecular clouds with total mass near $10^6 M_\odot$ and sizes 0.5–0.8kpc. Such complex must be surrounded by the extended neutral hydrogen envelope with column density greater then $5 \cdot 10^{20}$ cm^{-2} .The galactic structures of this type where observed by Grabelsky et al.(1987,1988).

It is very likely that the proposed process of giant complexes formation can explain why stars have tendency to create the large scale star complexes.

5. CONCLUSION

1. The physical processes leading to the dense cold core formation in superclouds having mases of the order $10^7 M$ are considered.

2. The relations for the core parameters estimations are derived. It is shown that cores have the masses in the range $(1-4) \cdot 10^6$ $M_\odot$, the equatorial radii 300–500 pc, and the axes ratio 0.4–0.5. The core is embedded into the extended HI envelope having a column density equal or greater then $5 \cdot 10^{20} cm^{-2}$ that is necessary for effective molecules formation.Therefore the core transforms into the complex of star forming giant molecular clouds. The parameters of star complexes in galaxes are very close to ones found in considered model.

3. Due to rapid cooling the supercloud subsonic random motions of the warm interstellar matter are transformed into the supersonic turbulence of the cold molecular core.

4. The properties of forming supercloud's cores are very sensitive to the level density n_{LD} which determines conditions for the gas cooling, and to the external pressure P_{ext} which take into the account the influence of the external forces. By variations of these parameters we can explain the observed diversity of giant molecular complexes and star forming regions in galaxies.

5. The existence of the critical column density $N_1 = 5 \cdot 10^{20} cm^{-2}$ necessary for molecules formation in the core explains why the active star forming galaxies have the surfase density greater or equal to N_1.

REFERENCES

Arshutkin, L.N.and Kolesnik, I.G. (1984) 'The structure of massive molecular clouds',Astrofisika 21, 147-161.

Efremov, Yu.N.(1989) Sites of stars formation in galaxies:star complexes and spiral arms(in russian),Nauka,Moscow.

Elmegreen, B.G.and Elmegreen, D.M.(1983) 'Regular strings of HII regions and superclouds in spiral galaxies: clues to the origin of cloudy structure', Mon.Not.R.astr.Soc. 203, 31-45.

Elmegreen, B.G. (1987) 'Supercloud formation by nonaxisymmetric gravitational instabilities in sheared magnetic galaxy disks',Astrophys.J. 312, 626-639.

Grabelsky, D.A., Cohen, R.S., Bronfman, L., Thaddeus, P. (1987) 'Molecular clouds in the Carina arm: large-scale properties of molecular gas and comparison with HI' Astrophys.J. 315, 122-141.

Grabelsky, D.A., Cohen, R.S., Bronfman, L., Thaddeus, P. (1988) 'Molecular clouds in the Carina arm: the largest objects; associated regions of star formation; and the Carina arm in the Galaxy' Astrophys.J. 331, 181-196.

Hachisu, I. and Eriguchi, Y. (1985) 'Equilibrium structures of rotating isothermal gas clouds.I.', Astron. and Astrophys. 143, 355-364.

Kaplan, S.A. and Pikel'ner, S.B. (1979) Physics of interstellar matter (in russian), Nauka, Moscow.

Kiguchi,M., Narita, S., Miyama, S.M., Hayashi, C. (1987) 'The equilibria of rotating isothermal clouds', Astrophys.J. 317, 830-845.

Kolesnik, I.G. (1977) 'Opacity effects in interstellar clouds and star formation', in I.G.Kolesnik (ed.), Early stages of stellar evolution (in russian), Naukova Dumka, Kiev, pp.10-13.

Kolesnik, I.G. (1987) 'Formation of giant molecular clouds in superclouds and the origin of supersonic turbulence', Kinematika i Fizika Nebesnykh Tel 3,no.6, 47-56.

Schmitz, F. (1983) 'An exact solution for an isothermal gas cloud with fast differential rotation', Astron. and Astrophys. 143, 355-364.

Spitzer, L.,Jr. (1978) Physical processes in the interstellar medium, John Wiley & Sons, Inc.

DISCUSSION

ZINNECKER: In addition to the original supersonic turbulence in GMC's which you described in your talk, kinetic energy input related to star formation (winds, bipolar outflous, supernova remnants) will further feed supersonic turbulence in these clouds.

KOLESNIK: Yes, of course. You put an attention to the very important processes. But in GMC's turbulent length scales occupy so wide range that for their maintainance some principle different sources. must operate.In the small length scale interval the main role is belonged to the sources that You have pointed. But in the large length scales the primordial slow decaying turbulence which have the life time equal about 10 million years will continue to operate.

Global Star Formation in the L1630 Molecular Cloud

Elizabeth A. Lada
Astronomy Department
University of Texas
Austin, Texas 78712

ABSTRACT. This paper presents the preliminary results of a study of star formation in the L1630 molecular cloud. We have completed an unbiased, well sampled, systematic CS (2→1) survey for density condensations and a 2.2 μm survey for embedded infrared sources in L1630. These surveys have provided us with a complete census of the dense cores and young stellar objects within this molecular cloud. As a result, four embedded stellar clusters have been identified in this cloud. These clusters are located near the most massive CS cores implying that the most active sites of star formation are located within the most massive dense cores.

1. Introduction

Over the last ten years, observations at radio and infrared wavelengths have revealed that stars form in molecular clouds. Little is known about the actual process of star formation, but it is becoming clear that knowledge of the density structure of molecular clouds is critical to understanding the evolution of the clouds and the formation of stars within them. Although the density structure of a number of individual molecular cores has been investigated, our knowledge of the large scale gas density distribution of star forming clouds is minimal.

Molecular line studies of regions with current star formation have shown that the youngest stars are invariably associated with dense molecular gas. Is the converse true? Is star formation always present in dense gas? Until recently, this question has not been adequately addressed since the dense regions studies were selected because they had signposts of star formation.

This paper presents the preliminary results of a study of dense cores and star formation in the L1630 (Orion B) Giant Molecular Cloud. For my thesis, I have completed an unbiased, well sampled systematic survey for density condensations and a survey for embedded infrared sources in the L1630 cloud in an attempt to understand the relationship between dense cores and young stars. The L1630 cloud was chosen for these surveys because it is one of the closest giant star forming complexes, having a distance of approximately 450 pc and because it is a well studied region. Several unbiased large scale molecular line

R. Capuzzo-Dolcetta et al. (eds.), Physical Processes in Fragmentation and Star Formation, 193–199.

surveys tracing lower density gas have been conducted in this cloud, such as CO surveys (Maddalena *et al.* 1983; Bally *et al.* 1990) and a ^{13}CO survey (Bally *et al.* 1990). In addition an IRAS biased CO survey for outflows has been done in this region (Fukui *et al.* 1986).

2. Survey for Dense Cores

To identify the dense cores, the L1630 cloud was surveyed in the J = 2→1 transition of CS using the AT&T Bell Labs 7-m telescope (Lada, Bally and Stark 1990). Thirteen thousand points were surveyed with one arc minute spacing to an rms noise level of 0.2 K. The total area covered by the survey was approximately 3.6 square degrees. The results of the CS (2-1) survey are presented in figure 1 in the form of an integrated intensity map. Emission was detected, at a 3 sigma level above the noise, over approximately 10 percent of the area surveyed. This emission is not uniformly distributed, but rather, it is clumpy. Individual velocity channel maps of the surveyed region indicate an even higher degree of clumping than that present in figure 1.

In order to study the clumping properties of the gas, we have identified individual clumps using channel maps, velocity - position maps and individual spectra. Forty two clumps were identified above a five sigma noise level. Clump radii range from $\leq$ 0.12 to 0.61 pc. The FWHM linewidths for the clumps range from 0.5 to 2.7 km s^{-1}. Virial masses were calculated for each clump and range from $\leq$ 8 to 528 $M_{\odot}$. Most clumps have masses less than 100$M_{\odot}$and only six clumps have masses greater than 200$M_{\odot}$.The clump mass spectrum is well fit by the power law

$$N = aM^{\alpha}$$

where N equals the number of clumps per solar mass interval, α = -1.1 and a = 564 $M_{\odot}{}^{-1}$. The power law index implies that a significant amount of the mass of the dense gas is contained in the most massive clumps. In fact, approximately 50 percent of the total clump mass is located in the 6 most massive clumps (M > 200 $M_{\odot}$).

3. Survey for Embedded Sources

A near infrared survey of a significant portion of the L1630 molecular cloud was carried out using the NOAO Infrared Array Camera on the Kitt Peak 1.3 m telescope (Lada *et al.* 1990). Three thousand $1'$ x $1'$ fields were surveyed at 2.2 μm, covering an area of approximately 0.8 square degrees. The regions surveyed included both areas containing CS emission and areas without CS emission The boundaries of this survey are shown in figure 1. The sensitivity of the survey is estimated to be 14th magnitude at K and the completeness limit of our counting statistics is estimated to be 13th magnitude. At the distance to L1630, this completeness limit would correspond to a main sequence star having a mass, M = 0.6 $M_{\odot}$.

As a result of the near infrared survey, 1185 individual sources having magnitudes brighter than 14th magnitude and 912 sources having m_K < 13 were identified. The embedded

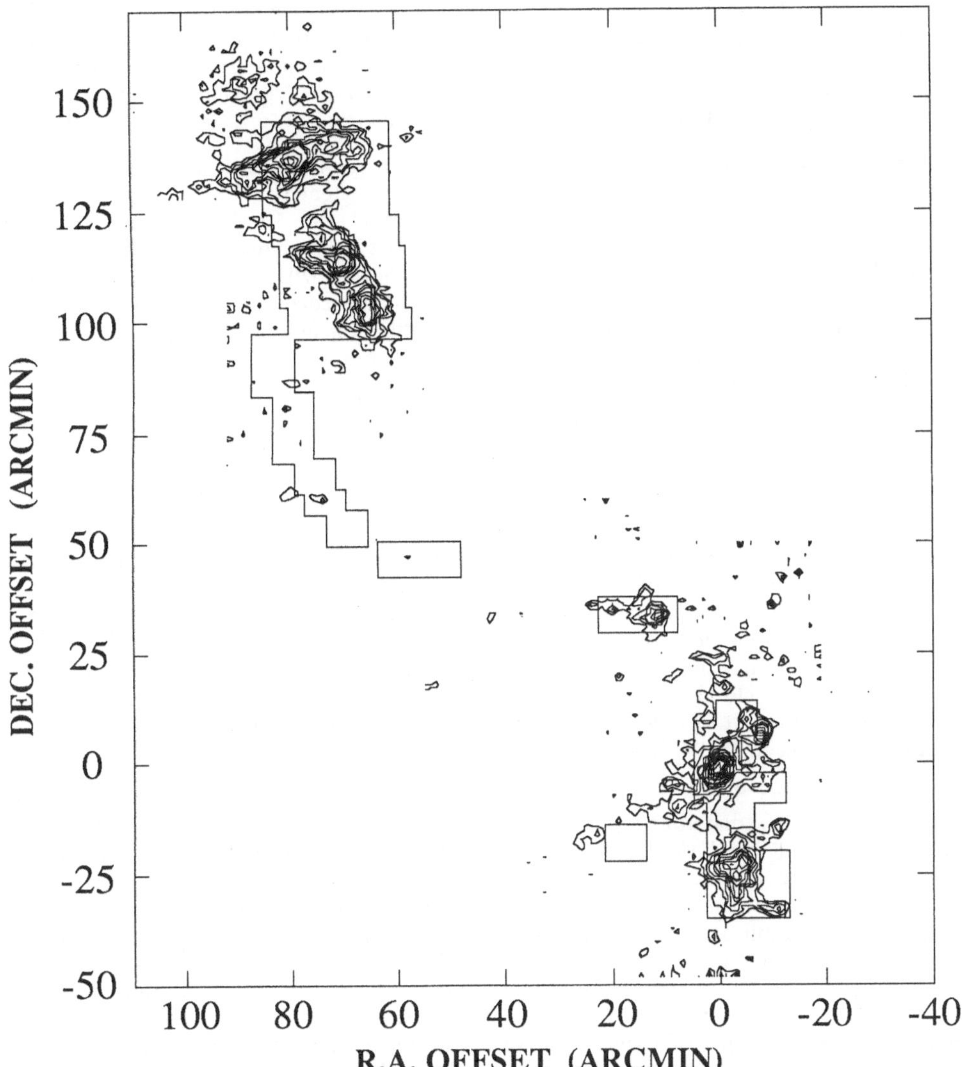

Figure 1. Distribution of CS(2-1) emission (integrated intensity) for the Orion B Cloud. The (0,0) position corresponds to R.A. $5^h\ 39^m\ 12^s$ Dec. -01° 55' 42" . The lowest contour level is at 0.8 K km/s corresponding to a 3 sigma detection above the noise with subsequent levels at 1.2, 1.6, 2.0, 3.2,..., 4, 5, 6,..., 15 K km/s. The solid lines represent the boundaries of the 2.2 micron survey.

sources are found throughout the region surveyed but the surface density distribution of these sources is not uniform. Rather many sources appear to be grouped or clustered. The clustering of sources is even more apparent at brighter magnitudes for which one might expect less contamination by background stars. Figure 2 presents the distribution of sources for $m_K < 12$. Here we clearly see sources clustered into at least four regions.

For purposes of discussion, we define a cluster to be a region in the sky where the star density significantly increases over the background star density. Using this definition, four individual clusters having surface densities greater than 10 times the background star density have been identified. The largest and most spectacular cluster is associated with the HII region NGC 2024. Three hundred nine sources with $m_K < 14$ are contained within this cluster. The remaining three clusters are also associated with well known star forming regions. The cluster associated with the reflection nebula, NGC 2071, contains 105 sources with $m_K < 14$ and the NGC 2068 cluster contains 192 sources. The smallest cluster is associated with NGC 2023 and contains 21 sources.

4. Star Formation in L1630

The CS and 2.2μm surveys have provided us with a complete census of the dense cores and young stellar objects within the L1630 molecular cloud. As a result, four embedded stellar clusters have been identified. Depending on the fraction of background stars, 58 – 96 percent of the embedded sources detected by the near infrared survey are located in these clusters. This fraction is surprisingly large and suggests that the majority of the stars forming in this cloud are forming in groups or clusters. In addition, the total area covered by the four clusters equals only 18 percent of the entire region surveyed indicating that star formation in L1630 is a highly localized process.

A comparison of the CS and 2.2μm surveys reveals that the distribution of infrared sources follows the dense gas distribution. 66 – 93 percent (depending on the background star distribution) of the total number of stars detected by the infrared survey are located within the 3 sigma CS emission boundaries. The four embedded infrared clusters are located near the most massive CS clumps ($M \geq 300\ M_\odot$). Apparently the most active sites of star formation are located within the most massive cores. Preliminary analysis indicates that only one massive clump ($M \sim 300\ M_\odot$) is not associated with a cluster. The lack of stars within this massive core is intriguing. Why is the star formation efficiency low in this massive core and how does this core differ from the four cores associated with stellar clusters? Is is possible that this core is at an earlier stage of evolution? Upon further analysis of my thesis data, I hope to address these questions by determining the level of star forming activity in this core and then comparing it to the four cores having stellar clusters.

This work was supported by a NASA Trainee Grant, NGT50320 to the University of Texas, at Austin. The author also acknowledges support from a Zonta Amelia Earhart Fellowship.

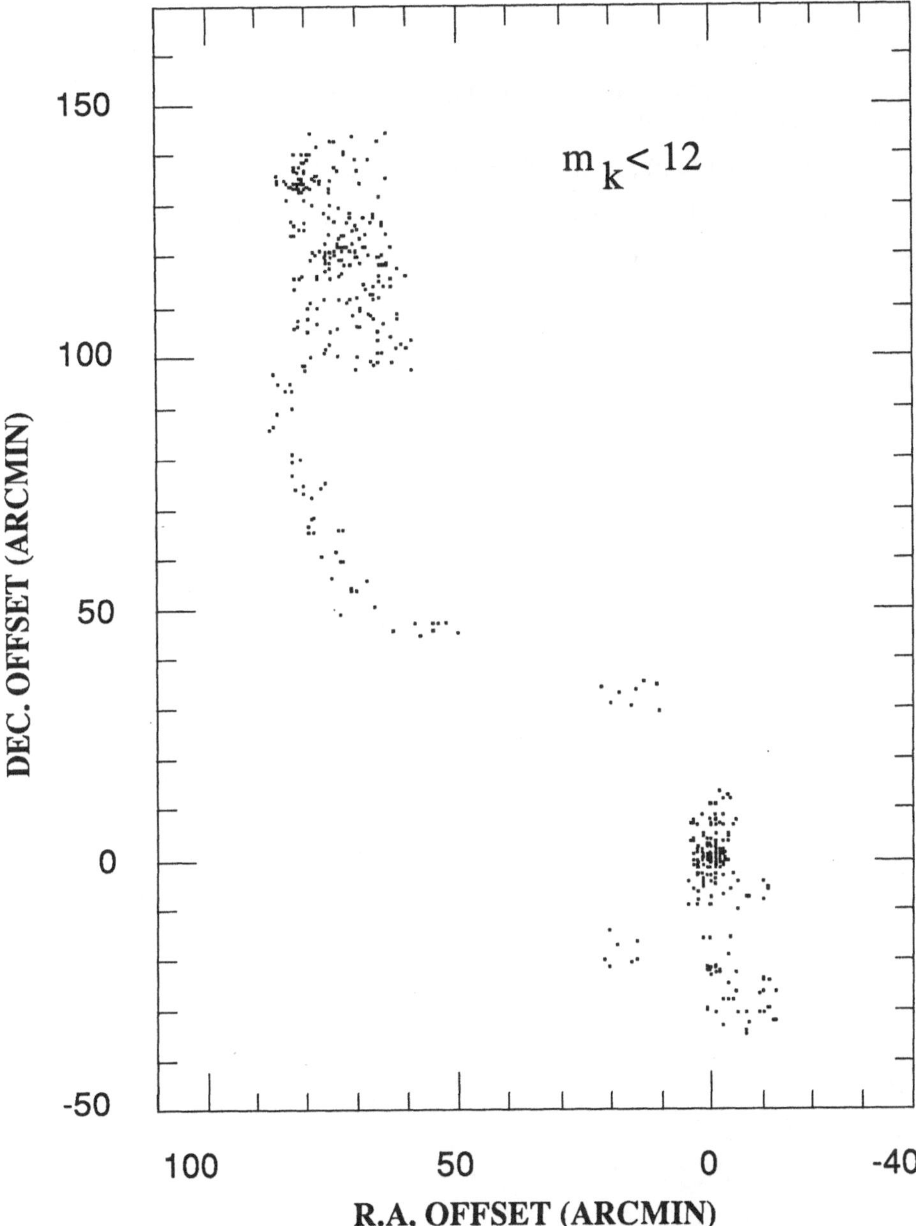

Figure 2. Spatial distribution of near infrared sources (2.2 micron) with m_k < 12.

References

Bally J. *et al.* 1990, in preparation.

Fukui, Y., Sugitani, K., Takaba, H., Iwata, T., Mizuno, A., Ogawa, H., and Kawabata, K. 1986 *Ap. J. (Letters),* **311**, L85.

Lada, E. A., Bally, J. and Stark, A. A. 1990, in preparation.

Lada, E. A., Evans, N. J., DePoy, D. and Gatley, I. 1990, *Ap. J.,* to be submitted.

Maddalena, R. J., Morris, M., Moscowitz, J., and Thaddeus, P. 1986, *Ap. J.,* **303**, 375.

Discussion

Mouschovias: Have you observed different lines of the same molecule or of different molecules sampling different densities to test a prediction I made a few years ago, namely, that if magnetic fields are responsible for the observed supersonic linewidths, you should see a narrowing and eventual thermalization of linewidth with increasing density in one and the same object? (significant narrowing should be observable by densities of 10^6 cm^{-3}).

Lada: At the moment, I only have CS(2→1) data for this region. However I have begun a program to obtain CS multi-transition data for the purpose of determining densities of these cores. I also plan to compare the CS survey to the already existing Bell Laboratory's CO and ^{13}COdata sets. Hopefully using these additional data sets, I will be able to address your question in the future.

Zinnecker: First, I'd like to congratulate you on your infrared array cluster data. Although the total net observing time is only 10 – 20 hours, I realize that it takes 1 – 2 years to reduce the data. Secondly, I have a question: When you speak of clusters, do you mean bound clusters? Some of them appear to be rather loose. Compared to the Trapezium cluster, their stellar densities seem to be quite low, aren't they?

Lada: Thank you Hans. In reply to your question, I have defined a cluster to be a region in the sky where the source density significantly increases over the background star density. By this definition, I did not mean to imply that the grouping I called a cluster necessarily corresponds to a bound open cluster. At this time I do not know whether these embedded infrared clusters will be bound systems or unbound associations. I hope to address this issue upon further analysis of my data. In regard to stellar densities, the stellar number densities of these clusters are much higher than those found in associations and are similar to stellar densities estimated for other young star forming regions such as the star forming core of Rho Ophiuchus. You are correct however when you say that the Trapezium cluster has much higher stellar densities. If we compare similar areas of the NGC 2024 cluster, which is the densest of the four L1630 clusters, to the Trapezium cluster (using the results of McCaughrean and collaborators) we find that the stellar density of the Trapezium cluster is 3.5 times higher than the density of the NGC 2024 cluster.

Kruegel: How do your CS clumps in NGC 2024 compare with the 1.3 mm dust continuum clumps?

Lada: This is an interesting question, but I have not made the comparison at this time.

Observations of Fragmentation

Hans Zinnecker

Max Planck Institut für Physik und Astrophysik,
Institut für extraterrestrische Physik, 8046 Garching, FRG

ABSTRACT. Observations of fragmentations are discussed, with a view towards star forming molecular clouds. After a brief outline of the motivation and the methods, we give a quick guide to the literature on the subject. We mention molecular line studies, dust continuum studies, and the impact of recent near-infrared images of embedded star clusters. We conclude with some remarks on how the observed clump mass spectrum in molecular clouds may relate to the mass spectrum of stars that form from these clumps.

1 Motivation

The primary motivation to observe the fragmented small-scale structure of molecular clouds is the conviction that these fragments represent the immediate progenitors of newly forming stars. Therefore the parameters that we observe and deduce for resolved clumps should give us a fairly good idea of the set of initial conditions for proto-stars (sizes, densities, temperatures, velocity dispersions, and masses etc). Observations of fragmentation can also reveal the structure of the fragments and the dynamics of the fragmentation process (hierarchical or otherwise). Recent advances in mm- and submm receiver technology (coupled with new large telescopes such as IRAM and JCMT) and the advent of infrared array cameras allow substantial progress to be made in the study of the last step "from clumps to stars".

2 Observational Methods

The most direct method to study fragmentation of molecular clouds is to map them with sufficiently high angular and velocity resolution in optically thin molecular transitions. This can isolate individual dense clumps in three dimensions (in two spatial coordinates and in velocity space). Then it becomes possible to determine the structure and the mass spectrum of the clumps, a vital step to understand the

R. Capuzzo-Dolcetta et al. (eds.), Physical Processes in Fragmentation and Star Formation, 201–209.

formation of stars and finally the stellar IMF. Besides molecular line studies, another important direct method is the mapping of molecular clouds in the submm continuum in order to find cold dust condensations. Compared to the line studies, continuum observations provide a somewhat less powerful tool, since such observations lack the 3rd dimension (i.e. velocity information); furthermore the continuum receivers are presently not sensitive enough to resolve anything but the densest clumps. The best indirect method to study fragmentation is to study the initial stellar population of star forming molecular clouds. Since these young stars are heavily embedded objects, infrared (1-5 μm) observations are required. Very recently, infrared imaging of large portions of molecular clouds has become possible (many square-arcmin), because 2D detector arrays similar to CCD cameras have become available. Therefore one can now investigate the spatial scales over which embedded objects are distributed (star-star distances in young embedded clusters, separations in binary and multiple systems). This allows one to infer a lower limit of the gas density when fragmentation into stars occurred.

3 A Guide to the Literature

Below we offer a quick overview of the literature. The description of the various references will be subdivided into three groups according to the observing methods employed. A detailed critical discussion of all the papers mentioned is beyond the scope of this short contribution.

3.1 Molecular line observations of fragmentation

One of the first studies was by Clark et al. (1977, ApJ 215, 511) who presented evidence for localized velocity components in a group of dust clouds in Taurus. Similarly, Crutcher et al. (1978, ApJ. 226, 839) found several distinct clouds in the NGC 2264 molecular cloud. Blitz (1987, in Physical Processes in Interstellar Clouds, eds. Morfill and Scholer) took up the subject and resolved many clumps in the Rosette molecular cloud. Very recently, Stutzki and Güsten (1990, ApJ in press) completed a very important study of the M17SW molecular cloud core, resolving more than a hundred clumps of various masses with the IRAM 30m telescope ($C^{18}O$ J=2-1 observations with an 11" beam). They were able to derive the clump mass spectrum (frequency per unit mass) which can be described as a power law in mass with an exponent close to -1.7 for masses between 1000 to 10 solar masses. Massi et al. (1988, A&A 194, 116) had previously studied small-scale clumping in M17 with the help of the VLA, and claimed the detection of dense ammonia condensations of the order of the Jeans mass (few solar masses here). VLA ammonia clumps of a few solar masses were also isolated in the Orion-KL region by Migenes et al. (1989, ApJ 347, 294), possibly compressed by the outflow from IRc2. Evidence for fragmentation in other regions of massive star formation is discussed in the recent review of Wilson and Walmsley (1989, A&A Rev. 1, 141), including W51, W49, W3(OH), DR21(OH), and SgrB2.

Fragmentation studies of cool molecular clouds without ongoing massive star formation are also pursued by a number of people. While Perault et al. (1985, A&A 152, 371) studied cool giant molecular clouds and found evidence for parsec-sized fragments, Loren (1989, ApJ 338, 902) analyzed the clumps in the much smaller Rho Ophiuchi molecular cloud. He found an exponent -1.1 for the clump mass spectrum. Thus the mass spectrum in this quiescent environment is shallower than that in the clumpy molecular core in M17 which is exposed to the penetrating UV radiation of an HII region (see above). In addition to the Rho Oph region, other more southern quiescent clouds are being investigated, owing to the advent of the SEST, a 15m radio dish at ESO/La Silla; see the work of Mattila et al. 1989 on the northern Chamaeleon cloud, especially their Figs. 5-8, published in the ESO Proc. "Low mass Star Formation and Pre-Main Sequence Objects", ed. Reipurth).

3.2 Dust continuum observations of fragmentation

The seminal paper on identifying dense protostellar clumps by their dust emission contrast with the surrounding medium is by Mezger et al. (1988, A&A 191, 44). They used the 30m IRAM telescope to probe NGC 2024 and S255 taking advantage of the small beam size (11") at the observing wavelength of 1.3mm. They detected several individual clumps and derived densities in excess of 10^8 cm^3 and masses of the order of several ten solar masses, but the masses could be an order of magnitude smaller depending on the uncertain dust opacity and the model fitting. (Follow-up studies at NIR wavelengths by Moore and Chandler 1989, MNRAS 241, 19p showed that at least one of these dust condensations contained an 2μm point source, most likely an embedded star; therefore not all of the these dense dust cores can be regarded as true protostars as suggested by Mezger et al.; see also Moore et al. 1989, MNRAS 237, 1p). Already in 1984, Jaffe et al. (ApJ 284, 637) had presented KAO far-infrared and submm observations of S255 and had discovered a double core, with masses 350 and 500 M . They also studied the previously known submm double cores in Orion (OMC1-N and OMC1-S; Keene, Hildebrand, and Whitcomb 1983, ApJ. 252, L11) and W3-IRS4/IRS5 (Jaffe et al. 1983, ApJ. 273, L89). The results of the IRAS satellite mission have added tremendously to the possibility of finding more candidates for fragmentation studies of cold dust condensations (Beichman et al. 1986, ApJ. 307, 337; Myers et al. 1987, ApJ. 319, 340; Wilking et al. 1989, ApJ. 345, 257).

Before these more recent submm and far-infrared dust observations, optical observations have long shown the existence of fragmented structure of a number of Lynds dark clouds (see the catalog compiled by Schneider and Elmegreen 1979, ApJ. Suppl. 41, 87; the catalog shows many examples of filamentary clouds with regularly spaced globular fragments). There are also the CO studies of Leung, Kutner, and Mead 1982, ApJ. 262, 583 and Myers, Linke, and Benson 1983, ApJ. 264, 517 on the structure and dynamics of isolated dark globules. Lately, an interesting HI absorption study bearing on fragmentation of the interstellar medium at the 25 AU scale appeared (Diamond et al. 1989, ApJ. 347, 302) using VLBI techniques.

3.3 Near-infrared array observations of fragmentation

Near-IR images bear on the fragmentation of young embedded star clusters and small groups of newly formed stars, including binaries. Here we will attempt to list all the activities of which we are aware. Some of the following papers that will be quoted are not yet published.

Probably the most widely recognized near-IR image of a young cluster is the one of the Orion Trapezium cluster, obtained by McCaughrean (et al., in preparation) using the infrared camera at UKIRT. The mosaiced image is reproduced in Sky and Telescope Vol. 77, p. 352 (April 1989) together with an optical photograph; see also Astronomy 17, p. 40 (August 1989). In the infrared the cluster appears round, whereas it looks irregular in the optical. The infrared image shows on the order of 500 stars in an area of 5'x5' or 0.7pc x 0.7pc. In the central region with a radius of 0.1pc there are about 100 stars, corresponding to a stellar number density of the order of 10^4 pc^3 . Optical images (Herbig and Terndrup 1986, ApJ. 307, 609) had revealed only about 120 stars. This highlights the fact that only near-infrared imaging at high spatial resolution (1"/pixel) can probe the stellar density unambiguously. This is because infrared imaging is largely freed from dust extinction and the competition from nebular emission. The high stellar density seen implies that fragmentation into stars must have taken place at gas densities in excess of 10^5 cm^3 , perhaps even much in excess of this value when we consider that the cluster is likely to have expanded somewhat after star formation took place. The 2μm luminosity function of the cluster stars keeps rising up to about K=11.5 and starts declining thereafter. This turnover occurs well before the detection limit (K=15) and thus in all likelihood is real, possibly implying a paucity of very low mass stars in the Orion Trapezium cluster.

An infrared image of the M17 embedded cluster, obtained at Kitt Peak by Gatley, DePoy, and Fowler (1988, Science 242, p. 1264 Dec 2 issue, front cover; also Sky and Telescope, Vol. 77, p. 352, April 1989) shows hundreds of mostly very red stars, hidden in the molecular cloud and previously undiscovered. The stellar density is not quite as high as that in the Orion Trapezium cluster, possibly because the expansion of the cluster, following gaseous mass loss, is more advanced here or because the initial gas density was lower. We do not know whether the 2μm luminosity function in M17 exhibits a similar turnover as that in the Trapezium cluster.

An infrared image of the young southern cluster NGC 3603, associated with the most luminous HII region in the Galaxy, was observed at CTIO by Moneti and Zinnecker (1989, shown in an article on high mass star formation by Melnick in ESO-Messenger 57, p. 4, his Fig. 4). Several sources not seen at 1.2μm become "visible" in the 2.2μm image. Nebulous infrared emission is also apparent indicating embedded ionizing stars. Some infrared objects were also found in the neighbouring molecular cloud, raising the question of further star formation triggered by the expansion of the HII region. The 2μm luminosity function in NGC 3603 does show a turnover at low luminosities but it occurs dangerously close to the detection limit, so that we are not convinced that it is real.

Other young embedded clusters, recently revealed by infrared imaging, include NGC 2024 (E. Lada, this meeting) and S106 (Hodapp and Rayner, in preparation); the latter has a 2μm luminosity function that keeps rising up to the completeness limit. Furthermore 2.2μm images exist of the following OB star forming regions (all of which lie at a distance of 2-3kpc) : W3-MAIN, W3(OH), S235, S255, S269, GL437 (Rayner and Zinnecker, in preparation). In each case it turned out that the OB stars are accompanied by a cluster of lower luminosity stars. These clusters all have diameters of around 1 arcmin (1pc at 3kpc) and contain approximately the same number of stars (50-100 to a limiting magnitude of about K=14.5). In some cases, but not in all, the brightest object appears to be near the projected centre of the cluster, as if the cluster was born with a core-halo structure..
Large-scale infrared images of the young cluster NGC 2264 have also been taken (Greene and C. Lada; McCaughrean, Rayner, and Zinnecker), to improve on the optical work by Adams, Strom and Strom (1983, ApJ. Suppl. 53, 893). However, these images still remain to be examined and no preliminary results are available. Some ground has been laid for the interpretation of the infrared data on this and many other clusters in a paper by Straw, Hyland, and McGregor (1989, ApJ. Suppl. 69, 99). Their paper deals with an infrared colour-magnitude diagram (K vs. J-K) of the partly obscured young cluster NGC 6334, but only for the more massive stars that have reached the Main Sequence; it ignores pre-Main sequence evolution of the lower mass cluster stars.
There are also some studies of small clusters without massive stars. One such cluster seems to be in the OMC2 molecular cloud core. JHK images of the region show a small cluster of intermediate-mass pre-Main Sequence stars (Rayner et al. 1989, MNRAS 241, 469). Furthermore, infrared array images around the Serpens source SVS4 reveal a dense "group" of 11 objects within a radius of 0.1pc of SVS4, indicating a rather high stellar number density (Eiroa and Casali 1989, A&A 223, L17). Similarly, a circular region of 0.16pc diameter on the Orion IRAS source 05338-0624 contains 20 stellar objects or semi-stellar knots (Strom, Margulis, and Strom 1989, ApJ. 346, L33), indicating a dense clustering, perhaps dense enough to emerge from its parent molecular core as a bound group after all the gas is dispersed. Even before infrared array images became available, candidate clusters of low-mass stars had been discovered, such as the embedded cluster around Elias 29 and WL16 in the core of the Rho Ophiuchi dark cloud (Wilking and Lada 1983, ApJ. 274, 698; Lada and Wilking 1984, ApJ. 284, 610). These authors patiently used the conventional and tedious method of raster scanning the suspected cluster area with a single beam detector (their search field comprised a 10'x10' area, sampled by a 12" beam). They found a total of 44 objects brighter than about K=12 within a volume of 0.125 pc^3 , corresponding to a mean separation of the objects of about 0.14pc and comparable with a Jeans length of 0.05pc for the observed temperature and density of the cloud core gas. Recently, a 10 times more sensitive survey of a subsection of the Wilking-Lada survey area was made using an infrared array (Rieke, Ashok, and Boyle 1989, ApJ. 339, L71). Surprisingly, no new sources were discovered between K=12 and K=14.5 that were not already known before (except a binary companion to WL20).

However, a larger 12'x12' infrared imaging survey by Barsony et al. (1989, ApJ. 346, L93) to a limiting magnitude of K=14 (totally covering the Wilking-Lada survey area and a little more) did not agree with the findings of Rieke and collaborators that very faint sources were apparently absent; of the 35 new sources they detected, 16 objects lie within the Wilking-Lada box, and Barsony et al. conclude that there are major variations in the initial luminosity function over size scales of 0.2-0.6 pc. Small number statistics and field star contamination may still affect these conclusions.

Zinnecker and Rayner (1989, unpublished) have recently identified a second cluster or subcluster in the Rho Ophiuchi dark cloud, about 1pc NW of the Wilking-Lada cluster, with the source Elias 21 (=GSS 30) as the dominant source of the second cluster. Their H and K images suggest the presence of about 25 infrared sources within a 5'x5' field (0.25pc x 0.25pc) near Elias 21. Possibly there are more in the wider surroundings. These images had a completeness limit of K=14-15 and were obtained at the Univ. of Hawaii 88" telescope using a 128x128 pixel array with an image scale of 0.6"/pixel. 2μm images of the entire Rho Oph cloud core (650 square arcmin) have been taken by Greene, Young and Meyers-Rice (1989, unpublished), also using a 128x128 infrared array camera (however with an undersampled image scale of 1.8"/pixel). About 250 sources appear in this mosaic to a detectable limit of K=14-15. (Even larger infrared arrays, viz. 256x256 pixel arrays, will soon be available, making infrared surveys of large areas more efficient than they are at present). Finally, we refer to the large scale (more than 1 square degree) IRAS study combined with near- and mid-infrared follow-up work of the Rho Ophiuchi region which produced a list of 78 young stellar objects associated with the cloud and 39 as yet unclassified sources (Wilking, Lada, and Young 1989, ApJ. 340, 823). Another rather tight cluster of infrared sources, discovered by the conventional method of raster scanning, is the "Coronet" cluster in the RCrA dark cloud (Taylor and Storey 1984, MNRAS 209, 5p). Here we have some 20 stellar objects within an area of 5'x 5' or 0.2pc x 0.2pc (limiting magnitude K = 14). This result was obtained at the Anglo- Australian telescope. An earlier effort using that telescope produced a large-scale (1400 square arcmin) infrared survey of the Chamaeleon dark cloud (Hyland, Jones, and Mitchell 1982, MNRAS 201, 1095). More than 100 sources were found, of which about half were identified as cloud members. This gave an overall stellar density of order 10 per pc^3 at a limiting magnitude of K = 10.5-11. Jones et al. (1985, Astron. J. 90, 1191) conducted a much smaller, but slightly deeper near-infrared follow-up study in the northern portion of the Chamaeleon cloud around HD 97300. They suggested that star formation had taken place preferentially on the perimeter of a dense core in the cloud, rather than in the core itself. They cite evidence that the stellar wind of HD 97300 may have triggered the formation of the 4 pre-Main Sequence stars in its vicinity. Studies of the embedded stellar populations of southern dark clouds will receive a major boost in the near future, as soon as deep infrared array imaging will be applied to them.

Apart from embedded star clusters and groups of young stars, mention must be made of pre-Main Sequence binary stars (for a summary of the available observations see Zinnecker 1989 in ESO-Proc. "Low-mass Star Formation and Pre-Main Sequence Objects", p. 447, ed. Reipurth). The separations and masses of these stars place a constraint on the gas density of the clumps out of which the components formed. If the components were to form too close together, tidal forces would disrupt such fragments (cf. e.g. Kumar 1972, Ap. Sp. Sci. 17, 453). We surmise that the filamentary structure of molecular clouds leads to elongated subcondensations the gravitational fragmentation of which should produce binary systems as a natural outcome (indeed ammonia cores with aspect ratios of 2:1 or more have been observed; Myers and Goodman, priv. communication).

4 Some Remarks on the Initial Mass Function

We have noted evidence that molecular clouds are clumpy, and that the mass spectrum of the clumps (dN/dM) of the clumps scales with the clump mass (M) to the -1.5 (or thereabouts) power. The basic idea to obtain the Initial Mass Function of stars born in these clumps is to translate the mass spectrum of the clumps into the mass spectrum of stars by virtue of a clump-star (i.e. initial-final) mass relation. If the final, stellar mass (m) scales with the p-th power of the initial clump mass M, one finds that the stellar mass spectrum dN/dm is a power law with index x = (p+0.5)/p; for example, for p=0.4 one gets x = 2.25, very close to the index of the Salpeter IMF. The physical justification for the power p in the final-initial mass relation could come from the theory of ambipolar diffusion (for low-mass stars) and from the theory of radiation pressure acting on dust grains (for high mass stars). We suppose that magnetic fields or radiation pressure prevent complete accretion of the clump mass onto a protostellar core and provide a clump-mass dependent efficiency factor in the accretion process (the higher the clump mass, the smaller the fraction that ultimately ends up in the star). Alternatively, protostellar winds from both low- and high-mass stars could truncate envelope accretion (in this case a single physical process would be responsible for the final-initial relation). The above reasoning neglects subfragmentation of the clumps which is likely to become increasingly important for larger and larger clumps. If we allow for an increasing probability of subfragmentation with increasing clump mass, this will deplete the high mass clumps at the upper end of the clump mass spectrum and populate more the lower parts of the clump mass spectrum, thereby steepening the slope of the original clump mass spectrum. The resulting new clump mass spectrum could then come very close to the stellar mass spectrum with a power law slope near -2.5 (i.e., subfragmentation of the original clump mass spectrum might account for steepening the slope of the mass spectrum by 1, from -1.5 to -2.5, as required for the IMF).

5 Acknowledgments

It is a pleasure to acknowledge stimulating discussions with Drs. Mark McCaughrean and Jürgen Stutzki.

6 Discussion

Cayrel:
How strong is the observational evidence for a space distribution of protostellar clumps along sheets or filaments?

Zinnecker:
Schneider and Elmegreen (1979) have shown convincing examples that fragmentation occurs along dark filaments. I understand that Güsten is investigating their fragmentation more closely via molecular line observations. Filamentary structure with dense condensations is also found in the Rho Ophiuchi cloud (Loren 1989, ApJ. 338, 902; Loren, Wootten, and Wilking 1989, ApJ., in press) as well as in the Taurus dark cloud (Cernicharo, Bachiller, and Duvert 1985, A&A 149, 273; Duvert, Cernicharo, and Baudry 1986, A&A 164, 349, their Figs. 1&2). The Orion molecular cloud, too, exhibits filamentary structure (Bally et al. 1987, ApJ. 312, L45). Bally has given a review of the structure and kinematics of star forming clouds in the ESO-Proc. on "Low-Mass Star Formation and Pre-Main Sequence Objects" (ed. Reipurth). Thus it would seem that the fragmentation of filaments plays a large role in defining the clumps out of which low-mass stars finally form. This is also supported by theoretical calculations (e.g. Bastien 1983, A&A 119, 109 and references therein; for fragmentation of sheets see Miyama et al. 1985 in IAU-Symp. 115 "Star Forming Regions", p. 451).

Lioure:
Is there any other observational evidence of the mass distribution of clouds inside molecular clouds? Is the slope of -1.7 seen in M17 a general feature or an isolated case?

Zinnecker:
There is. Stutzki et al. (in preparation) have high resolution data of the clumpy structure at the other end of the M17 cloud that is not affected by the HII region. I don't know which kind of mass distribution they find there. Also, as I mentioned, Blitz (1987) studied the Rosette molecular cloud and gets a clump mass spectrum with a slope of -1.5, while Loren (1989) finds the slope of the mass distribution of the filamentary fragments in the Rho Ophiuchi cloud to be -1.1. All these mass distributions are rather close to that predicted by models based on coalescence of

smaller clouds to form larger clouds (for an insightful discussion of the origin of a slope near -1.5 see the discussion given by Spitzer 1982 in his book "Searching between the Stars", p. 148).

Mouschovias:
Are all the fragments in your study self-gravitating? Any non-gravitating ones may have nothing to do with star formation.

Zinnecker:
The study of Stutzki and Güsten shows that the clumps are indeed virialized. Thus it seems they are all self-gravitating entities.

Di Fazio:
The fact that we see heavy stars in the centre of clouds, opposite to what one would expect by simple fragmentation, may have to do with turbulent wakes formed by the moving fragments. This raises the gas temperature and the characteristic mass in the centre of the cloud.

Zinnecker:
I understand what you are trying to say. There may be other reasons why the most massive stars seem to form in the middle. I am eager to hear what Richard Larson will tell us as a possible explanation for this.

EXTRASOLAR PLANETARY MATERIAL

E. KRÜGEL
Max-Planck-Institut fur Radioastronomie
Auf dem Hügel 69
53 Bonn, FRG

ABSTRACT. This is a shortened and slightly adapted version of a paper by Chini, Krügel and Kreysa to appear in Astronomy and Astrophysics. It reports the detection of four nearby main-sequence stars at mm/submm wavelengths and its interpretation that this emission is due to dust particles which are, at least, a hundred times larger than interstellar grains. If one assumes that their size distribution follows a power law of the form $n(a) \alpha a^{-3.5}$ then there is no limit on the upper boundary of the grain size and much larger bodies cannot be excluded observationally.

A fundamental question in astronomy is whether stars, other than the Sun, also possess a planetary system. Such planets will be extremely hard to detect because of their low mass and luminosity and their proximity to the central star. One must therefore look for indirect indications to their presence. In the solar system the existence of interplanetary dust today, five billion years after the formation of the Sun, can only be understood if the grains are continuously produced from the fragmentation and grinding down of larger bodies such as planetoids and comets.

The detection by IRAS of a dust cloud around the AOV star Vega (α Lyr), which is only 8.1 pc away, was a major step in the search for extrasolar planetary material (Aumann et al., 1984). A crude analysis showed the grains to be considerably larger than interstellar particles, a fact also known to hold for the interplanetary dust around the Sun. The conclusion was based on the width of the 60μm emitting region (about 20") and on the black-body temperature of 85K between 25 and 100μm: Only large grains can be that cool so close to the star.

This finding is in agreement with the idea that small grains (<10μm) are ejected from the stellar neighborhood by radiation pressure. Furthermore, the angular momentum of a grain circling the star is diminished as the photons from the star, due to the aberration of light, impinge on the grain slightly head-on. This is known as the Poynting-Robertson effect. It forces those particles, which are not expelled by radiation pressure, to spiral into the star. The life-time t_{PR} of a grain with respect to this effect is proportional to its radius. For particles greater than 1mm, t_{PR} becomes comparable to the main-sequence lifetime of Vega itself. Using

R. Capuzzo-Dolcetta et al. (eds.), Physical Processes in Fragmentation and Star Formation, 211–216.

such dynamical arguments, Aumann et al. (1984) claimed that the grains were millimeter-sized.

Aumann (1985) found in the IRAS Point Source Catalog a dozen of Vega-like stars, all very near and bright, suggesting that they are quite frequent, although difficult to observe. The presence of circumstellar material was again indicated by a far IR excess between 25 and 100μm. The associated luminosity is always a minute fraction ($\approx 10^{-4}$) of the bolometric brightness and so the visual optical depth must be negligible, unless the matter is concentrated in an extremely flat disk.

As it is unknown whether there exists, in addition to the dust detected by IRAS, a cold dust component, which would show up at longer wavelengths, we performed observations at 0.87 and 1.3mm of some of these stars. They were carried out on the IRAM 30m telescope on Pico Veleta with the MPIfR bolometer system (Kreysa, 1985) at 0.87 and 1.3mm. Our fluxes are given in Table 1. This table also presents the data at 2.2μm, at the IRAS bands and at 0.8mm as reported by Becklin and Zuckerman (1989). The latter are very broad-band measurements comprising two atmospheric windows and were obtained on the 15m JCMT on Hawaii with 17" resolution. The magnitude of the IR excess is expressed in Table 1 by the ratio a of observed emission to blackbody emission and is normalized at 2.2μm. In all cases β becomes significantly greater than 1 only for $\lambda \geqq 60\mu$m, reaches its maximum at 100μm and has already declined at 1.3mm.

We tried to interpret the complete spectra. We therefore continued previous modelling (Aumann et al., 1984; Buitrago and Mediavilla, 1985; Anandarao and Vaidya, 1986), this time, however, by including a size distribution for the grains and calculating the absorption efficiencies with more recent optical constants from Mie-theory. Such an improved model

TABLE 1. Observational results. Flux densities from 2.2 to 100μm are given in Jy, submm data are in mJy. Spatial resolution is 17" at 0.8mm (Becklin and Zuckerman, 1989), 9" at 0.87mm and 12" at 1.3mm. β is the ratio of the observed flux to the flux from a stellar photosphere emitting like a black body; it is normalized at 2.2μm. $\beta > 1$ thus indicates an IR excess due to dust. Not included is the 450μm flux density of 511±178 mJy for α PsA (Becklin and Zuckerman).

	Flux density [Jy]					Flux density [mJy]		
	2.2	12	25	60	100	800	870	1300
Vega	647	31.	10.1	10.7	8.8	22±5	22±9	4.5±1.5
β	1	1.1	1.4	8.7	20		3.7	1.7
α PsA	267	12.8	3.47	6.78	9.85	35±7	35±12	7.3±2.2
α PsA	1	1	1.2	13	52	12	14	6.6
τ_1Eri	31	1.33	0.47	1.57	4.42	<21	<30	<6.6
β	1	0.8	1.2	22	170	<51	<85	<41
ε Eri	145	6.73	1.91	1.27	1.67	<24	35±13	7.5±2.2
β	1	0.7	0.9	3.1	11	<11	18	8.4

should, in principle, yield better estimates for the size of the grains, although we are aware of the large uncertainties when dealing with circumstellar dust.

We modeled the emission of the dust by assuming the grains to be much larger than those in interstellar space with radii between 2μm and 4.4mm distributed in size according to n(a) α $a^{-3.5}$ (Mathis et al., 1977). Such a distribution, which is equivalent to dn(m) α $m^{-1.8}$ dm, has been shown to arise from collisional fragmentation (Hellyer, 1970). Absorption and scattering efficiencies Q are calculated from Mie theory; the optical constants of the dust material are taken from Draine (1985). The dust is distributed in the models in a spherical shell of constant density. It is fortunate that the angular distribution of matter around the star need not be known in these optically thin configurations: the temperature of a grain and its emission depend solely on its distance to the star. Distribution of the dust in a disk with the same inner and outer boundary as the shell would result in an identical spectrum as long as the disk absorbs the same fraction of stellar light as does the shell. Model parameters are given in Table 2.

TABLE 2. Description of models: spectral type SpT; distance D in pc; stellar temperature T_*; inner and outer radius R of dust cloud in AU; minimum and maximum grain radius a in μm (the latter is only a lower limit); dust mass M_d of model cloud in $M_\odot$ (again only a lower limit); grain temperatures T at inner and outer radius of cloud (they are approximate as they depend weakly on grain size); fraction of IR luminosity from dust.

	D	SpT	T_*	R_i/R_o	a_{min}/a_{max}	M_d	T_i/T_o	L_{FIR}/L_{tot}
Vega	8.1	A0V	9700	40/74	18/4374	1.8E-8	140/90	3 10-5
α PsA	6.7	A3V	8800	26/71	54/ 486	2.2E-8	120/60	6 10-5
τ_1 Eri	13.7	F6V	6400	20/52	6/4374	5.3E-8	95/55	4 10-4
ε Eri	3.3	K2V	4900	4/25	162/ 486	4.2E-9	135/45	8 10-5

The measured spectral points are depicted in Fig. 1 by asterisks. The values of the model are explicitly shown only if they deviate from the observations by more than 30%. Then they are marked by crosses. Smaller differences are irrelevant in view of the overall inaccuracies. The model always takes into account the observational beam width. We restrict the following discussion to the spectrum of Vega.

The model fit reproduces the observations very well. There is a cross only at 0.8mm, where the model gives a value four times larger than that observed by Becklin and Zuckerman. Note that the observed fluxes at 0.8 and 0.87mm coincide, although they were obtained with different beams of 17" and 9", respectively. For a source size of 20" at least one of them must be erroneous. The model in Fig. 1 and Table 2 fits our data point at 0.87mm.

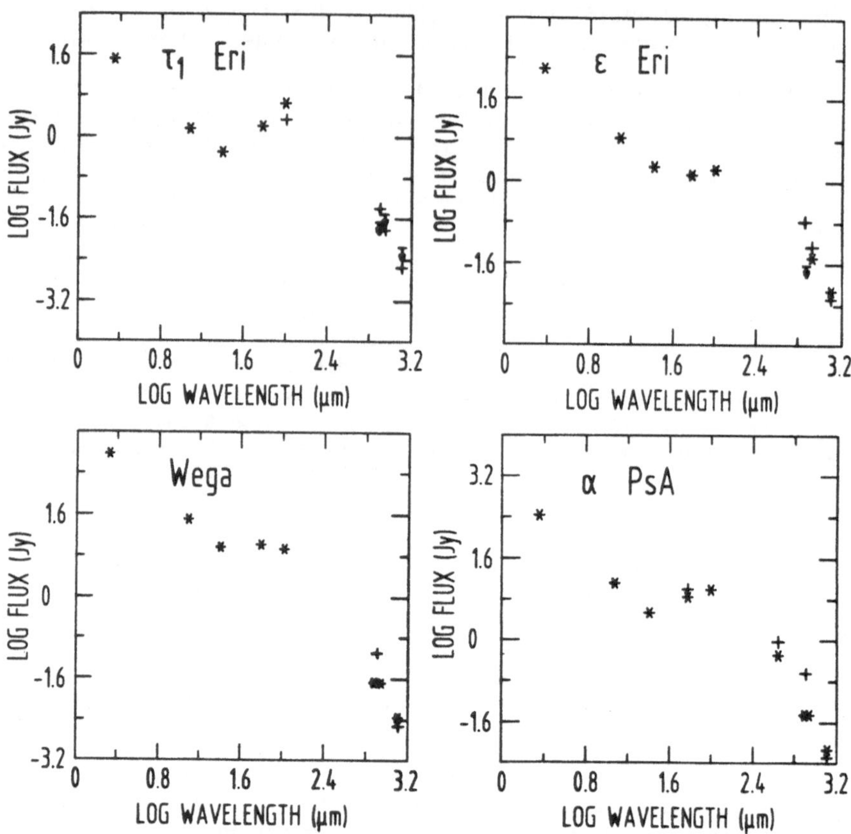

Figure 1. Observed spectra (asterisks; arrows for upper limits). Plotted are the fluxes at 2.2μm and the four IRAS bands (Aumann, 1985), at 0.45 and 0.8mm (Becklin and Zuckerman, 1989) and at 0.87 and 1.3mm from the present investigation. The results of the models described in Table 2 are presented separately by crosses if they deviate by more than 30% from the observations. As it is not clear from the figure, we mention that for Vega and α PsA the crosses at 0.8mm refer to the data of Becklin and Zuckerman.

The dust cloud lies between 40 and 74 AU from the star - an extension which is compatible with a size at 60μm of 20". The source is resolved at 0.87 and 1.3mm (10" correspond to 81 AU). The model for Vega contains grains with radii from 18 to 4374μm ≈ 4.4mm (Table 2). The grain temperatures are only mildly size dependent: at a given distance from the star the difference between the largest and smallest particles is less than 10%. The medium grain temperature falls from 140K at the inner edge to 90K at the outer. Our result about the sizes is in contradiction to Anandarao and Vaidya (1986) who claim that grains larger than 10μm cannot explain the fluxes.

To investigate the influence of the basic model parameters on the fit we varied inner and outer radius of the cloud, total dust mass and grain size. It turned out that the models were reasonably constrained. The extension of the dust cloud comes from the 60μm IRAS measurements, although the emitting region may be somewhat larger at submm wavelengths. When we varied in test runs the inner boundary of the dust cloud we found that if there are grains circling the star much closer than 40 AU they would produce excessive emission between 12 and 60μm. Smaller particles or even ordinary interstellar grains ($\ll 1\mu$m) also lead to unacceptably large fluxes in the IRAS bands at 12 and 25μm.

For a good agreement with the observations, the upper limit a_u of the grain size distribution may be reduced only slightly. We find reasonable fits down to $a_u = 486\mu$m. The corresponding mass is then 6.6 10−9 $M_\odot$. An open end to a_u, i.e. the existence of particles greater than 1cm, can, on the other hand, not be excluded from the models. They would not change the spectrum because they do not absorb much stellar light: For an $n(a) \propto a^{-3.5}$ distribution the bulk of the mass is locked up in the big particles, whereas the cross section is provided by the small ones. Therefore the dust mass and the upper grain radius given in Table 2 are only lower limits. Note that for interstellar dust at $\lambda \approx 1$mm the Rayleigh approximation is applicable ($a \ll \lambda$). Then $Q \propto a/\lambda$ and the observed flux density is proportional to the grain volume; this limits the total amount of dust M_d observationally. In the present measurements, however, a and λ are comparable and we cannot derive a limit to M_d.

The results of Table 2 can be summarized in the following way: The dust is distributed in a shell where the outer radius is two to three times larger than the inner radius. The total diameter is of order 100 AU and seems to be correlated with stellar luminosity in the sense that the dust cloud is larger for brighter stars. Grain temperatures range from 50 to 140K. If the grains, like those of the interstellar medium, have a size distribution $n(a) \propto a^{-3.5}$ or steeper then the diameter of the largest grains $2a_u$ cannot be determined. Our models, where $2a_u \approx 1$cm, give only lower limits to the total dust mass, typically 100 times less than the mass of the Earth. From the analysis of the spectra the presence of km-sized bodies is possible.

The smallest grains have radii of 10 to 100μm. The timescale t_{PR} for infall into the star due to the Poynting-Robertson effect is

$$t_{PR} = 4\pi c^2 \rho_g a R^2 / 3QL \tag{1}$$

($Q \approx 1$ is the efficiency for absorption and isotropic scattering in the visual, $\rho_g \approx 2.5$ g/cm^3 the grain material density, L the bolometric luminosity). Grains with $a = 10\mu$m are depleted in typically 10^5 yr. This probably accounts for the lack of small grains and the existence of the inner cavity. On the other hand, the grains may be replenished from grinding down larger bodies.

There is a curious relation between the far IR luminosity and the inflow of grains into the star due to the Poynting-Robertson effect. If there are N grains of radius a and mass m at distance R from the star, they produce a far IR luminosity $L'_{FIR} = L \cdot N\pi a^2/4\pi R^2$. They spiral into the star in a time given by Equ. (1). This leads to a mass loss rate

$$\dot{M} = Nm/t_{PR} = 4\ L_{FIR}/c^2 \qquad (2)$$

It is remarkable that $\dot{M}$ is independent of the distance of the particles and their diameter. L_{FIR}/L varies from 3 10-5 for Vega to 4 10^{-4} for τ_1 Eri. This implies a steady state mass loss rate $\dot{M}$ of order 10^{17} g/yr, which, over the lifetime of an A0V star (3 10^8 yr), is comparable with estimates for the mass of the Oort cloud around the Sun.

References:

Anandarao, B., Vaidya, D.: 1986, A&A **161**, L9.

Aumann, H., Gillett, F., Beichman, C., de Jong, T., Houck, J., Low, F., Neugebauer, G., Walker, R., Wesselius, P.: 1984, Ap.J. **278**, L23.

Aumann, H.: 1985, PASP **97**, 885.

Becklin, E., Zuckerman, B.: 1989, (preprint University of Hawaii).

Buitrago, J., Mediavilla, E.: 1985 A&A **148**, L8.

Draine, B.: 1985, Ap.J. Suppl. **57**, 587.

Hellyer, B.: 1970, MNRAS **148**, 383.

Kreysa, E.: 1985, Proc. URSI Intern. Symp. "mm- and submm-wave Astronomy", Granada, Spain, 11.-14.9.84, p. 153.

Mathis, J., Rumpl, W., Nordsieck, K.: 1977, Ap.J. **217**, 425.

H.W. YORKE: 1) Can you say anything about the geometrical distribution of large grains for these stars? 2) What would our solar system look like if viewed from the outside and how would it compare to the stars you observe?

E. KRÜGEL: 1) The fractional luminosity of the IR excess is very small and so is then the optical depth. The resulting spectrum is independent of the distribution of the grains and we cannot say anything about it. 2) Unfortunately, I don't know, although the answer to your question is contained in the zodiacal light measurements of IRAS, but it is not straightforward to extract it from there.

STAR FORMATION AND DISSIPATION IN GALACTIC DISCS

C.J. CLARKE
Institute of Astronomy
Cambridge CB3 0HA
U.K.

ABSTRACT. Disc galaxy evolution models may be classified according to the relative importance of star formation and dissipation in the disc gas. We discuss how the observed distribution of stars and of metals in galactic discs may be used to assess the degree of dissipation in the gas. We show that in order for viscously induced radial flows to have been important in the evolution of disc galaxies the timescale for the gas to dissipate its energy associated with local random motions must be $\sim 10^7$ years. Possible sources of dissipation in the disc gas are briefly discussed.

1. Introduction

The relative importance of dissipation and star formation in the interstellar gas is a crucial factor in determining the structure and evolution of galaxies (e.g Eggen, Lynden-Bell and Sandage 1962, Silk 1985, Lin and Pringle 1987a). This question can only be addressed from first principles, however, with a model for the gas that incorporates not only fragmentation and star formation but also the dissipation, on small scales, of energy extracted from the orbital motion of the gas. In the absence of such a model we are forced to turn the question around and ask: what may be learnt about the small-scale properties of the ISM from the global properties of galaxies? In this paper I discuss this issue in the context of galactic discs, considering, in turn, what are the constraints that the distribution of stars (Section 2) and of metals (Section 3) places upon the degree of dissipation in the disc gas. In Section 4 I briefly discuss possible viscosity mechanisms and their implications for models of galactic evolution.

2. Constraints On The ISM From The Mass Distribution In Galactic Discs.

Models for the formation of galactic discs may be broadly divided into those that envisage the *in situ* conversion of gas into stars (Larson 1976,Fall and Efstathiou 1986) and those in which radial gas flows in the disc plane play an important role (Silk and Norman 1981, Lin and Pringle 1987a). In models of the former type, the observed exponential decline of stellar surface density with radius (Freeman 1970) reflects the surface density of the gas on its arrival in the disc plane. Gunn (1981) showed that a roughly exponential profile results from the collapse of a uniformly rotating sphere, provided that every gas element preserves its angular momentum during the collapse (the so-called 'Mestel Hypothesis', Mestel 1963). If angular momentum exchange between gas elements is important during the collapse,

R. Capuzzo-Dolcetta et al. (eds.), Physical Processes in Fragmentation and Star Formation, 217–222.

however, a large amount of low angular momentum material may accumulate at small radii. In this case, the exponential stellar distribution must result from a combination of star formation and radial redistribution of gas in the disc plane. If t_* denotes the timescale on which a parcel of gas is entirely converted into stars and t_r is the timescale on which it changes its energy by order unity as a result of dissipation (thereby migrating over a radial distance of order its initial radius) then models of the former type correspond to $t_* << t_r$, whilst those of the latter to $t_* \geq t_r$.

The degree of viscous evolution undergone by galactic discs is however constrained by the present day value of the ratio R_J/R_M, where R_J and R_M are respectively the radii containing half the disc's angular momentum and mass. The effect of dissipation in a centrifugally supported medium is to redistribute angular momentum in such a way as to reduce the over-all energy: this involves an inward transport of mass and an outward transport of angular momentum (Lynden-Bell and Pringle 1974) and therefore causes R_J/R_M to increase with time. For an exponential disc R_J/R_M is of order unity for a wide range of rotation laws; Lin and Pringle (1987a) noted that this rules out the possibility that galactic discs are *extensively* viscously evolved and were thus led to reject the inequality $t_* >> t_r$.

If $t_* \sim t_r$, however, the combination of star formation and radial redistribution on comparable timescales conspires to produce an exponential stellar profile from *any* centrally condensed initial gas distribution (Lin and Pringle 1987a). This result is independent of the assumed rotation law and the prescriptions used to describe the star formation rate and the viscosity, provided only that $t_* \sim t_r$. Nor does the generation of an exponential disc require fine tuning between the values of these timescales (Clarke 1989); their rough equality however suggests that there is some physical connection between the processes of star formation and viscous dissipation in this scenario (Section 4).

3. Constraints On The ISM From The Metallicity Distribution In Galactic Discs.

The radial distribution of metals in galactic discs has often been used as a test of disc galaxy evolution models. Unbarred disc galaxies generally show a roughly exponential decline in primary element abundances with galactocentric radius (Shaver et al 1983, Janes 1979, Pagel and Edmunds 1981). This negative gradient has variously been explained in terms of preferential enrichment at small radii (e.g Rana and Wilkinson 1986, Wyse and Silk 1989) or else dilution of the outer disc by unenriched gas.

In the case of disc models with $t_* << t_r$ it is relatively easy to set up an abundance gradient by differential enrichment, because the degree of radial mixing of the interstellar gas is small. Moreover, in such models the outer disc is composed of high angular momentum material infalling from the halo, and it has been argued (e.g. Larson 1976) that the infall timescale for this gas may be comparable with the age of the disc, implying that unenriched gas is raining on to the outer disc at the present time. Matteucci and François (1989) have shown that this effect can produce an abundance gradient of the required magnitude.

If $t_* \sim t_r$ then radial flows tend to erase the effects of differential enrichment. Moreover, the outer disc is in this case formed from gas pushed out of the inner disc by viscous torques, which is pre-enriched by star formation at small radii. Consequently, such models predict, in their simplest form, rather shallow abundance gradients, with even an *increasing* metallicity in the outermost disc (Sommer-Larsen and Yoshii 1989).

Figure 1. Comparison between model Galactic metallicity distributions and observational data assembled by Lacey and Fall (1985). In each case a disc evolution model with $t_* \sim t_r$ and a star formation cut-off at 18 kpc is employed. The solid line corresponds to the effect of the advective flow only ($\nu_t = 0$). The dotted lines (i) to (iii) correspond to varying ratios of ν_t/ν: 1.,0.1 and 0.01 respectively.

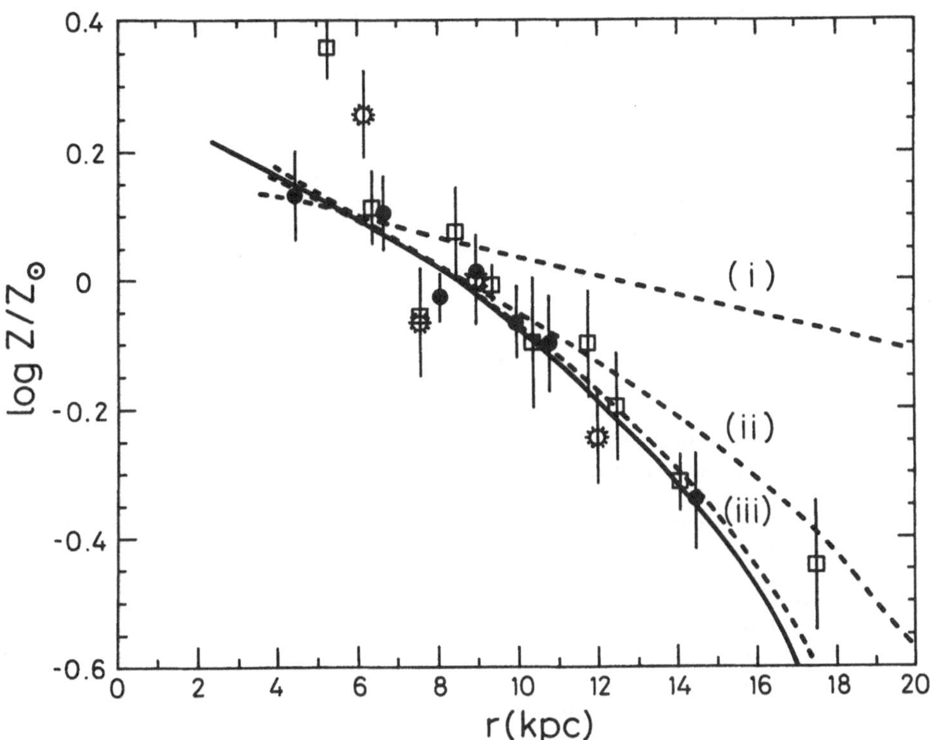

In all the models described above, it is assumed that star formation proceeds over the entire radial extent of the gas disc. This is probably incorrect, however. Star forming regions in disc galaxies are confined within a radius considerably less than that of the gas (Kennicutt 1989). An abrupt outer cut-off to the stellar disc is observed in many edge-on galaxies (Van der Kruit and Searle 1982) and there is indirect evidence for such an edge in more moderately inclined disc systems (Van der Kruit 1988) and also in the Galaxy (Djorgovski and Sosin 1989). The presence of a star formation cut-off implies a reservoir of low metallicity gas at large radii: Lacey and Fall (1985) first pointed out that radial inflow from such a reservoir would produce an abundance gradient, and demonstrated this by assuming an *ad hoc* radial flow velocity, constant in time and inwardly directed over the entire disc. The flow pattern is more complex if calculated self-consistently for a viscously evolving disc with $t_* \sim t_r$ (Clarke 1989): at first the flow is outwardly directed in the outer disc, but the point of flow reversal migrates outwards and after $\sim 5 \times 10^9$ years the flow is inwardly directed over the entire disc. The resultant metallicity gradient is shown by the solid line in Figure 1.

A word of caution is necessary here, however. Both Lacey and Fall (1985) and Clarke (1989) considered only the effects of the *mean* advective flow and did not treat the effect of diffusion of turbulent fluid elements against the mean flow. This latter effect is controlled by the turbulent diffusion coefficient ν_t which has the same dimensions as the kinematic viscosity, ν, that drives the mean flow. Morfill (1983) and Sommer-Larsen and Yoshii (1989) have argued that $\nu_t = \nu$ if the viscosity is turbulent in origin, although Clarke and Pringle (1988) dispute whether this is necessarily so. The dotted lines in Figure 1 show the effect of varying the ratio ν_t/ν from 0.01 to 1. Turbulent diffusion tends to erase the chemical gradient unless $\nu_t << \nu$.

In summary, therefore, there are numerous ways in which an abundance gradient can be produced if $t_* << t_r$. For models with $t_* \sim t_r$ a steep gradient results only if there is an outer cut-off to the stellar disc and if $\nu_t << \nu$.

4. Discussion

In the preceding sections we have classified disc galaxy models according to the ratio, t_r/t_*, of the radial flow timescale to the star formation timescale in the interstellar gas. We now briefly consider how the value of this ratio is related to the local properties of the disc gas.

The present day gas content of disc galaxies implies that t_* is of order the Hubble time. t_r is the timescale on which gas rotating at velocity v_{rot} can dissipate a specific energy of order v_{rot}^2. The *local* dissipation timescale, on which a specific energy $\sim \tilde{v}^2$ (where $\tilde{v}$ is the local gas velocity dispersion) can be extracted from the orbital motion of the gas is therefore $t_{diss} \sim t_r(v_{rot}/\tilde{v})^2$. Hence, employing typical observed values of $\tilde{v}$ and v_{rot} (6 km/s and 200 km/s respectively (Van der Kruit and Shostak 1984, Bosma 1981)):$t_{diss} \sim (t_r/t_*)10^7 years$.

Now, the existence of exponential discs rules out the possibility $t_* >> t_r$ and therefore one can reject models of the ISM in which $t_{diss} << 10^7$ years. It remains an open question, however, whether t_{diss} can be as short as 10^7 years, as required by disc models with $t_* \sim t_r$. Suggested sources of dissipation in the disc gas include shock heating in a non-axisymmetric potential (either generated by a stellar spiral or bar (Roberts and Shu 1972, Schwarz 1981) or by the development of self-gravitating modes in the gas (Lin and Pringle 1987b)) or else the effect of collisions between Giant Molecular Clouds (Fukunaga 1987). In each of these cases one can argue that regions of intense viscous dissipation are likely to be regions of intense star formation. However, the low value of t_{diss} is a potential problem for this class of model, especially in the outer disc where a value of t_{diss} of 10^7 years is only a few percent of the orbital period, t_{orb}. The condition $t_{diss} << t_{orb}$ implies that the radial length scale over which angular momentum is exchanged per orbit is much greater than the epicyclic excursion of the gas (or equivalently, in the vocabulary of viscous accretion discs, that the α -viscosity parameter (Shakura and Sunyaev 1973) $>> 1$). This suggests that collective effects, possibly involving the self-gravity of the gas, must be important.

5. Conclusions.

We have shown that two classes of disc galaxy evolution models result, depending on whether the gas can dissipate its energy associated with local random motions on a timescale as short as $\sim 10^7$ years. This issue will only be decided by further investigation of angular momentum transport associated with large scale fragmentation of the ISM. The value of the turbulent transport coefficient for the resultant flows will also determine whether such a model is compatible with the abundance gradients observed in disc galaxies.

References

Bosma,A., 1981. *Astr. J.*, **86**,1825.

Clarke, C.J., 1989. *Mon. Not. R. astr. Soc.*, **238**, 283.

Clarke, C.J. & Pringle, J.E., 1988. *Mon. Not. R. astr. Soc.*, **235**,635.

Djorgovski,S. & Sosin C. ,1989. *Astrophys. J. Lett.*, **341**,L13

Eggen, O.J.,Lynden-Bell, D. & Sandage, A.R., 1962. *Astrophys. J.*, **250**,262.

Freeman,K.C., 1970. *Astrophys. J.*, **160**,811.

Fukunaga, M., 1983. *Publ. astr. Soc. Japan*, **35**,173.

Gunn, J.E., 1981. In: *Astrophysical Cosmology, Proceedings of the Vatican Study Week on Cosmology and Fundamental Physics*,Sept. 28-Oct. 2, 1981,p. 233, eds Longair, M.S., Coyne,G.V. and Brück, H.A.. Pontifica Academia Scientarum, Citta del Vaticano.

Janes, K.A., 1979. *Astrophys. J. Suppl.*, **39**,135.

Kennicutt, R.C., 1989. *Astrophys. J.*, in press.

Lacey, C. & Fall, S.M., 1985. *Astrophys. J.*, **290**, 154.

Larson, R.B., 1976. *Mon. Not. R. astr. Soc.*, **176**,31.

Lin, D.N.C. & Pringle, J.E., 1987a. *Astrophys. J. Lett.*, **320**,L87.

Lin, D.N.C. & Pringle, J.E., 1987b. *Mon. Not. R. astr. Soc.*, **225**,607.

Lynden-Bell, D. & Pringle, J.E., 1974. *Mon. Not. R. astr. Soc.*, **168**, 603.

Matteucci, F. & François, P., 1989. *Mon. Not. R. astr. Soc.*, **239**,885.

Mestel, L., 1963. *Mon. Not. R. astr. Soc.*, **,186**, 479.

Morfill, G.E., 1983. *Icarus*, **53**, 41.

Pagel, B.E.J. & Edmunds, M.G., 1981. *Ann. Rev. Astr. Astrophys.*, **19** ,77.

Rana, N.C. & Wilkinson, D.A., 1986. *Mon. Not. R. astr. Soc.*, **218**, 497.

Roberts, W.W. & Shu, F.H., 1972. *Astrophys. J. Lett.*, **12**,L49.

Schwarz, M.P., 1981. *Astrophys. J.*, **247**,77.

Shakura, N.I. & Sunyaev, R.A., 1973. *Astr. Astrophys.*, **24**,337.

Shaver, P., McGee, R., Newton, L. & Danlo, A. 1983. *Mon. Not. R. astr. Soc.*, **204**, 53.

Silk, J., 1985.*Astrophys. J.*, **297**,9.

Silk, J. & Norman, C., 1981. *Astrophys. J.*, **247**,59.
Sommer-Larsen, J. & Yoshii, Y., 1989. *Mon. Not. R. astr. Soc.*, **238**, 133.
Van der Kruit, P.C., 1989. *Astr. Astrophys.*, **192**,117.
Van der Kruit, P.C. & Searle, L., 1982. *Astr. Astrophys.*, **110**,61.
Van der Kruit, P.C. & Shostak, G.S., 1984. *Astr. Astrophys.*, **134**,258.
Wyse, R.F.G. & Silk, J., 1989. *Astrophys. J.*, **339**,700.

Discussion

Boss: In order to explain the galactic abundance gradient, your model requires a viscous dissipation timescale, t_r, on the order of a hundred orbital periods. Can you estimate whether any of the mechanisms you suggested for generating turbulent dissipation (e.g. Giant Molecular Cloud collisions) could produce a timescale of this order?

Clarke: It's almost certain that GMC collisions won't do the job if you don't include self-gravity. The mean free path for a GMC is of the same order as its epicyclic excursion, implying that GMCs collide about once an orbit. This means a *local* dissipation timescale of an orbital period and a t_r a thousand times greater - far too long for our purposes. You need collective effects that ensure that angular momentum is transferred over more than the epicyclic radius each orbit. Lin and Pringle have suggested that self-gravity in the gas will do this, but nobody has checked this numerically.

Cayrel: There is evidence that old stars have less circular orbits than the recently formed stars. Is not the circularisation of the orbits a measure of the viscosity (by collisions between interstellar clouds) in your computations?

Clarke: Yes, it would be if you ascribe this effect to damping down of the vertical velocity of the clouds. But I think people prefer the explanation that the stars are *heated* because the increase in velocity dispersion with age is steepest for young stars, which is the opposite of what you'd expect from a gradually settling disc. The heating of the stellar population requires the presence of massive scatterers (e.g. GMCs, supermassive halo objects) but its not clear whether these control the dissipation in the gas. The timescale on which the disc can be torqued down by interaction with a population of halo objects is equal to the two-body relaxation timescale of the objects - far too long to be of interest.

STAR FORMATION IN STAR BURST GALAXIES

E. KRÜGEL
Max-Planck-Institut für Radioastronomie
Auf dem Hügel 69
D-5300 Bonn 1, FRG

1. Introduction

The present contribution does not aim at reviewing the subject, but rather intends to present some specific observational and theoretical results. For a summary on the topic of star bursts I may refer the reader to the almost up-to-date review papers by Scalo (1987) and Telesco (1988).

Star formation proceeds in some galaxies so vehemently that it is impossible to sustain this formation rate over the lifetime of the galaxy. The galaxy would run out of gas after a few 10^7 yr. This idea was published in 1975 by Rieke and Low and by Harwit and Pacini and later the term "starburst" was coined for this phenomenon. Modelling galaxy evolution, Larson and Tinsley (1978) found that peculiar galaxies with a recent history of vehement star formation cover a broad, but characteristic area in the UBV two-color diagram. The optical spectra of star burst galaxies, when low extinction allows them to be measured, are very similar to those of HII regions (Balzano, 1983) showing strong, narrow low-ionization emission lines. This is naturally explained if OB stars are the energy source.

Typically, star bursts occur in the center of a galaxy, although there are notable exceptions. The bulk of the luminosity is emitted in the far IR. The active nuclear region is marked by strong nonthermal radio emission from supernovae; it is well correlated with thermal $10\mu m$ radiation from dust around hot stars and has a linear size of 1 kpc or less (Condon et al., 1982). The $2\mu m$ radiation, on the other hand, may be an independent component arising not from recent star formation, but from an underlying stellar distribution. Details in the energy distribution depend on the parameters of the burst, like its strength, age and the amount of reddening. Successful models for the IRAS and optical fluxes from star bursts have been constructed by Belfort et al. (1987).

Much of our knowledge on star bursts comes from the study of the nearby galaxy M82 (reviews by Kronberg (1988) and Sofue (1988)), for which we possess an enormous wealth of data. There we can, at least indirectly, observe O stars on the main sequence by their ionizing effect through Brackett line emission, evolved massive stars (supergiants)

R. Capuzzo-Dolcetta et al. (eds.), Physical Processes in Fragmentation and Star Formation, 223–234.

through the CO absorption band at 2.3μm (Rieke et al., 1980), and exploded stars by a population of supernova remnants (Kronberg et al., 1985). Furthermore, the distribution of molecular gas is determined with high spatial resolution from the CO lines (Loiseau et al., 1989). Detailed modelling of M82 showed (Rieke et al., 1980) that a star burst can quantitatively explain the observed properties. Because of mass constraints it was also concluded that formation of massive stars is very much enhanced relative to the standard IMF. This seems to be a general finding as similar, only scaled-up results were obtained for the much more luminous (often called ultra-luminous) galaxies Arp220 and NGC6240 (Rieke et al., 1985). For these objects, however, an additional contribution to the total luminosity from a nonthermal compact source may be required. It should also be noted that a completely different mechanism for the production of their high luminosity has been proposed by Harwit et al. (1987). They suggest that two colliding galaxies radiate away part of their kinetic energy in the far IR.

Star burst galaxies cover a range of more than a factor of 1000 in bolometric luminosity L. At the low end are the blue compact dwarf galaxies with L of order 10^9 $L_\odot$. The main argument that they are bursting comes from their observed low metallicity, which would have to be much higher, had the star formation been sustained over an appreciable fraction of the Hubble time. At the high end, we find the "ultra-luminous infrared galaxies" discovered by IRAS with L $\gtrsim$ 10^{12} $L_\odot$, but an inconspicuous optical luminosity. The criteria for a galaxy to be classified as a star burst were never precisely defined and varied with the approach applied to study these objects. As shown above, high bolometric luminosity is not an appropriate way to characterize star burst galaxies. Any definition involving optical measurements (such as emission lines or a high L_{IR}/L_B ratio) is also only of limited use because it will break down for those galaxies which contain large amounts of gas and dust and are heavily obscured. We therefore look for another, hopefully better criterion.

2. L/M_g: an indicator of activity

As the interstellar gas is the material out of which stars are made and as a star burst dramatically increases the luminosity of the galaxy, bolometric luminosity over gas mass, L/M_g, should be a reliable indicator for the recent or present rate of star formation. This criterion is very simple because it involves only two basic quantities. Applying it to the Milky Way as a whole (L $\approx$ 5 10^{10} $L_\odot$, M_g $\approx$ 5 10^9 $M_\odot$), which is non-active, and to the central 1 arcmin of M82 (L $\approx$ 3 10^{10} $L_\odot$, M_g $\approx$ 10^8 $M_\odot$), we find a ratio L/M_g (solar units) of 10 and 300, respectively. This shows the range of values one may expect and we see that L/M_g is one or two powers of ten greater for star bursting regions. The criterion does, however, not discriminate against the way in which the gas produces the luminosity and may therefore also encompass activities other than star bursts.

So we wish to determine the stage of activity of a galaxy or its nuclear region by measuring bolometric luminosity and gas mass. After

IRAS it is straight forward to find L because the contribution to the luminosity from submillimeter wavelengths is not significant. Much more difficult is the measurement of the gas mass. M_g can be derived from the lines of trace molecules, such as CO, if hydrogen is molecular; from the 21cm line of HI, should hydrogen be mainly in its atomic form; and from the continuum emission of cold dust at wavelengths around 1mm. The latter procedure assumes that dust and gas are always mixed at a constant ratio. Once L and M_g have been determined, careful attention must be paid in extended sources to relate them to the same area on the galaxy, as they are usually obtained from observations involving different beam sizes.

Let us first recall how to derive gas masses from the dust continuum. Due to low optical depth, for a fixed mass ratio of dust to gas the measured flux S_ν is directly proportional to M_g: $S_\nu = K_\nu B_\nu(T_d) M_g D^{-2}$. The dust temperature T_d in the Planck function B is determined as a color temperature at two wavelengths, say, at 100μm and at 1300μm = 1.3mm. For this end, the frequency dependence of the emissivity is approximated by $\nu^m B_\nu(T_d)$. Fortunately, the particular choice of m is not crucial at a wavelength around 1mm. m = 2 is a good estimate as it was obtained as the mean value from submillimeter measurements of 13 galaxies (Chini et al., 1989a). K_ν is still a controversial quantity, but K = 0.003 cm^2 at 1.3mm per gram of interstellar matter is consistent with detailed modeling of galactic star forming regions (Chini et al., 1986a) as well as with our present knowledge of the optical properties of silicate grains coated with amorphous ice (Draine, 1985; Leger et al., 1983).

Clouds of atomic hydrogen are part of the total gas component. They are not the site of star formation, but constitute a reservoir for future molecular clouds. Whereas the CO surface density in extragalactic objects generally increases towards the center (however, central depressions also occur) (Young, 1987), the HI surface density is roughly uniform over the galactic disk and may therefore in the nucleus, to first order, be neglected with respect to CO.

The only molecule that can yield estimates for the total gas mass is CO. Both the main isotope CO and ^{13}CO may be used. One usually observes the lowest rotational transitions, (J=1−0) or (J=2−1) at 2.6 and 1.3mm, respectively, as they are easily excited. The ^{13}CO line suffers less from atmospheric attenuation, but is much weaker. Where it can be observed, ^{13}CO, although very likely optically thin, is not really superior to CO for deriving M_g because of the uncertainty in the abundance ratio $[^{13}CO]/[H_2]$.

For the main isotope, the CO-luminosity L_{CO} has been shown to be a good measure for the gas mass in the case of galactic clouds. L_{CO} is defined as the integrated line intensity $T_A dv$ times the beam area πR^2 on the source. If L_{CO} is given in [Kkm pc^2 s^{-1}] and M_g in solar masses then $M_g = 8\, L_{CO}$ (Young and Scoville, 1982). This result can be understood if the clouds are virialized, whence $dv = \sqrt{GM_g/R}$. For optically thick line emission filling the beam the antenna temperature equals the kinetic gas temperature, so $M_g = \sqrt{4\rho/3\pi G}\, T L_{CO}$. For T ≈ 20 K and $\rho \approx 10^{-20}$ g/cm^3 one obtains the above experimentally derived relation $M_g = 8\, L_{CO}$. This proportionality is also valid for sources smaller than the beam as the changes in T_A and πR^2 cancel each other. Can one apply it to galaxies? Their CO profiles are very broad (≈ 200 km/s) and the line width is determined by the stellar component and no longer by the gas. One now envisages the observed broad line as a superposition of many narrower lines, each representing one virialized cloud, which do not overlap in the phase space formed by the two spatial coordinates on the sphere and the radial velocity of the clouds (Dickman et al., 1986).

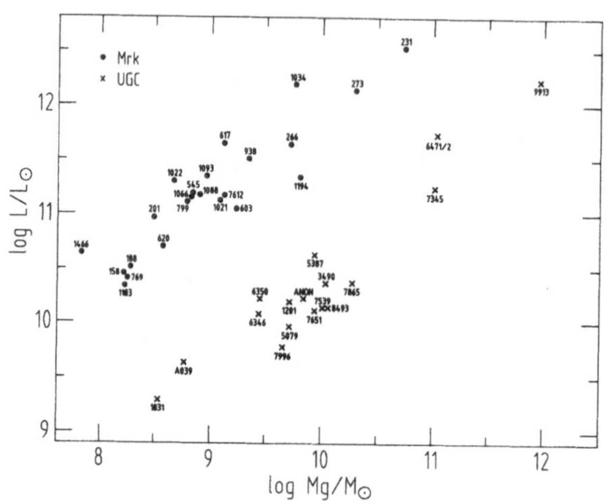

Figure 1. Gas mass from 1.3mm dust emission vs. luminosity for Markarians and for bright IRAS galaxies (UGC numbers). The bright IRAS galaxies contain all detections from the sample of Chini et al. (1986b). L refers to the wavelength band from 10 to 1300μm. Including the optical emission would raise their luminosity by typically 30%. For the Markarians, mainly starburst galaxies and Seyferts of both types, L denotes the bolometric luminosity (Krügel et al., 1988a,b).

The relation between L and M_g has been determined for a large number of galaxies from CO measurements (usually in the form: far IR luminosity vs. CO(J=1−0) luminosity) and for a smaller sample of about 50 objects from the dust continuum. CO data are presented for a wide range of galaxies, for example, in Fig. 3 of Young (1988) based on the FCRAO Extragalactic CO Survey; dust continuum data are displayed in Fig. 1. The results from these two methods are qualitatively similar, although discrepancies exist in detail. There is agreement that the bolometrically brighter galaxies are associated with greater gas masses and that the most luminous galaxies contain much more interstellar matter than the Milky Way. There is also consensus that the IRAS flux ratio S_{60}/S_{100}, or the corresponding dust color temperature, increases with activity.

However, dust continuum observations show a remarkable separation of galaxies in the L/M_g-plane into two modes. This is much less pronounced in the CO data, although present, as Solomon and Sage (1988;

see their Fig. 3) find a distinction between isolated and strongly interacting systems. The dust data of Fig. 1 contain, on the one hand, Markarian galaxies. They are all active in some sense, for example, according to the original definition of Markarian and coworkers (UV excess and emission line spectrum). Nowadays they are classified mainly as star bursts and Seyferts. Their L/M_g-ratio is around 100 in solar units and they were picked as the brightest 100μm sources ($S_{100} \geq 10$Jy) from the Markarian catalog. On the other hand, in Fig. 1 are plotted galaxies, mainly spirals, which were chosen from the IRAS Point Source Catalog. Their selection criterion was also strong 100μm emission ($S_{100} \geq 50$ Jy). They have an L/M_g-ratio of only 5 and their dust color temperature between 100 and 1300μm is only 16 K, instead of 32 K for the Markarians. These galaxies are nonactive.

For both samples in Fig. 1 gas mass is roughly proportional to luminosity and thus also to the star formation rate. Schmidt (1959) assumed that the star formation rate is related to some power n of the gas density; estimates for it range between 1 and 2. Fig. 1 suggests that the exponent in Schmidt's law is unity independent of the stage of activity.

It has been pointed out (for instance, Lequeux (1988)) that a correlation of the kind L vs. M_g can be deceptive: If one is dealing with a flux limited sample, distant objects have to be strong in both L and M_g in order to be detected, whereas near objects are intrinsically weak. This automatically creates a correlation. (This warning applies, of course, also to the CO data.) One should therefore try to enlarge the two samples of Fig. 1 in order to check whether the tightness of the dependence between L and M_g is confirmed or not.

Another point of concern is that the jump in L/M_g from the nonactive galaxies to the Markarians may be produced, at least partially, by an observational artifact. M_g was determined for the Markarians from measurements with an 11 arcsec beam, for the nonactive galaxies from measurements with a 90 arcsec beam. As both types of galaxies were extended, one might expect to obtain for the nonactive sample larger gas masses. This would imply smaller L/M_g-ratios, should the luminosity come from a compact source. However, the Markarians are more distant and the linear scale of the area observed was for them, on the average, still half as large as for the nonactive sample. This makes it unlikely that the drastic increase in L/M_g from the UGC to the Mrk sample is due to a beam size effect. Furthermore, if the relation L/M_g were sensitive to beam size one would expect both samples to show a larger scatter. Complementary measurements of these galaxies, either in the dust continuum or in a CO line, at the same wavelength, but with different beams would help clarify this point. But already now one can conclude from observations of some of the Markarians in Fig. 1, made at 870μm with a very small 8 arcsec beam (Chini et al., 1989a), that the interstellar matter in these objects is concentrated in a compact ($\leq$ 8 arsec) region.

Obviously, a critical evaluation of the trustworthiness of the gas masses is necessary. Furthermore, in some cases CO and dust measurements give very divergent numbers. For example, from the dust continuum one derives for UGC9913 = Arp220 (Fig. 1) from a 3.5-σ detection in a 90" beam a gas mass of 1.0 10^{12} $M_\odot$ (Chini et al., 1986b), from a map

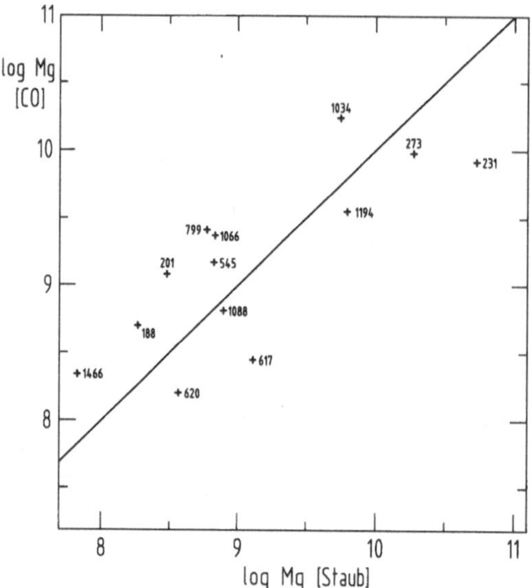

Figure 2. For some of the Markarian galaxies in Fig. 1 we measured the CO(2−1) line and determined the gas mass M_g[CO] from the CO luminosity. It can be directly compared to the gas mass derived from the 1.3mm dust continuum, M_g[dust], because the observations were made with the same 11" beam. The solid line is the locus of equal values for M_g[CO] and M_g[dust].

with the IRAM 30m over a 30" area 1.1 10^{11} $M_\odot$ (R. Zylka, priv. comm.), while Sanders et Mirabel (1985) obtain from CO in a 45" area 1.3 10^{10} $M_\odot$. With the goal of clarifying the discrepancy between the two methods we carried out a comparison of gas masses derived from CO lines and from dust emission (Krügel et al., 1990). To minimize errors, the measurements were performed using the same telescope (IRAM 30m) and beam for both the dust continuum and the CO(J=2−1) line. The resulting values for M_g agree within a factor of three (Fig. 2). This is encouraging considering the intrinsic uncertainties. There is a tendency for bolometrically very bright galaxies to show lower mass estimates from CO than from dust. For a number of reasons, such as that emission from grains is optically thin and that their abundance is not much affected by the environment, masses derived from dust are probably more reliable.

3. The physical origin of star bursts

The problem of finding the mechanism responsible for the violent star formation has not yet been solved convincingly. The majority of star burst galaxies shows clear evidence for tidal interaction such as long tails, a disturbed morphology or double nuclei. This was already noted by Larson and Tinsley (1978) and has been corroborated in all following studies. It is therefore at the basis for most theories that try to explain star bursts. A dynamical scheme that funnels the gas during the merging or interaction process into the nuclear region, where the bursts are generally observed, was first devised by Toomre and Toomre (1972). In a later study it was suggested that a decisive role in driving the gas into the nucleus plays the bar, which forms in the stellar component of the perturbed galaxy (Noguchi, 1988).

Despite the outstanding significance of interaction, it does not automatically lead to activity: For 30% of the objects from a large sample of strongly interacting spirals Bushouse (1986) did not find any sign of rapid star formation. More important, many active galaxies do not seem to be under the influence of a companion at all. For example, among the Markarians in Fig. 1 less than 60% show signs of interaction according to the catalog of Markarian galaxies by Mazzarella and Balzano (1986). Likewise, Sanders et al.(1988) conclude from a study of the IRAS Bright Galaxy Survey that the fraction of galaxies, which are interacting, decreases with luminosity. No isolated systems are found for $L > 10^{12}$ $L_\odot$. But at $L \leq 10^{11}$ $L_\odot$ three quarters of them are singles, and the majority of these are probably star bursts. Of course, a galaxy may reveal its gravitational interaction only on closer scrutiny; examples are the detection of close double nuclei or of a joint HI halo between two systems. Therefore the number of single galaxies cited by Sanders et al. (1988) may be overestimated. However, even if it is only approximately correct there must be a fair amount of systems where the star burst has not been triggered from outside. In these cases an intrinsic mechanism must operate in the center of the galaxy.

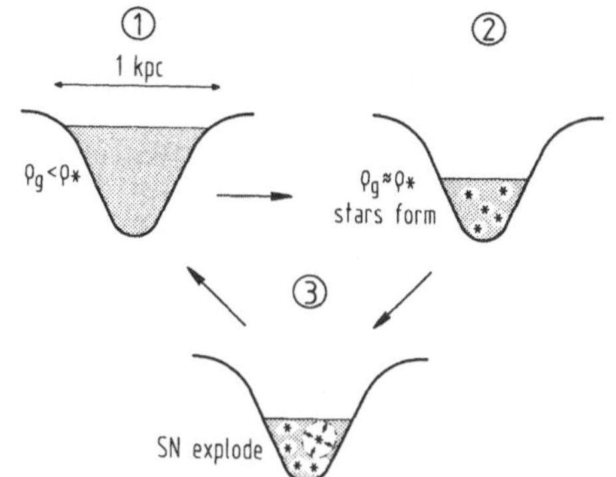

Figure 3. A schematic picture illustrating the scenario of repetetive bursts of star formation for an isolated galactic nucleus according to Loose et al. The shaded region denotes the distribution of gas. Stars and supernovae resulting from the burst are indicated. The underlying population of old stars is not shown, but it is responsible for the gravitational potential.

Such a mechanism was proposed by Loose et al. (1982) and worked out in one-dimensional, spherically symmetric hydrodynamical calculations. The study followed in time the dynamical evolution of the gas in a region slightly less than 1 kpc across in an environment similar to that of the center of the Milky Way. The center of the Milky Way was chosen because its fundamental parameters, like stellar mass and gas content, are known.

There are several particularities for the dynamics of the gas in a galactic center. First, the gas is trapped in the potential well created by the stars. Input of fresh gas can come either from mass loss from the old population of low mass stars (assumed to be of order 0.1 $M_\odot$/yr) or from outside. Second, virial and turbulent velocities of the gas are very high ($\gtrsim$ 100 km/s). Cloud collisions lead to dissipation of the turbulent energy and the gas will sink to the bottom of the gravitational potential well in a time given by nuclear size over characteristic velocity (about 10^7 yr). Third, star formation can only occur when the protostellar clouds are safeguarded against tidal disruption. This leads to a criterion more stringent than that of Jeans, namely that the gas density exceeds the smeared out stellar density (Kaplan and Pikelner, 1976). Fourth, the efficiency of a supernova for creating turbulence increases with the turbulent velocity of the medium, in which the supernova explodes. In a galactic nucleus we may expect that more than 10% ($\gtrsim 10^{50}$ erg) of the kinetic energy of a supernova goes into turbulence; for the galactic disk this number is several times lower. These particularities are incorporated in the gas dynamical equations of Loose et al.

The sequence of events for a galactic center is schematically pictured in Fig. 3. Starting at a time of inactivity, when there are no OB stars, the gas, due to dissipation of turbulence, sinks inward until its density reaches the critical value for star formation to set in. This happens on a time scale t_{dis}, say, of 10^7 yr. After several million years the first supernovae explode. If, as we may expect, the IMF is biased towards massive stars, there will be enough SN events to dispel the interstellar matter from the inner regions and stop the burst before the gas has been exhausted in star formation. The gas can reassemble again towards the inner regions, once the last supernovae have gone off. This happens 60 10^6 yr after the beginning of the burst if all stars with 6 $M_\odot$ or more explode. The cycle will be repetitive if the loss of gas in star formation is balanced by inflow of gas from old stars or from outside.

In a series of models calculated with varying parameters (Loose, 1981; Loose et al., 1982) periodic bursts were obtained whenever the lower limit m_l of the Salpeter IMF was raised to 1 $M_\odot$ or above. The burst enhanced the luminosity of the galactic center by about a factor of 20, which is also the luminosity ratio between active and nonactive galaxies in Fig. 1. The time between subsequent bursts was typically 10^8 yr, each burst lasting for about 2 10^7 yr. During the burst some 25% of the gas were converted into stars.

A severe shortcoming of the models is the neglect of rotation. Its presence favors the formation of a disk, and matter perpendicular to the rotation axis can flow to the bottom of the potential well only if angular momentum is transported to the outside. This introduces a new time-scale. Rotation would also prevent the outer zones of the galactic center from becoming highly devoid of gas during the compression phase. The latter point is the reason why comparison of the theoretical scenario with observations of M82 and NGC253 gave only rough agreement (Krügel et al., 1983). The IR spectra did not fit in detail because they were based on exactly the dust distribution of the dynamical models, which had, due to the absence of rotation, not enough obscuring matter in the outer regions.

The viability of the scheme by Loose et al. rests on a few parameters whose values are poorly known, such as time-scale for dissipation of turbulence, rate of gas supply, lower boundary of the IMF, efficiency for creating turbulence in supernova explosions. Nevertheless, with reasonable assumptions for these quantities it gives an explanation why a nucleus with a moderate gas content of an apparently isolated galaxy, like NGC253, can become active. It may, in principle, also be applied to the very bright star burst galaxies. But it does not provide an answer how the large amount of gas ($\geq 10^{10}$ $M_\odot$) observed in these objects has accumulated in their nuclear region of kiloparsec scale, nor has the model been tested under such extreme conditions.

4. Concluding remarks

Several attempts have been made to develop a coherent scheme or evolution ary scenario for the various kinds of activity connected with star bursts, Seyferts and quasars. For example, Weedman (1983) speculated that a star burst is the prerequisite for a Seyfert 1 nucleus, or even a quasar, in the sense that it produces a large number of relativistic objects, the stellar remnants of the massive stars, which partly accumulate in a very small volume (< 1pc) in the center of the galaxy and form the large mass accretor that then induces the activity.

A connection between high star formation rates and the Seyfert phenomenon was also demonstrated by Espinoza et al. (1987) who found, from comparing IRAS fluxes with those at shorter wavelengths, that in Seyferts the far IR emission generally originates from star formation and not from the "active" nucleus and is uncorrelated to the latter. That a high fraction of the far IR luminosity in Seyferts comes from star formation is corroborated by detailed studies of individual galaxies, such as NGC1068 (Telesco et al., 1984). Furthermore, all Markarians in Fig. 1, among them seven Seyferts, have clear thermal far IR and submillimeter spectra according to their spectral index between 0.1 and 1.3mm which, as it is greater than 2.5, makes synchrotron self-absorption unlikely.

The possibility that the different types of Seyferts can be unified and that our present classification is only due to the difference in viewing angle is discussed by Miller (1989). Norman et Scoville (1988), on the other hand, try to link the superluminous and very gas-rich galaxies, like NGC6240 or Arp220, to the quasars. They advocate that in the center of these galaxies a black hole forms as a result of mass loss from the stars created in the burst. The accretion of gas onto the black hole is then sufficient to explain the high luminosity of quasars. Here again the star burst is the precursor of the following activity. In a similar attempt of unification, Heckman et al. (1989) sketch the evolution from a star burst to a quasar.

The location of the Markarians in a well defined strip of the L/M_g-plane (Fig. 1) makes it plausible that there is a common origin among these galaxies of different activity class of converting interstellar matter into luminosity. The fact that even most radio-quiet quasars follow the same L/M_g-ratio and that their far IR emission is probably also thermal radiation by dust (Chini et al., 1989b) suggests a general underlying scheme.

All these ideas are very speculative. Nevertheless, they encourage us that the study of the star burst phenomenon, fascinating in itself, will provide insight into the riddles of nuclear galactic activity in general.

References

Balzano, V. (1983) Astrophys. J. 268, 602.
Belfort, P., Mochkovitch, R. and Dennefeld, M. (1987) Astron. Astrophys. 176, 1.
Bushouse, H. (1986) Astron. J. 91, 255.
Chini, R., Krügel, E. and Kreysa, E. (1986) Astron. Astrophys. 167, 315.
Chini, R., Kreysa, E., Krügel, E. and Mezger, P. (1986b) Astron. Astrophys. 166, L8.
Chini, R., Krügel, E., Kreysa, E. and Gemünd, H.P. (1989a) Astron. Astrophys. 216, L5.
Chini, R., Kreysa, E. and Biermann, P. (1989b) Astron. Astrophys. 219, 87.
Condon, J., Condon, M., Gisler, G. and Puschell, J. (1982) Astrophys. J. 252, 102.
Dickman, R., Snell, R. and Schloerb, F. (1986) Astrophys. J. 309, 326.
Draine, B. (1985) Astrophys. J. Suppl. 57, 587.
Espinoza, J., Rudy, R. and Jones, B. (1987) Astrophys. J. 312, 555.
Harwit, M. and Pacini, F. (1975) Astrophys. J. 200, L127.
Harwit, M., Houck, J., Soifer, B. and Palumbo, G. (1987) Astrophys. J. 315, 28.
Heckman, T., Blitz, L., Wilson, A., Armus, L. and Miley, G. (1989) Astrophys. J. 342, 758.
Kaplan, S. and Pikelner, S. (1976) in S. Pikelner (ed.), Formation and Evolution of Galaxies and stars, Moscow (in Russian).
Kronberg, P. (1988) in R. Pudritz and M. Fich (eds.), Galactic and Extragalactic Star Formation", Kluwer Acad. Publ., Dordrecht, p.391.
Kronberg, P., Biermann, P. and Schwab. F. (1985) Astrophys. J. 291, 693.
Krügel, E., Tutukov, A. and Loose, H. (1983) Astron. Astrophys. 124, 89.
Krügel, E., Chini, R., Kreysa, E. and Sherwood, W.A. (1988a) Astron. Astrophys. 190, 47.
Krügel, E., Chini, R., Kreysa, E. and Sherwood, W.A. (1988b) Astron. Astrophys. 193, L18.
Krügel, E., Steppe, H. and Chini, R. (1990) Astron. Astrophys. (in press).
Larson, R. and Tinsley, B. (1978) Astrophys. J. 219, 46.
Leger, A., Gauthier, S., Defourneau, D. and Rouan, D. (1983) Astron. Astrophys. 117, 164.
Lequeux, J. (1988) in R. Wolstencroft and W. Burton (eds.), MM and Submm Astronomy, p. 249.
Loiseau, N., Nakai, N., Sofue, Y., Wielebinski, R., Reuter, H.P. and Klein, U. (1989) Astron. Astrophys., in press.
Loose, H. (1981) Thesis, University of Göttingen.
Loose, H., Krügel, E. and Tutukov, A. (1982) Astron. Astrophys. 105, 342.
Mazzarella, J. and Balzano, V. (1986) Astrophys. J. Suppl. 62, 751.
Miller, J.S. (1989) IAU Symp. 134, 273.
Noguchi, M. (1988) Astron. Astrophys. 203, 259.
Norman C. and Scoville, N. (1988) Astrophys. J. 332, 124.
Rieke, G. and Low, F. (1975) Astrophys. J. 197, 17.
Rieke, G., Lebofsky, M., Thompson, R., Low, F. and Tokunaga, A. (1980) Astrophys. J. 238, 24.
Rieke, G., Cutri, R., Black, J., Kailey, W., McAlary, C., Lebofsky, M. and Elston, R. (1985) Astrophys. J. 290, 116.

Sanders, D. and Mirabel, I. (1985) Astrophys. J. 298, L31.
Sanders, D., Soifer, B., Elias, J., Madore, B., Matthews, K., Neugebauer, G. and Scoville, N. (1988) Astrophys. J. 325, 74.
Scalo, J. (1987) 10th European Regional IAU Astronomy Meeting, Vol.4, p. 101.
Schmidt, M. (1959) Astrophys. J. 129, 243.
Sofue, Y. (1988) in R. Pudritz and M. Fich (eds.), Galactic and Extra-galactic Star Formation, Kluwer Acad. Publ., Dordrecht, p. 409.
Solomon, P. and Sage, L. (1988) Astrophys. J. 334, 613.
Telesco, C. (1988) Ann. Rev. Astron. Astrophys. 26, 243.
Telesco, C., Becklin, E., Wynn-Williams and C. Harper, D. (1984) Astrophys. J. 282, 427.
Toomre, A. and Toomre, J. (1972) Astrophys. J. 178, 623.
Weedman, D. (1983) Astrophys. J. 266, 479.
Young, J. (1987) IAU Symposium No. 115, 557.
Young, J. (1988) in R. Pudritz and M. Fich (eds.), Galactic and Extra-galactic Star Formation, Kluwer Acad. Publ., Dordrecht, p. 579.
Young, J. and Scoville, N. (1982) Astrophys. J. 258, 467.

H. ZINNECKER: I would like to offer a physical speculation for the absence of low mass stars in the IMF of star burst nuclei, required in your model of star burst oscillations in galactic nuclei. As soon as the first OB stars form, their UV rardiation will penetrate the adjacent clumpy clouds leading perhaps to clump implosion (cf. M17SW). The high UV radiation field inside the clumpy clouds will tend to erase the low-mass clumps and only the dense, more massive clumps wil survive to form new massive stars by radiative implosion (cf. LaRosa 1983 ApJ; Klein et al. 1986 IAU Symp. 115; Zinnecker 1989 IAU Symp. 120).

R. CAYREL: According to your model one galaxy out of ten should be an active galaxy. How does that compare with statistics of active/normal galaxies?

E. KRÜGEL: For the nearby galaxies, where already fairly low levels of activity (a few 10^{10} $L_{\odot}$) can be observationally distinguished, there seems to be rough agreement.

C.J. CLARKE: I think there is a potential problem with sustaining a number of successive bursts in this way. It is not clear that the injection of thermal energy into the gas by supernovae will cause the re-expansion of the star forming region. Indeed, since the gas is mainly centrifugally supported in the nucleus, the increased turbulence induced by the heating may give rise to angular momentun redistribution and hence a net inflow. So an alternative, even probable, scenario would involve the formation of a massive central object.

E. KRÜGEL: The models did not include rotation. We obtained a massive central object in the series of models that were calculated whenever there were not enough SN explosion to dispel the gas. This happened, for example, for a normal IMF, not depleted in low mass stars.

J. SCALO: 1. You mentioned that a significant fraction of the Markarian galaxies are isolated, and so probably require some internal mechanism to trigger bursts. In fact, it appears more generally that most star burst galaxies (e.g. galaxies with small gas depletion times) show no signs of interaction or bars; it is only the most powerful star burst galaxies which have a large fraction of interactions, although even here the fraction is not 100%. It is also important to point out that a significant fraction of star bursts are non-nuclear.

2. If you look at the frequency distribution of star formation rates per unit gas mass for a basically magnitude limited sample of galaxies later than Sb, including star bursts, it does not appear that the distribution is bimodal, as for your two samples, although it is possible that the uncertainties in star formation rates and gas masses could be hiding it. I would take this to mean that there is a continuous range of burst amplitudes. Maybe the separation you find is due to the different selection criteria for the two sample. That does not affect the agreement with your model or others, in which I think a range of amplitudes in L/M_g is expected.

A. DI FAZIO: Endrik, you know that for a subset of active nucleus galaxies (i.e. the radio galaxies) we observe for the extended radio lobes, using the spatial separation and the estimated velocity, a time separation between the "puffs" of about 108yrs. Trying to think of these as the result of your bursts (maybe in the nucleus), we would like to know the order of magnitude of the burst emission power and the mean peak wavelength.

E. KRÜGEL: The bursts that we calculated were much more moderate in power and I hesitate to speculate that they could also be responsible for the puffs and the radio lobes.

THE NEW BRILLIANT BLUE COMPACT DWARF GALAXY: A TEST ON THE EARLY STAGES OF GALAXY EVOLUTION

YU. I. IZOTOV[1], V. A. LIPOVETSKII[2], N. G. GUSEVA[1], A. YU. KNIAZEV[2], J. A. STEPANIAN[3]

[1] Main Astronomical Observatory of the Academy of Sciences of the Ukrainian SSR, Kiev, USSR

[2] Special Astrophysical Observatory of the USSR Academy of Sciences, Nyzhny Arhyz, USSR

[3] Byurakan Astrophysical Observatory of the Academy of Sciences of the Armenian SSR, Byurakan, USSR

ABSTRACT. A blue compact dwarf galaxy SBS 0335 - 052 from the Second Byurakan Survey with extremely low heavy element abundance is found. The oxygen abundance in SBS 0335 - 052 is 77 times lower than the solar value and it is 1.7 times lower than that in I Zw 18. The electron temperature of gas is extremely high and equals 24800 K. SBS 0335 - 052 is a probable candidate for a young galaxy.

1. INTRODUCTION

The blue compact dwarf galaxies (BCDGs) attract attention by their active star formation processes and by their evolutionary youth. There are among the BCDGs the probable young, recently formed galaxies in the Galaxy vicinity, experiencing the first bursts of star formation. Hence, the search of extremely deficient BCDGs is very important. Unfortunately, the undertaken attempts to discover galaxy more heavy element deficient than I Zw 18 have failed (Augarde et al. (1987), Iovino et al. (1988)). Only recently Skillman et al. (1987) have reported the results of spectrophotometric observations of the faint galaxy GR 8, having the nearly I Zw 18 heavy element abundance.

The star formation processes in blue compact dwarf galaxies may proceed probably in different way in comparison to star formation processes in our Galaxy:

- because of lower metal content;
- dust properties in BCDGs may significantly differ from those in Galaxy;
- probably there are no giant molecular clouds in the BCDGs. The observations give us only the upper limits of CO luminosities of BCDGs.

R. Capuzzo-Dolcetta et al. (eds.), Physical Processes in Fragmentation and Star Formation, 235–240.

2. PHYSICAL CONDITIONS AND CHEMICAL COMPOSITION IN BLUE COMPACT DWARF GALAXY SBS 0335 - 052

The blue compact dwarf galaxy SBS 0335 - 052 from the Second Byurakan Survey was spectrophotometrically observed for the first time in November, 1988 by Izotov et al. (1989). This galaxy was found to have an exceptionally low value of heavy elements abundance as compared with known BCDGs. The equatorial coordinates of SBS 0335 - 052 are α_{1950} = $03^h\ 35^m\ 18^s$, $\delta_{1950} = -\ 05^{\circ}\ 12'\ 34''$, and distance is equal to d = (73.7 ± 1.5)$\cdot h_{50}$ Mpc, where h_{50} = H / 50, H is Hubble constant. The absolute photographic magnitude is $M = -\ 16.3^m$. On 25 January, 1989 spectrophotometric observations of the SBS 0335 - 052 on the echelle-spectrograph on the 6-meter telescope with high signal-to-noise ratio were carried out. The spectrum of the SBS 0335 - 052 is shown on Fig. 1.

Electron temperature T_e and number density N_e as well as the chemical composition of the galaxy SBS 0335 - 052 are given in Table 1. It is

TABLE 1. Chemical composition of the ionized gas in SBS 0335 - 052

	SBS 0335 - 052	I Zw 18		Sun (Gre- vesse (1984))
		Lequeux et al.(1979)	Dufour et al.(1988)	
T_e , K	24800			
N_e , cm^{-3}	250			
12 + log(He/H)	11.10	10.88	----	11.00
12 + log(O/H)	7.01	7.18	7.24	8.90
12 + log(Ne/H)	6.55	6.49	6.56	8.00
12 + log(Ar/H)	5.13	----	----	6.58
12 + log(S/H)	5.12	----	5.46	7.21
12 + log(N/H)	< 5.88	----	5.99	7.95

a matter of common knowledge that I Zw 18 and GR 8 are the most heavy element deficient galaxies, because of their oxygen abundances are about 1/40 (Lequeux et al. (1979)) and 1/30 (Skillman et al. (1988)) of solar value, respectively. As follows from Table 1, the oxygen abundance in SBS 0335 - 052 is only 1/77 of solar value and is considerably lower than that in I Zw 18 and GR 8. Hence the galaxy SBS 0335 - 052 is the most heavy element deficient blue compact dwarf galaxy among those hitherto observed. This result led us to conclude that SBS 0335 - 052 is

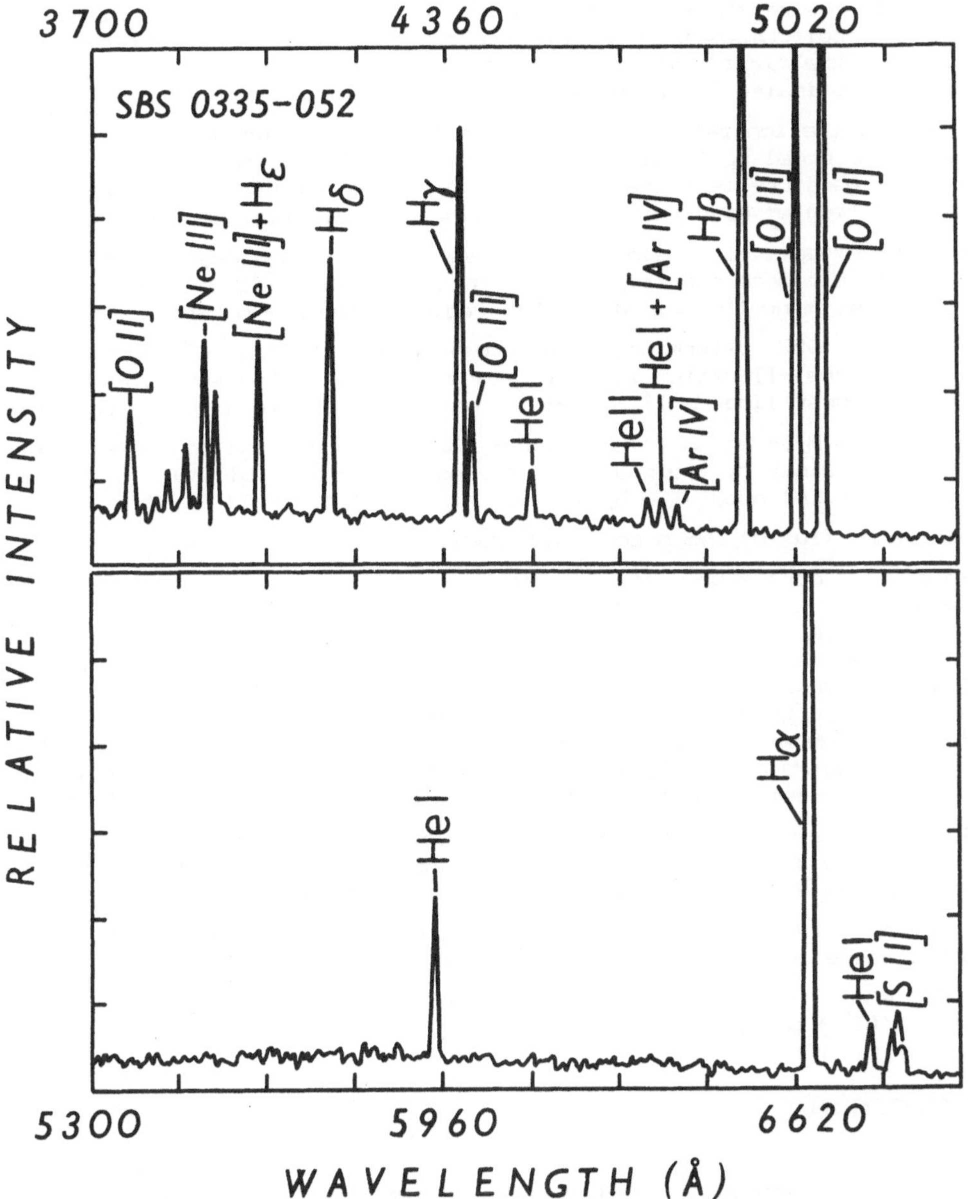

Figure 1. The spectrum of the galaxy SBS 0335 – 052.

probably one of the youngest galaxies formed from the matter with primeval chemical composition and experiencing the burst of the star formation for the first time.

We estimated the effective temperature of ionizing stars $T_{ion} \simeq 8 \cdot 10^4$ K, the ionization parameter $U \simeq 0.2$ and filling factor $f \simeq 1$ by using the model calculations of Campbell (1988). The value of ionization temperature is extremely high and can be satisfactorily explained by evolution of massive stars, which have the initial masses $M > 40\ M_\odot$ and are effectively losing their masses (Maeder and Meynet (1987)).

It is common practice to use the equivalent width of hydrogen line H_β to determine the age of the HII-regions. The age of the HII-region in SBS 0335 − 052, determined in such way, equals to $3.3 \cdot 10^6$ years. The youth of the HII-region is confirmed by lack of Wolf-Rayet star spectral features near line He II λ 4686 Å. The high value of the T_{ion} and small age of the HII-region in SBS 0335 − 052 in our opinion can be satisfactorily explained by the radiation of moderate number of stars with initial masses of about 100 $M_\odot$, existing now at the Wolf-Rayet stage. The estimated flux of Lyman continuum radiation is equal to $N_{Lyc} \simeq 2 \cdot 10^{53} s^{-1}$. This value corresponds to radiation of $\simeq 10^2$ stars with masses adopted above.

Massive stars with mass loss before reaching the Wolf-Rayet stage cause the ionized gas to be substantially enriched with helium. The helium mass fraction in matter with primeval chemical composition is usually adopted to be equal to $Y \simeq 0.25$. Then the additional mass fraction ΔY of the He at the age $\simeq 3.3 \cdot 10^6$ years is equal to 0.07 and the value of 12 + log(He/H) is 11.07, if we take into account the evolution of all stars with masses greater than 50 $M_\odot$. The estimated value of 12 + log(He/H) is usually observed in blue compact dwarf galaxies.

The observed abundance of oxygen and neon can be explained by mass loss of the estimated number of stars with the highest mass $\simeq 10^2\ M_\odot$ (Maeder (1983)), if the assumed mass of the HII-region equals to the usual value $\simeq 10^6\ M_\odot$. In this case the mass ratio Ne/O equals to about 0.42, which is in satisfactory agreement with observed value $(Ne^{2+} + Ne^{3+})/(O^+ + O^{2+}) \simeq 0.55$.

3.CONCLUSION

The carried out study has revealed that the BCDG SBS 0335 − 052 is, probably, a young galaxy experiencing the first burst of star formation. But in order to corroborate this conclusion, it is necessary to carry out observations in the near infrared range in search for a late star population. Ultraviolet observations of SBS 0335 − 052 are extremely important for studying the most massive stars in a galaxy.

REFERENCES

Augarde, R., Figon, P., Kunth, D. and Sevre, F. (1987) 'Spectroscopic survey of the case blue and emission line galaxies , Astronomy and Astrophysics 185, 4 - 8.

Campbell, A. (1988) ' Physical conditions in HII galaxies ', Astrophysical Journal 335, 644 - 657.

Dufour, R.J., Garnett, D.R. and Shields, G.A. (1988) The abundances of carbon and nitrogen in I Zw 18 ', Astrophysical Journal 332, 752 - 761.

Grevesse, N. (1984) ' Accurate atomic data and solar photospheric spectroscopy ', Physica scripta 8, 49 - 58.

Iovino, A., Melnick, J. and Shaver, P. (1988) ' The clustering of HII galaxies ', Astrophysical Journal 330, L17 - L20.

Izotov, Yu. I., Lipovetskii, V.A., Guseva, N.G. and Stepanian, J. A. (1989) ' SBS 0335-052 - the most heavy element deficient blue compact dwarf galaxy', Preprint of the Special Astrophysical Observatory of the USSR Academy of Sciences, No. 40.

Lequeux, J., Peimbert, M., Rayo, J.F., Serrano, A. and Torres-Peimbert, S. (1979) ' Chemical composition and evolution of irregular and blue compact dwarf galaxies', Astronomy and Astrophysics 80, 155 - 166.

Maeder, A. (1983) 'Evolution of chemical abundances in massive stars. I. OB stars, Hubble-Sandage variables and Wolf-Rayet stars. Changes at stellar surfaces and galactic enrichment by stellar winds', Astronomy and Astrophysics 120, 113 - 129.

Maeder, A. and Meynet, G. (1987) 'Grids of evolutionary models of massive stars with mass loss and overshooting. Properties of Wolf -Rayet stars sensitive to overshooting', Astronomy and Astrophysics 182, 243 - 263.

Skillman, E.D., Melnick, J., Terlevich, R. and Moles, M. (1988) 'The extremely low oxygen abundance of GR 8: a very low luminosity dwarf irregular galaxy', Astronomy and Astrophysics 196, 31 - 38.

DISCUSSION

DI FAZIO: Well, if this picture is confirmed, Yura, then we could say that the thought of high efficiency of galaxy formation - and concentration in time at some Gyrs ago ($\simeq 10^{10}$ yrs) is only nothing more than a prejudice. Are there programmed surveys to find similar objects ? Is this in the schedule of SAO ?

IZOTOV: I think that in the vicinity of our Galaxy it is difficult to search the massive young galaxies. We programmed survey of blue compact dwarf galaxies and hope that SBS 0335 - 052 not the unique young galaxy.

MATEO: What temperature is derived for the H II regions in I Zw 18 ?

IZOTOV: The temperature for the H II region in I Zw 18 equals about 19000 K.

CAYREL: This new blue compact galaxy could be very valuable for determining the primeval helium abundance. Am I right, that an He abundance fairly high compared to the one found in I Zw 18 on your transparancy?

IZOTOV: SBS 0335 – 052 is not unique galaxy in this sense. There are others blue compact dwarf galaxies with high values of He abundance. I estimated the enrichment by He, using calculations of Maeder (1983) and found that massive stars with $M > 50\ M_{\odot}$ with mass loss can substantially enrich the gas with heavy elements.

KRÜGEL: 1). Radio recombination lines would give additional electron temperatures and also line widths. Has this been done?
2). Can one speculate on a supermassive star ($> 10^{3}\ M_{\odot}$) which might explain your high temperature?

IZOTOV: 1). The observations have been made only in the optical range. From the line widths we can't determine the electron temperatures, because the line widths due to internal motions in galaxy are more smaller than instrumental width.
2). No. I think that physical properties and chemical composition can be explained by the evolution of usual stars with masses 50 – 100 $M_{\odot}$.

STAR FORMATION IN ELLIPTICAL GALAXIES

H.W. YORKE, R. KUNZE
Institute for Astronomy und Astrophysics
University of Würzburg
Am Hubland
D-8700 Würzburg
Federal Republic of Germany

ABSTRACT. Evidence for current low-level star formation in elliptical galaxies is briefly reviewed. Results of numerical 1-D and 2-D hydrodynamical calculations of the evolution of the interstellar gas in elliptical systems under various assumptions of the dependence of the star formation rate on the local environment are discussed and, when applicable, contrasted with observations. It is possible that elliptical galaxies with very little rotation can produce a single very massive gaseous object ($M > 10^5$ $M_\odot$) at the center. A rotating elliptical galaxy, however, invariably produces a disk out of which stars can form. Star formation under these conditions should be intermittent.

1. Introduction

Elliptical galaxies traditionally have been thought of as boringly simple objects. Their stars formed not long after the globular cluster formation epoch about 14 – 16 Gyr ago (*e.g.* Larson 1974). Little has happened since. The evidence in favor of this view appears to be quite convincing. Many of the following major points of argument have been summarized by O'Connell (1987a):

i) The resolution of Pop II giants in M32 and in the bulge of M31 was an important step towards the the formulation of the concept of stellar populations by Baade (1944). The immediate implication was that ellipticals and spiral bulges are old, metal-poor Pop II systems.

ii) The uniform (red) colors and spectra of E and S0 Galaxies imply that these systems are old and have similar star formation histories.

iii) The continuity of the sequence ranging from globular clusters to E and S0 systems in several photometric and spectroscopic properties implies that these are systems of similar age but varying metalicity Z (*e.g.* Faber 1973; McClure & van den Bergh 1978).

iv) Evolutionary synthesis models of ellipticals have shown that the colors are consistent with an age of 10 – 15 Gyr as long as $Z/Z_\odot \lesssim 1$ (*e.g.* Tinsley 1980).

v) Too little gas is present to support significant star formation. The theoretical explanation for this was the existence of supernova-driven galactic winds during an earlier evolutionary epoch (Mathews & Baker 1971).

R. Capuzzo-Dolcetta et al. (eds.), Physical Processes in Fragmentation and Star Formation, 241–255.

In contrast to the "classical" view, elliptical galaxies are now often regarded as extremely complicated objects with a complex star formation history. The cumulative effects of chaotic and intermittently interacting processes (*e.g.* Dressler 1984): mass infall, winds, stripping, mass loss (outgassing) by the constituent stars, mergers, episodes of star formation, *etc.*, produce a complex mixture of subsystems, which cannot be easily modelled by evolutionary synthesis calculations. The evidence in favor of this view is summarized in the following.

i) Since the work of Morgan & Mayall (1957) and Baum (1959) it has been known that the dominant component of ellipticals and spiral bulges are metal-rich ($Z/Z_\odot \gtrsim 1$). The Pop II giants in M32 and in the bulge of M31 are metal-poor and thus represent a minority component in these systems.

ii) In older galaxies the UBV colors evolve slowly (*e.g.* O'Connell 1987b).

iii) Newer studies of ellipical galaxies show significant spectral differences, especially in the UV (Burstein *et al.* 1988).

iv) The first evolutionary synthesis models assumed solar abundances, whereas elliptical galaxies are considered to be metal-rich. Indeed, the central regions could well have metalicities in excess of 2 to 3 $Z_\odot$. Newer synthesis models which take varying metalicity into account display significant star formation 5 – 8 Gyr ago and show some evidence for present-day star formation. It is interesting to note that the error in the estimate of age based on colors δt_{color} depends on the metalicity assumed in a simple manner (O'Connell 1986):

$$\delta \log t_{color} \simeq -\delta \log Z \quad .$$

v) Gas (and dust) are present in elliptical galaxies (see below).

1.1. GAS AND DUST IN ELLIPTICAL GALAXIES

Sandage (1957) pointed out that the total mass of gas ejected from aging stars in an elliptical galaxy should be about $M_{GAL}/200$ in 10 Gyr. Thus, the stars of a "typical" elliptical galaxy of mass 2×10^{11} $M_\odot$ will have outgassed $\sim 10^9$ $M_\odot$. Assuming a single star formation phase, supernova-driven winds can only effectively remove material for about the first few 10^8 yr — provided the supernova rate is sufficiently high and no massive halos are present. What happens to the remaining material? Can it be observed? As discovery has proceeded from HII gas to HI to X-ray emitting gas, the amount of gas observed in elliptical galaxies has increased dramatically (Schweizer 1987).

1.1.1. *Cool HI Gas* ($\lesssim 10^2$ *K*). Knapp (1987) has reviewed the 21 cm data for elliptical galaxies. HI has been detected in about 10% of nearby ellipticals. Typically, the mass is $\sim 5 \times 10^8$ $M_\odot$ distributed in a flattened disk-like manner. The orientation of the disks and the magnitude of angular momentum excludes the possibility that this gas has originated from the elliptical galaxies' own stars.

1.1.2. *Warm HII Gas* ($\sim 10^4$ *K*). In the central regions of 55 – 60 % of E galaxies the [NII]λ6584 line has been detected (Phillips *et al.* 1986; see also Veron-Cetty & Veron 1986). High luminosity ellipticals tend to have a LINER-type spectrum, whereas low luminosity ellipticals are generally HII galaxies. Typical masses are $M_{HII} \sim 10^3 - 10^5$ $M_\odot$ in a disk-like

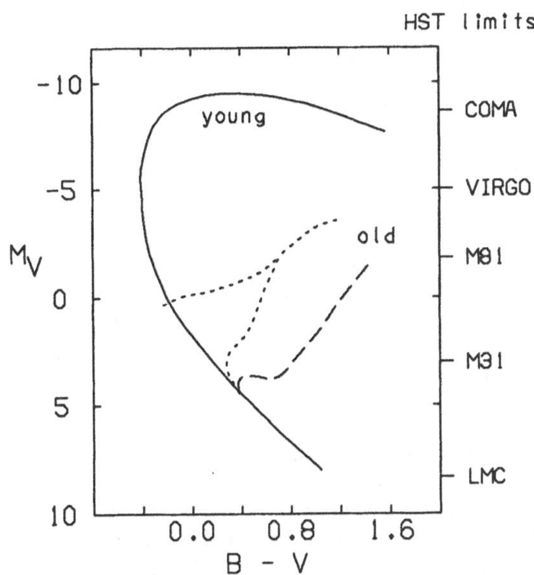

Figure 1. Schematic color-magnitude diagram of young (solid line) and old (dashed lines) stellar systems. The long (short) dashed line refers to a metal-rich (metal-poor) population. Limiting magnitudes (V = 27) for the Hubble Space Telescope at various distances are given on the right.

configuration of extent R $\sim$ 1 kpc and density $n_e \sim 10^3\,\mathrm{cm}^{-3}$.

1.1.3. *Hot Gas ($\sim 10^7$ K).* A large number of isolated elliptical galaxies exibit X-ray halos. The observation of spectral lines of highly ionized iron (*cf.* Sarazin 1986) indicates a metal enrichment of the halo gas of up to half solar values (*e.g.* Canizares 1981). Thus, the gas probably originates from the host galaxy. Mass estimates vary between $M_{X-ray} \sim 5\times10^8\,M_\odot$ to $\sim 5\times10^{10}\,M_\odot$ for the most massive ellipticals. The Gas distribution is almost spherical extending some R $\sim$ 100 kpc. Central densities have been estimated (assuming a radially constant temperature) to be $(n_e)_c \sim 0.01 - 0.1\,\mathrm{cm}^{-3}$. If this estimate were correct and if there were no significant heating sources, the hot gas should cool on a time scale $\tau_{cool} \lesssim 10^8$ yr. This is the typical argument used to postulate the existence of "cooling" flows (see below).

1.1.4. *Dust.* Dust is observable in some elliptical galaxies in the form of "obscuring patches" and dark lanes, the total mass of which is estimated to be $M_{dust} \sim 10^4 - 10^5\,M_\odot$. The dust appears to be associated with the HI gas.

1.2. THE STELLAR COMPONENT OF ELLIPTICAL GALAXIES

The careful analysis of the light from individual stars, applied in a statistical manner, has led to the concept of different "populations" of stars. However, these methods cannot be applied to elliptical galaxies due to the limiting magnitudes of present-day telescopes (see Fig. 1). Instead, one must use the integrated light and deconvolve it into its component parts by population synthesis. The basic ingredients of this process are described in the following.

Consider the integrated light to be expected from an aging system of stars formed in a δ-function burst:

$$F_\nu^{Burst}(t) = \int_{M_{min}}^{M_{max}} L_\nu(\mathrm{M}, t, \mathrm{Z}, ...) \; \Phi(\mathrm{M}, env.) \, d\mathrm{M} \qquad ,$$

where $L_\nu(\mathrm{M}, t, \mathrm{Z}, ...)$ is the spectrum to be expected from a single star of mass M, of age t, with metalicity Z. Perhaps further parameters are needed to characterize some stars (close binaries, etc.). The IMF, $\Phi(\mathrm{M}, env.)$, and the upper and lower mass limits, M_{max} and M_{min} are functions of the local environment. They may therefore vary as a galaxy ages. The integrated light of an entire galaxy is obtained by summing over all populations since the time of formation, t_{form}:

$$F_\nu = \int_{t_{form}}^{0} F_\nu^{Burst} \; \Psi(t, env.) \, dt \qquad ,$$

where $\Psi(t, env.)$ is the time (and environment) dependent star formation rate.

The theoretical problem of population synthesis therefore consists of combining extensive libraries for $L_\nu(\mathrm{M}, t, \mathrm{Z}, ...)$ with best guesses for the IMF, $\Phi(\mathrm{M}, env.)$, and the star formation rate $\Psi(t, env.)$. Ideally, one would like to specify $L_\nu(\mathrm{M}, t, \mathrm{Z}, ...)$ in the range: $0.1 \leq \mathrm{M}/\mathrm{M}_\odot \leq 100$, $0 \leq t \leq 20$ Gyr and $0.01 \leq \mathrm{Z}/\mathrm{Z}_\odot \leq 10$. This, however, will remain a dream for some time to come. Instead, one must construct a library by combining theoretical evolution isochrones with a combination of observed spectra and model atmospheres. The problems with this procedure are manyfold; they are based on our poor understanding of stellar evolution theory for certain short-lived (but very bright) phases and on our poor understanding of stellar atmospheres, particularly in the ultraviolet. The evolution isochrones are inaccurate due to unknown mixing and mass loss rates (*cf.* Chiosi & Maeder 1986). Problems in our understanding of the asymptotic giant branch (AGB), post-AGB (PAGB) and the horizontal branch (HB) are especially troublesome (bright UV sources!). Finally, stellar evolution isochrones are incomplete for $\mathrm{Z} > \mathrm{Z}_\odot$. No non-LTE, line-blanketed (important for the UV), and non-stationary model atmospheres exist today — one is forced to compromise. The effects of metalicity on model atmospheres has not been investigated in sufficient detail.

In spite of these problems, population synthesis still remains the principle means of investigation of the stellar content of E galaxies. To avoid uncertainties with the specification of $\Psi(t)$ from first principles, many investigators mix components from their spectral libraries in order to obtain a "best fit" to observed spectra (see *e.g.* Kjaergaard 1987 and Rocca-Volmerange & Guiderdoni, 1987 for some fits of M32). Of particular interest is the nature of the hot (UV) component observed in many ellipticals (Burstein *et al.* 1988), because this may shed light on the question of recent (< 6 Gyr) and present day star formation. The interpretation of the data is not unambiguous, however. Among the possible candidates for the UV light are young OB stars, accreting white dwarfs, HB stars, PAGB stars, and blue stragglers. Whereas Rocca-Volmerange (1989) and Kjaergaard have concluded that recent star formation is the best explanation for the UV excess in several ellipticals, Bertelli *et al.* (1989) argue the case in favor of PAGB stars. Clearly, the controversy has not yet been settled.

2. Star formation in cooling flows

Numerical models of steady cooling flows have been obtained for a wide range of galaxy parameters (*cf.* Sarazin & White 1987, Vedder *et al.* 1988 and references therewithin). Compared to the expected accretion rates ($\sim 1\ M_\odot\ yr^{-1}$) only modest amounts of warm gas at $T \sim 10^4$ K are observed in isolated ellipticals. The cooling, inflowing gas must therefore have a sink; it is very likely that gas condenses out of the flow in the form of stars. Another possible sink could be accretion onto a central (unobserved) disk or a black hole. Although some of the galaxies with cooling flows are strong radio sources there is no one to one correlation between cooling flows and active nuclei.

Gas in the temperature range of $10^5 - 10^7$ K is thermally unstable and therefore dense lumps can be expected to condense out of the accretion flow. White & Sarazin (1987a,b,c) found in a detailed instability analysis that these condensations are isobaric and comoving with the flow. Since the isothermal Jeans mass is reduced in the high pressure environment of the cooling flow many authors argue that only low mass stars will form from the cooling density pertubations in the accretion flow (Jura 1977, Fabian *et al.* 1982, Sarazin & O'Connell 1983, Thomas *et al.* 1986). Their argument of low mass star formation (with an upper cutoff mass for the IMF $m_{\rm cut} \lesssim 2\,M_\odot$) is supported at least indirectly by the fact that the presense of an X-ray halo does not affect the spectra or colors of an E galaxy (Hu *et al.* 1985).

Silk *et al.* (1986) argue that no straightforward inference can be drawn between a reduced Jeans mass in a high pressure environment and star formation. On the contrary, the formation of massive stars is to be expected on theoretical grounds (*e.g.* Silk 1985). Silk *et al.* (1986) modeled the effect of supernova heating on the cooling flow. The stars were assumed to form out of the flow with a normal solar-neighborhood IMF (Miller & Scalo 1979). The broad band colors are compiled with the population synthesis program of Bruzual (1983). They find the assumption of a Miller-Scalo IMF with a normal upper cutoff mass to be entirely adequate to account for the observed colors of elliptical galaxies surrounded by extended gaseous coronae. The inferred accretion rates are reduced by a factor of two, compared to the estimates without supernova heating.

2.1. EVOLUTIONARY MODELS OF GALACTIC FLOWS

Since the gaseous halos around elliptical galaxies are metal enriched, it most likely has originated from stellar mass loss. Though the stellar mas loss rate at the present epoch is about $1\,M_\odot\,yr^{-1}$ in giant ellipticals (Faber & Gallagher 1976) — sufficient to build up a gaseous corona with the observed X-ray luminosity within a Hubble time — stars have lost mass at a much higher rate during earlier epochs of galactic evolution (see *e.g.* MacDonald & Bailey 1981). Integrated over the lifetime of a galaxy, ellipticals may well have produced more gas than is observed in X-rays. Cold and warm gas cannot account for this "hidden gas" (see section 1.1).

Loewenstein & Mathews (1987; hereafter LM) calculated the evolution of hot galactic flows over the lifetime of an elliptical galaxy taking into account source terms — such as the stellar mass loss rate and the type I supernova rate (normalized to 25% of the observed present epoch rate (Tamman 1974)) — which are variable in time. They assumed that up to 90% of the total mass of the model galaxy is contained in a halo of dark matter. The

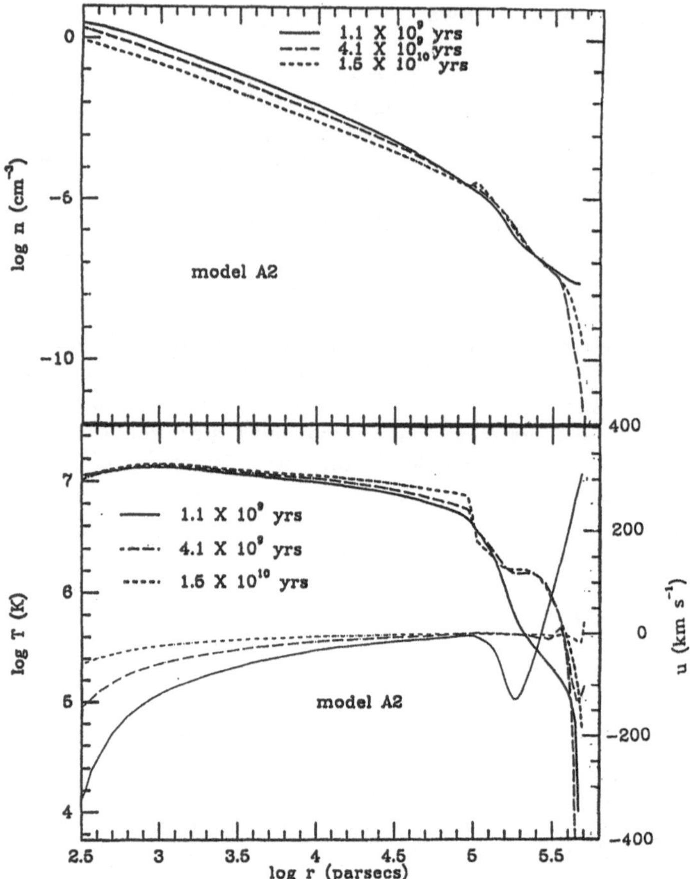

Figure 2. Evolution of density, velocity and temperature of model A2 from Loewenstein & Mathews (1987). Due to the increasing energy source term (supernova heating) with time relative to stellar mass loss, the inflow weakens during the evolution and the gas density profile becomes less steep. The temperature rises slightly during the evolution but remains nearly constant between galactic radii of 100 pc and 100 kpc.

total mass of the stars was $M_\star = 8.29 \times 10^{11}\,M_\odot$ in their reference case representing a giant elliptical galaxy. The calculations started at $t = 10^8$ yr when the type II supernova explosions declined (a delta function for the star formation rate is implicitely assumed).

In Fig. 2 we show the evolution of their reference model (A2). A partial wind develops in the early phases of evolution. In the first 2×10^9 yr about $2 \times 10^{10}\,M_\odot$ flowed into the core of the galaxy. Outside the stagnation radius, where the gas velocity vanishes, the gas is driven out. Ultimately, the expansion is stopped by gravity in the potential well of the dark matter halo. During the following 1.3×10^{10} yr the flow is quasi-steady. The stagnation radius moves outwards and after 1.5×10^{10} yr a nearly hydrostatic gaseous corona has developed. The resulting X-ray surface brightness profile was slightly steeper than the stellar surface brightness profile.

A higher supernova rate — the Tamman value at present epoch — does not drastically change the evolution of the flow. The X-ray luminosity is slightly lower than in calculation A2. More remarkable is the fact that the bulk of the gas is found to move outwards at the present epoch — while maintaining nearly hydrostatic equilibrium. If their source terms are allowed to evolve further, a galactic wind can be expected after 10^{10} yr. Thus, X-ray coronae around ellipticals may actually be a transient phenomenon.

Recently, D'Ercole *et al.* (1989) calculated the evolution of gas flows in elliptical galaxies with a different formulation of the source terms. In particular, the supernova rate was assumed to decline slightly *faster* than the stellar mass loss rate and to be normalized to the Tamman rate at $t = 1.5 \times 10^{10}$ yr, contrary to the assumptions of LM, whose rate was smaller and nearly constant in time. D'Ercole *et al.* find that at the present epoch the lower luminosity ellipticals ($L_B \sim 10^{10}$ $L_\odot$) exhibit supersonic galactic winds and only the most luminous galaxies with $L_B \gtrsim 10^{11}$ $L_\odot$ are in the phase of cooling inflow. Intermediate ranges in L_B have highly subsonic outflows. The X-ray luminosity of elliptical galaxies as well as the observed scatter in the L_X/L_B relation can be explained by their model.

2.2. THE EVOLUTION OF THE CORE REGION OF ELLIPTICAL GALAXIES

What happens to the gas flowing into the core of an elliptical galaxy, especially during the first few billion years of its evolution? The situation interior to the core radius of an E galaxy is unique since the stellar component can stabilize a large amount of inflowing gas against fragmentation (and star formation) — provided the core has an isothermal stellar velocity dispersion. For ionized gas Loose & Fricke (1980) find that masses of the order of 10^5 $M_\odot$ can be stabilized against collapse.

In order to investigate the influence of the clumpiness of the inflowing matter and the core's internal stellar mass loss Kunze *et al.* (1987; hereafter KLY) have calculated the gas flow in a non-rotating E0 galaxy during an early epoch. Contrary to the calculation of LM, who began at an age $t = 10^8$ yr, KLY start at $t = 10^9$ yr and assume initial gas free conditions. LM's calculated infall of more than 10^{10} $M_\odot$ within the first two billion years should certainly prolong the star formation phase in the core and strongly affect the inflow. In the scenario of KLY a type II supernovae-driven wind will have blown the galaxy gas free after 10^9 yr (see also Arimoto & Yoshii 1987).

KLY use a standard King function to model the galactic stellar distribution and assume no dark halo. Their models for an E galaxy's core (which is well-resolved by their implicit code) should be applicable even when a dark halo is present, because such a halo should influence the flow principally in the outer regions. KLY's model galaxy had a total mass of 1.1×10^{11} $M_\odot$, a central stellar density of $\rho_* = 6.5 \times 10^{-20}\,\mathrm{g\,cm^{-3}}$ and a core radius of 125 pc. A "kinetic bulk energy" E_T was introduced to account for the kinetic energy of clumpy, unthermalized matter. Dissipation of E_T was assumed to occur due to "clump-clump collisions", allowing E_T to decline exponentially. The dissipation time scale was identified with the local dynamical timescale of the system. The souce terms — stellar mass loss and supernova heating — were formulated in a manner similar to MacDonald & Bailey (1981), modified to allow source terms to E_T by means of efficiency factors. In Fig. 3 the results of case 1 (immediate thermalization; $E_T = 0$) are compared to the results

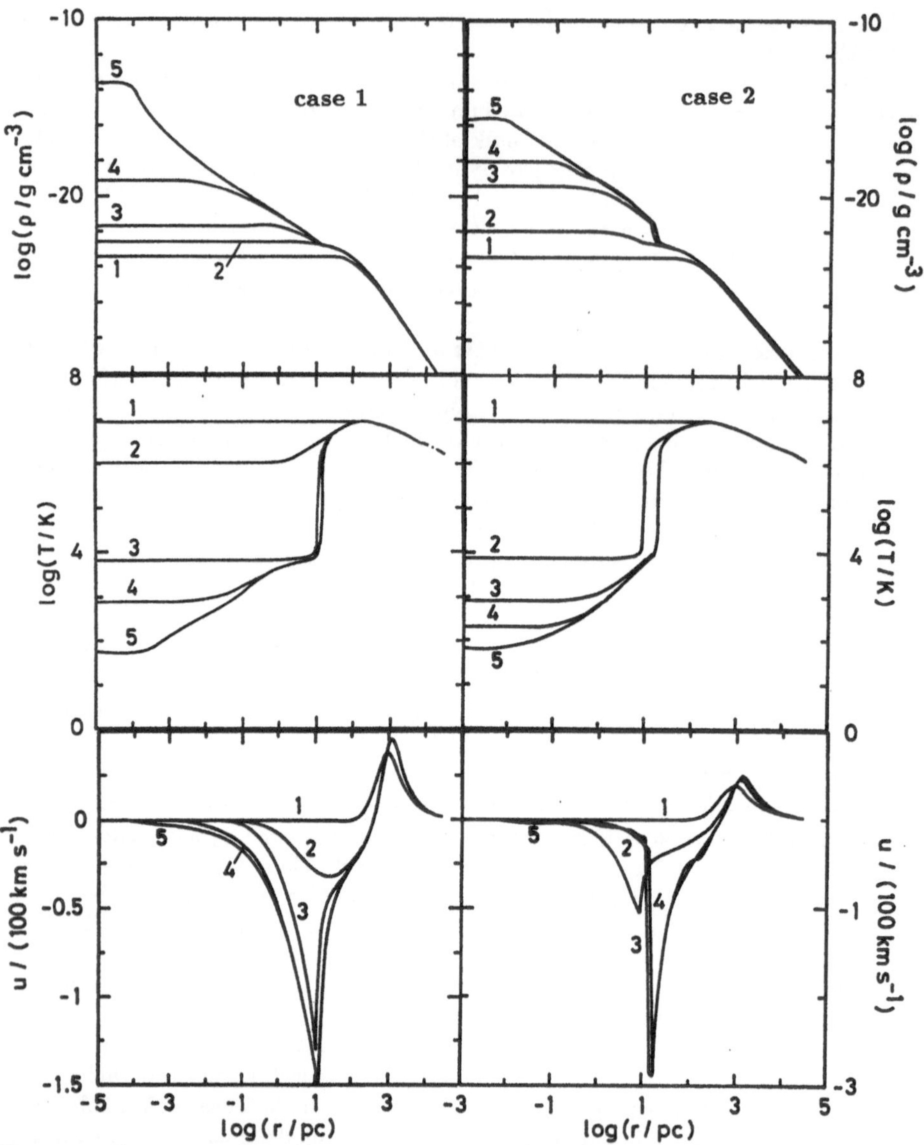

Figure 3. Density, temperature and velocity distributions for cases 1 and 2 (from Kunze *et al.* 1987). Numbers refer to the following evolutionary times (in 10^6 yr): case 1: [1]: 1.62, [2]: 2.58, [3]: 2.61, [4]: 2.62 and [5]: 2.63. Case 2: [1]: 1.06, [2]: 2.60, [3]: 3.00, [4]: 3.20 and [5]: 3.24.

of case 2 (10% of the kinetic energy of "normal" stellar mass loss — PN and red giant winds — and 28% of the supernova ejecta energy contribute to E_T).

Case 1 may be described as follows: after a short phase of gas accumulation a partial wind develops. Before the stagnation radius can move outwards significantly, the inflowing material begins to collapse at almost free-fall velocities as it cools to below 10^4 K. The gas at the center cools quickest; a "cooling" front moves outwards through the gas. Within a few 10^4 yr a low mass quasi-hydrostatic core forms. The subsequent evolution of this accreting core to stellar densities and nuclear burning was not followed. KLY estimat the mass of the resulting star to be of order of 10^3 $M_\odot$.

In the second case the evolution is similar until the gas in the central regions cools to $\sim 10^4$ K. Rather than collapsing in free-fall, the gas is initially stabilized by the "kinetic bulk energy". About 10^5 $M_\odot$ of cool neutral gas can accumulate interior to the cooling front. As the gas density increases the kinetic bulk energy can be dissipated more effectively and the central gas cloud can no longer be supported. The subsequent collapse proceeds until again a quasi-static nucleus forms. For this case the mass of the resulting central object is $\sim 10^5$ $M_\odot$; *i.e.* it will evolve into a supermassive star.

3. Star formation in early evolutionary phases of elliptical galaxies

The formation of massive stars can be expected during the formation of an elliptical galaxy. Thus, type II supernova with massive progenitors of $m_{SN} \gtrsim 8\,M_\odot$ could substantially contribute to the heating of the gas not consumed by star formation. Arimoto & Yoshii (1987) calculated the chemical evolution of ellipticals taking into account the complete ejection of the remaining gas by a supersonic supernovae-driven wind and successfully reproduced the structural and chemical properties of E-galaxies at the present epoch. They chose a criterion for the onset of a supersonic wind which compared the thermal energy content of the gas to the gravitational binding energy of the galaxy. Gas ejection occurred at $t \sim 10^9$ yr after onset of star formation in a typical giant elliptical with $M_\star \sim 10^{11}$ $M_\odot$.

However, the production of type II supernova progenitors can be self-regulative (see *e.g.* Hensler 1987). As gas is removed, star formation (and heating of the gas) ceases. Mass loss by evolving low mass stars is then able to accumulate again and — provided the type I supernova rate is not too large — to cool, thus allowing a new cycle of star formation. This process has been modelled in 2D hydrodynamical calculations by Kunze (1988) and Kunze & Yorke (1988). We shall describe the results of these calculations in the following.

Figure 4 (next page). Gas density (solid contours), temperature (dashed contours), and velocity (arrows) structure in the inner regions (3.5 × 5 kpc) of the model E3 galaxy at four evolutionary times as indicated in each quadrant. Initially the galaxy was gas free. Several values for $\log \rho$ and log T are given. The density contours are separated by $\Delta \log \rho = 0.25$ for $\log \rho > -26$; otherwise by $\Delta \log \rho = 0.5$. The velocity scale for all four quadrants is indicated at the upper right. Velocities exceeding 150 $\mathrm{km\,s^{-1}}$ are plotted as pointed triangles. The initial expansion is due to the gas coming from stars (which have an anistropic velocity distribution) readjusting to the galaxy's potential.

Figure 5 (second page). Same as Fig. 4 except the evolution at later times is shown and the velocity scale for all four quadrants is larger by a factor of five. Note the circulation pattern which has developed at $t = 5.93 \times 10^7$ yr.

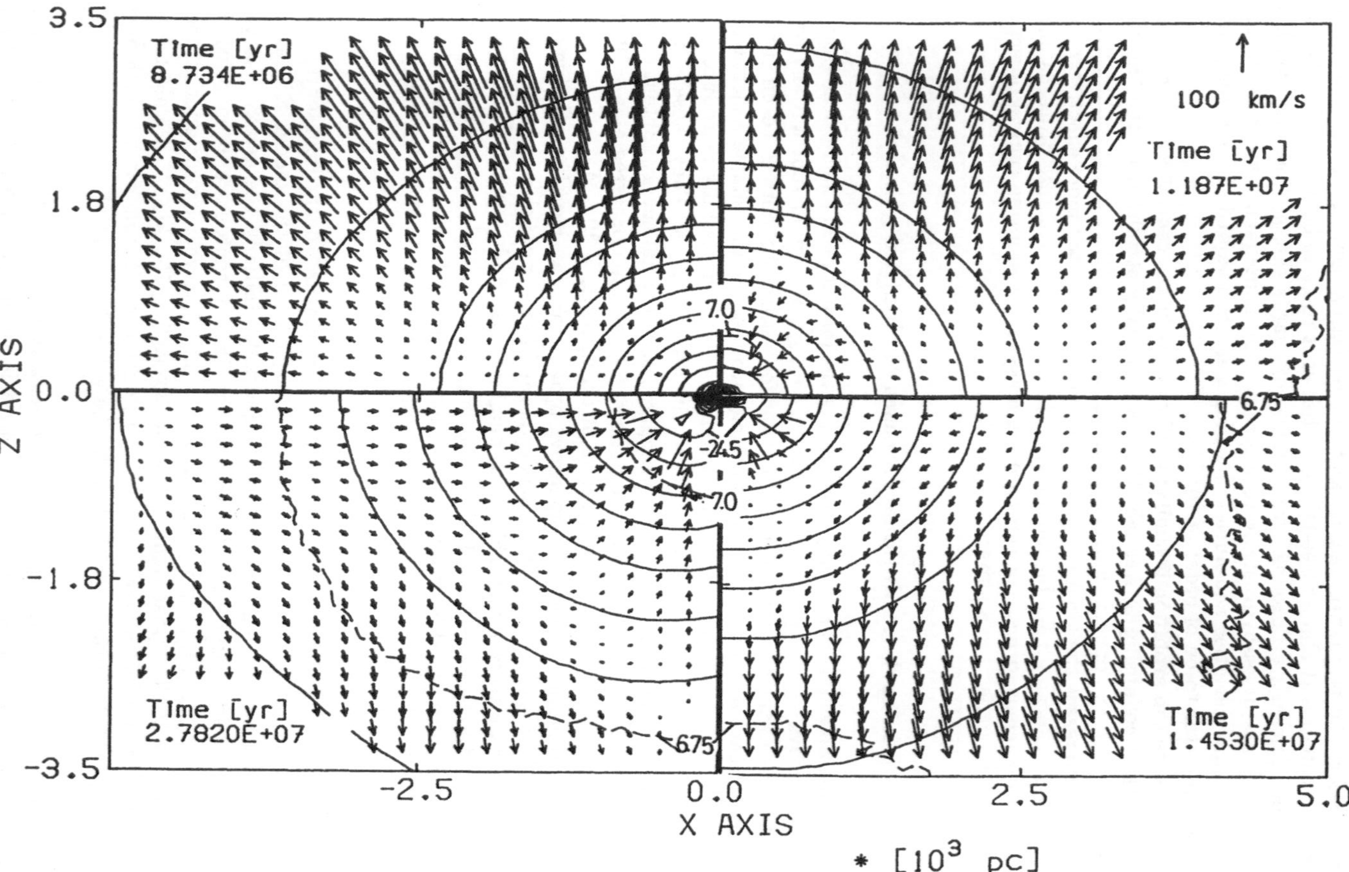

Time [yr]
8.734E+06
100 km/s
Time [yr]
1.187E+07
Time [yr]
2.7820E+07
Time [yr]
1.4530E+07
7.0
-24.5
7.0
6.75
6.75
3.5
1.8
0.0
-1.8
-3.5
Z AXIS
-2.5
0.0
2.5
5.0
X AXIS
* [10^3 pc]

Figure 4.

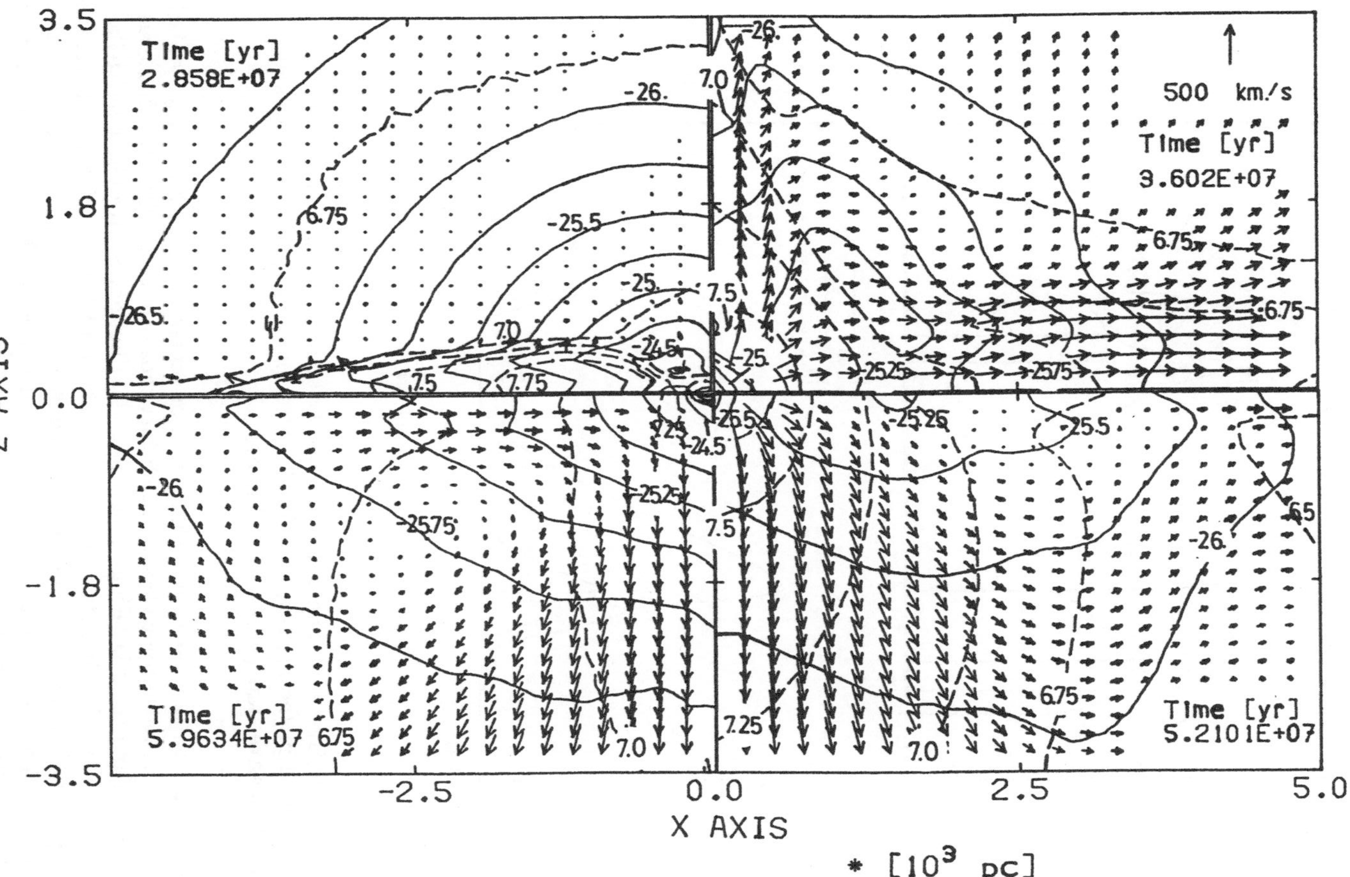
Time [yr]
2.858E+07
500 km/s
Time [yr]
3.602E+07
Time [yr]
5.9634E+07
Time [yr]
5.2101E+07
Z AXIS
3.5
1.8
0.0
-1.8
-3.5
-2.5
0.0
2.5
5.0
X AXIS
* [10^3 pc]

Figure 5.

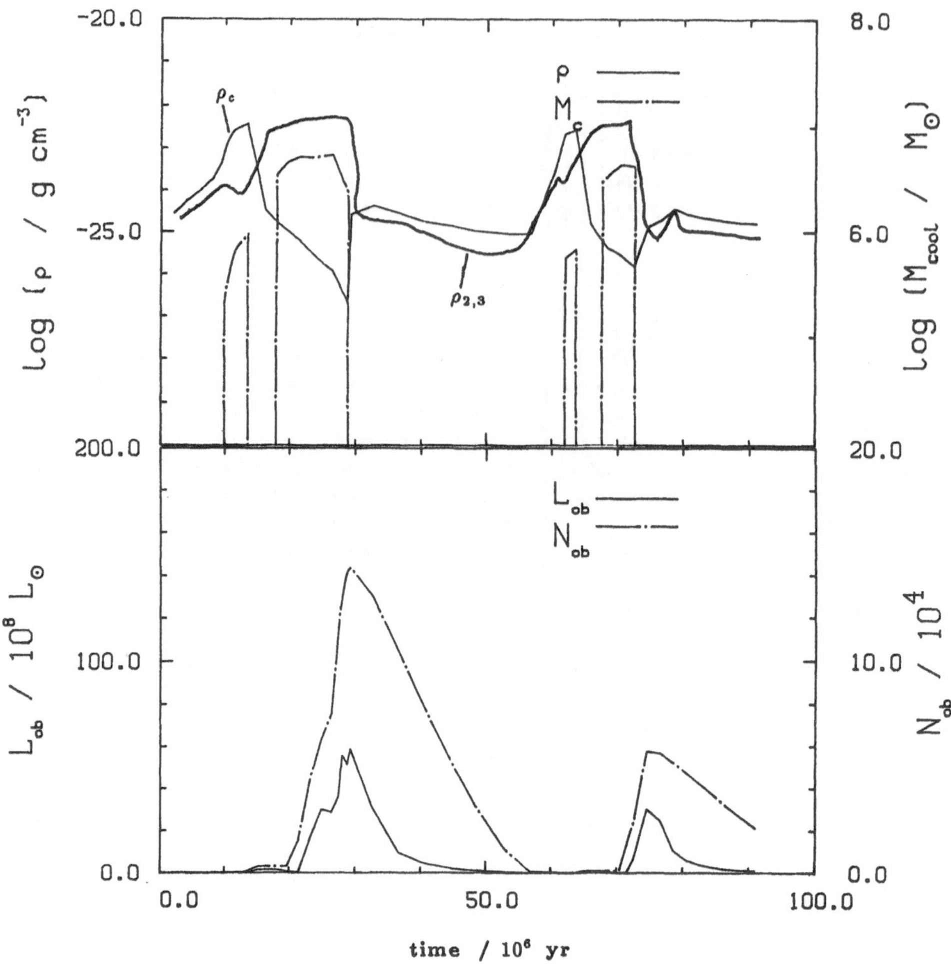

Figure 6. The time evolution of central density and the amount of material cooler than 10^4 K is plotted in the upper graph. In the lower plot the type II SN rate and the total number of massive stars ($m > 8\,M_\odot$) is shown.

The model galaxy was a slow oblate rotator representing an E3 galaxy. Ablation is maintained by an anisotropic velocity dispersion, chosen to be consistent with the assumed rotation curve and the 2-D "King"-type stellar distribution. Source terms for stellar mass loss and supernova heating (type I) were consistent with those of KLY. In contrast to KLY fragmentation and star formation were allowed explicitly, because, even for the slow rotation considered, the outgassed material could not reach the central regions where it might be stabilized. Instead, a gaseous disk was formed. Star formation was assumed to occur whenever the gas cooled below 10^4 K. The stars formed with a Salpeter IMF on a local dynamical timescale. The local type II SN heating rate is thus proportional to the star formation rate after a time lag given by the nuclear burning lifetime of massive stars.

In Fig. 4 and 5 we show the structure of gas density, temperature and velocity in the inner regions at 8 evolutionary times. The stellar component was assumed to rotate as a solid body for $R < 500\,\mathrm{pc}$ and at the maximum rotational velocity of $75\,\mathrm{km\,s^{-1}}$ exterior to this. The evolution was followed for 10^8 yr. Fig. 6 shows the evolution of central density, the total mass of cool gas and several star formation parameters. Cooling of the accumulated outgassed stellar material occurred after 10^7 yr and gas began to flow inwards primarily in the equatorial plane. Due to the small rotational velocity of this material centrifugal forces could halt the inflow only well inside the galactic core. A gaseous disk was formed out of which stars began to form. About 1.5×10^5 massive stars were formed during this first cycle. The energy input from their subsequent explosions was sufficient to reheat the gas to 10^7 K and reverse the flow. No supersonic wind was created. Instead, some of the hot gas circulated in eddies above and below the equatorial plane (see *e.g.* lower left quadrant of Fig. 5). After 6×10^7 yr cooling in the center dominated again and another star formation cycle started lasting for 2×10^7 yr. Although fewer stars were formed in the second cycle the heating of type II supernovae was sufficient to reheat the gas and cause its outflow.

Maximum color changes in the core due to the contribution of light from massive stars were estimated to be $\Delta(U - B) \sim 1.1$ and $\Delta(B - V) \sim 0.15$. During the subsequent evolution, periods of intense star formation in the core can be expected to alternate with periods of relative quiescence, since the energy provided by type II supernova explosions was not sufficient to establish a supersonic galactic wind. A steady-state situation could not be attained for the range of parameters considered. Clearly, a greater range of galactic parameters needs to be investigated.

Arimoto (1989) has also found recurrent star formation in giant elliptical galaxies with an extended version of the evolutionary population synthesis program of Arimoto & Yoshii (1987). Models of intermittent or recurrent star formation could explain observations of significant contributions of intermediate-age stars to the light of some elliptical galaxies.

4. Concluding remarks

From the results of hydrodynamical evolution calculations we conclude:

— We can expect many ellipticals to have had significant star formation as recently as ~6 Gyr ago.

— Ellipticals cannot support galactic winds at present. However, many can support hot coronae.

— Ellipticals experience low-level star formation at present (perhaps intermittantly).

— Slowly rotating ellipticals may have formed a supermassive object in the center.

ACKNOWLEDGEMENTS. We gratefully acknowledge support by the DFG (Deutsche Forschungsgemeinschaft) under grants Yo 5/5-1 and -2. Calculations were performed on the UNISYS 1180, the IBM 3090 (Gesellschaft für wissenschaftliche Datenverarbeitung in Göttingen), the CRAY XMP (Konrad-Zuse-Zentrum, Berlin), the Siemens 7.860L (Rechenzentrum der Universität Würzburg).

REFERENCES

Arimoto, N.: 1989, in "The epoch of galaxy formation", eds. C.F. Frenk et al. (Kluver Acad. Pub.) p. 205

Arimoto, N. and Yoshii, Y.: 1987, Astron. Astrophys. **173**, 23

Baade, W.: 1944, Astrophys. J. **100**, 137

Baum, W.A.: 1959, P.A.S.P. **71**, 106

Bertelli, G., Chiosi, C. and Bertola, F.: 1989, Astrophys. J. **339**, 889

Bruzual, G.A.: 1983, Astrophys. J. **273**, 105

Burstein, D., Bertola, F., Buson, L.M., Faber, S.M. and Lauer, T.R.: 1988, Astrophys. J. **328**, 440

Canizares, C.R.: 1981, in "X-Ray Astronomie with the Einstein Satellite", ed. R. Giacconi (Dordrecht: Reidel) p. 85

Canizares, C.R., Fabbiano, G. and Trinchieri, G.: 1987, Astrophys. J. **312**, 503

Chiosi, C. and Maeder, A.: 1986, Ann. Rev. Astron. Astrophys. **24**, 329

D'Ercole, A., Renzini, A., Ciotti, L. and Pellegrini, S.: 1989, Astrophys. J. **341**, L9

Dressler, A.: 1984, Ann. Rev. Astron. Astrophys. **22**, 185

Faber, S.M.: 1973, Astrophys. J. **179**, 731

Faber, S.M. and Gallagher, J.S.: 1976, Astrophys. J. **204**, 365

Fabian, A.C., Nulsen, P.E.J. and Canizares, C.R.: 1982, Monthly Notices Roy. Astron. Soc. **201**, 933

Hensler, G.: 1987, Mitt. d. Astron. Ges. **70**, 141

Hu, E.M., Cowie, L.L. and Wang, Z.: 1985, Astrophys. J. Suppl. **59**, 447

Jura, M.: 1977, Astrophys. J. **212**, 634

Kjaergaard, P.: 1987, Astron. Astrophys. **176**, 210

Knapp, G.R.: 1987, in *"Structure and Dynamics of Elliptical Galaxies", IAU Symposium 127*, ed. T. de Zeeuw (Dordrecht: Reidel) p. 145

Kunze, R.: 1988, Dissertation, University of Göttingen

Kunze, R. and Yorke, H.W.: 1988, Astr. Ges. Abstr. Ser. **1**, 40

Kunze, R., Loose, H.-H. and Yorke, H.W. (KLY): 1987, Astron. Astrophys. **182**, 1

Larson, R.B.: 1974, Monthly Notices Roy. Astron. Soc. **169**, 229

Loose, H.-H. and Fricke, K.J.: 1980, Astr. Letters **21**, 65

Loewenstein, M. and Mathews, W.G. (LM): 1987, Astrophys. J.

MacDonald, J. and Bailey, M.E.: 1981, Monthly Notices Roy. Astron. Soc. **197**, 995

Mathews, W.G. and Baker, J.G.: 1971, Astrophys. J. **170**, 241

McClure, R.D. and van den Bergh, S.: 1978, A.J. **73**, 313

Miller, G.E. and Scalo, J.M.: 1979, Astrophys. J. Suppl. **41**, 513

Morgan, W.W. and Mayall, N.U.: 1957, P.A.S.P. **97**, 205

Nulsen, P.E.J., Stewart, G.C. and Fabian, A.C.: 1984, Monthly Notices Roy. Astron. Soc. **208**, 185

O'Connell, R.W.: 1986, in "Stellar Populations", eds. C. Norman, A. Renzini, M. Tosi (Cambridge: Cambridge Univ. Press), p. 167

O'Connell, R.W.: 1987a, in "Starbursts and Galaxy Evolution", eds. Trinh Xuan Thuan, T. Montmerle, J. Tran Thanh Van (Gif Sur Yvette: Editions Frontieres), p. 367

O'Connell, R.W.: 1987b, in *"Structure and Dynamics of Elliptical Galaxies", IAU Symposium 127*, ed. T. de Zeeuw (Dordrecht: Reidel) p. 167

Phillips, M.M., Jenkins, C.R., Dopita, M.A., Sadler, E.M. and Binette, L.: 1986, Astrophys. J. **91**, 1062
Rocca-Volmerange, B.: 1989, Monthly Notices Roy. Astron. Soc. **236**, 47
Rocca-Volmerange, B. and Guiderdoni, B.: 1987, Astron. Astrophys. **175**, 15
Sanders, R.H.: 1981, Astrophys. J. **244**, 820
Sandage, A.: 1957, Astrophys. J. **125**, 422
Sarazin, C.L.: 1986, Rev. Mod. Phys. **58**, 1
Sarazin, C.L. and O'Connell, R.W.: 1983, Astrophys. J. **268**, 552
Sarazin, C.L. and White, R.E. III: 1987, Astrophys. J. **320**, 32
Schweizer, F.: 1987, in *"Structure and Dynamics of Elliptical Galaxies", IAU Symposium 127*, ed. T. de Zeeuw (Dordrecht: Reidel) p. 109
Silk, J.: 1985, Astrophys. J. **297**, 1
Silk, J., Djorgowski, S., Wyse, R.F.G. and Bruzual, G.A.: 1986, Astrophys. J. **307**, 415
Tamman, G.A.: 1974, in "Supernovae and supernova remnants", ed. C.B. Cosmovici (Dordrecht: Reidel), p. 371
Thomas, P.A., Fabian, A.C., Arnaud, K.A., Forman, W. and Jones, C.: 1986, Monthly Notices Roy. Astron. Soc. **222**, 655
Tinsley, B.M.: 1980, Fundam. Cosmic Phys. **5**, 287
Vedder, P.W., Trester, J.J. and Canizares, C.: 1988, Astrophys. J. **332**, 725
Veron-Cetty, M.-P. and Veron, P.: 1986, Astron. Astrophys. Suppl. **66**, 335
White, R.E. III and Sarazin, C.L.: 1987a, Astrophys. J. **318**, 612
White, R.E. III and Sarazin, C.L.: 1987b, Astrophys. J. **318**, 621
White, R.E. III and Sarazin, C.L.: 1987c, Astrophys. J. **318**, 629

Discussion following the paper presented at the meeting centered on the question of the origin of UV light in elliptical galaxies. However, only one question/comment sheet was handed in. The authors have attempted to include some of the main points of discussion in their manuscript.

M. Mateo: I should point out that recent photometric observations of M32 (Freedman, A.J. 1989, in press) show no evidence of a) young stars on the main sequence or b) a significant (if any) population of AGB stars. These observations seem to rule out not only very young stars, but any star formation in M32 in the last 3 – 5 Gyr — a significant fraction of that galaxy's age.

Yorke: I am happy to hear that some of questions we have discussed have been answered for one galaxy.

STAR FORMATION IN DISTANT GALAXIES

B. Rocca–Volmerange[1,2]
1 : Institut d'Astrophysique de Paris, CNRS, F–75014 Paris, France
2 : Université de PARIS XI, F–91405 Orsay-Campus, France

I. Introduction

High redshift galaxies are at present time detected along distance (up to $z \geq 3$) and time scales so extended that the look-back times reach until 90% of the age of the Universe. Among the striking facts, evidences of an intense stellar activity appear in most galaxies whatever their types or their environments. Signs of activity are essentially strong far-UV/optical/infrared stellar emissions and nebular emission lines. These spectral signatures of star formation processes are observed in extremely distant radiogalaxies as well as in clusters or in quasar environments: one of the firstly analysed effect (Butcher and Oemler, 1984) consists of an increasing number of blue galaxies with the redshift in distant cluster cores. The high density of blue galaxies in the cluster Abell 370 is interpreted as an excess of star formation activity stimulated by a ram-pressure effect (Mellier et al, 1987). Dressler and Gunn, 1983 observed strong absorption Balmer lines in spectra of elliptical galaxies. These so-called E+A galaxies are witnesses of a recent star formation activity. More recently, from a large sample of distant clusters (Gunn, Hoessel and Oke, 1986), Gunn, 1988 confirms a strong evolution of star formation for cluster galaxies until $z \simeq 1$.

Spectrophotometry of optical counterparts of the 3CR radiogalaxies until $z \simeq 2$ shows strong emission lines in the so-called Lyα galaxies (Spinrad, 1987 and Djorgovski, 1987 and references therein) with extreme Lyα equivalent widths ($\geq$ 1000Å) mainly attributed to a photoionisation process due to the massive stellar population. Also the optical and infrared counterparts of distant radiogalaxies observed in the VIJK bands from various catalogues (Laing et al, 1983, Lilly and Longair, 1984, Allington-Smith et al, 1985, Downes et al, 1986 and Dunlop et al, 1989a). The best example is the extremely distant (z=3.395) radiogalaxy discovered by Lilly,1988. The discovery of the galaxy 4C41.17 at a redshift z=3.8 (Chambers et al,1989) seems to show similar properties. A surprising fact is that in spite of an active research, no other $z \geq 3$ galaxy was discovered for two years. Is it due to the difficulty of detection or to the poorness of such a distant sample? the question is important for galaxy formation models which favor a bulk of galaxies preferentially at the redshift $z \simeq 2$.

R. Capuzzo-Dolcetta et al. (eds.), Physical Processes in Fragmentation and Star Formation, 257–265.

II. Star Formation Parameters

Observational constraints for testing the evolution of our Universe as well as for understanding formation of galaxies, open an hopeful era but it is tempered by a fundamental fact: the appearance of galaxies results from joined evolutionary and cosmological effects. It is not an easy work of separating their respective contributions. Moreover other effects such as amplification by gravitational lensing, absorption by dust, merging can be taken into account. Scenarios of galactic evolution, essentially based on our knowledge of nearby galaxies have been proposed. Star formation laws, initial mass function, metallicity are the main parameters. We shortly review the present status of these parameters in distant galaxies and give some deductive conclusions from a comparison with the most distant ($z \geq 3$) galaxies.

The star formation parameters are given in a classical form by the number of stars born by time and mass units:

$$d^2N(m,t) = \phi(m)\tau_*(t)\, d\, log\; m\; dt,$$

$\phi(m)\ \alpha\ m^{-x}$ is the Initial Mass Function (IMF) and τ_* (t) is the star formation rate.

i) the IMF distributed between two mass limits (M_{inf} and M_{sup}) is normalized to a fraction ς corresponding to the only visible component of forming stars, the other component being invisible stars (Bahcall, 1985). To respect the observed M/L ratio in nearby galaxies, we adopt a fraction ς=0.5.

Three different IMFs have been successively adopted: a) a *massive* IMF consists of a uniform slope x=0.9 and the two limits M_{inf}=2 $M_\odot$ and M_{sup}=120 $M_\odot$. b) the so-called standard IMF is given by x = 0.25 ($0.1 \leq m/M_\odot \leq 1$), 1.35 ($1 \leq m/M_\odot \leq 2$) and 1.7 ($2 \leq m/M_\odot \leq 80$) from Scalo 1986 instead of 2.3 from Miller and Scalo, 1979 or 2 from Lequeux, 1979. M_{inf} and M_{sup} are taken respectively as 0.1 $M_\odot$ and 80 $M_\odot$. c) a *low – mass* IMF with respective limits M_{inf}=0.1 $M_\odot$ and M_{sup}=1.5 $M_\odot$ is also adopted with the slope of case a). Fluxes and nebular lines resulting from a short efficient burst and the three IMFs are given in figure 1abc. Note for case a) the high far-UV luminosities and lines and the rapid evolution essentially due to the lack of low-mass stars.

ii) For simulations, the star formation rates (SFR) may follow two extreme scenarios strong initial bursts or continuous evolution laws (see Rocca-Volmerange and Guiderdoni, 1988 for details). A gas-SFR relation was adopted for spirals after the relation of the Hα equivalent width to the surface gas density in the Virgo cluster spirals (Guiderdoni and Rocca-Volmerange, 1985). Buat et al, 1989 and Kennicutt, 1989 independently confirm this relation and show from their samples that the molecular hydrogen density is not driving the star formation rate. A continuous law may be interpreted as an average of small bursts. In some cases (interactions, mergers, cooling flows) the origin of gas accreted by the star forming galaxy is easily explained. Also the UV-excess of elliptical galaxies in clusters with observational signs of mass deposited by cooling-flows is possibly explained with a gas-dependent scenario (the "UV-hot" model from Rocca-Volmerange,

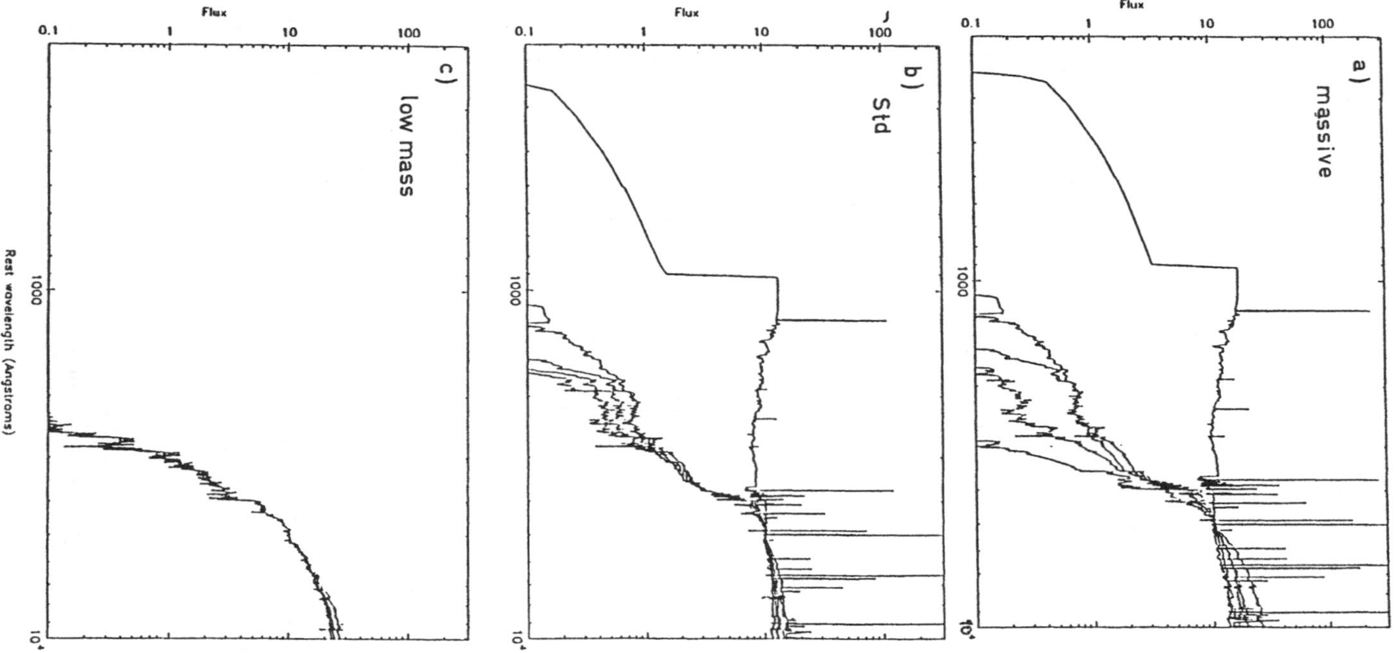

Figure 1abc. Fluxes and nebular lines during and after a star formation burst consuming all the initial content of gas through 0.1 Gyr. Three IMF as described in text are proposed. New stellar tracks according to Rocca-Volmerange and Guiderdoni,1989 have been adopted. Fluxes in arbitrary units normalized at 4990Åare shown each 0.5 Gyr from 0.1 Gyr. The wavelength range is 200Å to 10^4Å and the spectral resolution is 10Å. .

1989). But for field giant galaxies a clear explanation of the star formation time scale t_* defined as the main parameter driving the Hubble Sequence is not given. It will be one fundamental test for models of galaxy formation. Gas evolution is followed by:

$$\frac{d\sigma_T\ (t)}{dt} = -\tau_*\ (t) + \int_{m(t)}^{Msup} E(m)\phi(m)\tau_*(t - t_m)\ dm$$

E(m) is the gas fraction ejected by stars and σ_T (t) the gas density fraction normalised to the total mass.

iii) Metallicity is also a fundamental parameter to be measured in the distant galaxies, essentially to select between multiple solutions of IMF and SFR. Deficient metal tracks evolve more rapidly and are more luminous. In consequence of which, longer ages will be obtained with more metal-deficient tracks. Figure 2 shows the evolution of $[Z/Z_\odot]$=f(t) from the chemical evolution (Rocca-Volmerange and Schaeffer, 1989):

$$\frac{d\ Z\ \sigma_T(t)}{dt} = -\ Z\ \tau_*(t) + \int_{m(t)}^{Msup} E^Z(m)\phi(m)\ \tau_*(t - t_m)\ dm$$

$E^Z(m)$ is the fraction of metal ejected by stars of mass m. The integral term corresponds to ejecta from respectively non-exploding low and intermediate mass stars, type I and type II supernovae.

iv) Improvements of stellar evolutionary tracks
Several effects contribute to modify luminosity, effective temperature and lifetime duration of stars along their tracks in the Hertzsprung-Russell diagram. Among the most recent improvements of stellar models: overshooting, new opacities, helium abundances, mass loss, metallicity (VandenBergh, 1985 and references therein, Maeder and Meynet, 1988, Bertelli et al, 1989). Improved fits of the color-magnitude diagrams of the globular and open clusters. are obtained with these new sets of tracks which have to be taken into account.

III. Interpretation of the most distant Galaxies

To estimate these parameters in distant galaxies we calculate a synthetic atlas of galaxies (Rocca-Volmerange and Guiderdoni, 1988) from the model (Guiderdoni and Rocca-Volmerange, 1987). The basic principles of this approach of galactic evolution and some clues on cosmological parameters and evolution of the most distant radiogalaxies are given in Rocca-Volmerange and Guiderdoni, 1989. We also improved input data of our models by more recent set of tracks in stellar evolution (Vandenbergh, 1985, Maeder and Meynet, 1988). In the present paper, we only focus on the radiogalaxy 0902+34. This radiogalaxy was discovered by Lilly, 1988 at the redshift z=3.395, estimated from Lyα and CIV emission lines. These lines show evidences of a non-thermal component and metal enrichment. It is important to note that at this redshift the Lyα and Hβ lines coincide with the V and K bands, favouring the detection of the galaxy. Selected on its faint infrared emissivity associated to extreme red color J-K $\geq$2.75, its emissivity in Lyα

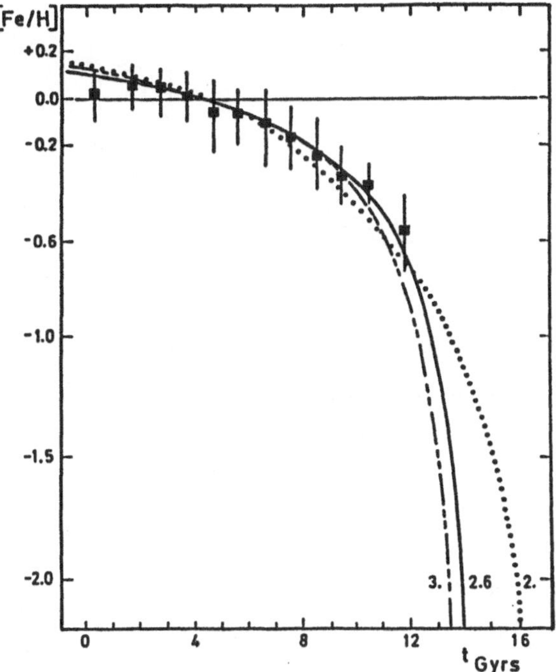

Figure 2. Relation metallicity-age in the solar neighborhood (Twarog, 1980) compared to our evolutionary models (Rocca-Volmerange and Schaeffer, 1989)

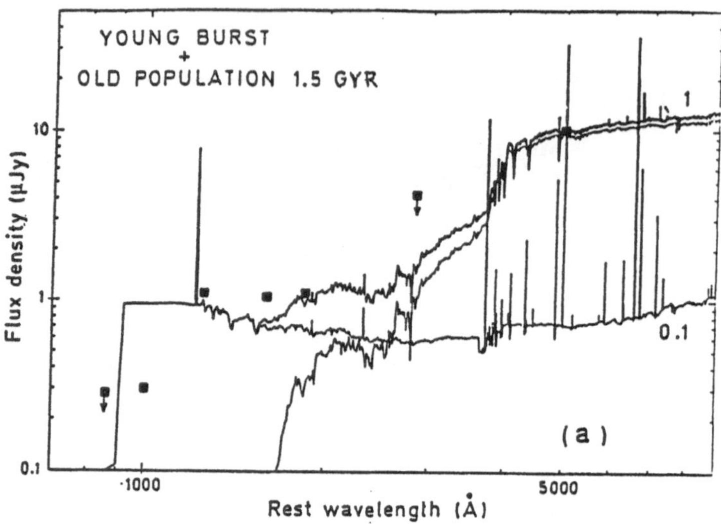

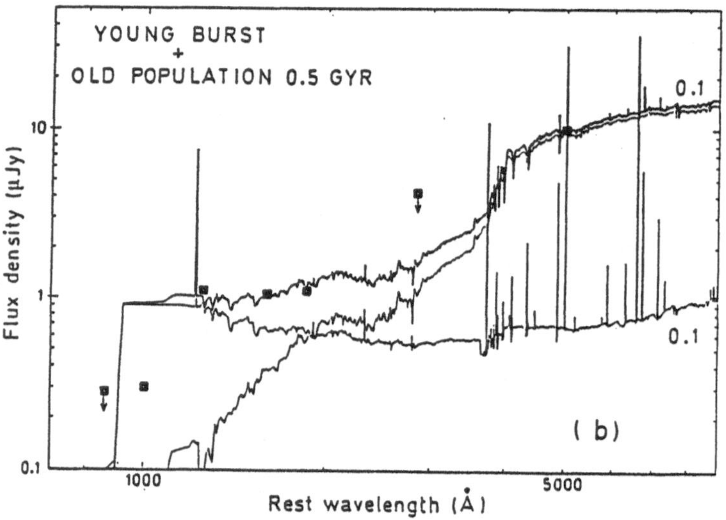

Figure 3ab Two fits of the U'BVRIJK photometry of the radiogalaxy 0902+34 (Lilly,1988,1989) with different histories of star formation. (a) 0.1 Gyr initial burst followed by passive stellar evolution for 0.4 Gyr and new burst. (b) 1Gyr initial burst followed by passive stellar evolution for 0.5 Gyr and new burst. The relevant time scale for the fit is the duration of the passive phase (0.4-0.5 Gyr). IMF is standard.

line ($\simeq 2.1 \times 10^{-18}$ W m^{-2}) and its equivalent width ($\simeq$1000 km s^{-1}) are about similar to the 3CR radiogalaxies.

The most striking feature of the radiogalaxy 0902+34 is the important gap of the apparent flux in the K band (initially 20μJy), down to the plateau observed in the $U'BVRI$ bands ($\sim$ 1μJy). After the redshift correction, this corresponds in the galaxy frame to a high visible emissivity compared to far-UV. The second feature concerns morphologies either of the extended and fairly condensed nebulosity ($\geq$ 100 kpc) emitting in the Lyman-α line at 1215Å (Spinrad,1989) and or of the rather compact emission in the K band as compared to 3C368 for instance. The first interpretation of the spectral energy distribution in the visible/infrared continuum in terms of stellar population is the strong emission of an evolved population at an age $\geq$1 Gyr (Lilly, 1988). This age may reach 2 Gyr when the red population is superimposed to a current burst explaining the far-UV emission and a part of the Lyman-α emission line (Rocca-Volmerange, 1988). The important fact is that if $H_0 = 50$ km s^{-1} Mpc^{-1} and $\Omega_0 = 1$, the Universe already has 1.42 Gyr at $z = 3.395$. So the condition $\Omega_0 = 1$ needs an age somewhere below 1.4 Gyr for this galaxy.

Recently, an improved aperture correction decreases the flux in the K band by a factor 2. The new flux now amounts to 10 μJy (Lilly, 1989). In a more refined version of our model (Rocca-Volmerange and Guiderdoni, 1989), including new sets of tracks from VandenBergh, 1985 and Maeder and Meynet, 1988 gives ages $\leq$ 0.5 Gyr and 1.5 Gyr (figure 3ab) to fit the visible/infrared continuum of 0902+34. (a) is a 0.1 Gyr initial burst started 0.5 Gyr earlier added to a short current burst (25 million years); (b) is a 1 Gyr initial burst started 1.5 Gyr earlier added to the same short current burst. The two bursting cases have 0.4 to 0.5 Gyr of passive evolution between the bursts. Thus the relevant time scale for the fit is not the duration of the initial burst, but the duration of the following phase of passive evolution, which has to be at least $\sim$0.4 Gyr for this continuum. If $\Omega_0 = 1$ and $H_0 = 50$, star formation begins at $z_{for} = 4.9$ for age 0.5 Gyr or at $z_{for} \geq 8.9$ for ages $\geq$1 Gyr. The mass of the old population derived from the rest–frame V band is 2.3 $10^{11} M_\odot$ (a) to 3.7 $10^{11} M_\odot$ (b), and the mass of the current burst from the 1200 Å rest–frame emission is only 5.9 $10^9 M_\odot$, after assuming that the current burst is going to last for 0.1 Gyr. In each case a dark mass fraction $\varsigma = 0.5$ is adopted. In order to select among the processes responsible for the efficiency of star formation, time scales as well as morphological arguments could be used. As previously shown, K band emissivity can be produced by young supergiant stars, AGB stars or a population of giants. We can hope to select the relevant ages from the spatial distribution of emissivity in the K band. An old population of giants must be relaxed and compact while a population of supergiants could be aligned along the radio axis (McCarthy et al, 1987). Lilly, 1989 shows that the reddest galaxies are not typically aligned. According to signs of star formation activity, bursts appear the best solutions to simultaneously reproduce the emission in the far–UV and visible light. However several comments can be done: only a better understanding of emission lines (thermal or non-thermal) as well as chemical evolution will allow to separate the various solutions which at the present time are not really constraining for separating the various parameters of star formation (IMF and rate). Also high resolution morphological data (Le Fèvre and Hammer, 1988) as well as a statistically sample of such distant galaxies would have to be instructive on the luminous mass distribution to constraint models of galaxy formation.

References

Allington-Smith, J.R., Lilly, S.J., Longair, M.S., 1985, *Mon. Not. R. Astr. Soc.*, **213**, 243

Bahcall, J. N., 1985, *Astrophys. J.*, **287**, 926

Bertelli et al, 1989, preprint

Buat, V., Deharveng, J.M., Donas, J., 1989, *Astron. Astrophys.*, **223**, 42

Butcher, H., Oemler, A.Jr., 1984, *Astrophys. J.*, **285**, 426

Chambers, K.C., Miley, G.K. & Van Breugel, W., submitted

Djorgovski, S.G., 1987, "Towards Understanding Galaxies at Large Redshift", Kron, R.G., Renzini, A. eds., Kluwer, p.259

Downes, A.J.B., Peacock, J.A., Savage, A., Carris, D.R., 1986, *Mon. Not. R. astr. Soc.*, **218**, 31

Dressler, A., Gunn, J.E., 1983, *Astrophys. J.*, **270**, 7

Dunlop, J., Peacock, J.A., Savage, A., Lilly, S.J., Heasley, J.N., Simon, A.J.B., 1989a, *Mon. Not. R. Astr. Soc.*, **238**, 1171

Dunlop, J., Guiderdoni, B., Rocca-Volmerange, B., Longair, M., Peacock, J., 1989b, *Mon. Not. R. Astr. Soc.*, **240**, 257

Guiderdoni, B., Rocca–Volmerange, B., 1987 *Astron. Astrophys.***186**, 1

Guiderdoni, B., Rocca–Volmerange, B., 1985, *Astron. Astrophys.*, **151**, 108

Gunn, J.E., Hoessel, J., Oke, J.B., 1986, *Astrophys. J.*, **306**, 30

Gunn, J.E., *The Epoch of Galaxy Formation*, C.S. Frenk *et al.* (eds.), NATO ASI Series C, vol 264.

Kennicutt, 1989, preprint

Laing, R.A., Riley, J.M., Longair, M.S., 1983, *Mon. Not. R. astr. Soc.*, **204**, 151

Le Fèvre, O., Hammer, F. & Jones, J., 1988, *Astrophys. J.* **331**, L73

Lequeux, J., 1979, *Astron. Astrophys.*, **80**, 35

Lilly, S., Longair, M., 1984, *Mon. Not. R. Soc.*, **211**, 833

Lilly, S., 1988, *Astrophys. J.* **333**, 161

Lilly, S., 1989, *Astrophys. J.* **340**, 77

Maeder, A., Meynet, G., 1988, *Astron. Astrophys. Suppl. Ser.* **76**, 411

Mc Carthy, P. J., Van Breugel, W., 1989, *"The epoch of Galaxy formation"*, Frenk et al, eds, NATO ASI Series C, vol 264, p. 57

Mc Carthy, P.J., Van Breugel, W., Spinrad, H., Djorgovski, S., 1987, *Astrophys. J.* **321**, L29

Mellier, Y., Soucail, G., Fort, B., Mathez, G., 1987, *Astron. Astrophys.*, **199**, 13

Miller, J., Scalo, J.M., 1979 *Astrophys. J.*, **41**, 513

Rocca–Volmerange, B., 1988, ESO *The Messenger* **53**, 26

Rocca–Volmerange, B., 1989, *Mon. Not. R. astr. Soc.*, **236**, 47

Rocca–Volmerange, B., Guiderdoni, B., 1988, *Astron. Astrophys. Suppl. Ser.* **75**, 93

Rocca–Volmerange, B., Guiderdoni, B., 1989, *The Quest for Cosmological Parameters*, Rencontres de Moriond, Eds J.Audouze, Tran Thanh Van., in press

Rocca–Volmerange, B., Schaeffer, R., 1989, *in press*

Scalo, J.M., 1986, *Fundamental of Cosmic Physics*, 11, 1

Spinrad, H., 1987, *High Redshift and Primeval Galaxies*, Ed. J. Bergeron, D. Kunth, B. Rocca-Volmerange, Tran Thanh Van, Ed Frontieres,p. 59

Spinrad, H., 1989, *The Epoch of Galaxy Formation*, C.S. Frenk *et al.* (eds.), NATO ASI Series C, vol 264, p. 57

Twarog, 1980, *Astrophys. J.*, **242**, 242
Vandenberg, D.A., 1985, *Astrophys. J. Suppl. Ser.* **58**, 711

THE EARLY EVOLUTION OF SPHEROIDAL STAR SYSTEMS

L. ANGELETTI and P. GIANNONE
Istituto Astronomico
Università 'La Sapienza', Roma
Via Lancisi, 29
I-00161 ROME (Italy)

ABSTRACT. Supernova–driven galactic wind and metal enrichment in spheroidal star systems are considered within the general framework of the closed one–zone–model approach. Three issues are addressed: i) protogalaxy binding energy, ii) non–uniform gas density distribution, iii) presence of dark matter. Models of a galaxy with a mass of $10^{10} M_{\odot}$ are discussed for illustrative purposes.

1. Galactic wind

A key observational outcome about galaxies is the mass–metallicity relation stating that the more massive the galaxy, the larger the mean metal abundance of constituent stars. A commonly accepted explanation of this fact is that diffuse gas in star–forming galaxies was metal enriched, heated, and finally ejected as a consequence of supernova (SN) explosions (Larson 1974). After the SN–driven galactic wind occurred, no substantial star formation has taken place, as matter further expelled from evolved stars is assumed to be lost to the intergalactic medium.

According to Larson (1974), a global galactic wind is assumed to occur at time t_{GW}, when the total residual thermal energy $E_{th}(t_{GW})$ of all supernova remnants (SNRs) reaches the binding energy $\Omega_g(t_{GW})$ of the remaining gas. Following this condition, Saito (1979b) considered the occurrence of a galactic wind in spheroidal star systems within the framework of the one–zone model. Further developments were later introduced by Arimoto and Yoshii (1987), Yoshii and Arimoto (1987), and Matteucci and Tornambè (1987).

In a previous paper (Angeletti and Giannone 1989a) we discussed some problems concerning the exact evaluation of t_{GW} and related quantities. Discrepancies between our results and those by the latter mentioned authors were found to be caused by a different way of computing E_{th} and a different choice of the initial protogalactic radii. Notwithstanding a claimed general agreement between theoretical results and observational data, initial protogalactic radii consistent with the model assumptions and an accurate evaluation of E_{th} lead to significant discrepancies with regard to observations. Namely, the computed spreads of star metallicities in low–mass systems

R. Capuzzo-Dolcetta et al. (eds.), Physical Processes in Fragmentation and Star Formation, 267–276.

($M_G \lesssim 10^8 M_\odot$) are at variance with the more or less pronounced chemical homogeneity observed in those systems. Moreover, the mean star metallicities in large–mass models ($M_G \gtrsim 10^8 M_\odot$) exceed the corresponding observed values. Finally, galactic wind and subsequent gas losses fail to explain masses and metallicities of the diffuse medium in clusters of galaxies: therefore, a large amount of primordial gas is required in order to match observations.

These facts suggest that the problem under discussion deserves further attention and current models need substantial implementation. On the basis of general considerations it may be expected that some improvement in the models might come from increasing the initial protogalactic radii or, equivalently, decreasing the initial gas density. In the present paper we focus on three issues relevant to the topic: i) the initial binding energy of a protogalaxy; ii) the gas density distribution and the related choice of the initial radii of 'equivalent' one–zone models, iii) the presence of dark matter as a major component of the galactic mass.

2. Initial binding energy

In the standard one–zone–model approach, a spherically symmetric star system is approximated by a uniform density sphere whose radius R_G is kept constant during the evolutionary phase before the occurrence of the galactic wind. Saito (1979b) adopted

$$R_G = 0.6\, G\, M_G^2/\Omega_G(M_G) \tag{1}$$

where M_G is the initial (total) galactic mass, $\Omega_G(M_G)$ the initial binding energy, and G the gravitational constant. Moreover, in the one–zone models the binding energy $\Omega_g(t)$ of the gas evolves in time according to

$$\Omega_g(t) = \Omega_g(0) f_g(t)[2 - f_g(t)] \qquad \text{for} \qquad 0 \leq t \leq t_{GW} \tag{2}$$

where $f_g(t) = M_g(t)/M_g(0)$ and $M_g(t)$ is the mass in the gas phase at time t. We assume here $M_g(0) = M_G$, hence $\Omega_g(0) = \Omega_G(M_G)$.

From observations of 'present epoch' spheroidal star systems (from globular clusters to giant elliptical galaxies, with the exception of dwarf spheroidals), the following relationship between binding energy Ω_a and mass M_a was derived by Saito (1979a)

$$\Omega_a(M_a) = k M_a^{\beta+1} \qquad \text{erg} \tag{3}$$

where $k = 3.41\ 10^{-6}$, M_a is in gr, $\beta = 0.45$, and subscript a refers to the present epoch systems.

The same relationship (3) was adopted in Saito (1979b), and in the subsequent papers mentioned above, between the initial binding energy and the initial mass in order to define the initial radius. Actually, this choice might not be fully correct if actual (observed) systems underwent a significant mass loss in the past (e.g. via galactic winds). Indeed, taking $\Omega_G(M_G) = \Omega_a(M_G)$

is equivalent to assuming that the initial and actual binding energies per unit mass are equal.

Let us now assume that a protogalaxy was initially (index i) in virial equilibrium and set $t = 0$ when such a protogalaxy began to cool rapidly (kinetic energy approximately equal to zero at $t = 0$) and to depart from equilibrium by collapsing, until a new stage (index v) of virial equilibrium was reached again. If energy and/or mass losses are negligible during collapse, one has (Silk 1977)

$$\omega_i(M_i) = 0.5\omega_v(M_i) \tag{4}$$

ω's being the binding energies per unit mass at the indicated stages.

In order to take into account the effects of a global galactic wind, we applied the findings by Hills (1980), assumed to be relevant here, in three particular cases: A) an impulsive wind after the initial collapse and subsequent virialization, B) a gradual wind after the initial collapse and subsequent virialization, C) an impulsive wind during the protogalaxy collapse. Moreover, we also considered gas losses from galaxies after the main wind episode; therefore, a gradual mass loss between t_{GW} and the actual epoch t_a was added to each one of the above cases, leading to cases a, b, and c, respectively.

When imposing condition $\omega_a(M_a) = kM_a^\beta$ [as required by observations, cf. (3)], it is found that the initial (i.e. at $t = 0$) binding energies per unit mass are given by

$$\begin{cases} \omega_i(M_i) = A_1 k M_i^\beta & \text{(case a)} \\ \omega_i(M_i) = A_2 k M_i^\beta & \text{(case b)} \\ \omega_i(M_i) = A_3 k M_i^\beta & \text{(case c)} \end{cases} \tag{5}$$

where $A_j = \phi_j(1 - f_t)^{\beta-1}$ (with $j = 1, 2, 3$), $f_t = f_w + f_{aw}$, $f_w = f_g(t_{GW})$ is the fractional mass lost in the main wind episode, and $f_{aw}(< 1 - f_w)$ is the fractional mass [mass in units of $M_g(0)$] of the gas lost between t_{GW} and t_a [obviously, $M_a = M_i(1 - f_w - f_{aw})$]. Moreover, it is $\phi_1 = q_1q_2q_5$, $\phi_2 = q_1q_3q_5$, and $\phi_3 = q_4q_5$, with q's expressed in terms of initial (at $t = 0$) radius R_i, radius R_v soon after virialization at the end of the initial collapse, final (at $t = t_a$) radius R_a, and the radii R_w^a, R_w^b, R_w^c attained by star systems after the main wind episode in cases a, b, and c, respectively.

Indeed, according to Hills (1980), one obtains: $q_1 = R_v/R_i = 0.5$; $q_2 = R_w^a/R_v = (1 - f_w)(1 - 2f_w)^{-1}$ with $f_w < 0.5$; $q_3 = R_w^b/R_v = (1 - f_w)^{-1}$ with $f_w < 1$; $q_4 = R_w^c/R_i = 0.5(1 - f_w)(1 - \lambda f_w)^{-1}$ with $f_w < \lambda^{-1}$ and $\lambda \geq 1$, where $\lambda = R_i/R_c$ and R_c is the protogalactic radius when the impulsive wind has taken place during collapse; and, finally, $q_5 = R_a/R_w = (1 - f_w)(1 - f_t)^{-1}$, where R_w refers to whichever radius R_w^a, R_w^b, or R_w^c is concerned.

Since values of R_i, f_w, and f_{aw} are unknown before solving the evolution equation for the gas mass in the star system and the pertinent equation (5), they have to be obtained by iteration. Actually, with starting conditions $A_j^{(0)} = 0.5$

(for $j = 1, 2, 3$) [i.e. with $f_w = f_{aw} = 0$], one has $R_i = 0.6\,GM_i/\omega_i(M_i)$ from (5). The evolution equation can then be integrated with a galactic radius R_i in case c and $R_v = 0.5\,R_i$ in cases a and b. The obtained values of f_w and f_{aw} give the new value $A_j^{(1)}$, and so on until convergency is achieved (provided the star system is in a bound state).

3. Non–uniform gas density models

The functional form of the r.h.s. of equation (2) follows from the factorization $\varrho_g(r,t) = f_1(r)f_2(t)$, where $\varrho_g(r,t)$ is the gas density, r the radial distance; $f_1(r)$=constant for a uniform density sphere. Moreover, Ω_G, as given by a relationship like (3) [i.e. $\Omega_G(M_G) = k\,M_G^{\beta+1}$], does not depend on R_G. Consequently, $\Omega_g(t)$ is independent of R_G, while $E_{th}(t)$ depends on it through the dependence of the complete (radiative) cooling time of the SNR shell on the density of the surrounding ambient gas [see e.g. Cox 1972]. Therefore, in order to improve the evaluation of $E_{th}(t)$ and to bring it as close as possible to its value in real star systems [in which $f_1(r) \neq$constant], a better selection of R_G was attempted.

In an order–of–magnitude estimate, we assumed (by neglecting gas recycling)

$$\frac{d\varrho_g(r,t)}{dt} = -\nu(r)\varrho_g(r,t) \tag{6}$$

with the star formation rate (per unit gas mass) $\nu(r) = \nu$ =constant. Hence, function $f_1(r)$ remains unchanged for $t \leq t_{GW}$. Then we adopted

$$\varrho_g(r,t) = \varrho_o(t)[1 + (r/r_o)^2]^{-p/2} \qquad \text{for} \quad r \leq r_t \tag{7}$$

so as to avoid more precise but also more complicated functional forms, which we thought to be unnecessary at this stage. From (6) it follows that the central density evolves according to $\varrho_o(t) = \varrho_o(0)\exp(-\nu t)$.

For selected values of M_G, p, and $\Omega_G(M_G)$, parameters $\varrho_o(0)$, r_o (structural length), and r_t (limiting radius) were obtained by constraining the model binding energy to be equal to $\Omega_G(M_G)$, and other model properties (e.g. core radius r_c and concentration parameter $c = \log r_t/r_c$) to be as close as possible to the corresponding observed values. For a first–order estimate, mean values of the relevant parameters can be taken from the samples of various classes of spheroidal star systems used by Saito (1979a) [for more details, see Angeletti and Giannone 1979b].

Finally, an 'equivalent' one–zone model was introduced: its radius R_E was chosen in such a way as to give the same $E_{th}(t)$ of the system with $\varrho_g(r,t)$ as in (7), $M_g(t)$ being the same. As typical results, we obtained $R_E \simeq 2.5\,R_G$ for $p = 3$, where R_G is given by (1) and (3) [i.e. by assuming $\Omega_G(M_G) = k\,M_G^{\beta+1}$]. On this ground, non–uniform gas density distributions occurring in real star systems suggest to enlarge the 'standard' radius (1).

It is worth noting that computations with $\varrho_g(r,t)$ as in (7) showed that more than half of $E_{th}(t)$ can be located in the outermost gaseous layers. This fact seems to suggest a continuous gas evaporation rather than an almost simultaneous global wind at t_{GW}.

4. Dark matter

Dark matter (DM) was included in the one–zone model under the following assumptions: i) DM is represented by a uniform and time–constant density sphere of radius R_D and total mass M_D; ii) non–dark matter (NDM) is represented by a uniform density sphere of radius $R_b(t) \leq R_D$ and total mass $M_b(t) \ll M_D$; iii) at $t=0$ it is: $M_g(0) = M_b(0)$, $\alpha_0 = M_b(0)/[M_b(0)+M_D]$, $\gamma_0 = M_b(0)/M_d(0)$, where $M_d(t)$ represents the DM mass within $R_b(t)$.

On the ground of the above assumptions, relationship (3) can be interpreted as referring to the total mass $M_T(t_a) = M_b(t_a) + M_d(t_a)$ within $R_b(t_a)$, as if there were no matter beyond $R_b(t_a)$. Therefore, quantity $\omega[M_T(t)] = 0.6\,GM_T(t)/R_b(t)$ is 'formally' defined and used here as in Section 2, though it has lost its physical meaning since the presence of matter beyond $R_b(t)$ is ignored. Moreover, one may set $R_b(t) = 0.6\,GM_T(t)/\omega[M_T(t)]$ and $\omega_a[M_T(t_a)] = kM_T^{\beta}(t_a)$ [cf. (3)].

We now take into account the same cases a, b, and c as in Section 2, with the only difference that virialized states are replaced by stationary states. The initial (i.e. at $t=0$; index i) values of $\omega[M_T(t)]$ are given by

$$(8)\quad \begin{cases} \omega_i[M_T(0)] = D_1\, kM_T^{\beta}(0) & \text{(case a)} \\ \omega_i[M_T(0)] = D_2\, kM_T^{\beta}(0) & \text{(case b)} \\ \omega_i[M_T(0)] = D_3\, kM_T^{\beta}(0) & \text{(case c)} \end{cases}$$

where $D_j = \phi_j(1+\gamma_0)^{1-\beta}[\gamma_0(1-f_t)+\phi_j^3]^{\beta-1}$ (with $j=1,2,3$), f_t as in Section 2, $\phi_1 = q_1q_2q_3$, $\phi_2 = q_1q_2'q_3'$, and $\phi_3 = q_4q_5$. Values of q's are now the 'real non–negative' roots of algebraic equations of third or fourth order appropriate to each case.

For case a, q_1 is the relevant root of equation

$$(9)\qquad q^3 + (\gamma_0 - 0.5)q - 0.5\gamma_0 = 0$$

where $q = R_b(t_s)/R_b(0)$, and t_s refers to the time when NDM reaches a new steady state at the end of the initial collapse (with negligible energy and/or mass losses). q_2 is the pertinent root of equation

$$(10)\qquad q^3 - [1 - 0.5\gamma_0 q_1^{-3}(1-2f_w)]q - 0.5\gamma_0 q_1^{-3}(1-f_w) = 0$$

with $q = R_b(t_{GW})/R_b(t_s)$, where $R_b(t_{GW})$ is the radius attained by NDM at the end of the impulsive (global) wind, and f_w is as in Section 2. Finally, q_3 is the appropriate root of equation

$$(11)\qquad q^4 + \left[\gamma_0(q_1q_2)^{-3}(1-f_t)\right]q - \left[1+\gamma_0(q_1q_2)^{-3}(1-f_w)\right] = 0$$

where $q = R_b(t_a)/R_b(t_{GW})$.

For case b, equation (10) is replaced by

$$q^4 + \gamma_0 q_1^{-3}(1 - f_w)q - (1 + \gamma_0 q_1^{-3}) = 0 \tag{12}$$

where $q = R_b(t_{GW})/R_b(t_s)$ and root q_2' of (12) replaces q_2 in (11) thus leading to root q_3'.

Finally, for case c, q_4 is the relevant root of equation

$$q^3 + [\gamma_0(1 - \lambda f_w) + 0.5\lambda^{-2} - 1]q - 0.5\gamma_0(1 - f_w) = 0 \tag{13}$$

where $q = R_b(t_{GW})/R_b(0)$, $R_b(t_{GW})$ being the radius attained by NDM in the steady state after the impulsive (global) wind during collapse. Moreover, $\lambda = R_b(0)/R_b(t_c)$, $R_b(t_c)$ being the protogalactic radius when the impulsive wind takes place during collapse. q_5 is the pertinent root of equation

$$q^4 + \left[\gamma_0 q_4^{-3}(1 - f_t)\right] q - \left[1 + \gamma_0 q_4^{-3}(1 - f_w)\right] = 0 \tag{14}$$

where $q = R_b(t_a)/R_b(t_{GW})$.

The time evolution of the gas potential energy $\Omega_g(t)$ (to be used in the one–zone models) is given by a relationship similar to (2), namely

$$\Omega_g(t) = \Omega_i[M_g(0)]f_g(t)[2I - f_g(t)] \qquad \text{for} \qquad 0 \le t \le t_{GW} \tag{15}$$

where $\Omega_i[M_g(0)] = 0.6\,GM_g^2(0)/R_b(0)$, $f_g(t)$ being as in Section 2, and $I = 1 + q_0(1 - \alpha_0)(5 - q_0^2)/4\alpha_0$ with $q_0 = R_b(0)/R_D = [\alpha_0/\gamma_0(1 - \alpha_0)]^{1/3}$.

In each of the above cases, $R_b(0), f_w$, and f_{aw} have to be determined by the iterated integration of the evolution equation for the gas mass and use of the pertinent equation (8), as outlined at the end of Section 2.

5. Comparison with observations

Comparing models with observations is made difficult because of the many free parameters to be constrained and the many observational evidences to be matched. From the theoretical side the key functions are: i) initial mass function $\varphi(m)$ (m is the stellar mass), ii) star formation rate ν (per unit gas mass), iii) stellar lifetime $\tau(m)$, iv) mass loss from stars, v) metal yields from evolved stars and SNe. From the observational side basic data concern: i) relations between mass, luminosity, metallicity, central velocity dispersion, and radius (or binding energy). Additional information comes from observations of clusters of galaxies and supports galactic wind models if the intracluster gas is from component galaxies (as conceivable in a first instance). Indeed, the intracluster mass and metallicity are $M_{gas} \approx (1 \div 2)M_{star}$ and $Z_{gas} \approx 0.5Z_{\odot}$

(Suchkov et al. 1987), respectively (M_{star} is the total stellar mass in the galaxies of the cluster).

As an illustration of above discussions, we present some results obtained for a protogalaxy with $M_i = 10^{10} M_\odot$. Parameters and assumptions entering into the problem were defined as follows [whatever non explicitly mentioned is as in Angeletti and Giannone 1989a]: i) $\varphi(m) \propto m^{-x}$ with $x = 0.95$ or 1.35; $0.05 = m_l \leq m \leq m_u = 60$ (solar units); ii) progenitors of type II SNe in the mass range 8 to 60 $M_\odot$, and progenitors of type I1/2 SNe in the mass range 5 to 8 $M_\odot$; iii) $\nu^{-1} = 67.81\ 10^6$ yr; iv) yields of the newly created metals from SNe and white dwarf precursors in the mass range 4 to 5 $M_\odot$ (zero metal yields from white dwarf precursors with $m < 4M_\odot$); v) mass loss from stars with $m < 5M_\odot$; vi) two formulations for $\tau(m)$, that is $\tau_1(m) = 11.7 m^{-2}$ (m in solar units), and $\tau_2(m) = 5m^{-2.7} + 0.012$ for $m \leq 8M_\odot$, or $\tau_2(m) = 1.2m^{-1.85} + 0.003$ for $m > 8M_\odot$ (all τ's are in units of 10^9 yr). Finally, we chose $Z_0 = 10^{-8}$ as the initial gas metallicity, $t_a = 15\ 10^9$ yr, and $Z_\odot = 0.02$, and computed E_{th} with cooling time t_c of a SNR shell at the time of the SN explosion.

In Table 1 we reported our 'standard' models for $M_i = 10^{10} M_\odot$, i.e. with $x = 0.95$, or $x = 1.35$, $\tau(m) = \tau_1(m)$, or $\tau(m) = \tau_2(m)$, $R_E = R_i, \omega_i(M_i) = k\, M_i^\beta$ according to Saito (1979b) and the other authors after him, and with no dark matter ($\alpha_0 = 1$) and no iterative procedure. The following definitions were also adopted: $f_t = f_w + f_{aw}$; $f_s(t_a) = M_s(t_a)/M_g(0)$ is the fractional mass of formed stars still shining at t_a and whose total mass is $M_s(t_a), f_r(t_a) = M_r(t_a)/M_g(0)$ is the fractional mass of the stellar remnants formed up to t_a and whose total mass is $M_r(t_a)$. Moreover, the actual mean star metallicity (weighted by mass) was denoted by $\overline{Z}_s(t_a)/Z_\odot$ and the metallicity reached by gas at t_{GW} by $Z_w/Z_\odot$; while the mean metallicity (weighted by mass) of the gas lost after t_{GW} and up to t_a, and from t_{GW} up to t_a, are indicated by $\overline{Z}_{aw}/Z_\odot$ and $\overline{Z}_g(t_a)/Z_\odot$, respectively.

From Table 1 one has $f_w \lesssim f_{aw}$, $f_t \simeq f_s(t_a)$ for $x = 0.95$ and $f_t \simeq 0.25 f_s(t_a)$ for $x = 1.35$, while metallicities at t_a are always larger than given by observations for the gas and for the stars as well (when $x = 0.95$). On the other hand, the differences coming from the two lifetimes $\tau_1(m)$ and $\tau_2(m)$ are not exceedingly large, so from now on we shall only consider $\tau_2(m)$ [which appears to be more realistic than $\tau_1(m)$].

Preliminary results, referring to Sections 2 and 3, are shown in Table 2 for standard ($R_E = R_v$) and non–uniform ($R_E \neq R_v$ and $p = 3$) gas density models. Mass loss as in case b was selected for the sake of convenience, and the virialization soon after collapse was chosen as the initial stage of galactic evolution (hence $\omega_v = 2\omega_i = C\, kM_i^\beta$, i.e. $C = 2A_2$). The iterative procedure, which also took into account that $\nu \propto C^2$ (it is $\nu^{-1} = 67.81\ 10^6$ yr for $C = 1$), led to the listed final values of C, from which R_v can be obtained. The comments already made about Table 1 with regard to fractional masses and star and gas metallicities at t_a also apply to Table 2. Comparison of the second (or fourth) row in Table 1 with the first (or third) row in Table 2 shows the effects of the appropriate determination of the initial radius (for the chosen models). Comparison of the upper (or lower) two rows in Table 2 illustrates

Table 1. Standard models for $M_i = 10^{10} M_\odot$ (for symbols see text).

x	$\tau(m)$	t_{GW} (10^6 yr)	f_w	f_{aw}	f_t	$f_s(t_a)$	$f_r(t_a)$	$\log \frac{Z_w}{Z_\odot}$	$\log \frac{\overline{Z}_{aw}}{Z_\odot}$	$\log \frac{\overline{Z}_g(t_a)}{Z_\odot}$	$\log \frac{\overline{Z}_s(t_a)}{Z_\odot}$
0.95	τ_1	216	0.13	0.29	0.42	0.43	0.15	0.792	0.797	0.796	0.290
0.95	τ_2	204	0.18	0.19	0.37	0.44	0.19	0.910	0.650	0.796	0.457
1.35	τ_1	243	0.05	0.14	0.19	0.72	0.09	0.443	0.534	0.513	-0.306
1.35	τ_2	261	0.05	0.11	0.16	0.72	0.12	0.600	0.114	0.326	-0.019

Table 2. Case b models for $M_i = 10^{10} M_\odot$ with $\tau(m) = \tau_2(m)$ (for symbols see text).

x	C	$\frac{R_E}{R_v}$	t_{GW} (10^6 yr)	f_w	f_{aw}	f_t	$f_s(t_a)$	$f_r(t_a)$	$\log \frac{Z_w}{Z_\odot}$	$\log \frac{\overline{Z}_{aw}}{Z_\odot}$	$\log \frac{\overline{Z}_g(t_a)}{Z_\odot}$	$\log \frac{\overline{Z}_s(t_a)}{Z_\odot}$
0.95	2.1	1.0	64	0.10	0.28	0.38	0.43	0.19	0.975	0.760	0.828	0.412
0.95	2.7	2.5	28	0.14	0.33	0.47	0.37	0.16	0.747	0.773	0.765	0.225
1.35	1.3	1.0	157	0.04	0.12	0.16	0.71	0.13	0.663	0.156	0.358	-0.056
1.35	1.6	2.5	66	0.12	0.14	0.26	0.63	0.11	0.374	0.401	0.389	-0.292

the effects of a change from a uniform to a non-uniform radial behavior of the gas density.

Effects of dark matter are under study; preliminary results show that dark matter can play a significant role in the problem. Sets of results obtained according to Sections 3 and 4 will be published elsewhere.

References

Angeletti, L., Giannone, P.: 1989a, Astron. Astrophys., in press.
Angeletti, L., Giannone, P.: 1989b, in preparation.
Arimoto, N., Yoshii, Y.: 1987, Astron. Astrophys., **173**, 23.
Cox, D.P.: 1972, Astrophys. J., **178**, 159.
Hills, J.G.: 1980, Astrophys J., **235**, 956.
Larson, R.B.: 1974, Monthly Not. R. Astron. Soc., **169**, 229.
Matteucci, F., Tornambè, A.: 1987, Astron. Astrophys., **185**, 51.
Saito, M.: 1979a, Publ. Astron. Soc. Japan, **31**, 181.
Saito, M.: 1979b, Publ. Astron. Soc. Japan, **31**, 193.
Silk, J.: 1977, Astrophys. J., **211**, 638.
Suchkov, A.A., Berman, V.G., Mishurov, Yu. N.: 1987, Sov. Astron., **31**, 371.
Yoshii, Y., Arimoto, N.: 1987, Astron. Astrophys., **188**, 13.

Discussion

P. SHAPIRO: Is it possible that the excess spread in metallicity predicted by this picture could be alleviated by adjusting the IMF? For example, perhaps the first stars were only the massive, short-lived stars, and the lower-mass, longer-lived stars formed only *after* the production of metals by the first stars.

P. GIANNONE: The form of the initial mass function is one of the constitutive conditions entering into the problem and, as such, various adjustments of it can be predicted to affect the model results. In particular, a change like the one you mentioned might reduce the excess spread in metallicity in low-mass star systems. However, in order to obtain the high degree of chemical homogeneity observed in globular clusters, rather restrictive (and perhaps not completely plausible) conditions on the values of transition mass and formation-time delay between massive and low-mass stars might be required.

M. MATEO: You note that having a large metallicity spread in low-mass systems is a problem, but my impression is that observationally there is a good evidence for a spread in metallicity in most of the dwarf spheroidals and it is interesting to note that there is a spread observed in the most massive globular cluster (ω Cen).

P. GIANNONE: The problem here concerns the amount of the metallicity spreads when model results are compared with observational data for dwarf spheroidal galaxies and most massive globular clusters. Models provide values of $\Delta[Fe/H]$ much in excess to observations.

A. DI FAZIO: i) How much mass percentage, in the dark mass case, did you allow for dark matter? ii) You showed a result by Arimoto and Yoshii according to which the mass loss was $\sim 80\%$ (in the max). Did they calculate the constraints in the mass–loss rates needed in order not to destroy the system?

P. GIANNONE: i) As I mentioned, among the cases we treated, we have concentrated on the one concerning 90% of the total galactic mass being in the form of dark matter; ii) A paper by Hills (see references), also mentioned by Arimoto and Yoshii, shows that a star system will disrupt if half or more of the mass of the system is ejected impulsively (i.e. in a time scale which is shorter than the dynamical time scale).

R. CAPUZZO–DOLCETTA: Is it possible that part of the difference among your and other authors' results is due to the treatment of the numerical time integration of the SN energy release around the discontinuity at $t_{SN} = t_{cooling}$?

P. GIANNONE: Discrepancies in the time of occurrence of galactic winds and related quantities between our results and those by the other mentioned authors are caused by a choice of larger initial protogalactic radii in Matteucci and Tornambè (1987), and by a (likely) overestimation of the total residual thermal energy (E_{th}) of all supernova remnants in Arimoto and Yoshii (1987) and Yoshii and Arimoto (1987). The discontinuity you mentioned can be relevant to the results, depending on how the numerical time integration is performed.

3. HYDRODYNAMICAL AND DYNAMICAL MODELS.

A.P. BOSS - Fragmentation of isothermal and non-isothermal protostellar clouds.

W.M. TSCHARNUTER - Protostellar core instabilities.

R. CAPUZZO–DOLCETTA, A. DI FAZIO, A.B. MEN'SHCHIKOV - A multifluid hydrodynamical radiative model for the evolution of a spherical protogalaxy.

A. PARRAVANO - Thermal instabilities in the warm interstellar gas as a regulating mechanism of star formation.

A.B. MEN'SHCHIKOV - Evolution of the first protostellar core.

R. BEDOGNI, P.R. WOODWARD - Numerical hydrodynamics of cloud implosion.

FRAGMENTATION OF ISOTHERMAL AND NONISOTHERMAL PROTOSTELLAR CLOUDS

A. P. BOSS
DTM, Carnegie Institution of Washington
5241 Broad Branch Road, N.W.
Washington, D.C. 20015
U.S.A.

ABSTRACT. A large number of fully three dimensional (3D) calculations of the collapse of low mass protostellar clouds, performed with both finite difference and smoothed particle hydrodynamics codes, have shown that isothermal protostellar clouds usually fragment during collapse into binary or higher order protostellar systems. This dynamical fragmentation requires, however, starting the collapse phase from an initial cloud that is less centrally condensed than $\rho_i \propto r^{-1}$ and that has appreciable specific angular momentum ($J/M > 2 \times 10^{20}$ cm^2 s^{-1}). 3D calculations of fragmentation in the nonisothermal regime ($\rho > 10^{-13}$ g cm^{-3}) of low mass protostellar collapse have shown that while fragmentation can still occur under certain conditions, the nonisothermal regime generally marks the beginning of the cessation of fragmentation. This occurs because rising temperatures result in increased ratios of thermal to gravitational energy, stifling thermally-driven fragmentation. The minimum protostellar mass formed by the collapse and fragmentation of Population I clouds appears to be $\sim 0.01 M_\odot$. Increased thermal support leads to the formation of rotationally flattened, triaxial, quasi-equilibrium protostellar cores that are the 3D analogues of the 1D cores found by Larson (1969). Protostellar cores may be either relatively large, outer cores, supported by molecular hydrogen, or relatively small, inner cores, supported by atomic and ionized hydrogen. Rotationally-driven fragmentation also appears to be stifled in these quasi-equilibrium cores, because angular momentum transport by spiral structure leads to orbital decay of any nascent binary structure, as found in calculations of the dynamic fission instability in rapidly rotating polytropes.

1. Introduction

This paper considers the simplest possible fragmentation problem, namely the formation of a binary (or multiple) protostellar system through the *dynamical fragmentation* of a collapsing, nonmagnetic, interstellar cloud. Because of the relative simplicity of this problem, compared to questions such as the fragmentation of magnetically-dominated molecular clouds or the formation of globular clusters, significant progress has been made in the last decade, and the claim can be made that we understand fragmentation during protostellar collapse well enough to have at least a basic outline for how binary stars forms (e.g., Boss 1988). More

R. Capuzzo-Dolcetta et al. (eds.), Physical Processes in Fragmentation and Star Formation, 279–292.

details on the context of this paper, the collapse of interstellar clouds to form protostars, can be found in the reviews by Shu *et al.* (1987) and Boss (1989a).

Spherically symmetrical (1D) models of cloud collapse were the first to discover the fundamental dynamics and thermodynamics of Population I protostar formation (Larson 1969). Starting from densities appropriate for dense molecular cloud cores ($\rho \sim 10^{-19}$ g cm^{-3}), a Jeans unstable $1M_\odot$ cloud initially collapses nearly isothermally, because dust grains radiate away the compressional energy liberated by collapse. This near-free-fall collapse is halted at the center by rising temperatures soon after the infrared optical depth reaches unity at $\rho \sim 10^{-13}$ g cm^{-3}. Thereafter the evolution is *nonisothermal*, that is, the optically thick inner region is roughly adiabatic, the outer cloud envelope is nearly isothermal, and intermediate regions have a more complicated thermal evolution. The halted collapse leads to formation of a quasi-equilibrium *outer core*, supported by molecular hydrogen and with a radius $R \sim 10$ AU, onto which the envelope continues to infall. When the central temperature reaches ~ 2000K, dissociation of molecular hydrogen removes thermal energy and leads to collapse of the outer core. This collapse is halted when dissociation is complete, producing the final, *inner core*, supported by atomic and ionized hydrogen, with $R \sim 10R_\odot$. The bulk of the cloud is still in the envelope, and is accreted onto the inner core over the time scale of a few free-fall times ($t_{ff} = (3\pi/32G\rho_i)^{1/2}$, where ρ_i is the initial cloud density). This classical picture may be complicated by the possibility of the inner core being periodically destroyed and reformed through a dynamical instability uncovered by Tscharnuter (1987, also this volume).

These 1D calculations have provided a successful theory of the formation of single stars, but have little to say about the formation of binary stars, or about single stars with planetary systems. The rest of this paper will briefly describe the methods and results of three dimensional (3D) calculations of protostellar cloud collapse, calculations that permit fragmentation into multiple bodies to occur.

2. Equations of Protostellar Formation

The time evolution of a 3D protostellar cloud is determined by the following set of equations and relations, assuming that magnetic fields do not dominate (i.e., initial densities high enough for significant ambipolar diffusion to have occurred; see Mouschovias, this volume). All numerical investigations of protostellar fragmentation solve at least the first five of the following equations, or mathematically equivalent equations.

The equations of hydrodynamics in spherical coordinates (r, θ, ϕ) include the continuity equation

$$\frac{\partial \rho}{\partial t} + \nabla \cdot (\rho \mathbf{v}) = 0,$$

and the three momentum equations

$$\frac{\partial(\rho v_r)}{\partial t} + \nabla \cdot (\rho v_r \mathbf{v}) = -(\rho \frac{\partial \Phi}{\partial r} + \frac{\partial p}{\partial r}) + \frac{\rho}{r}(v_\theta^2 + v_\phi^2),$$

$$\frac{\partial(\rho v_\theta)}{\partial t} + \nabla \cdot (\rho v_\theta \mathbf{v}) = -\frac{1}{r}(\rho \frac{\partial \Phi}{\partial \theta} + \frac{\partial p}{\partial \theta}) - \frac{\rho}{r}(v_r v_\theta - v_\phi^2 cot\theta),$$

$$\frac{\partial(\rho A)}{\partial t} + \nabla \cdot (\rho A \mathbf{v}) = -(\rho \frac{\partial \Phi}{\partial \phi} + \frac{\partial p}{\partial \phi}),$$

where ρ is the mass density, $\mathbf{v} = (v_r, v_\theta, v_\phi)$ is the Eulerian fluid velocity, $A = rsin\theta v_\phi$ is the specific angular momentum, and p is the pressure. The gravitational potential Φ is determined by Poisson's equation

$$\nabla^2 \Phi = 4\pi G \rho,$$

where G is the gravitational constant. When an isothermal or adiabatic approximation is made, this set of five equations for six unknowns (ρ, v_r, v_θ, A, Φ, p) is closed by specifying a pressure relation $p = p(\rho)$.

For nonisothermal phases of evolution, a full treatment of radiative transfer is required. Because of the extreme computational demands of 3D radiative transfer, so far only frequency-independent solutions in the diffusion or Eddington approximation have been attempted. In the diffusion approximation, one additional hydrodynamical equation must be solved, the energy equation

$$\frac{\partial(\rho E)}{\partial t} + \nabla \cdot (\rho E \mathbf{v}) = -p \nabla \cdot \mathbf{v} + \nabla \cdot (\frac{4}{3\kappa\rho} \nabla(\sigma T^4)),$$

where E is the specific internal energy, κ is the Rosseland mean opacity, and T is the gas (and radiation) temperature. In the Eddington approximation, the energy equation becomes

$$\frac{\partial(\rho E)}{\partial t} + \nabla \cdot (\rho E \mathbf{v}) = -p \nabla \cdot \mathbf{v} + 4\pi\kappa\rho(J - B),$$

where J is the mean intensity, and $B = \sigma T^4/\pi$ is the Planck function. In the Eddington approximation the radiation pressure can be determined from J, so the radiation temperature need not be the same as the gas temperature T. The mean intensity is determined by the equation

$$\frac{1}{3}\frac{1}{\kappa\rho} \nabla \cdot (\frac{1}{\kappa\rho} \nabla J) = J - B.$$

In regions of high optical depth, where $J \approx B$, alternative forms of the last two equations must be used, in order to avoid trying to subtract B from J. The nonisothermal system of equations is closed by specifying how the specific internal energy, pressure, and opacity depend on the density and temperature of the gas:

$$E = E(\rho, T),$$

$$p = p(\rho, T),$$

$$\kappa = \kappa(\rho, T).$$

Because of the nonlinear nature of these coupled partial differential equations, in order to solve either the isothermal or nonisothermal 3D equations of protostellar formation, a numerical solution is required. Solution methods are of two basic types: (1) finite differences (FD), where the differential terms are replaced by differences between variables defined on a grid with finite spacing, and (2) smoothed particle hydrodynamics (SPH), where a number of particles with finite size are used to represent fluid elements in the cloud, particle-particle interactions being defined through the fundamental equations. Specific FD techniques for the solution of the isothermal equations are described by Black and Bodenheimer (1975) and Boss (1980a). The SPH technique was originated by Lucy (1977) and has been developed most fully by Gingold and Monaghan (1977; see also Monaghan and Lattanzio 1985). FD techniques for the solution of the nonisothermal equations are described by Boss (1984; 1989b).

3. Isothermal Fragmentation

For protostellar clouds with densities between $\sim 10^{-19}$ g cm^{-3} and $\sim 10^{-13}$ g cm^{-3}, the isothermal approximation is very good, and this thermodynamical simplification undoubtedly encouraged the development of isothermal (or adiabatic) 3D hydrodynamics codes; at least 14 independent codes now exist. Roughly half the codes are FD codes (Narita and Nakazawa 1977; Norman and Wilson 1978; Cook and Harlow 1978; Różyczka, Tscharnuter, and Yorke 1980; Tohline 1980; Boss 1980a; Williams 1988) and the other half are SPH codes (Lucy 1977; Gingold and Monaghan 1977; Larson 1978; Wood 1981; Miyama, Hayashi, and Narita 1984). 3D isothermal codes have also been developed that include the effects of magnetic fields (FD: Dorfi 1982; SPH: Benz 1984). As we shall see, the basic agreement between all of these codes as to the outcome of isothermal collapse lends much credence to the results.

Boss and Bodenheimer (1979) presented the results of a successful comparison between two different FD codes, and suggested that their model be used as a standard test case for new 3D codes. In this model, the initial density perturbation was large enough, and the initial thermal energy low enough ($\alpha_i = E_{therm}/|E_{grav}| = 0.25$), that the cloud collapsed and fragmented directly into a binary system, without passing through the intermediate ring configuration first found in axisymmetric (2D) calculations of rapidly rotating cloud collapse by Larson (1972) and Black and Bodenheimer (1976). [A similar 3D model, also with $\alpha_i = 0.25$ and rapid rotation ($\beta_i = E_{rot}/|E_{grav}| = 0.20$), but with a very small initial density perturbation, formed a ring prior to fragmenting (Boss 1980b).]

However, when the standard test case was modeled by Gingold and Monaghan's (1981) SPH code, the binary that formed was found to undergo rapid orbital decay and merged into a single central object. Subsequent efforts to explain the discrepancy between the FD and SPH codes failed to yield a mutually acceptable explanation (Bodenheimer and Boss 1981; Gingold and Monaghan 1982), until certain refinements in the SPH technique (Monaghan and Lattanzio 1985) and a large increase in the number of SPH particles produced an SPH evolution (Monaghan and Lattanzio 1986) consistent with that of the FD codes (Boss 1988). While the brouhaha of 1981-82 may have discouraged other workers from trying the standard test case, perhaps now that agreement apparently has been reached, the other 3D codes should be tested on this problem.

Because of the existence of scaling laws, just two parameters are needed to characterize most of the physics of isothermal clouds: α_i and β_i, as previously defined. Figure 1 shows how the results obtained from all 3D calculations of the isothermal collapse of clouds starting from roughly uniform density initial conditions depend on these two parameters (Narita and Nakazawa 1977; Larson 1978; Boss and Bodenheimer 1979; Tohline 1980; Boss 1980b; Bodenheimer, Tohline, and Black 1980; Różyczka, Tscharnuter, and Yorke 1980; Boss 1981a,b; Wood 1982; Gingold and Monaghan 1983; Miyama, Hayashi, and Narita 1984; Monaghan and Lattanzio 1986). Figure 1 demonstrates the very good agreement between all of the codes, as well as with the fragmentation criterion ($\alpha_i \times \beta_i < 0.12$) of Hayashi, Narita, and Miyama (1982), derived from the stability of 2D rotating, isothermal equilibrium models.

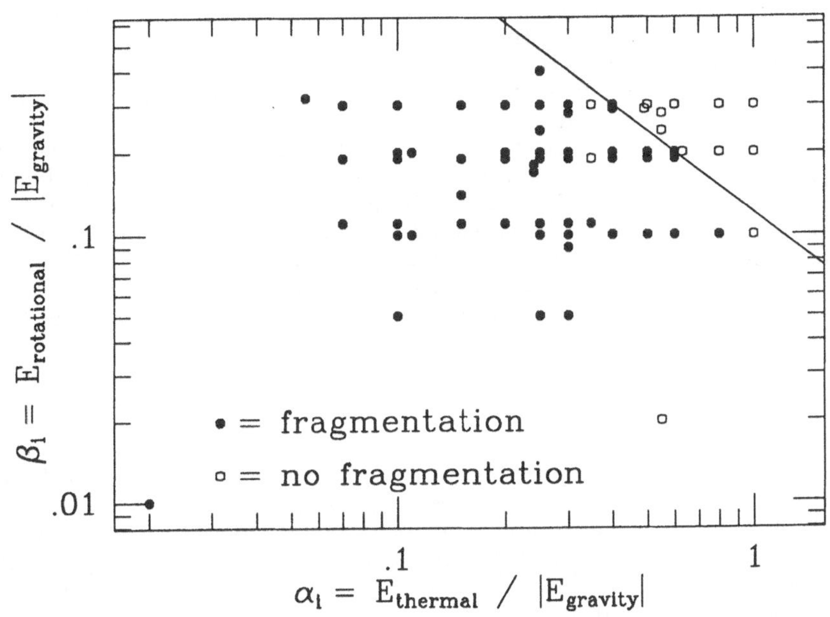

Figure 1. Summary of results of all 3D calculations of the collapse of isothermal protostellar clouds from uniform density initial conditions. The two parameters α_i and β_i characterize all possible initial conditions; α_i must be less than 1 for the cloud to be gravitationally bound, and β_i must be less than 1/3 for the entire cloud to start to move inward initially. For each initial condition, the result of the collapse is indicated, either fragmentation (formation of two or more clumps) or no fragmentation. Oblique line is the criterion for fragmentation advanced by Hayashi, Narita, and Miyama (1982).

Clouds with high α_i and β_i do not undergo collapse and fragmentation, but instead contract and reach a diffuse equilibrium state that is a triaxial analog of the Bonnor-Ebert isothermal sphere. Clouds with lower α_i and β_i undergo a sustained collapse to a denser configuration where self-gravity leads to fragmentation. Dynamical fragmentation can be either primarily rotationally-driven (high β_i) or primarily thermally-driven (low α_i). Fragmentation in the latter case is similar to that envisioned by Hoyle (1953) on the basis of Jeans mass arguments and by Hunter (1962) after calculating the growth of perturbations during the collapse of a uniform density sphere.

Cloud fragments in the isothermal regime tend to have masses and specfic angular momenta reduced by a factor of ten or more compared to their parent clouds. Their values of α are usually lower than that of their parent clouds as well, implying that they may sub-fragment themselves during what remains of their isothermal collapse phase. This suggests the possibility of a hierarchy of collapse and fragmentation, a likely way to explain the formation of hierarchical multiple stellar systems (Hoyle 1953; Bodenheimer 1978; Boss 1988). While hierarchical fragmentation remains to be demonstrated in a single numerical calculation, a strong circumstantial case can be made for at least a limited amount of hierarchical fragmentation.

While Figure 1 demonstrates the basic agreement of the 3D isothermal codes, it must be noted that important differences do exist, primarily between the FD and SPH codes. FD codes generally produce smaller numbers of fragments than SPH codes, as might be expected considering that the numerical viscosity associated with FD codes tends to smear out fragments, while the N (>> 1) particles used in SPH codes become N fragments if the particles cease to interact. Improving the agreement on this question is an important challenge for future work.

4. Nonisothermal Fragmentation

This section considers the possibility of fragmentation during nonisothermal collapse phases, that is, when the densest regions exceed $\sim 10^{-13}$ g cm^{-3}. We first consider fragmentation during the collapse to form the outer core, and then consider collapse leading to inner core formation. Because only one 3D nonisothermal hydrodynamics code has been developed to date, no code intercomparisons are possible, so the results of this section must be considered to be somewhat uncertain compared to the isothermal results.

4.1 FRAGMENTATION PRIOR TO OUTER CORE FORMATION

Clouds that collapse through the isothermal regime without fragmenting may still suffer fragmentation in the nonisothermal regime; also, fragments of clouds that fragmented in the isothermal regime may undergo sub-fragmentation as they enter the nonisothermal regime, i.e., hierarchical fragmentation may occur. Once an outer (or inner) core forms and rising thermal pressure halts the collapse, dynamical fragmentation must stop, leaving fission (a rotational instability that breaks a quasi-equilibrium body into several bodies) as the sole candidate for fragmentation. However, the fission instability (modeled in rapidly rotating polytropes by Durisen *et al.* 1986) does not appear to lead to two roughly equal mass

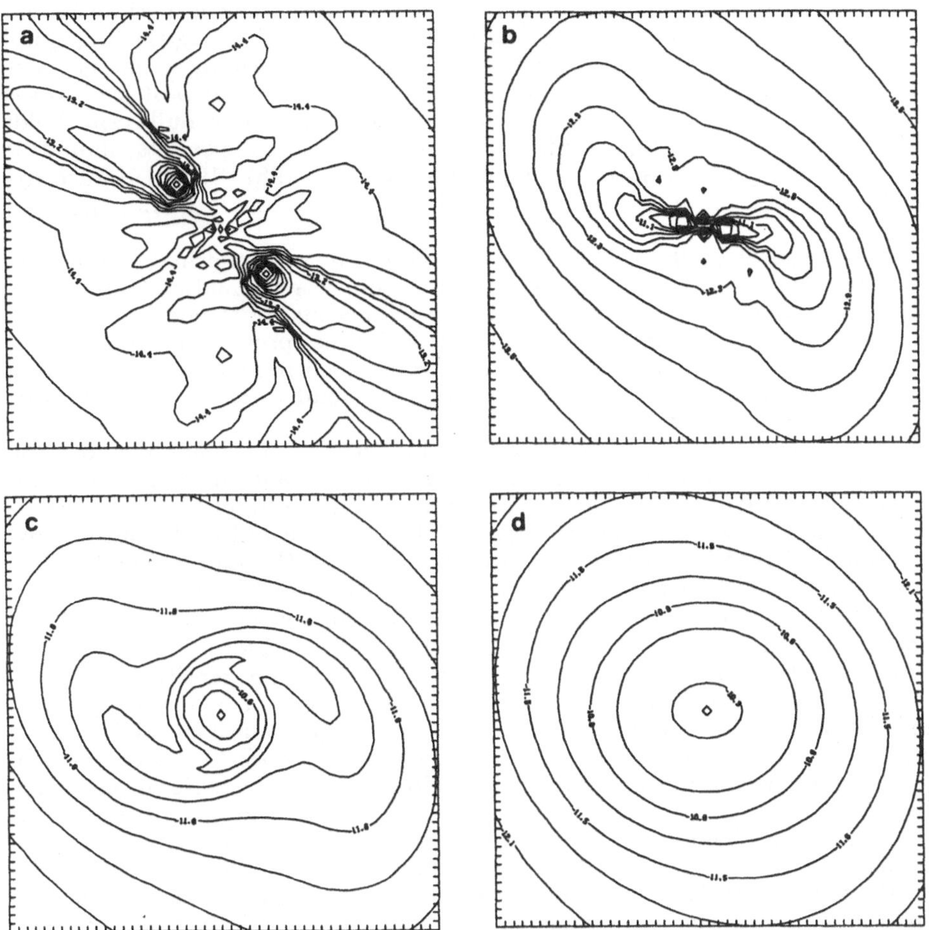

Figure 2. Density contours in the equatorial plane for four different 3D models of nonisothermal collapse prior to outer core formation (adapted from Boss 1986). Changes in contour levels represent changes in density by a factor of two. Each model started from $\alpha_i = 0.25$, $\beta_i = 0.04$, and $T_i = 10$K, but with varied initial masses: (a) $M_i = 2.0 M_{\odot}$, box size = 360 AU; (b) $M_i = 0.25 M_{\odot}$, box size = 96 AU; (c) $M_i = 0.10 M_{\odot}$, box size = 38 AU; (d) $M_i = 0.020 M_{\odot}$, box size = 15 AU. As the initial mass decreases, the initial density increases, the collapse starts farther into the nonisothermal regime, and fragmentation is stifled.

bodies, as is necessary to explain the formation of most relatively close binary stars. Instead, the fission instability leads to the growth of strong trailing spiral arms that prevent any nonaxisymmetric structure from forming a binary. The spiral arms produce rapid angular momentum transport, and at best lead to the ejection of a low mass ring of high angular momentum gas. This null result holds for polytropes with varied polytropic indices (i.e., adiabatic exponents simulating different thermal pressure laws; Williams and Tohline 1987, 1988). Hence dynamical fragmentation appears to be crucial for forming binary stars (e.g., Boss 1988).

Models of protostellar collapse of clouds into the nonisothermal regime have shown that fragmentation can still occur prior to outer core formation (Boss 1986), in exactly the same ways as in isothermal collapse, because at the beginning of the nonisothermal regime the increased influence of thermal pressure is modest. It is perhaps easiest to understand what can happen by focusing on those clouds that avoid fragmentation. There are three ways for a cloud to avoid fragmentation prior to outer core formation.

First, slowly rotating, nearly Jeans stable ($\alpha_i \sim 0.5$) clouds starting collapse in the isothermal regime can avoid fragmentation and instead form single protostars (Boss 1985). Note that one isothermal cloud in Figure 1 had $\beta_i = 0.02$ low enough and $\alpha_i = 0.55$ high enough to collapse yet avoid both rotationally- and thermally-driven fragmentation while still in the isothermal regime (Bodenheimer, Tohline, and Black 1980). This result is consistent with the nonisothermal models of Boss (1985), which implied that a solar mass cloud with $\beta_i < 0.02$ or specific angular momentum $J/M < 2 \times 10^{20}$ cm^2 s^{-1} could form a single star.

Second, clouds that begin their collapse phase at high densities encounter outer core formation before they have a chance to fragment. This is especially important for fragments that may have just formed at relatively high density and are now collapsing on their own. Because low mass clouds first become Jeans unstable at high densities, this means that as clouds masses decrease through fragmentation, eventually fragmentation is halted (Figure 2) and lower mass clouds cannot be formed. The lower bound on protostellar masses found in this manner is about 0.01 $M_\odot$ (Boss 1986). This minimum protostellar mass is slightly higher than early estimates based on the same concept (opacity-limited fragmentation) but with the analysis restricted to the Jeans mass derivation (Low and Lynden-Bell 1976; Rees 1976). Clearly this minimum protostellar mass of $0.01 M_\odot$ supports the possible existence of brown dwarf stars ($M < 0.08 M_\odot$).

Third, all of the models previously mentioned have assumed that a uniform density sphere is an adequate starting point. Of course, this is strictly incorrect, because such a cloud is initially far from equilibrium, whereas the dynamical collapse phase by definition starts from a quasi-equilibrium state, such as a magnetically supported cloud undergoing ambipolar diffusion. Such quasi-equilibrium states are likely to be centrally condensed. Boss (1987) examined the effects of varying the initial density profile from uniform to a gaussian to a power law profile ($\rho_i \propto r^{-1}$), and found that while a moderate gaussian profile had relatively little effect, the strong initial density singularity inherent in a power law profile effectively prohibited fragmentation (Figure 3). Hence clouds starting their collapse from a power law configuration do not appear to undergo dynamical fragmentation.

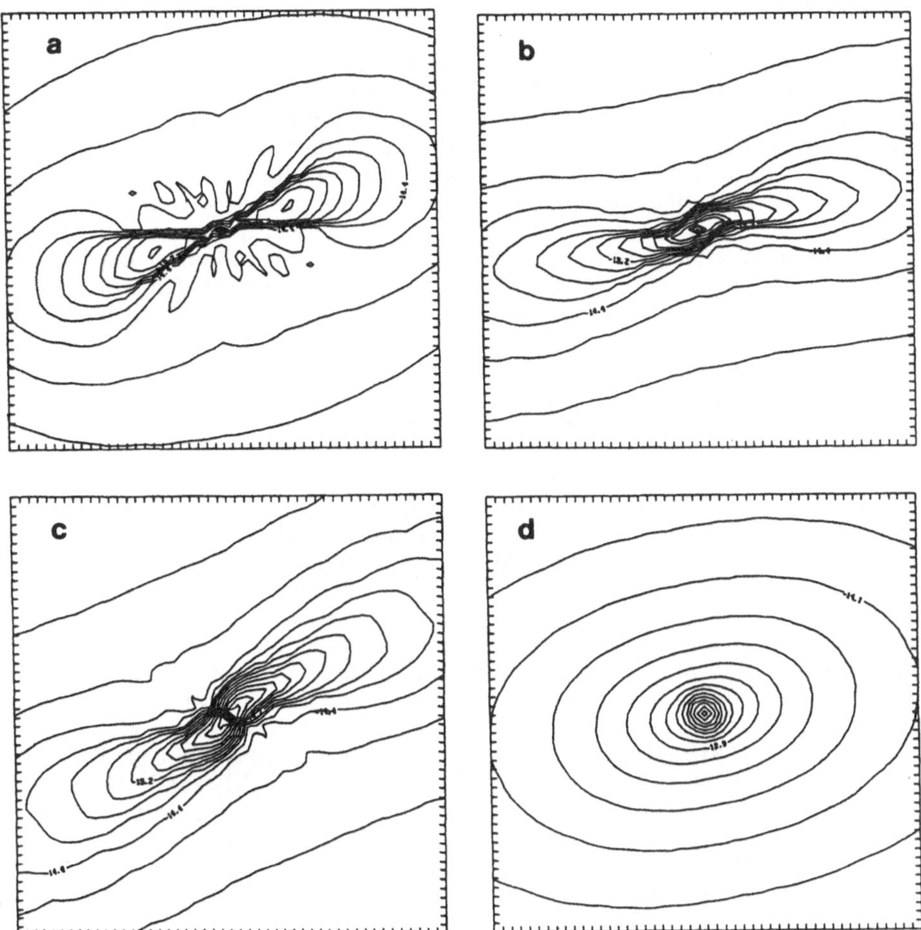

Figure 3. Density contours in the equatorial plane for four different 3D models of nonisothermal collapse that show the effects of varied initial density profiles (adapted from Boss 1987). Changes in contour levels represent changes in density by a factor of two. Each model started from $\alpha_i = 0.13$, $\beta_i = 0.0016$, $T_i = 10\text{K}$, $M_i = 1 M_\odot$, but with different initial profiles: (a) $\rho_i = \rho_o$, $\Omega_i = \Omega_o$ (box size 580 AU); (b) $\rho_i \propto exp[-(r/r_1)^2]$, $\Omega_i = \Omega_o$ (box size 300 AU); (c) $\rho_i \propto exp[-(r/r_1)^2]$, $\Omega_i \propto exp[-2/3(r/r_1)^2]$ (box size 270 AU); (d) $\rho_i \propto r^{-1}$, $\Omega_i = \Omega_o$ (box size 110 AU). Clouds with strong initial central density concentrations do not undergo dynamical fragmentation.

4.2 FRAGMENTATION DURING INNER CORE FORMATION

The final chance for dynamical fragmentation occurs during the collapse of the outer core to form the inner core. Boss (1989b) has modeled in 3D the formation of inner cores, starting from very idealized initial conditions similar to those in the outer core prior to its collapse; only the inner $0.01M_{\odot}$ is included in these calculations. A sequence of models with varied initial rotation rates showed that the subsequent evolution depended strongly on β_i. For models with $\beta_i < 0.1$, little rotational flattening occurred by the time the inner core began to form. Instead, a series of bounces ensued (similar to the situation when the outer core forms), but eventually these bounces led to the formation of several streams of outward moving gas that flowed out to $\sim 1/3$ AU. These streams appear to be the 3D manifestation of the inner core instability found by Tscharnuter (1987), and thus are driven by the energy liberated by the recombination of ionized and atomic hydrogen into molecular hydrogen. However, in 3D, gas continues to flow into the inner core region from directions other than where the outflows are located, so the inner core is not as severely depleted as occurs in a 1D calculation (cf. Tscharnuter 1987).

A model of inner core formation with $\beta_i = 0.1$, however, underwent a different evolution (Figure 4). In this case the inner core that formed was strongly flattened by rotation, and once the value of β exceeded 0.274 (the critical value for dynamical growth of nonaxisymmetry in an axisymmetric Maclaurin spheroid and more general fluid bodies), the inner core began to break-up into a binary. The binary did not survive, however, and merged into a bar. Continued evolution and accretion of high angular momentum gas drives the inner core into forming another binary, which also disappeared. As in the fission instability of axisymmetric polytropes (Durisen *et al.* 1986), the culprit that destroys growing binaries appears to be angular momentum transport by gravitational torques. Evidently a fission-like instability cannot form a stable binary in a fairly realistic model of a inner core; fragmentation of rapidly rotating inner cores is at best a transient event. While this model underwent considerable bouncing in its central regions, it did not form energetic, outward-moving streams of gas like the slower rotating models. This is consistent with previous 2D calculations by Tscharnuter, who found that rapid rotation stabilizes the inner core instability, by lowering the critical value of $\Gamma_1 = (\partial ln p/\partial ln \rho)_{ad}$ for dynamical instability from the value of 4/3 appropriate for spherical bodies to the value of 1 applicable to thin disks. Because protostellar cores are not isothermal (i.e., $\Gamma_1 > 1$), the instability is prevented when significant rotation is present.

In the slowly rotating models where outward streams developed, the amount of mass in the streams was small compared to the amount left in the inner core region, and thus these streams are unlikely to survive as separate entities, especially because they eventually will be swept back into the inner core by the continued infall of the cloud envelope. Hence significant fragmentation does not appear to be possible during inner core formation for slowly or rapidly rotating inner cores.

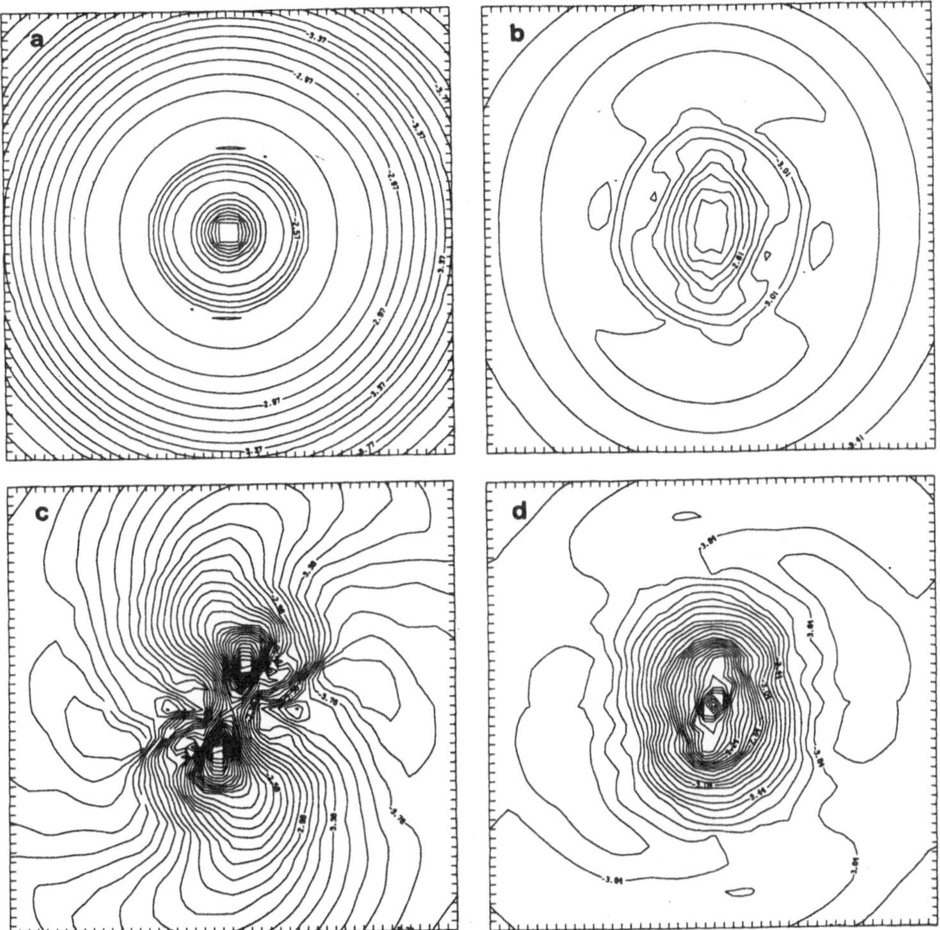

Figure 4. Time sequence of density contours in the equatorial plane for a 3D model of inner core formation through nonisothermal collapse (adapted from Boss 1989b). Changes in contour levels represent changes in density by a factor of 1.25. Region shown is 2 $R_\odot$ across. The binary-bar pattern rotates about 1/2 a rotation period between each frame. Starting nearly axisymmetric (a) with $\beta < 0.274$, nonaxisymmetry grows (b) once $\beta > 0.274$ and fragments the inner core into a binary (c), but the binary decays into a single bar-like object (d) because of angular momentum transport by the trailing spiral arms evident in (c) and (d).

5. Conclusions

Much of the exploratory phase of modeling 3D collapse and fragmentation of protostellar clouds has now been completed. A number of independent codes have been used to study the isothermal collapse problem, and a consensus has been reached that fragmentation into a binary or multiple system is the result of sustained isothermal collapse for most initial conditions. 3D models have also shown that while further fragmentation is possible in the nonisothermal regime, fragmentation ceases by the time that outer cores are formed, and furthermore, fragmentation does not appear to occur during collapse to form inner cores. While these results are all somewhat preliminary and unrefined to greater or lesser degrees, at present there does not appear to be any problem with explaining the formation of the entire range of binary star systems through dynamical fragmentation (e.g., Boss 1988). Future work needs to address more fully such issues as starting from overly idealized initial conditions (e.g., uniform density spheres), including the effects of any residual magnetic fields, overcoming numerical problems that may affect the number of fragments formed, and demonstrating hierarchical fragmentation in a single, self-consistent calculation.

Acknowledgment

The writing of this review was partially supported by U. S. National Science Foundation grant AST-8817334.

References

Adams, F., Ruden, S., and Shu, F. 1989, preprint.
Benz, W. 1984, *Astr. Ap.*, **139**, 378.
Black, D. C., and Bodenheimer, P. 1975, *Ap. J.*, **199**, 619.
—— 1976, *Ap. J.*, **206**, 138.
Bodenheimer, P. 1978, *Ap. J.*, **224**, 488.
Bodenheimer, P., and Boss, A. P. 1981, *M.N.R.A.S.*, **197**, 477.
Bodenheimer, P., Tohline, J. E., and Black, D. C. 1980, *Ap. J.*, **242**, 209.
Boss, A.P. 1980a, *Ap. J.*, **236**, 619.
—— 1980b, *Ap. J.*, **237**, 866.
—— 1981a, *Ap. J.*, **246**, 866.
—— 1981b, *Ap. J.*, **250**, 636.
—— 1984, *Ap. J.*, **277**, 768.
—— 1985, *Icarus*, **61**, 3.
—— 1986, *Ap. J. Suppl.*, **62**, 519.
—— 1987, *Ap. J.*, **319**, 149.
—— 1988, *Comm. Ap.*, **12**, 169.
—— 1989a, *Publ. Astr. Soc. Pac.*, **101**, 000.
—— 1989b, *Ap. J.*, **346**, 000.
Boss, A. P., and Bodenheimer, P. 1979, *Ap. J.*, **234**, 289.
Cassen, P. M., Smith, B. F., Miller, R. H., and Reynolds, R. T. 1981, *Icarus*, **48**, 377.
Cook, T. L., and Harlow, F. H. 1978, *Ap. J.*, **225**, 1005.

Dorfi, E. 1982, *Astr. Ap.*, **114**, 151.
Durisen, R. H., Gingold, R. A., Tohline, J. E., and Boss, A. P. 1986, *Ap. J.*, **305**, 281.
Gingold, R. A., and Monaghan, J. J. 1977, *M.N.R.A.S.*, **181**, 375.
—— 1981, *M.N.R.A.S.*, **197**, 461.
—— 1982, *M.N.R.A.S.*, **199**, 115.
—— 1983, *M.N.R.A.S.*, **204**, 715.
Hayashi, C., Narita, S., and Miyama, S. M.1982, *Prog. Theor. Phys.*, **68**, 1949.
Hoyle, F. 1953, *Ap. J.*, **118**, 513.
Hunter, C. 1962, *Ap. J.*, **136**, 594.
Larson, R. B. 1969, *M.N.R.A.S.*, **145**, 271.
—— 1972, *M.N.R.A.S.*, **156**, 437.
—— 1978, *M.N.R.A.S.*, **184**, 69.
Low, C., and Lynden-Bell, D. 1976, *M.N.R.A.S.*, **176**, 367.
Lucy, L. B. 1977, *Astron. J.*, **82**, 1013.
Miyama, S. M., Hayashi, C., and Narita, S. 1984, *Ap. J.*, **279**, 621.
Monaghan, J. J., and Lattanzio, J. C. 1985, *Astr. Ap.*, **149**, 135.
—— 1986, *Astr. Ap.*, **158**, 207.
Narita, S., and Nakazawa, K. 1977, *Prog. Theor. Phys.*, **59**, 1018.
Norman, M. L., and Wilson, J. R. 1978, *Ap. J.*, **224**, 497.
Rees, M.J. 1976, *M.N.R.A.S.*, **176**, 483.
Różyczka, M., Tscharnuter, W. M., and Yorke, H. W. 1980, *Astron. Ap.*, **81**, 347.
Shu, F. H., Adams, F. C., and Lizano, S. 1987, *Ann. Rev. Astr. Ap.*, **25**, 23.
Tohline, J. E. 1980, *Ap. J.*, **235**, 866.
Tscharnuter, W. M. 1987, in *Physical Processes in Comets, Stars, and Active Galaxies*, eds. E. Meyer-Hofmeister, H. C. Thomas, and W. Hillebrandt (Berlin: Springer-Verlag), p. 96.
Williams, H. A. 1988, Thesis, Louisiana State University, Baton Rouge.
Williams, H. A., and Tohline, J. E. 1987, *Ap. J.*, **315**, 594.
—— 1988, *Ap. J.*, **334**, 449.
Wood, D. 1981, *M.N.R.A.S.*, **194**, 201.
—— 1982, *M.N.R.A.S.*, **199**, 331.

Discussion

Zinnecker: Can you elaborate on your statement that a peaked gas density distribution prevents fragmentation? For example, ambipolar diffusion seems to produce a $\rho \propto r^{-2}$ profile; does your statement imply that binary stars cannot form from such an initial density profile?
Boss: The 3D calculations show that binary formation by fragmentation during the cloud collapse phase cannot occur in a cloud initially as centrally condensed as $\rho_i \propto r^{-1}$, and hence fragmentation cannot occur with $\rho_i \propto r^{-2}$ either. Binary fragmentation can still occur in more modestly centrally condensed clouds, such as gaussian density profiles with 20:1 density contrasts. If one wishes to account for binary formation in a cloud with $\rho_i \propto r^{-2}$, one needs to invoke some mechanism other than fragmentation, which I think is dubious. On the other hand, $\rho_i \propto r^{-2}$ is a plausible initial condition for the formation of single stars like the sun.

Lizano: Do you know about the recent work of Adams, Ruden, and Shu (1989) on disk instabilities that could produce a binary system through the $m = 1$ mode? Could you use your 3D code to study this problem?
Boss: Hans Zinnecker has kindly loaned me a copy of this preprint. Adams, Ruden, and Shu (1989) have performed a linear stability analysis for the growth of $m = 1$ modes in a thin, relatively massive disk surrounding a solar-mass star, and found that the $m = 1$ mode is favored for growth. They suggest that further growth of the $m = 1$ mode could result in formation of a second protostar. Substantial growth of $m = 1$ modes has been found previously in 3D calculations of the break-up of isothermal rings (Norman and Wilson 1978). Also, preferential growth of $m = 1$ modes has been encountered during 3D calculations of protostellar collapse (Boss 1980b), and an analytical reason has been given for this $m = 1$ growth (Boss 1980b), so I agree that $m = 1$ modes are likely to undergo growth in such a disk. However, my guess is that these modes will saturate in amplitude prior to forming a binary companion, for the same reason that the $m = 2$ mode in the dynamic fission instability (Durisen *et al.* 1986) failed to produce a stable binary: the growth of trailing spiral arms on a time scale close to that of $m = 1$ growth (i.e., the rotation period) will cause the nascent binary member to spiral inward (through loss of angular momentum by gravitational torques) and merge with the central primary. This is just a guess, and this question could and should be answered by further 3D code calculations of the nonlinear evolution of $m = 1$ modes in a massive disk orbiting an equally massive protostar (see also Cassen *et al.* 1981).

PROTOSTELLAR CORE INSTABILITIES

W. M. Tscharnuter
Institut für Theoretische Astrophysik
Universität Heidelberg
Im Neuenheimer Feld 561
D-6900 Heidelberg
F.R.G.

ABSTRACT. An axisymmetric collapse model for the formation of the presolar nebula and a spherically symmetric, Larson-type protostellar evolution is discussed. Particular attention is paid to the dynamical behavior of the star-like core. It has been found that, subsequent to their formation, protostellar core embryos containing only a few percents of a solar mass tend to undergo oscillations of large amplitudes. Dynamical oscillations and even disruptive core expansion ('hiccups') have also ben observed to occur in spherically symmetric models. On the basis of N. Baker's (1966) one-zone-model, it can be argued that protostellar cores are vibrationally unstable and may even become dynamically unstable. Thus, oscillations of protostellar cores are to be expected, but their amplitudes, as observed in the model sequences, could be very sensitive to the accuracy achieved by the numerical solution of the discretized structure equations.

1. Introduction

Since Larson's (1969) pioneering work on spherically symmetric protostellar collapse, it has become clear that calculations of this type are extremely difficult to carry out, unless simplifying assumptions are made, e.g., the stellar core being hydrostatic and the accretion flow being stationary (cf. Stahler *et al.*, 1980). In particular, it turned out that the modelling of the tansition from the overall collapse to the main accretion phase, which ensues subsequent to the central bounce and core formation, requires a very robust numerical code which allows one to use large time steps. Only very recently a numerical scheme has been developed which seems to satisfy all demands on numerical stability, accuracy, and efficiency (Winkler and Norman, 1986). Results presented in Sect. 2 and critically discussed in Sect. 3 are the first to be obtained with the new numerical procedure, and to incorporate a substantially improved equation of state (Wuchterl, 1989a).

Collapse models of *rotating* protostars need to be, at least, axially symmetric. In most calculations attention is focussed on the formation of a disk-like structure, whereas the core region is treated in a rather qualitative way. For the central region a very crude, global book-keeping for the total core mass M_c and angular

R. Capuzzo-Dolcetta et al. (eds.), Physical Processes in Fragmentation and Star Formation, 293–301.

momentum J_c is made according to the *inflow* rate of these two quantities ($\dot{M}$ and $\dot{J}$, respectively) into the central 'hole' across a rather arbitrarily chosen inner boundary (cf. Morfill *et al.*, 1985). From M_c and J_c, Bodenheimer (1989) derives the respective MacLaurin spheroid, with an equatorial radius R_e, and determines the accretion luminosity $L = GM_c\dot{M}/R_e$ (G is the gravitational constant) which he takes as the inner boundary condition for radiative transfer, at any instant of time. Star-like cores will form only if the angular momentum problem is solved in one way or another, i.e., if angular momentum is removed from the central regions of the protostellar cloud and transported outward, while mass is moving inward.

Three-dimensional (3-D) models (cf. Boss, 1986, for a review; Boss, 1989; Boss, this conference) have never addressed the problem of stellar core formation because of the very stringent Courant-Friedrichs-Lewy (CFL) condition for numerical stability. Such calculations, however, are very important for studying angular momentum transport by gravitational torques, which become efficient if non-axisymmetric structures (bars, spirals) develop in (self-gravitating) disks. In this context model calculations dealing with the *fission problem* of rapidly rotating protostellar cores should also be mentioned (Durison and Tohline, 1985; Durison *et al.*, 1989).

At present, the formation process of star-like cores within collapsing protostellar fragments can be studied in detail only with axial (2-D) and spherical (1-D) geometry. However, hydrodynamical models covering the main accretion phase during which the material from the free-falling (1-D) or disk-like, centrifugally supported (2-D) envelope is accumlated in the central object (e.g., the 'Proto-Sun' within the 'solar nebula') are still restricted to the 1-D case (Appenzeller and Tscharnuter, 1974, 1975; Bertout, 1976; Tscharnuter and Winkler, 1979; Winkler and Newman, 1980ab; Balluch, 1988). This is because dynamical oscillations of protostellar cores have inhibited further progress, particularly with 2-D model sequences which require an unreasonably large amount of computer time, if the evolution is dominated by the very short dynamical timescale (at most several tenths of a year) rather than by the much longer lasting ($10^3 - 10^5 yr$) 'free-fall' accretion or the 'viscous' timescale ($10^5 - 10^7 yr$), e.g., for α-accretion disks. To my knowledge, only two model sequences pertaining to core formation in rotating protostars have been published as yet (Morfill *et al.*, 1985; Tscharnuter, 1987b).

In the sequel, I shall present two examples for protostellar core models, one for the 2-D and the other one for the 1-D case, and comment critically on the protostellar 'hiccup'-problem — oscillations of protostellar core embryos — that has been the main issue for the last couple of years (Tscharnuter, 1987a, 1989).

2. Results

The main results relating to protostellar core instabilities are very briefly summarized below. As technical tools I have used a fully implicit numerical scheme with 1st order donor-cell advection in conservative form (cf. Winkler and Norman, 1986) and an implicitly defined, fully adaptive numerical grid in the radial direction (Dorfi and Drury, 1987). Dependence on the polar angle in the 2-D case is accounted for by expansion into Legendre polynomials. The radiation field is described by the Eddington approximation. For more details see Tscharnuter (1987b).

2.1. AXIALLY SYMMETRIC MODELS

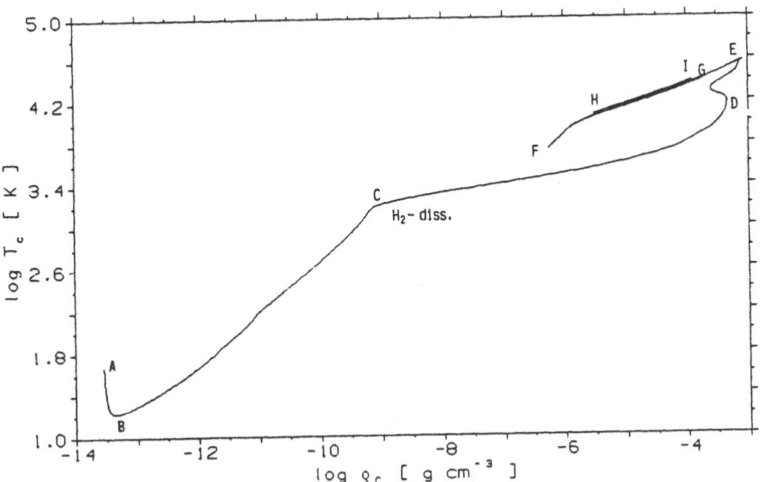

Figure 1: Central density ($\log \rho_c$) versus central temperature ($\log T_c$). Labels A through I mark important evolutionary stages. A-B: cooling and beginning of the collapse; B-C: adiabatic collapse and formation of the first core, i.e., a thick disk extending to about the dimensions of the solar system; C-D: second collaps after angular momentum has been transported away from the central regions by turbulent friction; D-E: formation of the 'protosun' and a subdisk extending to a few a.u.; E-F-G-H-I: quasi-adiabatic oscillations of rather large amplitude (2-D 'hiccups').

Fig. 1 shows the evolutionary path of a rotating low-mass protostar in the central density — central temperature graph. The following initial conditions were chosen (*non-homogeneous* starting model): total mass $M = 1.2$ $M_\odot$, specific angular momentum $J/M = 1.26 \cdot 10^{20} cm^2 s^{-1}$, radius $R = 4.06 \cdot 10^{15} cm$; mean density $\overline{\rho} = 8.51 \cdot 10^{-15} gcm^{-3} \rightarrow$ mean free-fall time $\bar{t}_{ff} = 721.3 yr$; central density $\rho_c = 2.81 \cdot 10^{-14} gcm^{-3}$ and temperature $T_c = 47K$; central angular velocity $\Omega_c = 2.53 \cdot 10^{-11} s^{-1} \approx$ mean angular velocity; internal energy / gravitational energy $= 0.050$, rotational energy / gravitational energy $= 0.054$.

Angular momentum transport is included as an α-disk-like turbulent viscosity after Shakura and Sunyaev (1973) with α ($= 0.1$ in this calculation) as a free parameter. This is why the second collapse can take place in a fashion similar to that of the 1-D models (labels C-D in Fig. 1). The final outcome (labels E-I) is a pressure-supported "protosun" containing about 5-6% of a solar mass within a sphere of 20-30 solar radii. The bulk of the core material is partly ionized and the mass-weighted mean of the adiabatic exponent $\Gamma_1 = (\partial \log P/\partial \log \rho)_S \approx 4/3$, which indicates that the core is on the verge of dynamical instability. The sequence is a good candidate for modelling the formation of the solar nebula.

2.2. SPHERICALLY SYMMETRIC MODELS

The initial conditions for the 1-D sequence are (*homogeneous* starting model): $M = 3\ \mathrm{M}_\odot$, $\rho = 10^{-20} gcm^{-3}$, $R = 5.22 \cdot 10^{17} cm$, $t_{ff} = 6.66 \cdot 10^5 yr$, $T = 8.4K$ (Jeans temperature). Constant external pressure is assumed.

Fig. 2 shows the variations of the density and temperature at the center. The collapse proceeds isothermally until $\rho_c \geq 10^{-13} gcm^{-3}$ during slightly more than one free-fall time t_{ff}. Then the first optically thick core forms within a few $10^2 yr$. The second stellar collapse commences when $T_c \geq 2 \cdot 10^3 K$. This is to be compared with the graph B-C-D in Fig. 1.

As is seen from Fig. 2, the four bounces and ensuing hiccups found in the calculation occur at rather high values of ρ_c. The accretion shock is pushed outward, eventually catching up with the shock that is surrounding the optically thick inner part of the protostar. The expansion ceases, and the reverse shock detaches, after about the same mass as was contained in the original stellar core embryo (a few $10^{-2}\ \mathrm{M}_\odot$) has flown across the outward travelling shock, which corresponds to about $10^{15} cm$ for the terminal shock distances.

Fig. 3 shows the huge (more than 20 orders of magnitude) density variations during the hiccup phase. The evolution has been followed until more than 99% of the total mass was accumulated.

3. Discussion

The results presented above are non-standard and need further investigation. A problem related to protostellar collapse is the formation of giant gaseous planets with a solid core. The core is assumed to accrete small planetesimals, while its mass begins to exceed Mizuno's (1980) critical value ($\approx 12 - 13\ \mathrm{M}_\oplus$) above which collapse is supposed to set in. Wuchterl (1989a) succeeded in following the evolution beyond that critical point into the non-linear regime by using a modified version of the 1-D radiation hydrocode for protostars. He found out that after a Kelvin-Helmholtz contraction of the gaseous envelope containing about 6 $\mathrm{M}\oplus$ pressure waves are excited which steepen to *outgoing* shocks. Most of the envelope's mass is lost and a Uranus-like object is left over in the first place. In applying Baker's (1966) one-zone-model (OZM), originally developed for Cepheïd pulsations, Wuchterl (1989ab) was able to show that certain opacity features pertaining to ionisation (dissociation) of hydrogen (molecules), H^- and dust absorption, give rise to

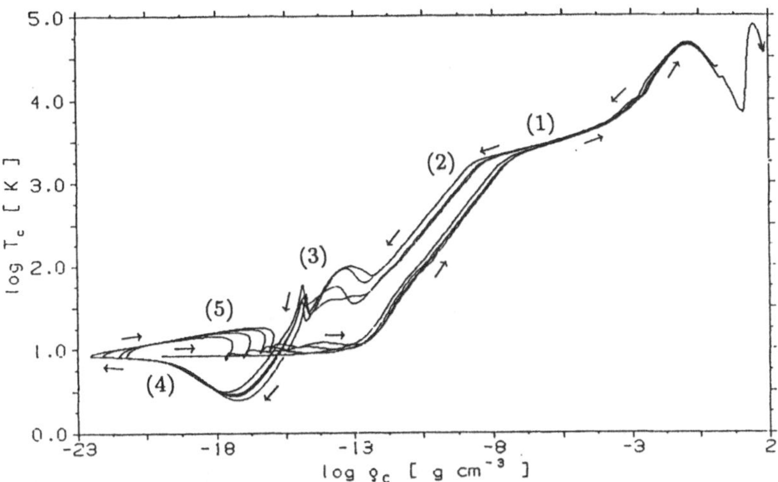

Figure 2: 1-D 'hiccups' displayed in the central density ($\log \rho_c$) - central temperature ($\log T_c$) diagram. ρ_c and T_c at the bounce are about $0.5 gcm^{-3}$ and $3 \cdot 10^4 K$, respectively. Pressure ionization is the dominant process, and the free electrons are partially degenerate (decreasing temperature with increasing density!). For densities above $10^{-2} gcm^{-3}$ at temperatures below $10^5 K$ the equation of state becomes rather uncertain. Important events during the expansion phase are labelled from (1) to (5): At point (1): The opacity gap reaches the center and heat is quickly transported inward, away from the temperature maximum that has developed off-center $\Rightarrow$ (1)-(2): change of the adiabat. (3): The core as a whole becomes optically thin, heat generated by the outward travelling "accretion shock" (together with the reverse shock) diffuses into the core remnant, thereby again raising the entropy. (4): Now the fragment as a whole is optically thin. Radiative heating from outside becomes more efficient than adiabatic cooling. (5): The reverse shock hits the center; after reflection the collapse starts anew. In this calculation (constant external pressure) 4 hiccups occurred. Then the mean density has increased to such an extent that, due to the high mass infall rate $\dot{M} \approx 10^{-3} - 10^{-4}$ $M_{\odot} yr^{-1}$, further hiccups are suppressed and, hence, within a few thousand years a substantial fraction of the total mass is accreted.

a very effectively acting κ-mechanism. Radiative energy, generated by disspation of kinetic energy of the infalling planetesimals in a thin layer above the solid core, is pumped into mechanical energy, i.e., pulsations of the gaseous envelope. There is a striking similarity between the structure of Wuchterl's giant protoplanets and highly evolved, shell-burning stars on the AGB, but of course, scaled down to very low masses (several ten earth masses) and temperatures less than $1.5 \cdot 10^4 K$.

Since the OZM makes use of local quantities only (luminosity, pressure, ...), application to protostellar cores is straightforward. In doing so we are immediately led to the conclusion that stellar core embryos are also vibrationally unstable. But the big caveat here, is the fact that the energy input (by gas accretion) is performed "from above". This is contrary to what happens in giant planet formation or evolved stars (Cepheïds, Mirae, ...) where the layers of energy production — dissipation of

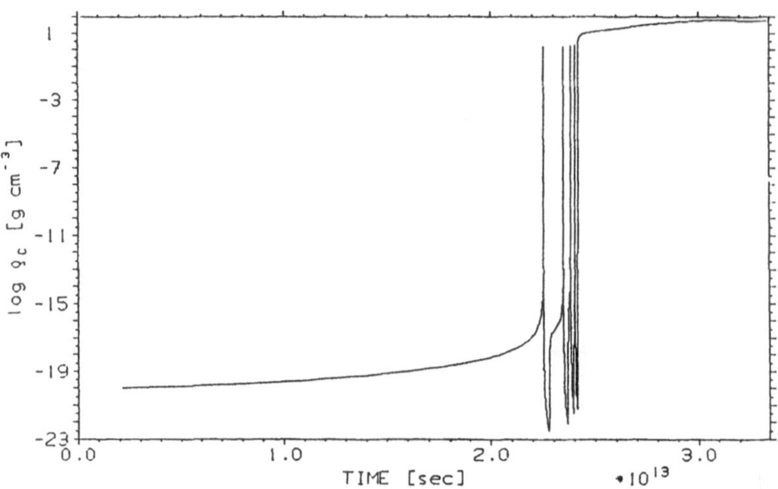

Figure 3: $\log \rho_c$ versus time for the 1-D sequence.

kinetic energy or thermonuclear reactions, respectively — are located in the deep interior.

Another mechanism, currently being investigated (Balluch, private communication), refers to the stability behaviour of protostellar accretion flows with radiative cooling downstream from the accretion shock. Under certain conditions the location of the shock becomes unstable, and the hydrostatic quasi-equilibrium of the core is disturbed. Whether or not the sources of instability mentioned will lead to hiccups, as found in the non-linear calculations, depends decisively on the amount of energy available for expansion (Shu private communiacation).

Unfortunately, the *total* energy cannot be taken as a primary variable in the calculations, because the mechanical energy balance (kinetic energy E_{kin} and gravitational energy W) is already *implicitly* defined by integrating the momentum equation(s). Otherwise the internal energy E_{int} would be a derived quantity, which could be ill-defined *locally* in zones where the velocity is highly supersonic. So, the degree of non-conservation of total energy is a good criterion for the *global* accuracy of the numerical models. In the present model calculations total energy conservation is guaranteed to about 8-15% relative to the sum $E_{kin} + E_{int} + |W|$. It cannot be ruled out that, given a fixed number of zones, say, a few hundred, the *local* refinement of the numerical grid is still not sufficient to imprive on *global* energy conservation to a degree necessary for establishing a 'correct' solution of the structure equations.

In this sense the comparative study of Boss and Tscharnuter (1989) is, in the first place, nothing but a consistency proof of numerical results obtained by entirely differently structured codes. The *physical* problem, however, will remain unsolved so long as the common initial model — the collapsing first, optically thick core at a certain instant of time $t = 0$ — might already be affected by the uncertainty in the total energy. The first core, which has always been regarded as an unimportant transient structure, is most probably the clue for the final solution of the hiccup problem. If so, the only remedy for the energy disease, protostellar models might have been suffering from to date, is to increase the number of gridpoints by at least one order of magnitude — or even more.

References

Balluch, M.: 1988, *Astron. Astrophys.* **200,** 58

Bertout, C.: 1976, *Astron. Astrophys.* **51,** 101

Bodenheimer, P.: 1989, *Formation and Evolution of the Solar Nebula,* in: NATO ARW *Theory of Accretion Disks,* eds. F. Meyer, W.J. Duschl, J. Frank, E. Meyer-Hofmeister, Kluwer, Dordrecht, p. 75

Boss, A.P.: 1986, *Theory of Collapse and Protostar Formation,* in: *Proceedings from the Summer School on Interstellar Processes,* eds. D. Hollenbach and H. Thronson, D. Reidel, Dordrecht

Boss, A.P.: 1989, *Evolution of the Solar Nebula. I. Nonaxisymmetric Structure during Nebula Formation,* preprint, to appear in *Astrophys. J., October 1st issue*

Boss, A.P, Tscharnuter, W.M.: 1989, *Protostellar Core Instabilities: Verification of Dynamical Hiccups,* (prepint)

Dorfi, E.A., Drury, L.O'C.: 1987, *J. Comput. Phys.* **69,** 175

Durison, R.H., Tohline, J.E.: 1985, *Fission of Rapidly Rotating Fluid Systems,* in: *Protostars and Planets II,* eds. D.C. Black and M.S. Matthews, Univ. Arizona Press, Tucson, p. 534

Durison, R.H., Yang, S., Cassen, P., Stahler, S.W.: 1989, *Numerical Models of Rotating Protostars,* Indiana Astronomy Publ. No. 89–101 (preprint)

Larson, R.B.: 1969, *Monthly Notic. Roy. Astron. Soc.* **145,** 271

Mizuno, H.: 1980, *Prog. Theoret. Physics,* **64,** 544

Morfill, G.E., Tscharnuter, W.M., Völk, H.J.: 1985, *Dynamical and Chemical Evolution of the Protoplanetary Nebula,* in: *Protostars and Planets II,* eds. D.C. Black and M.S. Matthews, Univ. Arizona Press, Tucson, p. 493

Shakura, N.I, Sunyaev, R.A.: 1973, *Astron. Astrophys.* **24,** 337

Stahler, S.W., Shu, F.H., Taam, R.E.: 1980, *Astrophys. J.* **241,** 637

Tscharnuter, W.M.: 1987a, *Models of Star Fomation,* in: *Physical Processes in Comets, Stars, and Active Galaxies, Lecture Note in Physics,* eds. E. Meyer-Hofmeister, H.C. Thomas, and W. Hillebrandt, Springer Verlag, p. 96

Tscharnuter, W.M.: 1987b, *Astron. Astrophys.* **188,** 55

Tscharnuter, W.M.: 1989, *Formation of Viscous Protostellar Accretion Disks,* in: NATO ARW *Theory of Accretion Disks,* eds. F. Meyer, W.J. Duschl, J. Frank, E. Meyer-Hofmeister, Kluwer, Dordrecht, p. 113

Tscharnuter, W.M., Winkler, K.-H.A.: 1979, *Computer Physics Communications* **80,** 171

Winkler, K.-H.A., Newman, M.J.: 1980a, *Astrophys. J.* **236,** 201

Winkler, K.-H.A., Newman, M.J.: 1980b, *Astrophys. J.* **238,** 311

Winkler, K.-H.A., Norman, M.L.: 1986, *WH80s: Numerical Radiation Hydrodynamics,* in: *Astrophysical Radiation Hydrodynamics,* eds. K.-H.A. Winkler and M.L. Norman, NATO ASI Series C: Mathematical and Physical Sciences Vol. **188,** D. Reidel, Dordrecht

Wuchterl, G.: 1989a, *Zur Entstehung der Gasplaneten: Kugelsymmetrische Gasström auf Protoplaneten,* Ph.D. Thesis, Univ. Wien

Wuchterl, G.: 1989b, *Hydrodynamics of Giant Planet Formation I: The κ-Mechanism in Proto Giant Planets,* (preprint)

Questions from the Audience

MOUSCHOVIAS: I am trying to understand physically the oscillations in your central density by more than 20 orders of magnitude! How much mass is contained in the central zone of your code, and is the "gravitational pressure" $\sim GM^2/r^4$ indeed as large as the thermal pressure nkT ($\sim 10^{23} \times 10^{-16} \times 10^4 = 10^{11} dynes/cm^2$)? Is sufficient care taken to prevent (artificial) mass advection in the central zone, which would give rise to a large density increase?

TSCHARNUTER: These 'oscillations' are indeed puzzling, and it cannot be completely ruled out that numerical inaccuracies, e.g., accumulation of discretizing errors, could drive the outward motion. However, there is *no* advection into the Lagrangean central zone containing 10^{-9} $M_\odot$. In case of equilibrium the pressure difference between the center and a zone at radial distance r or mass $M_r \equiv \Delta M_r$ is given by $\Delta P = P_c - P(r) = \frac{1}{4\pi} GM_r^2/r^4$; hence the"gravitational pressure" can be much smaller than the central pressure P_c, depending on $r > r_c$, where r_c is the radius of the innermost sphere. As an example, from $r = 2.97 \cdot 10^9 cm$, $M_r = 3.629 \cdot 10^{27} g$, $P(r) = 7.77 \cdot 10^{10}$, $P_c = 7.91 \cdot 10^{10} dyn/cm^2$ follows $\Delta P = 1.4 \cdot 10^9$, which is of the same order as the "gravitational pressure" $= 0.9 \cdot 10^9 \ll P_c$.

MOUSCHOVIAS: Have you run any models with angular momentum one or two orders of magnitude smaller than the model you described? That initial condition is allowed by our magnetic braking calculations.

TSCHARNUTER: No, I have not done so. The model I presented in my talk refers to the smallest amount of angular momentum > 0 I have ever considered in my calculations.

PALLA: I have two questions:

1. Does the amplitude of the oscillation change by varying the zoning of the grid?
2. Could you explain why previous studies did not show these huge oscillations?

TSCHARNUTER: Ad 1.: No, as long as the total number of grid points vary in between 150 and 512. Test calculations with several thousand gridpoints are underway.

Ad 2.: In all previous calculations there was a tendency for protostellar cores to undergo oscillations. However, since the codes used were less robust, one always tried to damp oscillations artificially, mostly by rather dirty numerical tricks.

CAYREL: How dissipative are the shocks which are produced in your computations of a spherically symmetric collapse?

TSCHARNUTER: Shocks are smeared out artificially over a typical length scale of 1 % of the local radial distance by means of a pseudo tensor viscosity. The corresponding entropy generation is taken into account by adding the viscous energy generation term in the equation of internal energy balance.

BEDOGNI: At a certain phase of the collapse strong shocks form moving out and back into the nebula. Is it possible that at the contact discontinuities between this pair of shocks a RAYLEIGH-TAYLOR instability develops stopping the collapse?

TSCHARNUTER: Due to radiative heat transport the contact discontinuity does not appear.

DI FAZIO: 1) What is the velocity of the density oscillations that you showed in the $\log T_c - \log \rho_c$ graph: is it supersonical?

2) What is the minimum time step in units of the free-fall time, in the slowest (computationally) points of the calculations? Is it $\Delta t/\tau_{ff} \approx 10^{-7}$? Or less?

TSCHARNUTER: Ad 1): Yes, the velocities become supersonic.

Ad 2): The minimum time step $\Delta t \approx 3 \cdot 10^3 sec$. This is to be compared with the (initial) free-fall time $\tau_{ff} \approx 3 \cdot 10^{10-12}$, yielding $\Delta t/\tau_{ff} \approx 10^{-9} - 10^{-7}$.

ZINNECKER: What are the possible observable signatures of your protostars?

TSCHARNUTER: No detailed (non-grey) radiative transfer calculations have been made so far. But in general, variations of several magnitudes in the bolometric luminosity on timescales of months or years are predicted.

BOSS: Wuchterl's mechanism for losing an envelope from a Jupiter-sized protoplanet provides an interesting new way for making a Uranus-like planet. Do you have any idea how you could, e.g., allow this instability at $\sim$ 20 AU but prevent it at $\sim$ 5 AU, and so account for the composition of the Jovian and outer planets?

TSCHARNUTER: The mechanism seems to depend on the dust opacity features. The location of these features in the envelope and their presence (or absence) depend on the outer boundary condition for the protoplanet, i.e., on the location in the solar nebula. In this way the efficiency of the ejection mechanism could be modulated by the spatial variations of the physical conditions in the solar nebula.

A MULTIFLUID HYDRODYNAMICAL RADIATIVE MODEL FOR THE EVOLUTION OF A SPHERICAL PROTOGALAXY.

R. CAPUZZO-DOLCETTA
Istituto Astronomico, Università di Roma "La Sapienza", Via Lancisi 29, I-00161 Roma, Italy

A. DI FAZIO
Osservatorio Astronomico di Roma,
Viale del Parco Mellini 84, I-00136 Roma, Italy

A. B. MEN'SHCHIKOV
Astronomical Council, Academy of Sciences of the USSR, Ul. Pyatnitskaya 48, SU-109017 Moscow, USSR

ABSTRACT. In this paper we present a new radiative, multifluid, hydrodynamical model of protogalactic evolution, using the fragmentation law by Di Fazio (1986). A treatment for radiation transport suitable for non-equilibrium conditions was included. The most important results are: strong shock waves are formed at the time of the bounce of the system, causing a mass loss from the protogalaxy both in fragments and gas; the initial quasi-isothermal phase at ≈ 300 °K is abruptly abandoned when the collapse heating overwhelms the molecular hydrogen radiative cooling, and the second quasi-isothermal phase at ≈ 3500 °K is also abandoned when the center of the system reaches a bremsstrahlung phase. The X-luminosity is, in this phase, similar to that of QSO's. After this transient violent radiative phase, the protogalaxy's radiation activity seems to slowly decline and degrade towards longer wavelengths. In the stellar mass range, two peaks are attained in the IMF at ≈ 20 and ≈ 100 $M_{\odot}$. The slopes of the declining part of the IMF in that range is similar to Salpeter's. Globular clusters seem to form in two bursts, in regions r > 7 Kpc and r > 20 Kpc, with radial metallicity distribution implications. The slope of the globular cluster family IMF is ≈ -1.8.

1. Introduction

Much attention has been devoted, in the literature, to the physics underlying the evolution of protogalaxies. One of the main problems, in attempting to describe and to follow with time the evolution of a protogalaxy, is that a protogalactic cloud is a very complex object, and that many basic physical processes concur to determine its evolution. Moreover, a very remarkable characteristic of said processes is that they are not separable. The interactions and the feedbacks among the most important processes are very strong, and it is almost impossible to aim to obtain a realistic model if we ignore some of the processes, to concentrate only on some other one. Of course, the first

R. Capuzzo-Dolcetta et al. (eds.), Physical Processes in Fragmentation and Star Formation, 303–317.

models, indeed, attempted some of the mentioned strong simplifications (i.e., one-zone, only chemical or only dynamical models, e.g. Gott and Thuan, 1976) and obtained nevertheless very useful informations. The far more complete models by Larson (1969, 1970) succeeded in giving good spatial dependence descriptions, and even multifluid (gas-stars) population-distinguished simulations. Nevertheless, we are tempted, but also forced, to try the construction of more advanced, more complete models of protogalactic evolution, due to the theoretical advancements: i) in the field of radiative properties of the ISM (opacities and emission functions at low densities in NLTE, in particular effects due to the cooling function in presence of molecular hydrogen, see. e.g. Hirasawa, 1969, and Capuzzo-Dolcetta, Di Fazio, and Palla (CDP), 1989, this workshop), as well as: ii) in the field of multifluid-interacting, non-equilibrium models, e.g. Di Fazio, Vagnetti, Wilson (1980), iii) regarding molecule formation in gaseous phase, e.g. Izotov and Kolesnik, 1984, and iv) in the field of radiation transport in hydrodynamical environment (see e.g Hummer and Rybicki, 1971, Tscharnuter and Winkler, 1979, Yorke, 1980); v) in the field of NLTE equations of state; vi) concerning the analytical fragmentation theory due to gravitational instability, able to follow in time the mass spectra of formed fragments (Di Fazio, 1986); vii) about the turbulent phenomena in a multi-phase environment (gas with orbiting objects, see e.g. Battinelli et al., 1989, this workshop), and many other ones. On the other hand, the construction of more advanced and more detailed models of protogalactic evolution is also stimulated by: i) the far larger amount and better quality (precision) of the observations nowadays available on the inner structure of galaxies -thanks also to the satellite based observations in the IR and UV bands- as well as ii) the notable quantity of observations of very distant galaxies, QSO's (maybe connected with, or even present in, protogalaxies), AGN's, intergalactic matter at low and high redshifts, iii) the impressive observative material now available on the mass functions of galaxies (not only derived from luminosity functions: see e.g. the Karachentsev, 1981, 1985, and Peterson, 1979, catalogues of binary galaxies and their dynamically determined masses), and of the globular clusters (GC's) of nearby galaxies (the GC's are known to be good tracers of the early history of galaxies), and iv) the large quantity of spectro-photometrical data on medium and large distance galaxies. Moreover, the now advanced projects for the construction of the multi-mirror very large telescopes also influence the opportunity of construction of more deep and detailed protogalactic evolution models, which should make large use of base-physics, and should try to restrict the parametrization of the phenomena to the minimum possible. In fact, the observations planned for these large telescopes include surveys of distant objects and searches for protogalaxies. In this line of thought, we made a first step towards the construction of a more complete protogalactic evolution model. This step consists in including in a multifluid model (gas of primordial composition + gas of elements heavier than helium + N collisionless pseudo-fluids of fragments) the following new physical inputs: i) the treatment of radiation transport suitable for the overall non-equilibrium dynamics; ii) the analytical fragmentation theory by Di Fazio (1986), able to give in each instant and point the mass spectrum of the fragmentation rate, as well as the instantaneous and cumulative initial mass function of fragments; iii) the new opacities and emission functions in NLTE (including molecular hydrogen) by CDP, 1985, 1989. In this paper, we show the preliminary results of the work, which will appear in a more extended, enriched and detailed version in a forthcoming article.

2. Outline of the multifluid radiative spherical model

We consider a spherical protogalactic cloud, just after the separation from the Hubble flow, or alternatively, from a previous larger protocloud. We divide the gas into 2 components (for the sake of better following chemical enrichment), namely primordial composition and elements heavier than helium ("metals"). The fragments are divided into 15 pseudo-fluids (of collisionless particles: the single fragments are assumed to have a negligible collision rate, considering the model's density and the fragments' typical orbital velocities). Each of these pseudo-fluids, the "k-fragment fluid", is characterized by having fragments of mass in the interval $[m_K, m_K+\Delta m_K]$. The fragment pseudo-fluids are coupled to the gas through the mutual gravity, the fragmentation, and the mass-loss processes (which transfer back mass from fragments to gas). The mass transfer due to the latter two processes causes the coupling in the momentum and energy equations, through the energy and momentum conservation terms. The fragmentation rate spectrum (see Di Fazio, 1986, p. 53, Eq. (14)), depending on the local and instantaneous conditions, distributes mass in the various k-fragment channels.

The used equations are, for $i \in [1,2]$:

$$\begin{cases} \dfrac{\partial\rho_i}{\partial t} + \mathrm{div}(\rho_i \mathbf{v}) = -B_i + D_i \\[2ex] \dfrac{\partial S_i}{\partial t} + \mathrm{div}(S_i \mathbf{v}) = -\dfrac{\partial p_i}{\partial r} - \dfrac{GM(r)\rho_i}{r^2} + \dfrac{\partial S_{iBD}}{\partial t} \\[2ex] \dfrac{\partial \epsilon_i}{\partial t} + \mathrm{div}(\epsilon_i \mathbf{v}) = -p_i \mathrm{div}\, \mathbf{v} - \Lambda + k^a \rho c \epsilon_{rad} + \dfrac{\partial \epsilon_{iBD}}{\partial t} \\[2ex] \dfrac{\partial M(r)}{\partial r} = 4\pi r^2 (\rho + \Sigma_K \rho_{FK}) \\[2ex] \rho = \rho_1 + \rho_2 \\[2ex] S_i = \rho_i v \; ; \end{cases} \qquad (1)$$

$$\begin{cases} \mu\frac{\partial I}{\partial r}(r,\mu) + \frac{(1-\mu^2)}{r}\frac{\partial I}{\partial \mu}(r,\mu) = -k^{a+s}\rho I(r,\mu) + \Lambda + k^s\rho J(r) \\ J(r) = \frac{1}{2}\int_{-1}^{1} I(r,\mu')\Phi(\mu')d\mu' \\ \epsilon_{rad} = J/c \quad ; \qquad T_{rad} = (\epsilon_{rad}/a)^{1/4} \\ k^{a,s} = k^{a,s}(\rho, T, T_{rad}, \{n_i\}) \\ \Lambda = \Lambda(\rho, T, T_{rad}, \{n_i\}) \\ p = p(\rho, T, T_{rad}, \{n_i\}) \\ n_i = n_i(\rho, T, T_{rad}, \{n_j\}) \end{cases} \tag{2}$$

$$\begin{cases} B_i = X_i\Sigma_K(\frac{\partial\rho}{\partial t}_{FK})_{frag} \\ D_i = Y_i\Sigma_K(\frac{\partial\rho}{\partial t}_{FK})_D \\ (\frac{\partial\rho}{\partial t}_{FK})_{frag} = 3AG^{1/2}\rho^{3/2}\ [F(M_K)-F(M_K+\Delta M_K)+ \\ \qquad - \log_e\frac{1+F(M_K)}{1+F(M_K+\Delta M_K)} - \frac{1}{3}\log_e\frac{M_K}{M_K+\Delta M_K}\ ] \\ F(\xi) \equiv [1-(M_J/\xi)^{2/3}]^{1/2} \\ \frac{\partial\Phi}{\partial t}(m,t) = \frac{\pi^{3/2}}{36}\ G^{1/2}\rho^{3/2}m^{-2}F(m) \end{cases} \tag{3}$$

$$\begin{cases} \frac{\partial\rho}{\partial t}_{FK} + \mathrm{div}(\rho_{FK}\mathbf{v}_{FK}) = -D_K + B_K \\ \frac{\partial S}{\partial t}_{FK} + \mathrm{div}(S_{FK}\mathbf{v}_{FK}) = -(\gamma_F-1)\frac{\partial\epsilon}{\partial r}_{FK} - \frac{GM(r)}{r^2}\rho_{FK} + \frac{\partial S}{\partial t}_{FKBD} \\ \frac{\partial\epsilon}{\partial t}_{FK} + \mathrm{div}(\epsilon_{FK}\mathbf{v}_{FK}) = -(\gamma_F-1)\epsilon_{FK}\mathrm{div}\ \mathbf{v}_{FK} + \frac{\partial\epsilon}{\partial t}_{FKBD} \\ S_{FK} = \rho_{FK}v_{FK} \end{cases} \tag{4}$$

The group of equations (1) refers to gas dynamics and thermodynamics. Group (2) contains the equation of transport, the related functions, the equation of state, and the relations giving the equilibrium chemical abundances n_i, for atoms, ions, and molecules. We consider the following species: H, H^+, H^-, H_2, He, He^+, He^{++}, and heavier elements put together in one group. Group (3) is the set of equations related to the processes of fragmentation and mass-loss. B_i is the birth rate function per unit volume, and D_i is the mass-loss rate. $X_{1,2}$ are, respectively, the cumulative abundance of hydrogen and helium, and the heavy element abundance. Y_i are the yields relative to each chemical group. The 3rd, 4th, and 5th of (3) describe the fragmentation process (fragmentation rate and mass function) due to gravitational instability in the cloud, as in Di Fazio (1986, p. 52, Eq.(12); p. 53, Eq.(14)). Group (4) are the dynamical and thermodynamical adiabatic equations for the ensamble of K-fragments. In particular, ϵ_{FK} is the K-fragment fluid internal energy in orbital random motions.

3. Results and discussion

3.1 INITIAL CONDITIONS AND SOME FEATURES OF THE MODEL

As initial conditions for the spherical protogalactic gas cloud we choose: $R_0 = 30$ Kpc, $M_\bullet = 2.5 \cdot 10^{11} M_\odot$, $T_{0c} = 100$ °K, $T_{rad0} = 30$ °K, $Y = 0.75$, $Z = 10^{-9}$, $v(r,0) \equiv 0$, $\rho_{FK}(r,0) \equiv 0$. With the latter data, the initial average free-fall time is $\tau_{ff} = 161$ My. The initial gas density profile has been taken as quasi-uniform throughout the cloud, except in a narrow region on the boundary, where $\rho \propto r^{-15}$. This is in order to simulate the fact that the cloud has just separated itself from the background (the Hubble flow or, alternatively, a larger mother-cloud), and thus it feels the effect of the density perturbation mainly near the boundary. Said previous process of fragmentation from the background is likely to have imported some entropy in the cloud. Thus we choose a hydrostatic-like profile for the internal energy density $\epsilon(r,0)$, but scaled to have a central value corresponding to the assumed value of the central temperature, T_{0c}. This yields a ratio (actual central pressure/hydrostatic central pressure) ≈ $3 \cdot 10^{-5}$. The initial radiation flux is that of a black body background radiation at $T_{rad} = 30$ ·K.

The chosen profile for the initial energy density is not critical: in fact, in only ~ $0.3\tau_{ff}$ (hereafter τ_{ff} is the free-fall time) the initial profiles are forgotten due to the non-reversible thermodynamics. The chosen initial value for T_{rad} corresponds, in the case of a scenario in which the protogalaxy forms directly from the Hubble flow, to a formation redshift $z^f \approx 10$ in an Einstein-De Sitter model.

The model follows the formation, evolution, and dissipation of shock waves in the gas, as well as collisionless shocks in the fragment fluids. When the outward velocity is greater than the escape velocity the model accounts automatically for a mass loss.

3.2 NUMERICAL RESULTS

Figs. 1, 2, 3 are snapshots of the variables v, v_F, the escape velocity v_e, T, M_J, the optical depth τ, the pho-

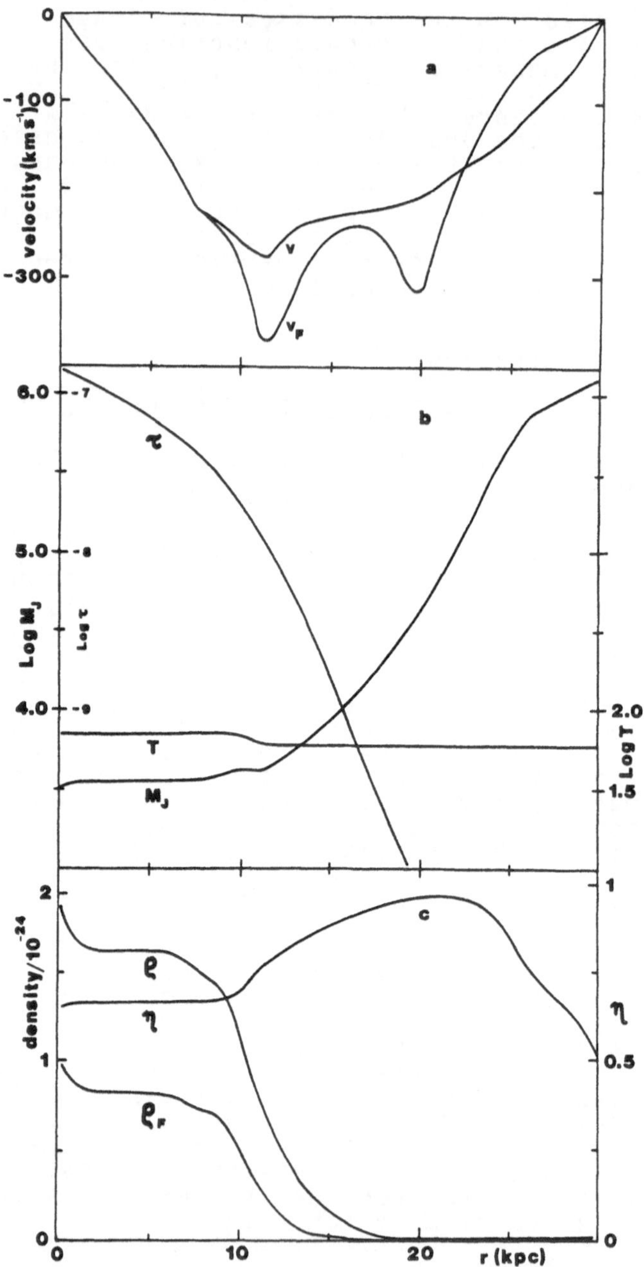

Fig. 1 Radial behaviour of some relevant quantities (see text) at $t = 0.907 T_{tt}$. The Jeans mass M_J is in solar masses. Densities, in panel c, are in 10^{-24} g cm^{-3}.

ton number density n_γ, ρ, $\rho_F \equiv \Sigma_k \rho_{Fk}$, the mass fraction in gas, η. The shown velocity for the fragment fluids, v_F, is actually a mass-averaged velocity: $v_F \equiv \Sigma_k S_{Fk}/\Sigma_k \rho_{Fk}$. This is for the sake of making the plot of the kinematics more compact. The times for the snapshots are, in units of τ_{ff}, 0.907, 1.06, 1.13, i.e. around one free-fall time, as the most interesting activity occurs in this period.

In Fig. 1a we can see a violent collapse of the spherical structure. Note the double wave at ~ 10 and at ~ 20 Kpc in the fragment velocity. The structure is shown to be throughout transparent and uniform in temperature [58 ÷ 70] °K (Fig. 1b). The behaviour of M_J with radius shows how different is the typical mass of the fragments formed throughout the structure: globular cluster-like objects form in an external region (r>~20 Kpc), while the inner zone is populated by lighter objects. For r<~10 Kpc we note a zone where the latter have a typical open cluster-like mass, of ~ 3500 $M_\odot$. Furthermore, said ensamble of globular cluster-like objects is falling toward the center of the protogalaxy following nearly radial orbits, as in that zone $\epsilon_{Fk} \ll 1/2\rho_{Fk}v_{Fk}^2$. An indication on the relative abundance of fragments is given by the mass fraction in gas η in Fig. 1c, showing that fragmentation, at that time, was more efficient in an inner region (r<~10 Kpc), while the scarcity of gas in the outer zone is due to a kinematic segregation effect due to the difference in velocity between fragments and gas (see upper right end of Fig. 1a).

Fig.2a shows a bounce wave in the fragment fluid at r ≈ 8 Kpc, with a rising velocity, but still less than the escape velocity; the gas is still collapsing at high velocity (of the order of 1000 Km/sec). This fast collapse causes the heating of the structure as shown by the temperature curve in Fig. 2b. This increase of the temperature is due to the fact that the sum of the various contributions to the heating has overcome the peak of the cooling function in the molecular region, thus causing first the dissociation of the molecules, and then the gas ionization. The system is still transparent but the optical depth shows an increasing trend with time. The most relevant feature shown by the Jeans mass curve in Fig. 2b is that no more proto-globular clusters appear to be forming at this time. The formation activity is around masses of ~ $2 \cdot 10^4$ $M_\odot$ in the shell between 1 and 10 Kpc, and a few thousand $M_\odot$ above 10 Kpc. In the central protogalactic region the fragmentation activity is negligible, because the gas left is in not relevant quantity (see Fig. 2c). The fragmentation process and the collapse has caused an increase of 5.5 orders of magnitude in fragment density. At the same time, the smaller increase in gas density in the central zones is due to the gas deplenishment caused by the fragmentation process.

Fig. 3a (at the time t = $1.13\tau_{ff}$) shows that the fragment bounce has reached the boundary with velocity significantly higher than the escape velocity (of the order of 300 Km/s). An outgoing shock is present in the gas phase near the boundary. The central region is beginning to become opaque (τ~0.1). Such an optical depth causes a partial

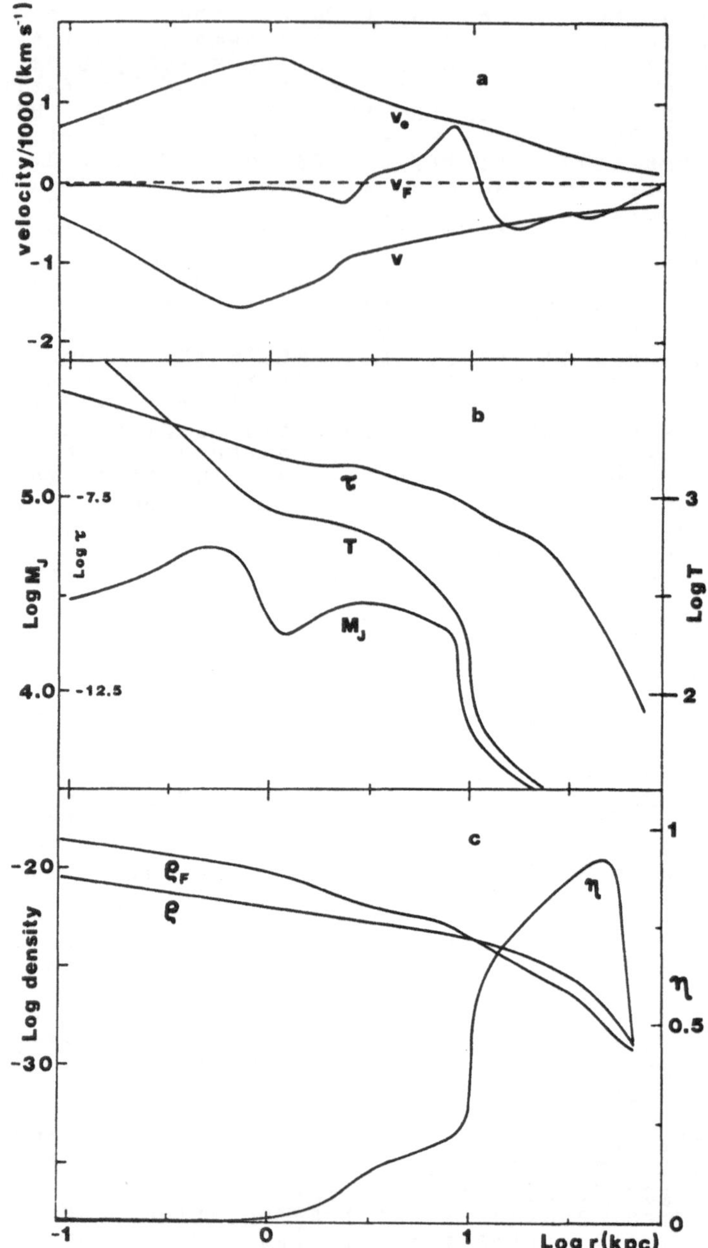

Fig. 2 Plot of the same quantities of Fig. 1 vs. radius at the time $1.06\tau_{ff}$. In panel a is also shown the escape velocity v_e. The logarithmic densities (panel c) are in gcm^{-3}.

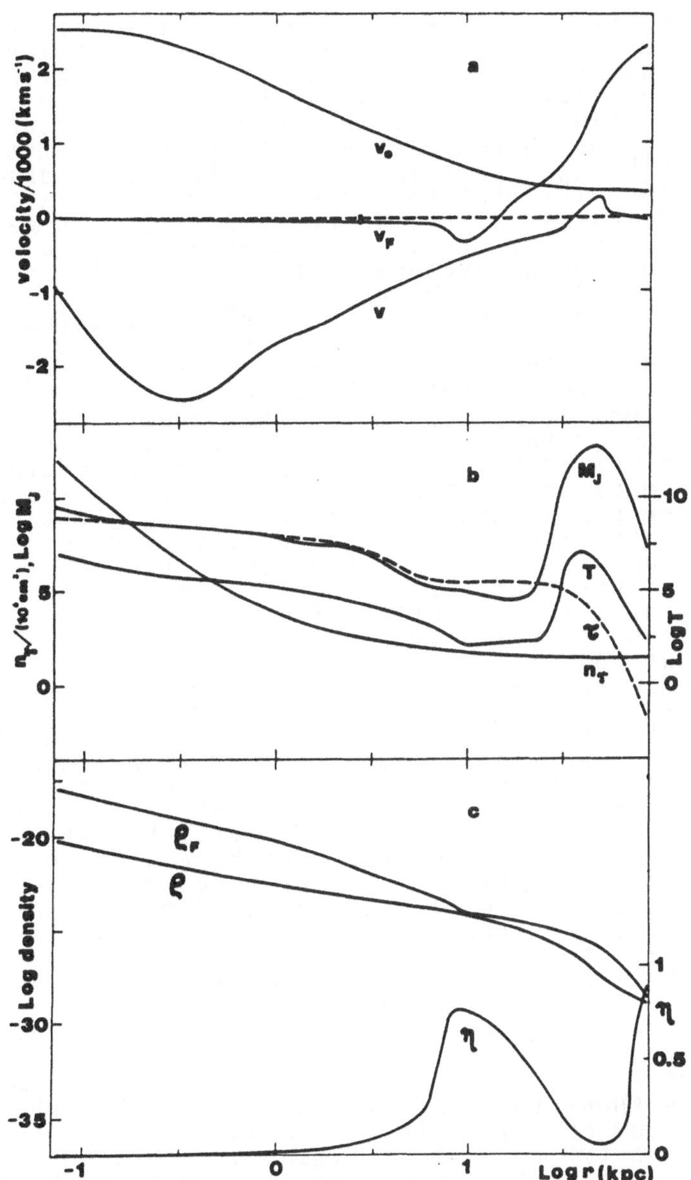

Fig. 3 Same variables shown in Fig.2, with the addition of the photon number density n_γ (panel b, in $10^6 cm^{-3}$). The τ curve in panel b is dashed. The time is 1.13 τ_{ff}. The scale for the optical depth, Logτ, is the same as for the temperature, but - 10 units: i.e., Logτ = 0 is marked 10.

photon retention, as seen from the central pile-up of photons (see in Fig. 3b the n_γ profile). This corresponds to an increase of the central radiation temperature of about a factor 10 ($T_{rad} \approx 320$ °K) with respect to the actual background value of 18 °K. The matter temperature profile implies that the thermal luminosity, whose bolometric value is $\sim 2.3 \cdot 10^{10}$ $L_\odot$, peaks in quite different regions of the spectrum in different points of the protogalaxy. In the nuclear region (r<1 Kpc) there is a strong X-ray emission (the central temperature is of the order of $2 \cdot 10^7$ °K and the space average peak wavelength is ~2 Å), while in the region 1 < r/Kpc < 2 we have a UV core ($\langle\lambda_{max}\rangle \sim 1000$ Å). The region 2÷10 Kpc off the center emits mainly in the infrared ($\langle\lambda_{max}\rangle \sim 2\mu$), even though the inner part of this region emits also in the optical. In the outward-moving shocked gas, the temperature rises to $3.5 \cdot 10^7$ °K, so defining another X-ray emitting region ($\langle\lambda_{max}\rangle \sim 30$ Å). The Jeans mass curve in Fig. 3b shows that practically the only fragmentation activity is now in 6 < r/Kpc < 22, as in the central regions the Jeans mass is greater than the available gas mass, and in the outer region the Jeans mass is even greater than the system's mass. These conditions determine a cutoff in the fragmentation rate (see Di Fazio, 1986, p.53 Fig. 2). In the remaining zones, the left-over gas is in mass only a small fraction of the existing fragments, and thus the fragmentation process can change the fragment density by non-relevant quantities.

It is interesting to remark that the quantity $x_S \equiv \frac{2GM(r)}{c^2 r}$ (x_S indicates how close we are to the need of GTR), that at the beginning was $\sim 10^{-11}$, at $t=1.13\tau_{ff}$ reaches the value $\sim 3 \cdot 10^{-4}$ in the center. The time scale of its increase is $0.2\ \tau_{ff}$. The mass involved in this fast collapse is $\sim 10^8\ M_\odot$. The calculations were stopped at $t/\tau_{ff} \approx 1.5$. At such time we could not yet determine whether the core will or not become a black hole. For this it will be necessary to push further the calculations, and also to add more mesh points in the inner region. A speculative consideration is that the situation in the center of our model protogalaxy could possibly be that of a forming QSO, in that the formation conditions of a QSO are often thought to be those of a violently collapsing hot object in the center of a protogalaxy.

At the time 1.5 τ_{ff}, the gas left is 16% in mass.

Fig. 4 shows the time evolution of the protogalaxy's bolometric luminosity, with a remarkable peak of $\sim 10^{45}$ erg/s (similar to a typical QSO's luminosity) at the time $t/\tau_{ff} \approx$ 1.08. Just before the peak (at $t/\tau_{ff} \approx 0.9$, corresponding to the first arrow in Fig. 4) the space-averaged emission has $\langle\lambda_{max}\rangle \approx 43\mu$. At the time of the luminosity peak, the averaged wavelengths are approximately the same as described in Fig. 3b, while just after (third arrow, at t~ 1.19τ_{ff})$\langle\lambda_{max}\rangle \approx$ 3Å, 10Å, 6μ, 10Å, averaged in the same zones as when describing Fig. 3b. The following evolution cools the system down progressively, and at $t/\tau_{ff} \approx 1.5$ $\langle\lambda_{max}\rangle \approx$ 6Å for r<3 Kpc, and 7μ for the external zone. This evolution of the system luminosity to higher wavelengths is in agreement with the energy-redshift relation for extragalactic active objects (see Giovannelli and Polcaro, 1986). Incidentally, the duration of the high luminosity phase is ~30 My, which is compatible with the believed duration of QSO activity (see e. g. Komberg, 1990).

In Fig. 5 we display the cumulative initial mass function of fragments at time 1.05 τ_{ff}.
Two well defined peaks are noted at ~20 $M_\odot$ and at ~100 $M_\odot$ Moreover, different tail slopes are present. If we approximate the declining parts of the spectrum with a power law α $M^{-\alpha}$, the low-mass peaked distribution has α = 2.18, as indicated by the first straight line. The decline in the region 3.5 < Log $M/M_\odot$ < 4.1 (which corresponds to massive open clusters) has α= 2.33, as shown by the second straight line; the globular cluster mass range has α = 1.85. For M > 560 $M_\odot$ (i.e. a mass value which is believed to be of the order of the upper limit for the initially stable first burning stars, see e.g. ElEid et al.) α = 1.82. Note that the globular cluster zone has a slope sensibly flatter than the rest.

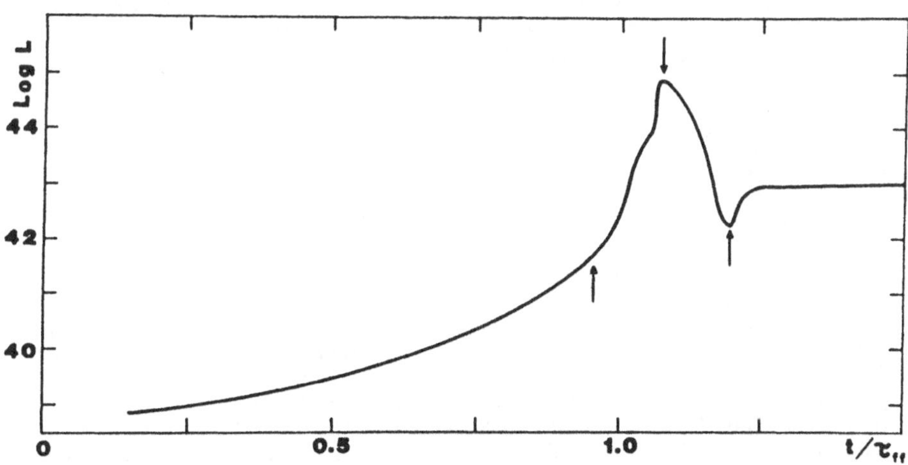

Fig. 4 Time dependence of the protogalaxy's bolometric luminosity in erg·sec^{-1}. In coincidence with the arrows, the spectrum is described in the text.

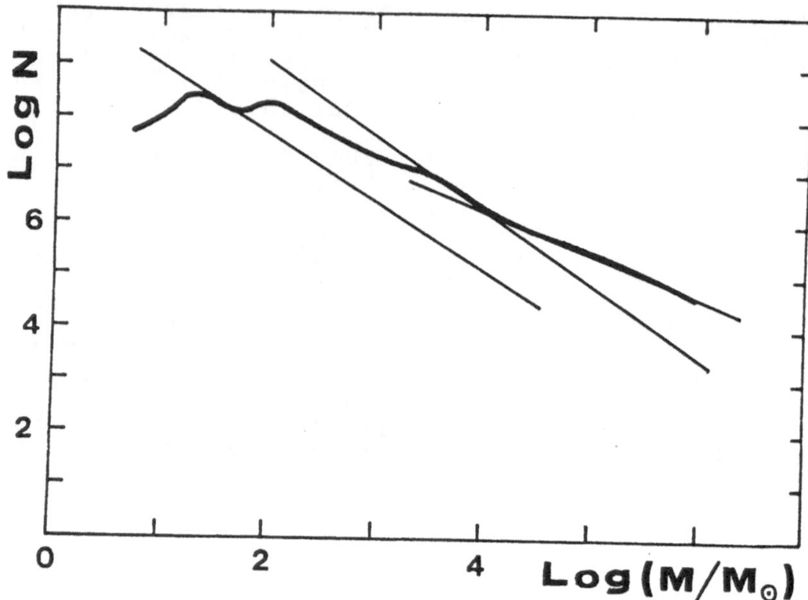

Fig. 5 The mass function integrated on even spaced logarithmic mass intervals (of width 0.25) of the fragments formed up to $t = 1.5\tau_{ff}$. The slopes in some interesting regions are indicated by straight lines.

4. Conclusions

The presented preliminary results have shown some interesting features of the evolving protogalaxy. There is a transient when the luminosity is very similar to that of many QSO's and AGN's. The trend in time is to a spectrum degradation to higher wavelengths, which qualitatively corresponds to what is observed for high redshift objects. At the time ($\sim 1.5\ \tau_{ff}$) when the model was stopped, we note the tendency to the formation of a massive black hole in the nucleus. To check this possibility, we need to extend in time the calculations. A notable feature is the formation of a central region with an optical depth sufficiently high (0.86) to cause a retention of photons and a consequent local increase of the radiation temperature.

An interesting result is that there seems to be at

least two bursts of protoglobular cluster formation, separated in time by about 40 My: the first around 1 free-fall time and in an external zone, r > 20 Kpc, and the second occurring in a zone 6 < r < 18 Kpc. This may have implications on the metallicity-galactocentric distance distribution of globular clusters (e.g. Zinn, 1978) in that the second generation forms in an enriched environment gas.

The stellar and open cluster-like zones of the obtained initial mass functions have slopes compatible with Salpeter's value 2.35, while the globular cluster IMF is quite flatter.

Acknowledgements

This work has been accomplished thanks to the Scientific Exchange Agreement between the Academy of Sciences of the USSR and the Italian National Council of Researches (CNR), contract No. 2.16 between the Astronomical Council (Moscow) of the USSR Academy and the Astronomical Observatory of Rome. Many thanks to Dr. B. M. Shustov for useful discussions on the radiation transfer problem and for offering computing facilities, ospitality and the collaboration of the Moscow Institute.

REFERENCES

Battinelli, P., Capuzzo-Dolcetta, R., Di Fazio, A., Urpin, V.A., Vereshchagin, S.V.: 1989, this volume

Capuzzo-Dolcetta, R., Di Fazio, A., and Palla, F.: 1989, this volume

Di Fazio, A.: 1986, Astron. Astrophys. **159**, 49

Di Fazio, A., Vagnetti, F., and Wilson, J.R.: 1980, Astrophys. Space Sci. **72**, 223

ElEid, M.F. Fricke, K.J., and Ober, W.: 1982, in *"The Most Massive Stars"*, ESO Workshop, eds. S. D'Odorico, D. Baade, K. Kjar, p.303

Giovannelli, F., Polcaro, V.F.: 1986, Monthly Notices Roy. Astron. Soc. **222**, 619

Gott, J.R. III, Thuan, T.X.: 1976, Astrophys. J. **204**, 649

Hirasawa, T.: 1969, Progr. Theor. Phys. **42**, n.2, 523

Hummer, D.G., Rybicki, G.B.: 1971, Monthly Notices Roy. Astron. Soc. **152**, 1

Izotov, Yu.I., Kolesnik, I.G.: 1984, Sov. Astron. **28**, 15

Karachentsev, I.D.: 1985, Astron. Zhurn. **62**, n.3, 118

Karachentsev, I.D.: 1981, Astrofizika **17**, 429

Komberg, B.V.: 1990, Sov. Astron., in press

Larson, R.B.: 1969, Monthly Notices Roy. Astron. Soc. **145**, 405

Larson, R.B.: 1970, Monthly Notices Roy. Astron. Soc. **147**, 323

Peterson, S.D.: 1979, Astrophys. J. Suppl. Ser. **40**, 527
Yorke, H.W.: 1980, Astron. Astrophys. **86**, 286
Zinn, R.: 1978, Astrophys. J. **225**, 790

Discussion :

Palla: does the model you described predict the existence of a hot phase, say $T > 10^5$ °K, in the intercloud medium?
Di Fazio: yes, in the collapse phase around one free-fall time; after, when the heating mechanisms overwhelm the (very efficient) cooling, the temperature, in the center, temporaneously leaps to bremsstrahlung values. In other zones of the protocloud, instead, we have continuously decreasing values down to 10^4 °K, and then, from 3 to 10 Kpc, a steep decline down to 100 °K. From 13 Kpc outward the temperature temporaneously re-encreases due to the passage of a strong shock. The interesting feature that we can underline is that, even when the environment gas is locally hot or very hot, the fragments are relatively cold in their interior: 50÷500 °K, due to the much denser matter inside them.

Cayrel: 1) What is the largest optical thickness you obtain from the most efficient transition? 2) Are you also computing the optical thickness of the fragments?
Di Fazio: 1) The largest τ is (up to the time at which we temporaneously stopped the computations) 0.86 for the average nucleus peak wavelength, but it is from 20 to 50 in the lines of molecular hydrogen rotational jumps. The model is able to account for line transport, but this computation was done with grey average coefficients. Nevertheless, 0.86 is an average optical depth which causes the retention (accumulation) of photons due to the said rotational transitions, but also to bound electron scattering, and, in the center, to free-free absorption and to Thomson scattering. 2) No, we are not yet computing the optical thickness of the fragments. This can for now be estimated only approximately making auxiliary runs at the appropriate lower masses and initial conditions.

Bedogni: Which Mach numbers do you have for the formed fragments moving in the gas? Do they interact with the bounce shock wave that travels outward from the system's center?
Di Fazio: The Mach numbers range from 3 to ~40. The fragments' motion is highly supersonic, and they will create strong shocks. Maybe it will be interesting to use your work, and ours in collaboration with Urpin and Vereshchagin, to account for the energetics of the interaction of the fragment wakes with the protogalactic gas. No interaction with the system's shock wave is accounted for. The latter, in our model, interacts (heats, dissipates, and radiates) only with the gas, but not with the fragments. I believe

that the interaction with the fragments is negligible, in its effects, with respect to the interaction gas-shocked wakes of the fragments (see Battinelli *et al.*, this workshop).

THERMAL INSTABILITIES IN THE WARM INTERSTELLAR GAS AS A REGULATING MECHANISM OF STAR FORMATION

Antonio Parravano
Universidad de Los Andes
Facultad de Ciencias
Departamento de Fisica
Merida 5101, Venezuela

Abstract

The role of a self-regulating mechanism of star formation based on the sensitivity of the process of condensation of small cool clouds in the warm gas upon the UV radiation produced by massive stars is reviewed in three applications: a) in isolated galaxies, b) in a simplified one-zone model of mass exchange between three phases (diffused gas, clouds, and stars), and c) in the galactic warm interstellar gas. The results indicate that thermal instabilities as a condensation mechanism of the interstellar diffused gas, play a central role in the self- regulated process of stellar formation.

1 The Self-Regulating Mechanism

Many self-regulated mechanisms of star formation have been proposed (i.e. Franco and Cox, 1983; Franco and Shore, 1984; Dopita, 1985; Struck-Marcel and Scalo, 1987; Wang and Cowie, 1988). Here attention is concentrated in a self-regulating mechanism based on the sensitivity of the condensation of small cool clouds upon the radiation density in the band 912 – 1100 Å (Parravano 1987). This mechanism is capable of affecting the large scale structure of galaxies due to the fact that it acts at a large scale in a very short time.

The interstellar warm gas always condenses when the critical pressure $Pmax$ is lower than the gas pressure. Here, $Pmax$ is the pressure in the marginal state of stability of the isobaric mode for the transition warm gas $\longrightarrow$ small clouds due to thermal instabilities. The critical pressure $Pmax$ depends mainly on the gas metallicity ($[Xi/H]$) and on the Fuv parameter. As a first approximation, $Pmax$ can be expressed as (Parravano 1988):

$$\log(Pmax/k) = 3.18 - 1.6[Xi/H] + 1.35 \log Fuv \quad (1)$$

R. Capuzzo-Dolcetta et al. (eds.), Physical Processes in Fragmentation and Star Formation, 319–324.

where

$$Fuv = \left[\frac{Y_e}{\rho_s/(3\,g\,cm^{-3})}\right]\left[\frac{Q_a}{a/(2\,10^{-6}cm)}\right]\left[\frac{160 M_d \overline{U}_\lambda}{M_H 7\,10^{-17} erg\ cm^{-3}\ \mathring{A}}\right] \tag{2}$$

and Y_e is the mean efficiency for photoelectric emission, ρ_s the solid density of the grains, Q_a the efficiency factor for absorption, a the mean radius of the grains, M_d/M_H the dust-to-hydrogen mass ratio, and $\overline{U}_\lambda$ is the mean energy density in the band 912 – 1100 Å. The Fuv parameter is defined in such a way that $Fuv = 1$ for the average conditions in the Galaxy.

According to equation (1), $Pmax$ is a positive slope function of the UV energy density. Then, when $Pmax$ is near the warm gas pressure, an increase (decrease) of the UV energy density tends to inhibit (promote) cloud formation. Thereby, star formation can be regulated by the massive star sub-system if small clouds condensed from warm gas are the main source to incorporate diffuse gas into molecular clouds which evolve into giant molecular clouds and form stars. This hypothesis might be expresed as: *The star formation rate is self-regulated at a value that it maintains Pmax close to the gas pressure Pg.*

2 Applications

2.1 The condition $Pmax = Pg$ in isolated galaxies and the global star formation rate

If the condition $Pmax = Pg$ is present in the warm gas then, as a first approximation, isolated galaxies will satisfy the following relation (Parravano 1989):

$$\log M_H = P_o + P_1 \log \frac{M_{wd} L_B d_1}{M_\odot 10^8 L_\odot d_2} + P_2 \log d_1 - P_3([Xi/H] - [Xi/H]_o) \tag{3}$$

where M_H is the hydrogen mass, M_{wd} the mass of warm dust, L_B the blue luminosity, d_1 and d_2 the major and minor diameters respectively, and $[Xi/H]_o$ is a reference metallicity. If hydrostatic equilibrium is assumed, then $P_1 = 0.35$, $P_2 = 0.47$, and $P_3 = 0.42$. The parameter P_o depends upon the physical and chemical properties of the gas and dust, and on the morphology of the galaxy. Nevertheless, because of the lack of observational data, it is supposed that P_o is constant.

Under the condition $Pmax = Pg$, the star formation rate can be expressed as (Parravano 1989):

$$SFR(M_\odot/yr) = 1.4610^{-15}\,\frac{10^{0.53([Xi/H]-[Xi/H]_o)}}{d_1^{0.67}\sqrt{d_1/d_2}}\,T_d^2\,M_H^{1.43} \tag{4}$$

where T_d is the warm dust temperature.

Relations (3) and (4) have been tested for a sample of 70 isolated galaxies of different types and dimensions (ranging over almost four orders of magnitude in gas mass). When $logMg$ is plotted versus the three last right hand terms of equation (3), the sample shows a high correlation coefficient (0.95) and a regression line with a slope equal to 1.01. The agreement between the observed and predicted slopes, together with a high correlation coefficient, reinforces the hypothesis and makes it possible to adjust the sample with only one free parameter ($P_o = 6.2$). It is important to notice that P_1, P_2, and P_3 are completely given by the model.

When the SFR inferred from the blue and infrared luminosities (Parravano 1989) is plotted versus the right hand side of equation (4), the correlation coefficient is good (0.93) and the adjustement to the hypothesis $Pmax = Pg$ is satisfactory. This result indicates that the proposed expression for the SFR (eq. 4) is more realistic than the law of Schmidt because it contains important galactic parameters such as the mass and properties of the dust, the metallicity, and the galactic diameter.

2.2 The stabilizing effect of the self-regulating mechanism

With the purpose to study the stabilizing effect of the self-regulating mechanism presented above, Parravano, Rosenzweig, and Teran (1989) included it in a simple model of mass exchange among three phases (diffused gas, clouds, and stars). With the absence of the regulating mechanisn, the system shows a strong tendency to develop periodical bursts of star formation. But, when the regulating mechanism is included, the non-linear oscillations dissapear and the system maintains a stellar formation rate approximately constant for long periods of time. Moreover, the system tends towards the state of marginal stability for the condensation of small clouds. The results show a pattern of evolution that agrees with the observational evidence in the sense that, in most of the isolated galaxies, the SFR appears to have remained remarkably constant for long periods of time. The present model also alleviates the gas consumption which makes it easier to explain the presence of gas in evolved galaxies.

2.3 The state of the galactic warm gas

Given $[Xi/H]$ and Fuv, and if the condition $Pmax \simeq Pg$ is present in the warm gas, then equation (1) permits us to calculate its pressure Pg. Thus, for each set of values $([Xi/H], Fuv)$ the thermodynamical state $(n_{wg}, T_{wg}, \xi_{wg})$ is univocally determined. Here, n_{wg}, T_{wg}, and ξ_{wg} are the density, temperature, and degree of ionization of the warm gas, respectively. Moreover, it is sufficient to know the value of two of the five variables $([Xi/H]$, Fuv, n_{wg}, T_{wg}, $\xi_{wg})$ in order to calculate the values of the remaining ones.

In order to estimate the thermodynamical state of the warm gas as a function of the galactic radius (R), the data of Güsten and Mezger (1982) for the metallicity, and the linear fit of Bohigas (1988) to data points for the warm gas density (n_{wg}) have been used. These data permit us to estimate the radial dependence between 6 and 13 Kpc for the gas pressure Pg and for the parameter Fuv (Figure 1), and for the temperature T_{wg} and ionization degree ξ_{wg} (Figure 2) in the galactic plane.

The linear fit to data points of Bohigas (1988) is only useful to give a crude picture of the radial dependence of n_{wg}. Hence, the results in Figures (1) and (2) are preliminary and have to be taken only as indicators of the radial dependence of the variables Pg, Fuv, T_{wg}, and ξ_{wg}. Nevertheless, the results in Figures (1) and (2) agree with the expected dependence and show the potentiality of prediction of the hypothesis $Pmax = Pg$.

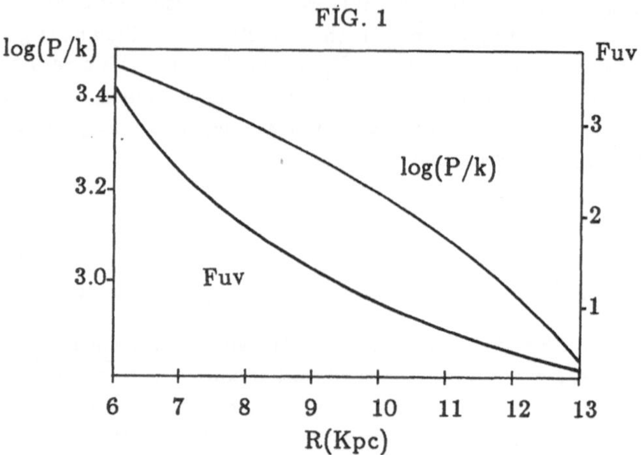

Figure 1. The radial dependence for the gas pressure Pg and the Fuv parameter in the galactic plane under the condition $Pmax = Pg$.

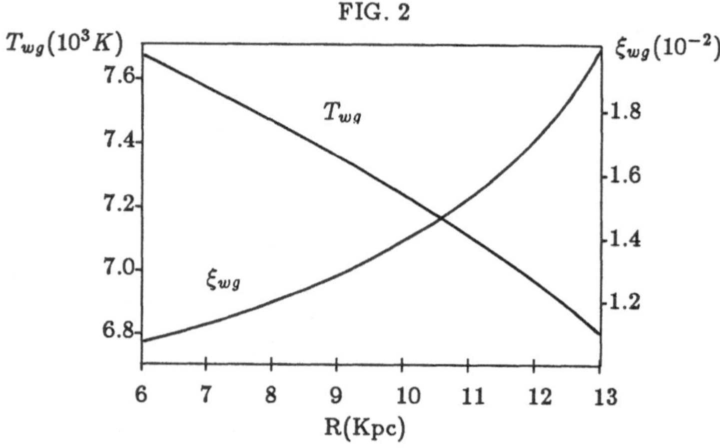

Figure 2. The radial dependence for the temperature T_{wg} and the ionization degree ξ_{wg} of the warm gas in the galactic plane under the condition $Pmax = Pg$.

3 Conclusions

The self-regulating mechanism presented above is capable to affect the large scale structure of the galaxies due to the fact that it acts at a large scale in a very short time. Additionally, the radiation in the band $912 - 1100\,Å$ is produced by massive stars ($m \geq 6\,M_{\odot}$). Thereby, the stellar formation is controlled by the star formation rate in the near past.

The results indicate that the condition $Pmax \sim Pg$ is present in isolated galaxies. This fact allows to propose an analytical formula for the star formation rate as a function of global galactic parameters. The hypothesis also permits to estimate the radial dependence of the thermodynamical state of the warm gas in the galactic plane. Finally, the self-regulating mechanism used here is capable to inhibit intrinsic bursts of star formation.

All these results indicate that thermal instabilities as a condensation mechanism of the interstellar diffused gas, play a central role in the self-regulated process of stellar formation.

The author is very grateful to the local organizing commitee of the workshop for the generous financial support.

REFERENCES

Bohigas, J.:1988, Astron. Astrophys. **205**, 257
Dopita, M.A.: 1985, Astrophys. J. Letters **295**, L5
Franco, J., Cox, D.P.: 1983, Astrophys. J. **273**, 243
Franco, J., Shore, S.N.: 1984, Astrophys. J. **285**,813
Güsten, R., Mezger, P.G.: 1982, Vistas Astron **26**, 159
Parravano, A.: 1987, The interaction of Supernova Remnants with the Interstellar Medium, IAU 101, University of Cambridge Press
Parravano, A.: 1988, Astron. Astrophys **205**,71
Parravano, A.: 1989, Astrophys. J. (in press)
Parravano, A., Rosenzweig, P., Teran, M.: 1989, Astrophys. J. (submited)
Struck-Marcell, C., Scalo, J.M.: 1987, Astrophys. J. Suppl. Series **64**, 39
Wang, Z., Cowie, L.L.: 1988, Astrophys. J. **335**, 168

QUESTIONS

T. Ch. Mouschovias: I have two questions:

1) The same stars that produce the UV radiation you are using also are likely to become supernovae, which are more effective for damping energy in the warm intercloud medium. Have you considered this effect?

2) Field (1965) showed that even a magnetic field as small as 1 microgauss renders the thermal instability one-dimensional. Since the resulting cloudlets have masses $\leq 0.1\,M_{\odot}$, can you really build molecular clouds and giant molecular clouds out of the one-dimensional motion of these cloudlets?

A. Parravano:

1) The transition warm gas $\longrightarrow$ small clouds by thermal instabilities is expected to occur far from the places of recent star formation. In these regions the radiation density in the band 912 – 1100 Å is maintained by the large scale distribution of OB stars. The contribution of supernova to this radiation field is negligible. Supernova are important in pressurizing the interstellar medium. Therefore, many of the parameters used in the present model are sensitive to the galactic rate of supernova. Moreover, supernovae shell fragmentation is an important source of neutral clouds. Nevertheless, the results presented in (2.1) show that thermal instabilities in the warm gas play a central role in the process of self-regulation of stellar formation.

2) At the time when the two-phase model for the ISM was proposed (Field et al. 1969) the dominant heating mechanism was cosmic rays. It is now accepted that photoejection of electrons by dust grains is the dominant heating mechanism and that the primary cosmic ray ionization rate is 40 times lower. Therefore, in the model used here, the ionization degree is very low and heat conduction is dominated by neutral hydrogen. Consequently, condensations that are nearly spherical are expected and their motions are not constrained by the magnetic field. In order to build up molecular clouds we can evoke accretion of warm gas by small clouds and coalescence.

EVOLUTION OF THE FIRST PROTOSTELLAR CORE

A.B. MEN'SHCHIKOV
Astronomical Council of the USSR Academy of Sciences
Pyatnitskaya 48
109017 Moscow
USSR

ABSTRACT. Numerical calculations of the structure and evolution of the first protostellar core in hydrostatic equilibrium are presented, which were performed using stellar evolution code. It is shown that the core with the lifetime $\sim 10^3$ yr always passes through the three stages: convective, radiative and that of dust evaporation. At the end of its life the core becomes dynamically unstable due to molecular hydrogen dissociation. Dependence of the results on mass accretion rate, shock luminosity, dust grain opacity is studied.

1. Introduction

The first important theoretical investigation of star formation was made by Hayashi *et al.* (1962) who studied the evolution of the convective protostars that were rather dense initially. In these models the initial state was far from the typical conditions in star forming regions, and therefore the first important stage of dynamical collapse was not followed in that calculations. A significant step forward was made by Larson (1969) who first solved numerically in hydrodynamical calculations the full problem of star formation, starting with the initial collapse of a fragment of molecular cloud. Discrepancies between quantitative results of similar calculations which were made by different authors later, have been analyzed by Winkler and Newman (1980a,b). They showed that the results depend on the specific numerical method and presented hydrodynamical calculations with the method which allowed them to avoid numerical difficulties. An important contribution was made by Stahler *et al.* (1980a,b; 1981) who used completely different numerical technique. Dividing the whole protostar into several physically defined regions (the dust envelope, the opacity gap, the accretion shock and the hydrostatic core), they were able to calculate the total structure and evolution of a protostar by matching solutions at boundaries of these regions. Their results agree well with the observed birthline of low-mass stars (Stahler, 1983).

A simplified general evolutionary scheme of a transition from molecular clouds to the newly formed star can be described in few words as follows. After about one free-fall time ($\sim 10^5$ years) the nearly isothermal collapse leads to the formation of a small (nearly adiabatic) first hydrostatic protostellar core at the center of the cloud. The lifetime of this dust core is much shorter than initial free-fall time. Dissociation of molecular hydrogen in its central parts leads to the collapse of the first core, during which the second (ionized, and stable) protostellar core forms.

R. Capuzzo-Dolcetta et al. (eds.), Physical Processes in Fragmentation and Star Formation, 325–332.

It accretes surrounding dust envelope containing most of the mass of the initial fragment until all the envelope is accreted by the core, and the infrared protostar becomes a star which radiates energy mainly at the optical wavelengths.

In this picture, which follows from a number of hydrodynamical calculations of protostellar collapse, not all stages are important for the final observable properties of a single star. For this reason Stahler *et al.* (1980a) did not consider the very short stage of the first hydrostatic core in their calculations. We decided to study this stage in the present work because it could give some additional information about the star formation process and because it can be important for the formation of binaries (Tutukov, 1983).

2. General Approach and Equations

The results of hydrodynamical calculations of protostellar collapse show that after the moment of the first core formation, the cloud is naturally divided into two different regions: the dust hydrostatic core itself and the freely–falling envelope. The first protostellar core forms due to increased optical depth of central parts of the cloud, having mass $\sim 0.01 M_\odot$ almost independently of the initial cloud's mass. Mechanical equilibrium of the core implies that it is possible to construct its model and to follow its evolution by solving usual equations of stellar structure instead of hydrodynamical equations. In this case the optically thin envelope may be described by a single parameter, the mass accretion rate $\dot{M}$. Interaction between the core and the envelope should be taken into account while defining the outer boundary conditions.

We consider a spherically-symmetrical opaque protostellar core in hydrostatic equilibrium. It is assumed that the chemical composition is $X = 0.7$, $Y = 0.27$, and that 1% of the core mass is in the form of silicate grains with the radius $5\ 10^{-6}$ cm ; the density of the grain material is $3\ \mathrm{g/cm^3}$. For temperatures $T < 10^3$K the dominant contribution to the opacity comes from the dust grains, and the opacity depends on temperature only. When $T > 10^3$K , the grains begin to evaporate, and the opacity depends both on gas density and temperature. We may restrict our consideration to the temperature range 50 K$< T < 2\ 10^3$K , because at higher temperatures the first protostellar core appears to be dynamically unstable due to H_2 dissociation. Therefore, only the density-independent grain opacity was used (Tutukov and Shustov, 1981). To take into account dust grain evaporation, the opacity artificially decreases by several orders of magnitude at the temperatures $T > 10^3$K .

For our calculations it is sufficient to use the simplest equation of state, i.e. that of the perfect gas, because the gas never becomes degenerate at such low densities. Saha equation for number densities of H_2 , H , H^+ , He , He^+ , He^{++} was solved to obtain the vibrational, rotational and the dissociation energies of H_2 , and ionisation energies of H and He . Mean vibrational and rotational energies of an H_2 molecule was calculated by the summation over its energy levels assuming equilibrium distribution of their populations (Lang, 1978).

The structure and evolution of the first protostellar core, with the above assumptions, are described by the following equations:

$$\frac{dP}{dm} = -\frac{Gm}{4\pi r^4} \tag{1}$$

$$\frac{dT}{dm} = -\frac{GmT}{4\pi r^4 P}\nabla \tag{2}$$

$$\frac{dr}{dm} = \frac{1}{4\pi r^2 \varrho} \tag{3}$$

$$\frac{dL_r}{dm} = \varepsilon_g \tag{4}$$

where mass $m(r)$ is chosen as an independent variable. Here P , T , r , L_r are pressure, temperature, radius, and luminosity respectively; ϱ is density, ∇ is logarithmic temperature gradient, ε_g is the rate of gravitational energy release, and G is the gravitational constant. The equation of state for perfect gas and the Schwarzshild criterion of convection are used. The convective gradient was calculated according to the mixing length theory (Paczynski, 1969).

Outer boundary conditions for the accreting protostellar core are different from that without accretion. The boundary conditions were treated in the following way. Let us assume that the fraction α of the shock luminosity is freely radiated away from the shock front. The remaining fraction $(1\text{-}\alpha)L_{sh}$ is going into the increasing internal energy of the surface layers. The accreting core feels also external dynamical pressure, which should be added to the boundary pressure. The resulting outer boundary conditions are:

$$P_s = \frac{2}{3}\frac{GM}{\kappa R^2} + \frac{\dot{M}}{4\pi R^2}\left(\frac{2GM}{R}\right)^{1/2} \tag{5}$$

$$T_s = \left(\frac{L + (1-\alpha)GM\dot{M}/R}{4\pi R^2 \sigma}\right)^{1/4} \tag{6}$$

where M is the total mass of the core and σ is the Stephan-Boltzman constant.

In the present work we have considered the following model parameters. The mass of the first protostellar core was taken to be $0.01\ M_\odot$, while the mass accretion rate $\dot{M}$ for the main model is $2\ \ 10^{-5} M_\odot/\mathrm{yr}$. The case with the accretion rate $2\ \ 10^{-6} M_\odot/\mathrm{yr}$ was also calculated to check the influence of its variation. The fraction α of freely radiated shock luminosity was usually set to 0.5 , but one our model had $\alpha = 0.1$.

3. Results and Discussion

In all cases considered the evolution of the first hydrostatic protostellar core proceeds in a qualitatively similar way. The evolution of the core without accretion was also calculated, because it is useful for comparison with all other cases.

The evolution of the protostellar core's center on the density-temperature diagram is shown on Fig. 1. Initial models were constructed with central temperatures in the range $50 \div 80$ K, having optical depths ~ 100 only. At the temperatures about 100 K all the models attain hydrostatic and thermal equilibrium. Their subsequent evolution is simply gravitational contraction with monotonically rising central temperatures and densities. As expected, the models with higher mass accretion rates have more steeply rising temperatures relative to densities, because they need to have higher thermal pressure to balance external dynamical pressure. When central temperature becomes higher than 10^3K dust grains begin to evaporate and opacity rapidly decreases with temperature. This forces the central parts of the protostellar core to contract more rapidly, with temperature rising less steeply relative to the density. Of course, this behaviour is usual for other self-regulating hydrostatic configurations, like ordinary stars. At the end points of the tracks on Fig. 1 the core becomes dynamically unstable due to H_2 dissociation.

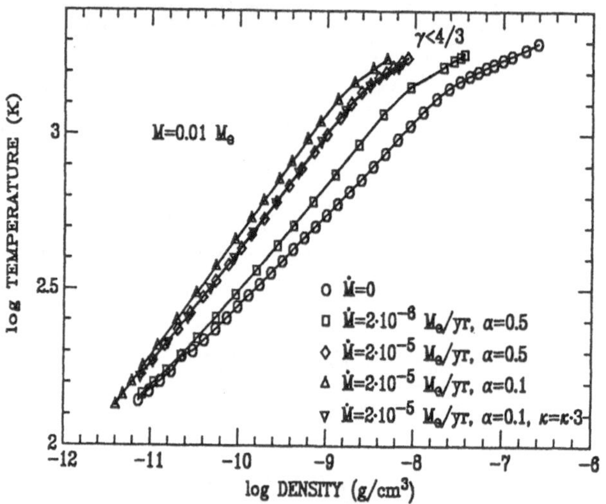

Figure 1. Density-temperature diagram for the first protostellar core. At the end points of the tracks the central zone with $\gamma < 4/3$ covers $\sim 30\%$ of the total core's mass.

The convective structure of the protostellar core is displayed on Fig. 2 as a function of time, for the main evolutionary sequence. The core forms as nearly fully convective configuration, at the middle of its life becoming fully radiative one. The transition takes $\sim 10\%$ of core's lifetime which is about 800 yr. At the end of evolution its central parts contract more rapidly, and the increased luminosity is the reason of the formation of a thin intermediate convective zone, which moves outward. At the same time the adiabatic index of the central parts of the core becomes less than $4/3$, due to the dissociation of H_2, and that region also expands rapidly.

The H-R diagram for all the sequences calculated is presented on Fig. 3, including for comparison the case without accretion. Generally accretion increases the effective temperatures and luminosities, as one can expect from the Eqs. (5),(6) and from Fig. 1. Inspection of Fig. 2 and Fig. 3 shows that there exist three following evolutionary phases during the lifetime of the core.

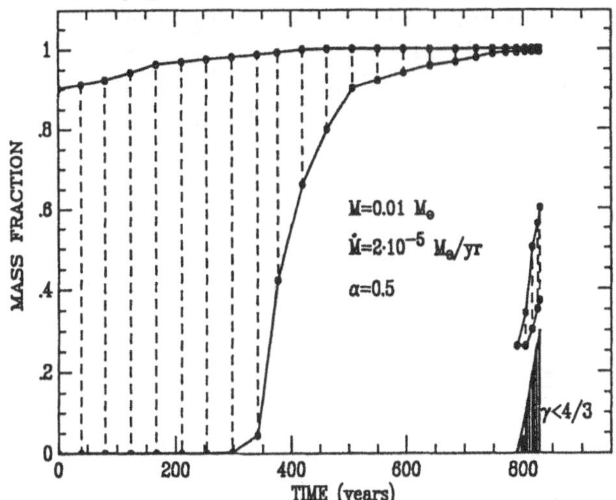

Figure 2. Convective structure of the first protostellar core for our main run. The central region where $\gamma < 4/3$ is also shown.

1. The convective stage. It corresponds to the descending part of the evolutionary track without accretion on the H-R diagram. This stage is very similar to the well-known Hayashi track for ordinary stars. Contracting along the track, protostellar core heats up to $3 \; 10^2$K at the end of the stage (~50% of its total lifetime).

2. The radiative stage. It corresponds to the ascending part of the evolutionary track without accretion. When the convective zone is decreasing, the radiative transfer of energy becomes dominating and the core contracts more slowly, reaching at the end of this stage temperatures about 10^3K (~45% of the total lifetime). It is easy to see that this stage looks also like the ordinary radiative pre-main-sequence stage.

3. The stage of dust grain evaporation. This is the relatively short stage at the end portions of the tracks (~5% of the core's lifetime). Evaporation of grains leads to the more rapid contraction of central parts, which in turn makes dissociation of molecular hydrogen faster. Adiabatic index recomes less than 4/3 in the progressively larger central zone, and at the end of this stage the central parts of the core become dynamically unstable. From the computational point of view, our program could not find solution when the central 30% of core's mass had $\gamma < 4/3$, apparently because of the hydrostatic equilibrium break down. A dynamical collapse should follow, forming the second, ionized stellar core.

Some evolutionary sequences with variations of the main parameters are also shown on Fig. 3. These include runs with $\alpha = 0.1$, and with increased opacity (by the factor 3). The evolutionary tracks show dependence on the parameters which are rather uncertain.

Of course, we cannot compare the calculations of the dust protostellar core's evolution with observations immediately. But it is interesting to compare our H-R diagram with the results of Stahler *et al.* (1980a). Their calculations started after collapse of the first core when the second ionized core has already formed and the flow has become stationary. We can compare location of the dust photosphere from

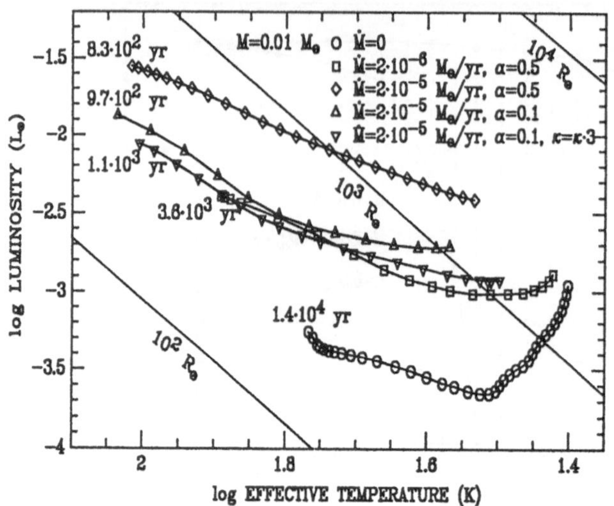

Figure 3. H-R diagram for the first protostellar core. At the end points of the tracks the total evolutionary times for each sequence are shown.

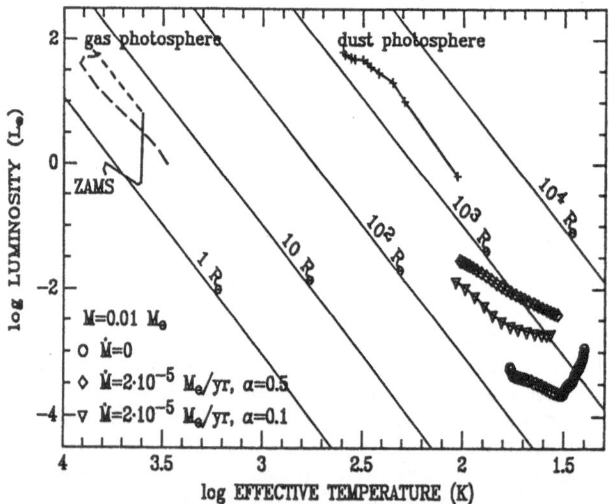

Figure 4. Comparison of the calculations made by Stahler *et al.* (1980a) with the present results. The curve labelled as the "gas photosphere" is the pre-main sequence track of the second protostellar core (initially not observable). After most of the dust envelope has been accreted, it appears at the Hayashi track, as an optically visible star.

their calculations and the photosphere of the first protostellar core on the H-R diagram showing both their and our results (Fig.4). One can conclude that there is a reasonably good agreement. The radius of the first protostellar core which has just formed is comparable with that of the dust photosphere at the main accretion stage, as it should be. At the end of the first protostellar core's life, when its central parts have collapsed in a very short timescale, the resulting luminosity produced in the shock front around the ionized stellar core is $\sim 1L_{\odot}$. In other words, there should be a quick transition from the end point of the protostellar core's track to the first point of the track of the dust photosphere.

4. Conclusions

We have studied the structure and evolution of the first protostellar core appearing at early stages of star formation using numerical stellar evolution program. One can construct hydrostatic models of the core and follow its evolution up to the end phase of increasing dynamical instability in its central parts leading to the second collapse, after which the second protostellar core forms and the main accretion phase begins. Subsequent stages of single protostar evolution calculated by Stahler *et al.* (1980a), are probably only weekly dependent on the properties of the first protostellar core. But the present results extend in some respects their calculations and show how evolve such hydrostatic configurations and may be useful when considering binary star formation.

References

Hayashi, C., Hoshi, R.,Sugimoto, D. (1962) 'Evolution of the stars', *Progr. Theor. Phys. Suppl.* **22**, 1-183.

Lang, K.R. (1978) *Astrophysical Formulae*, Springer-Verlag, Berlin Heidelberg New York, p.148

Larson, R.B. (1969) 'Numerical calculations of the dynamics of a collapsing proto--stars', *Mon.Not.Roy.Astr.Soc.* **145**, 271-295

Paczynski, B. (1969) 'Envelopes of red supergiants', *Acta Astr.* **19**, 1-22

Stahler, S.W., Shu, F.H., Taam, R.E. (1980a) 'The evolution of protostars. I. Global formulation and results', *Astrophys. J.* **241**, 637-654

Stahler, S.W., Shu, F.H., Taam, R.E. (1980b) 'The evolution of protostars. II. The hydrostatic core', *Astrophys. J.* **242**, 226-241

Stahler, S.W., Shu, F.H., Taam, R.E. (1981) 'The evolution of protostars. III. The accretion envelope', *Astrophys. J.* **248**, 727-737

Stahler, S.W. (1983) 'The birthline for low-mass stars', *Astrophys. J.* **274**, 822-829

Tutukov, A.V. (1983) 'The formation of binaries', *Pis'ma Astron. Zh.* **9**, 160-165

Tutukov, A.V., Shustov, B.M. (1981) 'The evolution of dusty envelopes around young massive stars', *Astron. Zh.* **58**, 109-118

Winkler, K.-H.A., Newman, M.J. (1980) 'Solar-type star formation with a spherical symmetry. 1. The key role of the accretion shock', *Astrophys. J.* **236**, 201-211

Winkler, K.-H.A., Newman, M.J. (1980) 'Solar-type star formation with a spherical symmetry. 2. Effects of detailed constitutive relations', *Astrophys. J.* **238**, 311-325

Discussion

Tscharnuter: You treat the first core as a hydrostatic configuration, but dynamical calculations show that the first core undergoes oscillations of large amplitudes, before the second collapse stes in. Can you comment on that?

Men'shchikov: Previous dynamical calculations that I referred to showed that the pulsations were not large, and in the following evolution the first core has a hydrostatic configuration. So we treat the core as if it is a hydrostatic one.

Di Fazio: What type of extinction did you use for the grains?

Men'shchikov: The grains were assumed to be silicates, and their Rosseland mean opacities were used for calculations.

NUMERICAL HYDRODYNAMICS OF CLOUD IMPLOSION

R. BEDOGNI
Osservatorio Astronomico di Bologna
Bologna
Italy

P. R. WOODWARD
Minnesota Supercomputer Institute
Minneapolis, MN
USA

ABSTRACT. The implosion of a spherical cloud under the action of an external shock is investigated by means of high resolution two dimensional numerical hydrodynamics. The Parabolic Piecewise Method (PPM) has been used to simulate numerically the cloud evolution. Numerical models have been computed setting two adimensional parameters: the density contrast of the cloud with respect to the ambient medium ρ_c/ρ_a and the Mach number M of the shock, for a cloud in pressure equilibrium with the unperturbed gas. Two models are computed and discussed for the same pairs of parameters in the cases of adiabatic and isothermal approximations.

1. Introduction

The interaction of a blast wave with a cloud plays a crucial role as in the understanding of the general properties of the ISM medium as well for studying the conditions under which shock compression can induce gravitational collapse (McKee 1988). In this paper we shall focus our attention in describing some of the dynamical conditions to form dense globules by the action of the hydrodynamical instabilities. Woodward (1976,1979) performed numerical simulations, using a mixed Eulerian-Lagrangean code, to reproduce the interaction of a shock of Mach number $M \sim 2.6$ against a spherical cloud of density contrast $\chi = \rho_c/\rho_a \sim 80$, with respect to the unperturbed medium, in both adiabatic ($\gamma = 5/3$) and isothermal ($\gamma = 1$) cases. Here we show the results of the numerical models, for the same pairs of M and χ of the old computations, obtained with a new numerical technique. The PPM method

R. Capuzzo-Dolcetta et al. (eds.), Physical Processes in Fragmentation and Star Formation, 333–339.

(see Colella and Woodward 1984) provides a second order accuracy in the smooth part of the grid and, making use of the parabolic representation and steepening constrains, permits an accurate treatment of the cloud-intercloud interface. As a consequence of that the method maintains the integrity of the sharp gradients and avoids the spurious small scale oscillations, so the shocks and the constant discontinuities are very well represented. The new generation of supercomputers, like the CRAY X-MP, offers the potential of high resolution simulations for compressible gas dynamics showing in great details the behaviour of the dynamical instabilities. The outputs of the code result in forms of hundreds of compressed "dumps" to have, with a continuous displaying on local IRIS workstation, animations showing the dynamical evolution of the whole phenomena. Here we present only some of the most significant frames to partially describe the main physical characteristic of the shock-cloud interaction.

2. Numerical Models

The hydrodynamical equations are solved in cylindrical geometry using uniform square computational zones in Z and R directions, for both adiabatic (γ=5/3) and isothermal (γ=1) cases. The same γ for the unperturbed medium is chosen for the cloud too : $\gamma_a = \gamma = \gamma_c$. Solving the hydro's equations we neglected radiative terms, thermal conduction, magnetic fields and self-gravity.

The cloud is initially in pressure equilibrium p_a=p_c and at rest with respect to the ambient medium, so $v_{za} = v_{zc} = 0$. The sound velocity and density of the ambient medium are scaled to one i.e., c_a=1=ρ_a so pressure becomes p_a=1/γ. This requires that the density ratio between cloud and ambient gas ρ_c/ρ_a is equal to the factor χ. The Mach number of the incident shock is given by the ratio between the shock velocity v_{zs} and the sound velocity of the unperturbed gas c_a : $M = v_{zs}/c_a$. The shocked variables ρ_s,p_s,v_{zs} are fixed by the standard Rankine-Hugoniot conditions (see Landau and Lifshitz, 1959):

$$\rho_s = \frac{(\gamma+1)}{((\gamma-1)+2/M^2)},$$

$$p_s = \frac{M^2((2\gamma+(1-\gamma)/M^2)}{\gamma(\gamma+1)},$$

$$v_{zs} = \frac{M(1-((\gamma-1)+2/M^2)}{(\gamma+1)},$$

and v_{zs} is the normal component of the shock velocity. The R component of the velocity is equal to zero (v_r=0) for both shocked and ambient media. The subscripts a,c and s refers to the ambient, cloud and shocked quantities.

In fig. 1. we show the initial configuration before the strong shock strikes the cloud. At the top of the R-grid and at the right side of the Z-grid we set

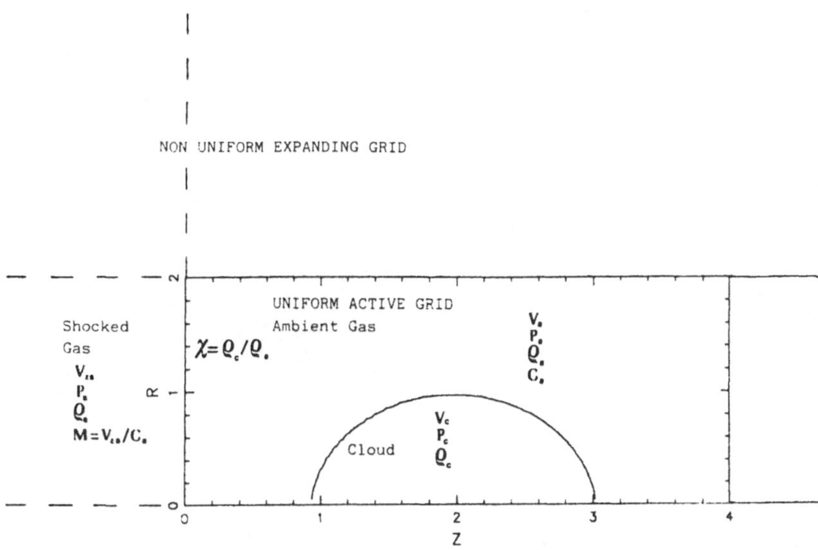

Figure 1. Initial conditions. The shock moves from Z=0=R against the spherical cloud centered in Z=2 and R=0. The active grid extends from Z=0 to Z=12 but it is shown only to Z=4.

flow-out boundary conditions (b.c.) using an exponential non-uniform expanding grid extended from -10 to 0 in Z-direction and from 2 to 20 in R-direction. At the left side, where the shock is coming from, are flow-in b.c. and, finally, at the bottom of Z-grid reflecting b.c. An axial symmetry is enforced at the Z=0 axis, so the plots appear perfectly symmetric with respect to the longitudinal direction. The numerical models are computed with the same resolution taking a grid size $\Delta_{R-Z} = 0.025$.

3. Results

3a. Adiabatic Model

This first numerical model is compututed chosing the same parameters of the old Woodward's simulation (Woodward 1976), but setting $\gamma_c = 5/3$ to get an adiabatic cloud. The simulation confirms the description of the shock-cloud interaction proposed by Heathcote and Brand (1983) and McKee (1988). In fig. 2a the external shock S_{ex} has surpassed the cloud and sits in $Z \sim 5$ while the transmitted one (in $Z \sim 2$) has crossed almost half of the compressed cloud. So the first frame represents the *shock compression phase*, using the terminology of McKee (1988). No standing bow shock forms in front of the face of the cloud because the low Mach

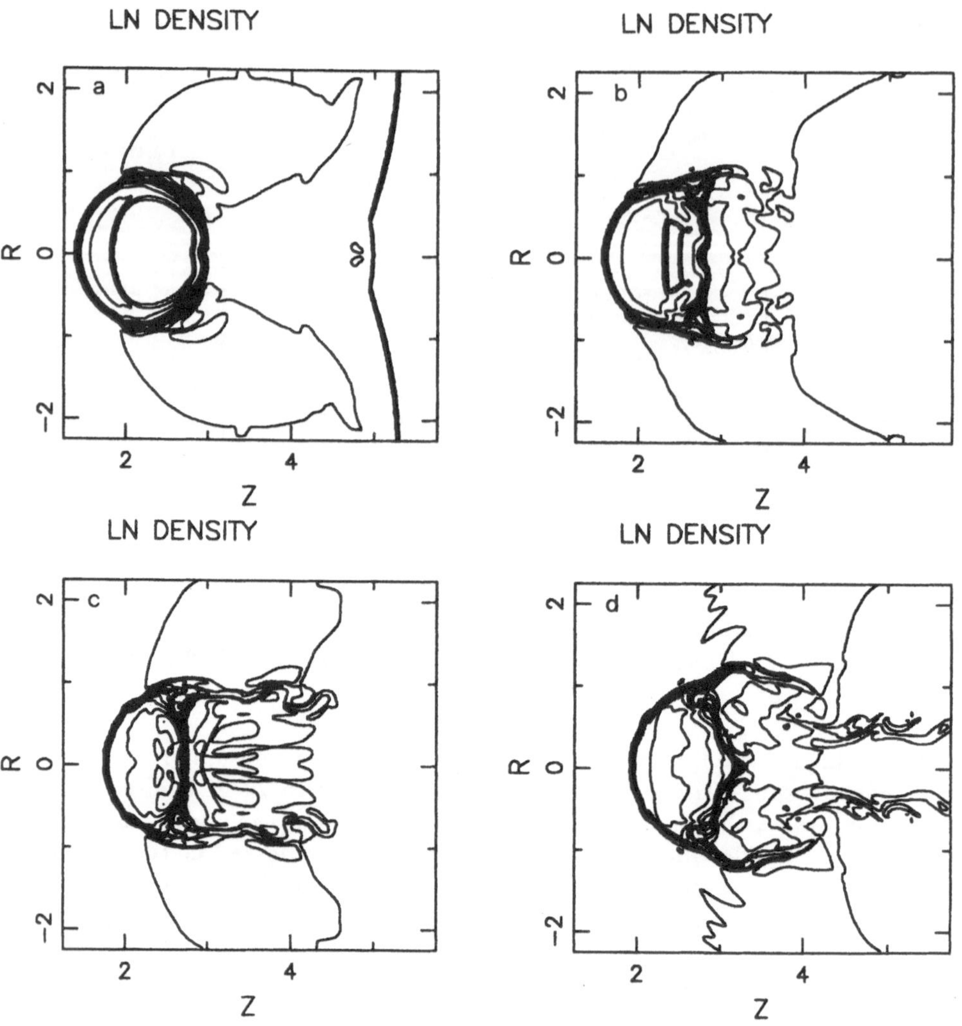

Figure 2.Close up view of lnρ case A: M=2.6, $\chi = \rho_c/rho_a = 80$ *and* $\gamma = 5/3$. *Range : 15 contours from -0.534 to 6.405; time sequence a)t=2.008, b)t=3.003, c)t=4.004, d)t=5.011*

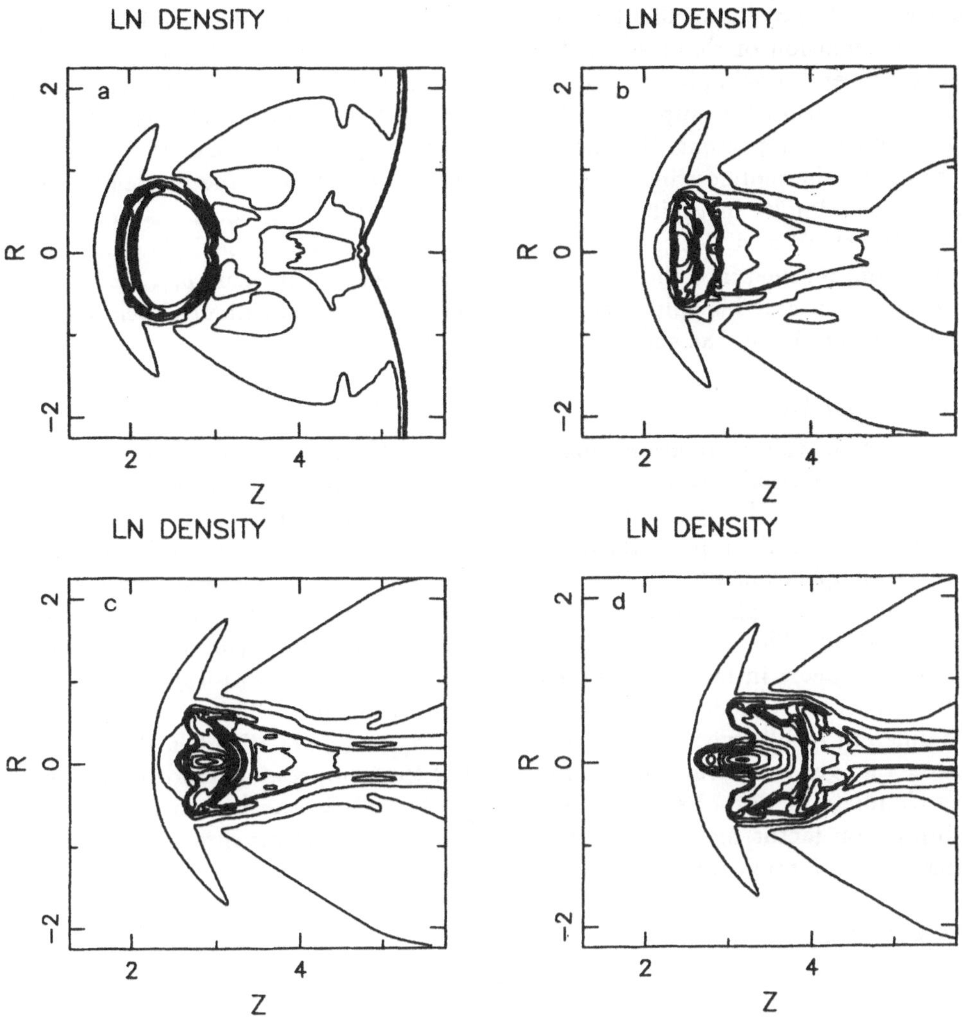

Figure 3. Close up view of $\ln\rho$ case B: M=2.6, $\chi = \rho_c/rho_a = 80$ and $\gamma = 1$. Range : 15 contours from -0.431 to 9.988; the same time sequence of Case A.

number (< 2.76) of the incident shock S_{ex}. In fig. 2b we can see the situation just before the collision of cloud shock S_{tr} with the shock reflected S_{re} from the back of the cloud and moving in the opposite direction of the previous one. Also lateral compressions, from the upper and lower sides, of the cloud are clearly visible; so the second frame represents the end of the *shock compression phase*. In figs 2c. and 2d are shown the plots during the *re-expansion phase* in which the Rayleigh-Taylor and Kelvin-Helmoltz instabilities tends to develop tongues of dense material at the sides of the cloud and to segment the front face of the cloud itself. The maximum compression, reached into the shocked cloud, is given by the value $\rho_{max}/\rho_c \sim 7.56$. The cloud maintains his spherical shape because the weak incident shock does not blow away completely the shocked cloud.

3b. Isothermal Model

The outputs of this simulation look like those of the isothermal model of Woodward (1979) and similar results are obtained by Rozyczka and Tenorio-Tagle (1986) including in the hydrodynamical equations the cooling term, not present in our code. Fig. 3a clearly represents the same situation of fig. 2a at the beginning of the *shock compression phase*. The *re-expansion phase* starts immediately after the end of the previous phase (fig. 3b) and it is well represented by the last two plots figs. 3c and 3d. The cloud is partially flattened toward the symmetry axis and a bow shock appears in front of the dishomogeneity. The Rayleigh-Taylor instability results apparent from the distortion of the cloud front face. This simulation gives quite different results with respect to the old computation of Woodward (1979) but we need a longer temporal evolution to check if the compressed cloud fragments in many cloudlets or not. Because the incident shock is isothermal arises naturally a compression factor higher than the adiabatic case. The maximum compression during this simulation results $\rho_{max}/\rho_c \sim 272$.

4. Conclusion

The numerical models show a compression in the shocked cloud that, in both cases, seems to favour a gravitational collapse under the action of self-gravity. The role of the magnetic field, as demonstrated by Nittman (1981), would not limit the density reached in the shocked cloud but reducing the shear which led to the Kelvin-Helmoltz instability. Further investigations are required to check the role of the dissipative terms like thermal conduction and cooling effects.

References

Bedogni, R, Woodward, P.R.:1989, *Astron. Astrophys.* (submitted)
Colella, P., Woodward, P.R.:1984, *J. Computational Physics* **54**, 174.

67
Landau, L.D, Lifshitz E.M.:1959,*Fluid Mechanics*, Pergamon, New York
McKee, C.F.: 1988, in *Supernova Remnants and the Interstellar Medium*, IAU Coll. **101**, eds R.S. Roger and T.L. Landecker, Cambridge University Press, p. 205.
Nittman, J., 1981, *Monthly Notices Roy. Astron. Soc* **197**, 699
Rozyczka, M., Tenorio-Tagle, G.:1987, *Astron. Astrophys.* **176**, 329
Woodward, P.R.: 1976, *Ap.J.* **207**, 484.
Woodward, P.R.: 1979, In IAU Symp 84, *The Large Scale Structure of the Galaxy*, ed. W.B. Burton (Dordrecht: Reidel), p. 159.

DISCUSSION

BOSS: Your simulation shows, very nicely, that initially a significant cloud compression occurs because of the shock passage. However, toward the end of the simulation the compressed regions appeared to be blow apart. If self-gravity had been include could these compressed cloudlets have collapsed to form stars, or might they be disrupted first?

BEDOGNI: The cases presented here do not show that the shocked cloud is disrupted in many cloudlets while other computations, with different values of M and χ (Bedogni and Woodward 1989), clearly suggest the cloud is blow apart.

SHAPIRO:How do your results differ from the earlier simulations by Woodward in 1976?

BEDOGNI: The isothermal models results quite different with respect to the old simulation while a direct comparison with the adiabatic one is not directly possible because the γ_c is not the same.

Di FAZIO: Did you attempt evaluating the supersonic resistivity coefficient? This being $c(v) = F/(\rho\ S\ V^2)$, where F is the total force exerted by the shock, S is the cross section offered by the cloud, and V is the relative cloud-shock velocity. It is useful to compute $c(v)$ in your computation of supersonic turbulent wakes where the situation is spectacular; the fragment moves supersonically in the gas, which has inertial-system velocity somewhat smaller than the fragments.

BEDOGNI: We are doing that selecting, for this particular purpose, more than two computational models in terms of the pairs of the critical adimensional number M and χ.

4. THE INITIAL MASS FUNCTION.

A REVIEW ON IMF THEORIES

R. CAYREL
Observatoire de Paris
61, avenue del'Observatoire
F-75014 PARIS, France

ABSTRACT. The theories of the IMF are reviewed, starting from those essentially based on stochastic arguments, to those involving more Physics. It is concluded that until there is a fairly well accepted theory for star formation itself, theories of the IMF will remain very fragile.

1. Introduction

The oldest and maybe the safest approach to a theory of the initial mass-function (IMF in short) is to base such a theory on the recognition of our ignorance. Ignorance can lead to definite statistical predictions if it can be shown that the mass function results from elementary stochastic processes. These models are the first ones that we shall survey in this talk.

Other theories derive from simplified models for the process of star formation. But the process of star formation, as it will certainly appear in the rest of this meeting, is a tremendously difficult problem, which is very far from being amenable to a quantitative description. As a consequence, each theory has a merit of its own, but none can pretend to give a full answer to the subject. Hopefully the rest of this conference will bring more light on this difficult but crucial problem.

2. Hierarchical and related stochastic theories

2.1 AULUCK AND KOTHARI SCHEME

One of the most elementary "theory" of the mass distribution resulting from the fragmentation of a cloud of interstellar matter (ISM) is due to Auluck and Kothari (1960,1965). Let first define the so-called IMF. If newly born stars are formed in swarms and if m_1, m_2, ... m_n are the masses of the objects in a swarm ,the number of objects with masses between m and $m+dm$ should be called the initial mass function. But the tradition is that it is the number of objects with a logarithmic mass between $\log m$ and $\log m + d\log m$ which is called the mass function.

R. Capuzzo-Dolcetta et al. (eds.), Physical Processes in Fragmentation and Star Formation, 343–355.

Therefore one has:

$$dN = \Phi(\log m)\ d \log m$$

The normalization of $\Phi(\log m)$ is a matter of taste. One choice is to normalize in such a way that the integral of the IMF on all masses is equal to one. The IMF gives then the probability for a star of having a mass between two limits. However some prefer to normalize Φ at one solar mass, especially to compare several IMFs, because it is a well defined quantity, not depending on a large number of stars of small masses very poorly known (brown dwarfs!).
The IMF has another physical meaning which is worth mentioning here.
The fraction $d\mu$ of the total mass M_{tot} which has gone into pieces of masses between m and $m+dm$ is:

$$d\mu = mdN/M_{tot} = 0.43429 m\Phi dm/mM_{tot} = 0.43429\Phi dm/M_{tot}$$

Therefore Φ does describe also ,with proper normalisation, the *mass* distribution, and not the *number distribution* ,but now with a *linear* scale on m.

Auluck and Kothari have obtained the mass distribution generated by a stochastic process consisting in subdividing randomly each edge of a cube (with a uniform probability law) and finding the volume (or mass) distribution of the pieces. The problem has an analytic solution in one dimension but is far from trivial in 3D. Only an asymptotic law for large pieces is simple. The fraction of pieces with masses m' larger than m (m being itself much larger than the average mass m_0 of the pieces), is given by:

$$f(m'>m) \cong \frac{2\pi}{\sqrt{3}}\left(\frac{m}{m_0}\right)^{\frac{1}{3}} exp\ \left\{-3\left(\frac{m}{m_0}\right)^{\frac{1}{3}}\right\}$$

Kushwaha and Kothari (1960) have compared their distribution with the mass-function of the Hyades, and found a rough agreement, taking m_0 equal to one tenth of a solar mass.

Nobody can seriously consider this as a "physical theory" of the initial mass function. But it illustrates the point that starting with a simple uniform one-dimensional law one can get something resembling the actual mass function.

2.2 HIERARCHICAL THEORIES

Hoyle (1953) can be considered as the father of hierarchical theories. The basic assumption is that fragmentation does not occur in a single step, but in several successive steps. If each step contains some amount of randomness, then, after a large enough number of steps, some definite statistical properties appear in the mass distribution. Let us first illustrate this by an example borrowed to Larson(1973). The assumptions are that a cloud of mass M suffer a succession of *binary* fragmentations with, each time, an equal probability for the fragment of subdividing into two equal pieces, or not fragmenting at all. After p steps one gets pieces of masses $2^{-q}M$, with probabilities to find the original mass under the form of such masses as shown by table 1. Very clearly the probability after p steps of getting the mass as $2^{-q}M$ fragments is the binomial law:

$$P_p = 2^{-p}\frac{p!}{q!(p-q)!}$$

It is well known that the binomial coefficients ,near their maximum , are approximatively described by a normal law with a standard deviation $\frac{\sqrt{p}}{2}$ and a maximum at $2^{-\frac{p}{2}}M$. This is a *log-normal law* ,the abcissae being equally distributed on a logarithmic scale.

Table 1

p	M	$\frac{M}{2}$	$\frac{M}{4}$	$\frac{M}{8}$	$\frac{M}{16}$
0	1				
1	$\frac{1}{2}$	$\frac{1}{2}$			
2	$\frac{1}{4}$	$\frac{1}{2}$	$\frac{1}{4}$		
3	$\frac{1}{8}$	$\frac{3}{8}$	$\frac{3}{8}$	$\frac{1}{8}$	
4	$\frac{1}{16}$	$\frac{4}{16}$	$\frac{6}{16}$	$\frac{4}{16}$	$\frac{1}{16}$

The value of p cannot be chosen arbitrarily. Because $2^{-\frac{p}{2}}M$ is the most frequent stellar mass, the numerical factor $2^{-\frac{p}{2}}$ must be of the order of the ratio between a stellar mass and the typical mass of a cloud of

ISM. This leads to a value of p near 20 and a dispersion of 2.2 in a logarithmic scale of base 2, or 0.67 in a log scale of base 10. This is after all not bad in comparison of the Miller and Scalo (1969) empirical log normal law, which has a σ of 0.68!

The log-normal law can be defended on more general arguments. Elmegreen (1985) or Elmegreen and Mathieu (1983) and Zinnecker (1984) have studied the problem in great detail.
We shall only remind their most general argument . If a final stellar mass is the result of a multiplicative process (as successive fragmentations), then the stellar mass is a product of random variables:

$$m=f_1 f_2 \cdots \quad f_n$$

and the log of the mass is a *sum* of random variables. According to the central limit theorem, such a sum obeys a normal law if n is great enough. In practice, the quoted works have shown that "great enough" means about 5. Elmegreen (1985) has shown that power laws can also be sometimes obtained by statistical models.

We shall stop here, noting that there are theoretical arguments in favour of log-normal laws, but sending the reader to the references for a deeper insight of these statistical approaches.

3.Opacity limited fragmentation

3.1 THE OPACITY LIMIT

Silk (1977a) has developped the concept of opacity limited fragmentation . The Jeans mass M_J:

$$M_J \sim \rho^{-1/2} T^{3/2}$$

actually decreases during a gravitational collapse if the temperature T remains constant, as the density ρ increases. But if the cloud, or a fragment becomes opaque to its own radiation it cannot contract isothermally anymore and if it contracts adiabatically (never completely the case), the temperature rises as $\rho^{\gamma-1}$ and the Jeans mass goes up if $\gamma > \frac{4}{3}$. Interestingly, this value is just below the value of γ for a mono-atomic gas (5/3) and for a di-atomic gas (7/5). . A more detailed study of the thermodynamics of the contracting mixture of gas and dust leads, according to Silk, to a effective $\gamma \cong 1.6$ inhibiting further opportunity to fragment for an optically thick fragment.
For a gas with the standard proportion of dust, the opacity limit is set by the absorption of the dust grains. The two conditions:

a) τ(*optical depth*) = 1 over a Jeans length;

b) radiative cooling rate= rate of energy acquisition by the compressional work in the collapse;

actually set the fragment mass, temperature and gas density at which the minimum mass is reached. Computations based on the infrared optical properties of silicate interstellar grains show that $M_{min} \cong 0.01\mathfrak{M}_{\odot}$, $T \cong 10$ *to* 40 K , $\rho \cong 10^{-13}$ *cgs*. This limit is only weakly dependent on the chemical composition of the medium.

3.2 THE PREDICTED IMF

In a subsequent paper (Silk 1977b) Silk has been able to study the feedback of the already formed protostellar fragments on the course of further fragmentation. Basically the feedback is caused by the absorption of radiation coming from the already formed protostars by the remaining gas. A very simplified theory leads to a power-law mass function with an index depending upon the luminosity law of the protostars and of the temperature dependence of the grain opacity.
With:

$$L \sim m^{\eta}$$

and:

$$\kappa \sim T^{\delta}$$

where κ is the grain opacity, one gets:

$$dN \sim m^{-(1+x)} dm \quad \text{or} \quad dN \sim m^{-x} d\log m$$

with:

$$x = \eta - 1 - \frac{2}{2+\delta}$$

The numerical value of x for $\eta=3$ and $\delta=2$ is 1.5 fairly close to the empirical Salpeter value of 1.35.

4. Coalescence Theories

Although coalescence may be more relevant to the formation of planets and giant molecular clouds than to stellar formation, it has been studied and computed in the frame of numerical models. Several references exist on the subject. A good recent one is Pumphrey and Scalo (1983). Former ones by Nakano (1966), Arny and Weissman (1973), Silk and Takahashi (1979) are important contributions to the field. The basic equation is an integro-differential equation expressing that the number of clouds (or protostars) per unit volume of mass m has a 'sink' term which are the encounters with coagulation with other clouds of mass m', and a source term which are the creations of clouds of mass m, by coagulation of two clouds of masses m'' and $m-m''$. A critical parameter is the rate of coagulating encounters $\alpha(m,m')$. One assumes this rate to be of the form:

$$\alpha = n_1 n_2 v_{12} \sigma_{12}$$

where the notations are obvious ,and where the cross-section σ_{12} is usually of the form :

$$\sigma_{12}= \pi(r_1+r_2)^2$$

These models predict IMFs of the type:

$$n(m) \sim m^{-\beta}exp(-am)$$

However the coalescence process works never alone (for example contraction of the fragments quickly modify the cross-sections) and we shall postpone the study of this process to the next section which combines coalescence and opacity-limited fragmentation ,just the opposite process.

5. Fragmentation and coalescence

The competitive effects of opacity-limited fragmentation and coalescence have been investigated by Yoshii and Saio (1985). They take into account the heating of the gas by protostars, the fast decrease of coalescence cross-section on a Helmotz-Kelvin time-scale and the condition that the local Jeans length must be smaller than the distance between preexisting protostars.They assume also that the protostars retain the virial velocity of the gas from which they formed.Their conclusions can be summarized as follows:

a) the coalescence process is very unefficient in most practical cases and can be neglected. A caveat however: if stars are born with the high densities found by Herbig and Terndrup (1986) in the Trapezium cluster of the Orion nebula, this conclusion may not hold.

b)the IMF is, between a minimum and a maximum mass, a power law with an index controlled by the mass-luminosity relation of the protostars, as found by Silk.

there is a minimum mass independent of the size and mass of the cloud of the order of $0.01\mathfrak{M}_\odot$ as already found by Silk.

c) there is a maximum mass critically depending on the size of the cloud (for a given mass of the cloud).

The first stars form always with the minimum mass. Then the gas is heated by the first protostars and the Jeans mass increases. This is the beginning of building up a mass-function. The process can stop in two ways: either the gas get exhausted, or the Jeans length become larger than the distance between already formed protostars.For the standard case considered by the authors (M_{cloud}=$10^5\mathfrak{M}_\odot$, R_{cloud}=7 pc , and η =3, see above section for the definition of η) the maximum mass is about 20 $\mathfrak{M}_\odot$ and the gas is not exhausted (limit set by the existence of protostars within a Jeans length).The slope of the IMF is set by the value of η, and is corresponding to an index 1.5 as compared to 1.35 for the Salpeter law. For a less dense cloud (R=14 pc) the result is strikingly different: the feedback from the protostar radiation is so less efficient that all protostars have virtually all masses very near the minimum mass when the gas gets exhausted. Varying η gives IMF slopes very far from acceptable values. For example η=1.5 gives a totally flat IMF, truncated at about 50$\mathfrak{M}_{\odot}$. In all these cases coagulation is not contributing to the shape of the IMF.

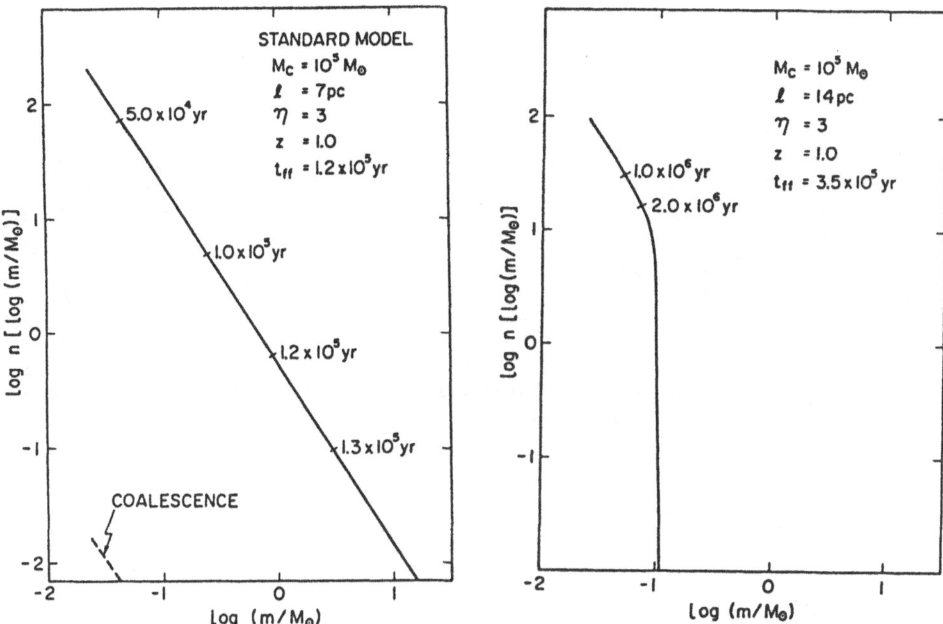

Figure 1. IMF resulting from fragmentation and coalescence (from Yoshii and Saio 1985).

Figure 2. Same as fig. 1 but for a larger cloud of the same mass. Note the drastic change in M_{max}.

6. Di Fazio's theory

Very different from the former theories is the one developped by Di Fazio (1986). Again an interstellar cloud of density ρ and temperature T is considered as subject to Jeans instability. But the spectrum of perturbations is taken to be given by:

$$dN=4\pi k^2\, dk/\tau(k)$$

where k is the wave-number, and $\tau(k)$ is the time-scale of growth of perturbations of wave-number k. This last quantity can be readily obtained from the dispersion relation:

$$\omega^2 = k^2 c_s^2 - 4\pi G\rho \quad (\ c_s\text{=speed of sound, G= Newton constant})$$

and is given by:

$$\tau(k) = (\pi/G)^{1/2}\rho^{-1/2}/(1-(M/M_J)^{2/3})^{1/2}$$

where M_J is the Jeans mass.This leads to the 'instantaneous' mass function:

$$dN \sim (M/M_J)^{-2}\sqrt{1-(M/M_J)^{-2/3}}$$

This mass function does evolve with time because the density of the cloud decreases as fragments condense out of the gas.
Di Fazio follows the dynamical evolution of the cloud which produces fragments with an evolving mass-function of the type described. The end-product is a mass-funtion of a slightly different form , but very similar qualitatively.The process stops when the density of the remaining gas is low enough to bring M_J up to the total mass of the remaining gas. Nothing prevents each fragment to fragment in turn, and this leads to the interesting following scenario: starting from a $10^{11}\mathfrak{M}_\odot$ mass one gets fragments of typical masses $10^8\mathfrak{M}_\odot$(first generation of massive clouds). Each of these clouds do fragment into $10^5\mathfrak{M}_\odot$pieces, which are considered as protoglobular clusters. Each protoglobular cluster does fragment into 100 $\mathfrak{M}_\odot$pieces which are identified with the first generation of population III stars.

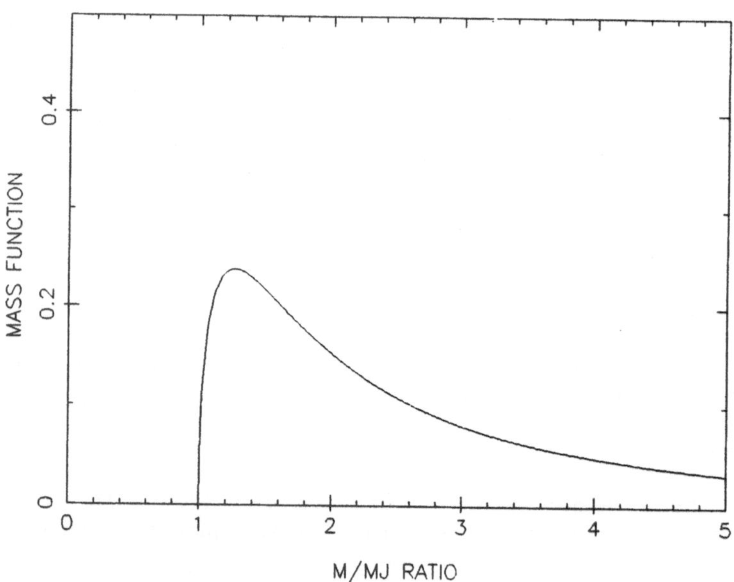

Figure 3. Instantaneous IMF in Di Fazio's approach. The maximum of the IMF occurs at M/M_J= 1.26.

Afterwards the injection of heavy elements in the gas brings the next generation down to a tenth of a solar mass. So we are almost back to a hierarchical scheme, but with some more physical ground this time. Refinements to the 1986 paper can be found in Di Fazio 1987.

7. The big question

All the theories exposed so far rely on the assumption that a cloud suffering a gravitational collapse *does* fragment down to the minimum local Jeans mass as this last one decreases during the collapse.However there are serious doubts that this is true. Layzer (1963) made this point very clearly. From the formula giving $\tau(k)$ in the former section it clear that the time scale of growth of subcondensations is increasing for smaller scales and reaches infinity at Jeans mass. Because of the non-linear behaviour of gravitational collapse one can think that the fastest mode is taking over the slower modes, and as the fastest mode is for the full size of the cloud, the whole idea of fragmentation is shaken. The problem has been recently investigated by Blottiau and al.(1988).Whereas an initially uniform sphere evolves as shown by fig.4 ,with a central uniform core of increasing density, and a shell with a power law outwards decreasing density, a perturbed sphere may develop a substructure, or may not. When it does, it is at a so advanced stage of the collapse that the Layzer argument seems very serious. So attempts to find other characteristic scales in a gravitational collapse have been looked for.

8. Fragmentation from sheets and filaments: bimodal star formation

The observation of star formation regions in the sky suggests that the ISM is more or less organized in sheets or filaments (Silk 1986).
If so, this introduces new length-scales of interest. Following Larson (1985) we shall consider the dispersion equation for a thin sheet of surface density μ.This equation is:

$$\omega^2 = \gamma c_s^2 k^2 - 2\pi G\mu k$$

The occurrence of k in the last term introduces a very important change in the growth rate variation with k. The critical value corresponding to the Jeans mass (ω=0) is:

$$k_c = 2\pi G\mu/\gamma c^2$$

and the growth rate:

$$1/\tau(k) = i\omega = \sqrt{2\pi G\mu k(1-k/k_c)}$$

has a maximum at $k=k_c/2$.
So there is now a preferred scale equal to twice the Jeans-scale. The corresponding mass:

$$M_{2c} = 4.67\gamma^2 c_s^4/G^2\mu$$

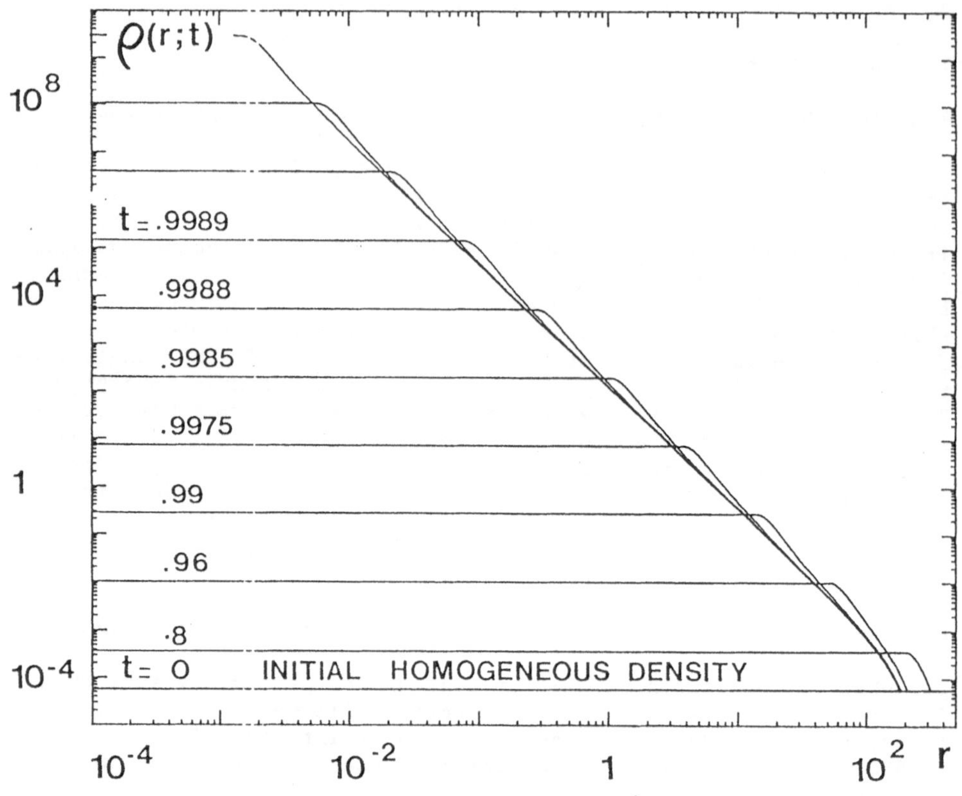

Figure 4. Evolution of density profiles during a gravitational collapse (according to Blottiau and al. 1988). The sphere is initially at rest and homogeneous.The equation of state is a polytropic one, with a power γ = 1.2.

For a typical region of star formation like the Taurus cloud, the temperature is about 10 K, and $\mu \cong 160\ \mathfrak{M}_\odot pc^{-2}$, the mean molecular weight is 2.3 and one has:

$$M_{2c} \cong 2.0\ \mathfrak{M}_\odot$$

The fit with the most frequent stellar masses in the Taurus region ($0.6\mathfrak{M}_\odot$) is not too bad. More formulae are available in Larson (1985) for thick sheets in gravitational equilibrium, for filaments and for rotating or magnetic discs.

An interesting feature of this approach is that stellar masses are strongly dependent upon temperature via the c_s^4 factor. In a molecular cloud near a typical H II region μ is more like $300\mathfrak{M}_\odot pc^{-2}$ and $T \simeq 40$ K. This leads to:

$$M_{2c} \cong 25\ \mathfrak{M}_\odot$$

This is the basis for supporting the bimodal stellar formation scenario. Cold clouds form stars of small masses ,like those seen in T-associations, and hot clouds produce stars of large masses, like those observed in O-associations. However it is not clear that O-associations contain only large- mass stars (see, for example Walker 1954).

In conclusion, a mechanism is proposed here, which links stellar masses to properties of the gas ,itself in the form of sheets or filaments , favouring a definite mass scale, function of gas temperature and surface (or linear) densities, the temperature factor being predominant.

9.What determines stellar masses?

So far we have implicitely admitted that stellar masses were determined by a fragmentation process involving mostly the properties of the molecular cloud out of which the stars have formed. But as noticed by Shu,Adams and Lizano, in a remarkable review paper (1987), such an approach will never explain why most of the mass found in self-gravitating bodies is just in the range where these bodies become nuclear-burners. There is little dispute that the maximum stellar masses are not determined by the formation process but by the physics of the body itself . A star cannot be more massive than about 60 $\mathfrak{M}_\odot$ because its surface layers are blown out by its *own* radiation. Why its minimum mass should not be looked for in the fact that a self gravitating body becomes a nuclear burner ? There are indeed arguments coming from theoretical computations (Larson 1978) that cores form by gravitational collapse with a mass near the opacity limit (near $0.01\mathfrak{M}_\odot$) and that these cores grow by accretion to much larger masses, in the range of observed stellar masses (unfortunately very little is known about the frequency of masses in the brown dwarf range because of selection effects in the study of binary stars). Then the problem is to know why a core ceases to grow at some mass. The answer by Shu and al.

(1987) is that the core ceases to grow because once the star has a mass large enough to become a nuclear burner it also generates a wind reversing the accretion process. In order to generate a mass spectrum (and not a universal stellar mass) it is of course necessary to involve some random variables which can be initial rotation, magnetic fields, and the like. Although this approach is quite appealing it must be noted that a few counter-arguments have been developped (Zinnecker 1989).What does actually occur in Nature will be soon found out by the observation of star forming regions at higher and higher spatial resolution.We shall not further develop this interesting concept because is has not predicted a quantitative IMF so far.

10. Conclusion

At this point we should summarize the status of "theories of the IMF". Theories of the IMF are numerous. Several do contain a part of truth, but as long as there is not a well accepted theory for star formation a theory of the IMF is necessarily very precarious.We have stood away in this talk from star formation processes involving angular rotation and magnetic fields ,although they seem unavoidable ingredients. The reason is that this type of work rarely goes as far as a prediction of the IMF. It must be however admitted that they are a necessary step in the prediction of the IMF. It is our feeling that a few concepts have a good chance to survive:

-The general statistical arguments supporting log-normal laws, or power-laws

-The opacity-limited fragmentation concept

-The concept that the IMF is not a pure product of cloud fragmentation processes but does depend also of internal properties of the object itself.

Hopefully the rest of the conference will shed more light on this difficult subject!

REFERENCES

Auluck, F.C. and Kothari, D.S. 1960, Nature,**174**, 565
Auluck, F.C. and Kothari, D.S. 1965, Z.f.Astrophys. **63**, 9
Arny, T. and Weissman, P. 1973, Astron.J. **78**, 309
Blottiau, P., Bouquet, S., Chièze, J.P. 1988, Astron.Astrophys. **207**, 24
Di Fazio, A. 1986, Astron.Astrophys. **159**, 49
Di Fazio, A. 1987, in 'Starbursts and Galaxy Evolution',eds.Trinh Xuan Thuan,Montmerle, Tran Than Van, éditions Frontières,Gif sur Yvette, France
Elmegreen, B.G. 1985, in 'Birth and Infancy of stars',eds. Lucas, Omont and Stora, Les Houches session XLI, Elsevier Science Publishers,Amsterdam,p.257

Elmegreen B.G. and Mathieu R.D. 1983,Mon.Not.Roy.Astron.Soc.**203**, 305
Herbig G.H. and Terndrup D.M. 1986, Astrophys.J. **307**, 609
Hoyle, F. 1953, Astrophys.J. **118**, 513
Kushwaha, R.S. and Kothari, D.S. 1960, Z.f.Astrophys. **51**, 11
Larson R.B. 1973, Mon.Not.Roy.Astron.Soc. **161**, 133
Larson, R.B. 1985, Mon.Not.Roy.Astron.Soc. **214**, 379
Larson, R.B. in IAU Colloquium No. 120,'Structure and Dynamics of the Interstellar Medium', Granada,Spain April 1989, eds. Moles, Tenorio-Tagle, Melnick, Springer-Verlag,Berlin
Layzer, D. 1963, Astrophys.J. **137**,351
Miller, G.E. and Scalo, J.M.1979, Astrophys.J.Supp. **41**, 513
Nakano, T. 1966, Prog.Theo.Phys. **36**, 616
Pumphrey, W.A. and Scalo, J.M. 1983, Astrophys.J. **269**, 531
Silk, J. 1977a, Astrophys.J. **214**, 152
Silk, J. 1977b, Astrophys.J. **214**, 718
Silk J. 1986, in 'Luminous Stars and Associations in Galaxies',eds. De Loore and al.,p.301
Silk, J. and Takahashi, T. 1979, **229**, 242
Shu, F.H.,Adams, F.C., Lizano, S. 1987, Ann.Rev.Astron.Astrôhys. **25**, 23
Yoshii, Y. and Saio, H.,1985, Astrophys.J. **295**, 521
Zinnecker, H. 1984, Mon.Not.Roy.Asron.Soc. **210**, 43
Zinnecker, H. 1989, in 'Evolutionary Phenomena in Galaxies' ed. J.Beckman,Cambridge University Press

DISCUSSION

R.B. LARSON (comment) : You placed some emphasis on stochastic fragmentation and log-normal mass spectra, but I am not sure that present data provide much support for the log-normal shape. The most recent analysis by Scalo does not show an increasing slope of the IMF with increasing mass, but, if anything, the slope seems to decrease.

Fragmentation theories and the IMF

Federico Ferrini
Istituto di Astronomia, P.zza Torricelli 2, 56100 Pisa, Italy
E.S.O., Karl–Schwarzschild–Str. 2, D–8046 Garching, W.Germany
Francesco Palla
Osservatorio Astrofisico di Arcetri, L.E.Fermi 5, 50125 Firenze, Italy
Umberto Penco
Istituto di Astronomia, P.zza Torricelli 2, 56100 Pisa, Italy

ABSTRACT. We investigate the problem of the origin of the Initial Mass Function for field stars. The IMF, computed following the nonlinear approach by Ferrini, Marchesoni and Vulpiani (1983), has been extended to take into account the simultaneous presence of several instability criteria, applied to the fragmentation of molecular clouds. The two basic properties of the observed IMF are reproduced: the region of the peak and the power law that describes the behaviour of the IMF for masses larger than the characteristic mass. Small scale features are found in some of the computed IMFs, in the form of small departures from a smooth, monotonic behaviour.

1. Introduction

The understanding of the fragmentation process of molecular clouds leading to star formation and the origin of the IMF still remains largely incomplete (cf. Zinnecker, 1987, for a recent review). Present observational evidence on the shape of the IMF is convincingly clear only for the high mass end of the distribution function, that shows the characteristic power–law form originally discussed by Salpeter (1955). A deviation from a steep slope is indicated at smaller masses, but the controversy remains as to whether the mass spectrum at the lowest end stays constant or turns over (Reid, 1987; Hawkins, 1986). The reality of other features, like the apparent peaks at 1.2 and 3 $M_{\odot}$ and the consequent bimodal shape of the IMF, is also an open question (e.g., Güsten and Mezger, 1982; Scalo, 1987). The far reaching implications of such properties on models of galactic evolution have been fully analyzed by various authors (Larson, 1986; Wyse and Silk, 1987). However, in a recent study Rana (1987) has shown how the derivation of the mass function of

R. Capuzzo-Dolcetta et al. (eds.), Physical Processes in Fragmentation and Star Formation, 357–366.

stars in the solar neighborhood, by using new data on the luminosity function and scale heights of main sequence stars, fits well the original idea of an IMF described by a simple power law, without necessarily invoking bimodality. The feature around $\log(m/M_{\odot}) \simeq 0$ that has been interpreted as suggestive of a second mode of star formation, according to Rana, is just one of several humps in the distribution, that occur at various masses.

It has been suggested that the strong temperature dependence of the Jeans mass naturally tends to produce a double–peaked IMF in regions of different temperatures (cf. Larson, 1987; Silk, 1987); however, the same high sensitivity of the Jeans mass combined with the spread of observed cloud temperatures will prevent any clear–cut bifurcation into two well defined modes. That is, any critical mass where the transition between the two modes might occur will be wiped out, and in that case the resulting shape of the IMF would result from the superposition of many individual IMFs, each peaked to a different value of the characteristic mass. It will be more correct, then, to think in terms of multimodality rather than bimodality: it is the purpose of this paper to test quantitatively this conclusion, in the light of current theories of fragmentation.

In particular, we will follow the scheme suggested by Ferrini, Marchesoni and Vulpiani (1983, hereinafter referred to as FMV) who treated the problem of the fragmentation of molecular clouds using the tools of nonlinear, nonequilibrium dynamics: the key physical variable of the model being the expression of the nonlinear potential that describes the clumpy nature of the observed clouds. The model, originally developped for a single isolated molecular cloud, showed how naturally a mass spectrum evolves, once a local instability criterion for fragmentation is specified. We expand the analysis of FMV in two ways: (i) by modelling the molecular cloud system as composed by a large number of clouds, whose mass distribution and star formation efficiency are taken from presently available compilation of molecular line observations of galactic molecular clouds. and (ii) by considering various instability criteria in the presence of magnetic field and turbulence for different cloud geometries (spherical, filaments, sheets). We explore if a non–monotonic IMF may arise and discuss the relevant cases. As a test to our approach we also examine the sensitivity of the results to the assumed mass spectrum, by using the expression given by Fleck (1983), derived under the assumption that turbulence plays a dominant role in determining the IMF.

2. Derivation of the IMF

Here we briefly summarize the necessary steps used by FMV to derive a theoretical IMF, while we refer to that paper for a rigorous mathematical discussion. (1) Describe the global dynamics of the molecular cloud system, subject to some external perturbation, by using a statistical mechanics approach, namely by writing the Lagrangian and the corresponding equation of motion. (2) The solutions of the equation of motion are determined and the energy spectrum is computed, together with the distribution of the perturbations in the various modes. This is the major quantitative difference with respect to other fragmentation schemes proposed by other authors, mainly following Hoyle's suggestion (1953) of hierarchical fragmentation of gas clouds: the non–linear nature of the Lagrangian determines a non uniform excitation of the different modes resulting in a non trivial spectrum of fluctuations that can be identified *a posteriori* with the clumpy structure of the parent cloud. (3) Fragmentation of the gas system, and hence star formation, can occur if an instability criterion, to be specified, is satisfied inside the individual clumps. A mass spectrum of the newly formed stars develops, that can then be easily converted to an IMF.

The expression of the mass spectrum, i.e. the number of stars of mass m in the mass interval $m, m + dm$, that FMV derived according to the previous steps is:

$$\nu(m)\ dm \equiv \nu_0\ \overline{\nu}(m)\ dm = \nu_0 \left\{ 1 - \exp\left[-0.1 \left(\frac{m}{m_{\rm cr}} \right)^{-Am^{\delta}} \right] \right\} dm, \qquad (1)$$

where ν_0 is a normalization constant, fixed by the total mass $M_{\rm cl}$ of the cloud and by the fraction of gas locked into stars, $\varepsilon_{\rm cl}$, directly related to the star formation efficiency, as commonly defined in observational studies of star forming regions. $m_{\rm cr}, A$, and δ are functions that depend on the adopted instability criterion.

As instability criteria that define $m_{\rm cr}$, we consider (1) an infinite uniform medium, (2) an infinite sheet geometry, (3) presence of magnetic field, (4) a homogeneous turbulent medium.

To make our analysis more general, we have searched for other expressions of the mass functions based on a theoretical approach different from that given by FMV to be used to test the model. Among various possibilities, the one suggested by Fleck (1983) is particularly suitable for our purposes in that the dependence of the mass spectrum on the variables that describe the physical state of the clouds is explicitly stated. In fact, following his definition, the mass spectrum obtained assuming a gaussian velocity distribution of interstellar turbulence takes the form:

$$\nu(m)\ dm = \nu_0\ m^{\phi}\ \exp\left[-\frac{G(m,T)}{\sigma^2} \right] dm, \qquad (2)$$

where ϕ and $G(m,T)$ are defined for the supersonic and subsonic regimes.

Having specified the mass spectrum according to either Eq. (1) or Eq. (2), the IMF resulting from the fragmentation of a cloud of mass $M_{\rm cl}$ characterized by a star formation efficiency $\varepsilon_{\rm cl}$ is given by (cf. Scalo, 1986):

$$\xi_{\rm cl}(m) = \frac{m\ \nu(m)}{0.434} = \frac{m}{0.434} \left[\frac{\varepsilon_{\rm cl}\ M_{\rm cl}\ \overline{\nu}_{\rm cl}(m)}{\int m\ \overline{\nu}_{\rm cl}(m)\ dm} \right]. \qquad (3)$$

3. The model

We now apply the derivation of the IMF of a single cloud described in the previous section to a more general system, that we will call the "cloud ensamble". In this model the ensemble represents a sufficiently large volume of the ISM comparable to the size of a typical galactic giant molecular cloud complex (GMC), with mass of the order $10^6 - 10^7 M_\odot$. The ensamble presents a clumpy structure with clouds distributed in mass and characterized by different physical conditions, namely temperature, density, magnetic field strength, etc. To specify the mass distribution of the clumps, $N(M_{\rm cl})$, we rely on the compilations of Solomon et al. (1987) and of Feitzinger and Stüve (1986). In addition, several other relations between the observed variables need to be introduced, namely: (1) the gas density–temperature relation; (2) the relation between the magnetic field strength and the gas density, and (3) the variation of the gas temperature with the cloud mass. The knowledge of these relations is instrumental in defining properly the critical mass that enters in the computation of the IMF.

We then suppose that these clouds experience a perturbation that triggers the onset of gravitational instability and let the physical process controlling the critical mass for the instability be of different nature (i.e., thermal, turbulent, magnetic). The star formation activity within each cloud can be characterized by an efficiency factor, $\varepsilon_{\rm cl}$, that specifies the fraction of the available gas that is actually converted into stars. Finally, we can compute the resulting IMF as being due to the superposition of the individual IMFs of the subunits of the ensamble. The important point that we want to stress here is that, on the theoretical side the only *Ansatz* of the model is in the specification of the instability criterion, all the relations necessary to compute the ensemble IMF ($\xi_{\rm ens}$) can be quantitatively determined thanks to the wealth of observational data presently available.

Using the definitions of Sect. 2, what we want to compute is:

$$\xi_{\rm ens} = \int \xi_{\rm cl}(m) N(M_{\rm cl})\, dM_{\rm cl}, \tag{4}$$

where $N(M_{\rm cl})$ is the number of clouds of mass between $M_{\rm cl}$ and $M_{\rm cl} + dM_{\rm cl}$ and the integral is evaluated by counting all the stars of equal mass produced in each cloud, and by summing them over the whole cloud mass spectrum.

So far, we have not specified the critical mass that enters in $\xi_{\rm cl}(m)$. In principle, one could consider the effect of adopting a unique definition of $m_{\rm cr}$ in the computation of the IMF. However, since we feel that in reality the critical mass for fragmentation is determined by a cooperative effect of thermal as well as non-thermal forces, what we have actually computed instead of Eq. (4) is:

$$\xi_{\rm ens} = \int \sum_{k=1}^{6} \mu_k(M_{\rm cl}) \xi_{\rm cl}(m) N(M_{\rm cl})\, dM_{\rm cl}, \tag{5}$$

where $\mu_k(M_{\rm cl})$ has the meaning of a statistical weight of the k-th critical mass introduced in Sect. 2 (the values 5 and 6 refer to the Fleck's IMF in the case, respectively, of supersonic and subsonic regime) and is such that $\sum_{k=1}^{6} \mu_k = 1$ for each M_{cl}. Since we have introduced two expression for the mass spectrum, ξ_{cl} can be either $\xi_1(m)$ or $\xi_2(m)$. According to Eq. (5), the expression in Eq. (4) would represent the case where μ_k is equal to 1 only for one particular value of m_{cr} and 0 for all the others.

4. Results

In Fig. 1 we present the results of the integration of Eq. (4) for each of the critical masses and for a particular choice of the input relations.

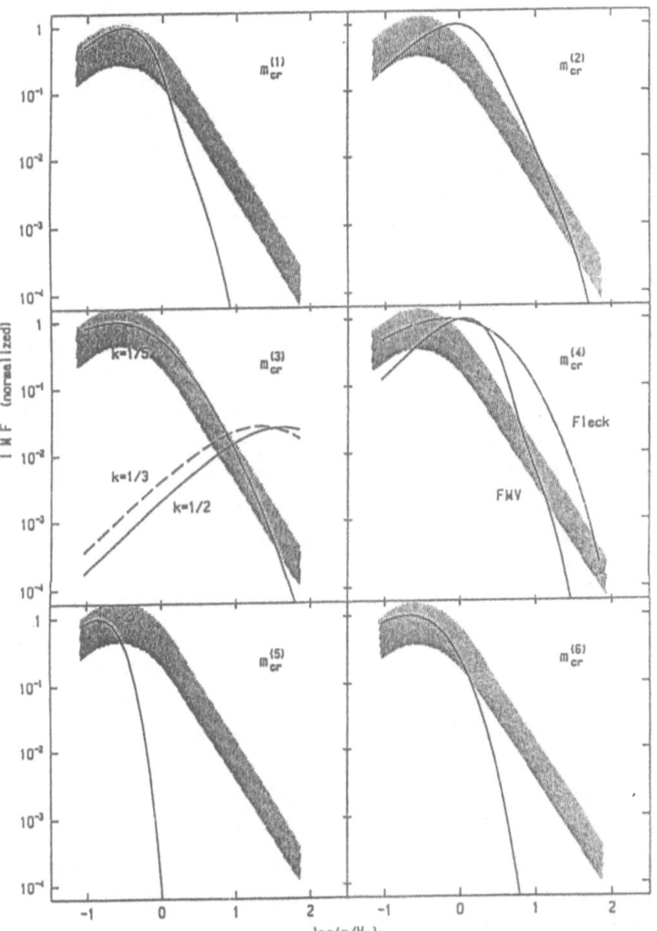

Fig. 1 The resulting IMF of each m_{cr}. All the curves have been normalized to the peak value of the distribution, except for the two curves of the case $m_{cr} = m_{cr}^{(3)}$

with $k = 1/2, 1/3$. The shaded area superposed to each curve has been derived from the IMFs of Scalo and Rana.

In the panel for $m_{cr}^{(3)}$, we explore three different choices for the exponent k in the relation between the magnetic field strenght and the gas density ($k = 1/2$ and $k = 1/3$ refer to the analysis of Myers and Goodman (1988), while $k = 1/5$ is suggested by Fleck (1988)). We superpose to each computed IMF a dotted area corresponding to the position of the peak, to the slope, and to the dispersion around the analytic fit obtained from the Scalo's and Rana's IMFs.

It is clear from Fig. 1 that we cannot reproduce the observed IMF, as given by either Rana or Scalo, nor we can obtain some of the fine details discussed by Rana, if the fragmentation process in the whole system is governed *only by one specific mechanism*. In fact, the property of the derived IMF common to all of the choices of the critical mass is that the slope of the curves in the high mass tail of the distribution is too steep. In addition, the position of the peak is a sensitive function of m_{cr}: this is already indicative of the fact that, if the assumption of a unique fragmentation criterion is relaxed in favor of multiplicity, then it becomes possible to replenish the high mass mode so as to obtain a better agreement with the observed IMF.

We refer the reader to Ferrini, Palla and Penco (1989) for a detailed discussion of the constitutive relations, the analysis of the single fragmentation modes and the modifications introduced by a different choice of the input relations, as well as a presentation of the results obtained by including several mode mixings.

Here, we relaxe the hypothesis that only a unique fragmentation mode has interested all the clouds and consider that the various instability modes described in Sect. 2 act at the same time or that, viceversa, only a certain range of cloud masses are subject to a particular mode, but not to the other ones; as outlined in Sect. 3, the individual ξ_{cl} must be combined to form the global ξ_{ens} by weighting each of them by the proper coefficients $\mu_k(M_{cl})$. These are constrained by the observations, in the sense that the measurements of observables such as the gas temperature, density, magnetic field strength etc., yield a direct estimate of the relative importance of each force in the balance that determines the cloud dynamics.

A proposal for a gradual change in the importance of the fragmentation mode in the range of cloud masses may be deduced from Larson's analysis (1981) on the effects of turbulence on star formation in molecular clouds. Following Larson, we assume that for the low mass clouds ($M_{cl} \leq 300 M_{\odot}$) the principal mechanism is that of supersonic turbulence, while for higher masses both gravitational fragmentation modes (Jeans and polytropic sheets) are of interest, and we assign equal weight to them. The analysis can be modified if we introduce the subsonic turbulence in view of the fact that in the hierarchy of clumps of various sizes, the internal velocities of the smaller ones are no longer supersonic. For instance, we consider that subsonic turbulence acts only for $M_{cl} \leq 30 M_{\odot}$ (cf. Leung et al., 1982).

From the analysis done by Shu and collaborators of the role of the magnetic field on the star formation process (Shu et al., 1987; Lizano and Shu, 1989), we may consider that the critical mass for fragmentation is influenced by the magnetic field for cloud masses smaller than $10^3 M_\odot$.

We incorporate all the fragmentation modes in the computation of ξ_{ens}. The results are shown in Fig. 2. The first remarkable result is the excellent agreement with the observed IMF. The peak of the mass distribution is largely due to the effect of the gravitational fragmentation mode, but a contribution from turbulence and/or magnetic field must be present to account for the high mass end of the spectrum.

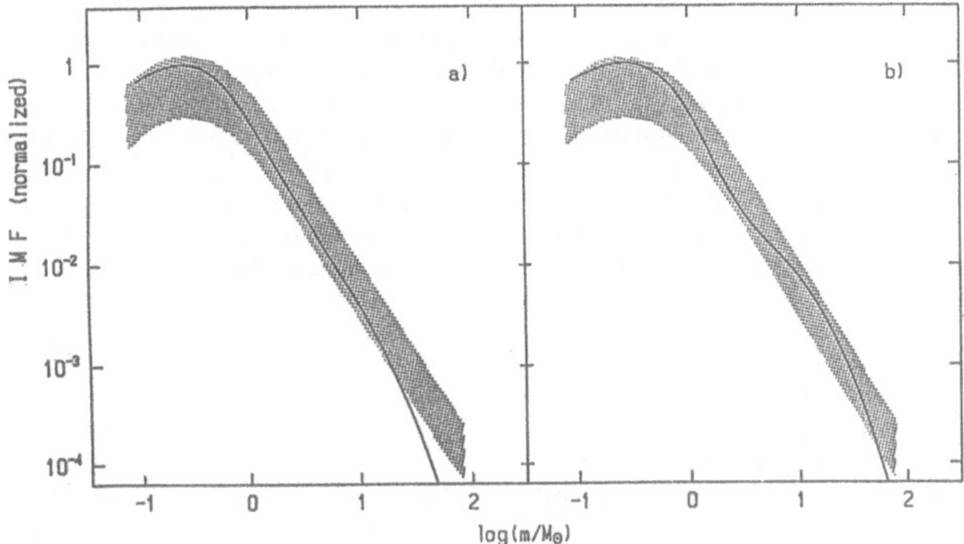

Fig. 2: The IMF in the case of magneto–gravitational–turbulent mixing for a constant SFE and for a cloud mass distribution given by $N_1(M_{cl})$ (panel a), and by $N_2(M_{cl})$ (panel b).

The other point of interest is the non–monotonic behavior of the IMF, that shows a hump at $\log(m/M_\odot) \sim 0.5$. In a certain way, this curve is the best illustration of what could be considered as indicative of a bimodal IMF, even though its derivation rests on a *multimodal* fragmentation scheme. Nevertheless, it would be tempting to make such a claim, since the solution found in the curve of panel b) is rather insensitive on the assumptions used for the input relations. If we change either the SFE or the $T-n$ relation, the shape of the curve and the presence of the hump are still present, but horizontally shifted to higher masses. The strongest dependence of these results is only on the cloud mass function: as clearly seen by comparing panel a) and panel b), the presence of a secondary hump would be washed out even though the agreement with the observations would be still considered satisfactory. This last point clearly urges for a better estimate of the exponent of the realtion $N(M_{cl})$:

if the concept of bimodality of the IMF has to be grounded on theoretical facts, then it is not only the physical conditions of the molecular clouds (temperature, density, magnetic field) that must be determined with great care, but, and perhaps more importantly, the frequency of the occurrence of low mass clouds within GMC complexes.

5. Conclusions

We have computed the IMF resulting from the fragmentation of molecular clouds, based on the theoretical approaches given by FMV (1983) and Fleck (1983), but extended to take into account the simultaneous presence of several instability criteria. In addition, the physical quantities that define the characteristic mass of fragmentation (temperature, density, magnetic field) have been taken from observations.

The comparison with the observed IMF shows that, in order to reproduce both the shape and the position of the peak of the distribution, the instability criteria must include the effects of turbulence and magnetic field. The peak of the IMF, that defines the characteristic mass, can indeed be obtained by using only gravitational instability, but then the upper end falls short by unacceptably high value of the slope.

The presence of small scale structures in the IMF is suggestive of the cooperative effects of several fragmentation modes.

In relation to the working hypothesis of a bimodal IMF either in space, (Güsten and Mezger, 1983) or in time (Larson, 1986), the present calculations have shown that such a possibility might occurr only under strict constraints on the physical conditions of molecular clouds. Namely, a rigorous bimodality in the distribution of the observed properties may indeed produce a bimodal IMF, which is instead is ruled out in all the other cases. The suggestion by Güsten and Mezger that the high mass mode of star formation operates preferentially in the spiral arms encounters some difficulties, considering that in no case we found an IMF deficient of low mass stars, using the observed relations for molecular clouds in spiral arms.

As for the time dependence of the IMF, this is more diffucult to assess on the basis of the present calculations, which are aimed mainly at reproducing the observed IMF in the solar neighborhood. In general, the properties of molecular clouds may change dramatically during the life of a galaxy, as a consequence of the star formation history (thermal and mechanical energy inputs; phase consumption and restitution) and of the chemical evolution (grains populations, chemical reactions, heating and cooling processes). Then, different states of the clouds may favour or suppress a particular mode of fragmentation and the resulting IMF may show some properties that differ from the present ones. Here, we remind that the theory of a bimodal IMF has been formulated partially on the basis of some indirect observations, but more importantly as an attempt to reconcile with observations some models of galactic evolution (e.g. Larson 1986, Wyse and Silk 1987). However, it has been shown elsewhere (Shore et al., 1987) that global properties, such as the stellar age–metallicity relation and the G–dwarf problem, can be satisfied in the framework of multi–phase galactic models in which the IMF does not vary in time and the star formation rate results from the coupling between the evolution of the halo and the disk, without invoking a bimodal IMF.

References

Chandrasekhar, S.: 1951, *Proc. Royal Soc. London A*, **210**, 26

Feitzinger, J.V., Stüve, J.A.: 1986, *Astrophys. J.*, **305**, 534

Ferrini, F., Marchesoni, F., Vulpiani, A.: 1983, *Monthly Notices Roy. Astron. Soc.*, **202**, 1071 (FMV)

Ferrini, F., Palla, F., Penco, U.: 1989, *Astron. Astrophys.*, in press

Fleck, R.C.: 1983, *Monthly Notices Roy. Astron. Soc.*, **201**, 551

Fleck, R.C.: 1988, *Astrophys. J.*, **328**, 299

Güsten, R., Mezger, P.: 1983, *Vistas in Astronomy*, **26**, 159

Hawkins, M.R.S.: 1986, *Monthly Notices Roy. Astron. Soc.*, **223**, 845

Hoyle, F.: 1953, *Astrophys. J.*, **118**, 513

Larson, R.B.: 1981, *Monthly Notices Roy. Astron. Soc.*, **194**, 809

Larson, R.B.: 1986, *Monthly Notices Roy. Astron. Soc.*, **218**, 409

Larson, R.B.: 1987, in *Stellar Populations*, eds. C. Norman, A. Renzini, and M. Tosi, Cambridge Univ. Press, p. 101

Leung, C.M., Kutner, M.L., Mead, K.N.: 1982, *Astrophys. J.*, **262**, 583

Lizano, S., Shu, F.H.: 1989, *Astrophys. J.*, **342**, 834

Myers, P.C., Goodman, A.A.: 1988, *Astrophys. J. Lett.*, **326**, L27

Ostriker, J., Thuan,T.X.: 1975, *Astrophys. J.*, **202**, 353

Rana, N.C.: 1987, *Astron. Astrophys.*, **184**, 104

Reid, N.: 1987, *Monthly Notices Roy. Astron. Soc.*, **225**, 873

Salpeter, E.E.: 1955, *Astrophys. J.*, **121**, 161

Scalo, J.M.: 1986, *Fund. Cosm. Phys.*, **11**, 1

Scalo, J.M.: 1987, in *Starbursts and Galaxy Evolution*, eds. T. Montmerle and T.X. Thuan, Edition Frontieres, p.445

Shore, S.N., Ferrini, F., Palla, F.: 1987, *Astrophys. J.*, **316**, 663

Shu, F.H., Adams, F.C., Lizano, S.: 1987, *Ann. Rev. Astr. Astrophys.*, **25**, 23

Silk, J.: 1987, in *Star Forming Regions*, IAU-Symp. 115, eds. M. Peimbert and J. Jugaku, Reidel Publ. Co., p. 663

Solomon, P.M., Rivolo, A.R., Barrett, J., Yahil, A.: 1987, *Astrophys. J.*, **319**, 730

Wyse, R.F.G., Silk, J.: 1987, *Astrophys. J. Lett.*, **313**, L11

Zinnecker, H.: 1987, in 10th IAU European Regional Meeting, J. Palous ed., vol. 4, p.77

Questions

H.Zinnecker: "Why does the sum of many IMFs produce the smooth overall IMF? This is probably due to only a small variation in the dispersion of the individual IMFs. Why, physically, is that variation so small?"
F.Ferrini: "Most of the computed IMFs show a remarkably similar width: this is an outcome of the non–linear nature of fragmentation, that results in an approximate log–normal distribution."

W.M.Tscharnuter: "How can supersonic turbulence be maintained in low mass fragments?"
F.Ferrini: "We consider turbulence for two aspects: (a) following Chandrasekhar (1951), in a homogeneous turbulent medium, it is possible to recognize a critical mass; (b) assuming a gaussian velocity distribution of interstellar turbulence it is posssible to derive a mass spectrum (Fleck, 1983). On the smaller scales, supersonic turbulence has not been considered, as it was previously discussed about the adopted mixing. You may also notice from Fig. 1 ($m_{cr}^{(4)}$) that supersonic turbulence produces an IMF with a peak strongly shifted towards high stellar masses.

P.Lenzuni: "Is it possible that you get (at least approximately) the same results using different combinations of the ingredients you include in your model?"
F.Ferrini: "We explored the sensitivity of the resulting IMF on the adopted values of the input quantities, i.e. SFE, cloud mass spectrum, temperature etc.. For example, the position of the peak is quite sensitive to $T(M_{cl})$ and in particular to the value of T for low mass clouds. For what concerns the multi–humps structures, whose appearance is due to the concurrent contribution of several fragmentation modes, the position of the humps is reminescent of the position of the peaks of the individual modes. At the same time, the qualitative structure of the IMFs remains constant."

THE INITIAL MASS FUNCTION OF VERY LOW MASS STARS AND THE SIGNIFICANCE OF BROWN DWARFS

Francesca D'Antona
Osservatorio Astronomico di Roma
I-00040 Monte Porzio
Italy

ABSTRACT. In recent years we have learned how tricky is the determination of the IMF at the low luminosity end of the main sequence. I recall here the main reasons why the stellar structure computations produce a functional form of the mass-luminosity relation for $M \leq 0.5 M_{\odot}$ which, coupled with the observationally determined luminosity function in the visual band, produces a smoothly increasing IMF down to $M{=}0.1M_{\odot}$.
I discuss then the updates of the theory and the large errors which still weigh on the derivation of luminosity and effective temperature from the observational data.
I further consider whether the recent detection of several infrared (substellar?) companions to intrinsically faint stars may give us further hints on the mass-luminosity relation close to the end of the main sequence, and, second, to the number density expected for single brown dwarfs in the galactic disk.
I finally review the problems emerging from the consideration of the luminosity function based on stellar counts in the infrared bands.

1. Introduction

From the point of view of star formation, Brown Dwarfs (BDs) pose two interesting problems:

1. do BDs form at all?
2. if BDs exist, are they an important constituent of the mass of the Galaxy?

Answer to the first question would put constraints on the theories of fragmentation, while answer to the second one is necessary first to understand whether BDs are responsible for the dynamically inferred missing mass in the local system and /or in the galactic disk (Oort 1960, Bahcall 1984), second to model properly the chemical evolution of the Galaxy, as BDs and low mass stars act as reservoir of matter which does not partecipate in the nucleosynthesis.
Renewed interest in the study of BDs came out abruptly in 1984 with two concomitant scientific achievements: i) discovery of the first candidate BD (VB8B, McCarthy et al. 1985); ii) new derivation of the dynamical mass of the disk of the Galaxy by Bahcall (1984) which confirmed Oort's (1960) result that we do not see about half of the mass of the galactic disk.
Later on, the very existence of VB8B was seriously questioned (Perrier and Mariotti 1987, Skrutskie, Forrest and Shure 1987), and different models of the galactic disk (Bienaymè et al. 1987), or use of differently selected samples of stars to be used as tracers of its dynamics (Kuijken and Gilmore 1989, but see also Gould

R. Capuzzo-Dolcetta et al. (eds.), Physical Processes in Fragmentation and Star Formation, 367–380.

1989) now lead to new doubts on the Oort limit.
In any case, the observational effort put into the study of very low luminosity stars and in searching for BDs is producing interesting results and it will, at least, let us better understand the interesting and difficult physics of the objects close to the end of the main sequence and derive correct constraints on the IMF of low mass stars.

2. The mass luminosity relation

2.1 ROLE OF STELLAR STRUCTURE FOR THE LOW MASS STARS ML RELATION

The theory of low mass stars is a fundamental ingredient to derive correctly the IMF of low mass stars. If fact, it is necessary to accurately know the mass-luminosity (ML) relation, whose derivative enters as a factor multiplying the luminosity function (LF).
The observational points in the ML relation are too few (Popper 1980, Scalo 1986) and have large error bars to be useful here: they can be fitted with a power law relation, which has, in any case, scarce relation with the behaviour which we expect on theoretical grounds.
In fact, theory predicts three main features in the ML relation:

1. flattening (dM/dL increases) at $M \sim 0.5 M_{\odot}$.
2. steepening (dM/dL decreases) at $M \leq 0.2 M_{\odot}$.
3. close to the main sequence minimum mass, the steepening becomes more and more pronounced, and dM/dL approaches zero.

2.2 THE ML RELATION AT $\sim 0.5 M_{\odot}$ AND GLOBULAR CLUSTERS IMF

Flattening of the ML relation around $M \sim 0.5$ $M_{\odot}$is related to the fact that the stellar structure enters the region of molecular recombination, where the adiabatic gradient in the external convective layers becomes smaller, and the decrease of central temperature (T_c) with decreasing the stellar mass produces then a smaller variation of T_{eff} and L with respect to the upper mass range. This fact is well known as the *break* in the main sequence, and Copeland et al (1970) have shown that it is due to the inclusion of H_2 in the equation of state.
It is interesting to notice that *the lower the metal content is, the larger the luminosity at which the flattening occurs* (D'Antona 1987). This is due to the fact that for smaller opacity the structures enter into the partial dissociation region 'before'. A somewhat different but equivalent way of understanding this behaviour is the following: molecular dissociation increases the opacities in the envelope and the stars become more and more convective. Finally, at $M \leq 0.4 M_{\odot}$ the stars become fully convective (D'Antona and Mazzitelli 1985, D'Antona 1987). In convective main sequence stars, which are scarcely superadiabatic, the surface luminosity depends mainly on the atmospheric opacity, thus a comparable production of nuclear energy (the internal structure does not depend too sensitively on the metal content) provides larger luminosity the smaller is the opacity.
This theoretical result has its observational confirmation: the LFs of the main

sequence of Globular Clusters (GCs) of different metallicities present a steepening, due to the flattening of the ML relation, at a magnitude ($M_v \sim 7-8$) which depends on the clusters metallicity, as predicted.
Adopting the correct ML relation for each cluster, we can derive the mass function of main sequence stars in GCs, and so obtain information on the IMF. Mc Clure et al (1986a) found a strict correlation of the slope of the IMF with the GCs metallicity. Actually, this correlation is probably not so strict, as different fit of the theory to the observational data (D'Antona 1987) and larger samples of GC main sequence luminosity functions (Mc Clure et al. 1986b, see also the discussion in Renzini and Fusi Pecci 1988) do not confirm it. In any case, study of the LFs of GCs main sequence and the conversion to mass function by means of the appropriate ML relation is fundamental for a complete understanding of star fragmentation in isolated, relatively simple systems such as the GCs, and consequently, it is fundamental to our understanding of fragmentation in more complex systems such as galaxies.

2.3 THE ML RELATION AT $M \leq 0.2M_{\odot}$

The ML relation changes shape again at $M \leq 0.2M_{\odot}$, with an increase of slope which becomes a drop of luminosity while we reach the minimum mass fo hydrogen burning. This behaviour is due to the following reasons:

i) The influence on the structure of the region of molecular dissociation, where the adiabatic gradient is much smaller than 2/5, becomes smaller when decreasing T_{eff}, also because pressure ionization becomes important;

ii) the p-p reaction rates decrease faster with decreasing temperature for $logT \leq 6.6$.

Both these effect determine an increase in dT_c/dM and thus in dL/dM for a given variation of the atmospheric parameters (D'Antona and Mazzitelli 1982).
Both the flattening in the ML relation at M=$0.5M_{\odot}$and its steepening at $M \leq 0.2M_{\odot}$are relevant if we want to derive from knowledge of the LF the IMF of the sample stars. Being, for low masses (practically unevolved):

$$\frac{dN}{dM} = \frac{dN}{dL} \times \frac{dL}{dM} \tag{1}$$

the IMF depends on the *derivative* of the ML relation. This has been pointed out in D'Antona and Mazzitelli 1983, 1986 and by Liebert and Probst 1987.
In particular, D'Antona and Mazzitelli 1986 show that the IMF derived from the sample of stars within 20 pc (Wielen et al 1983) can be considered continuous and it is increasing down to $M_v = 17$, corresponding to M=0.1 $M_{\odot}$.

2.4 CONSEQUENCE: A SMALL OBSERVATIONAL SAMPLE

Although the slope of the mass function derived in D'Antona and Mazzitelli 1986 is not very large (the mass function can be described by a power law with index x=1, so that we should not expect a large number of BDs) any plateau or decrease in the *visual* LF at $M\sim 0.2M_{\odot}$ can be completely ascribed to the ML relation! This implies, on the one hand, that we *must not* expect an increasing LF *in the visual* down to the hydrogen burning minimum mass, but also that it is difficult to get a good, numerous sample of VLM stars in order to better define *observationally* the

shape of the IMF in this important region.
In fact, *very few* are the nearby stars with *mass* determination (astrometric binaries, see Popper 1980, Liebert and Probst 1987, and the recent recompilation by McCarthy et al 1988).
A clean cut evidence for masses smaller than the hydrogen burning limit (or, better, smaller than the transition masses) is the recent determination of strictly periodic radial velocity variations in the star HD 114762 (Latham et al 1989), for which the orbital solution provides:

$$M_c \times sin\ i\ = 0.011 \pm 0.001 M_\odot \tag{2}$$

Thus, unless the inclination i is very small, the companion of this nearby star is either a BD or even a giant planet of Jupiter type.
It is highly probable that improvement in the tecniques for the determination of radial velocity changes will increase this sample considerably in the future.
Another probable case of substellar masses is provided by the system Wolf 424, for which Heinz (1989) derives components of 0.059 and 0.051 $M_\odot$.

3. The H-R diagram

Determination of photometric masses for the lowest mass stars is often doubtful, as the *age- luminosity* relation produces a large spread for a given mass (see the discussion in Liebert and Probst 1987).
Figure 1a shows the location of observed low mass stars in the theroretical H-R diagram, from which a further complication in the comparison theory-observations appears: there are very few M dwarfs for which the distribution of emergent flux with frequency is known reasonably well, but even for these stars the error bars for placing them in the theoretical HR diagram must be considered very large. Let us only consider the case of VB8 and of LHS2924, according to the determinations by Zuckerman and Beckin (1988) and by Berriman and Reid (1987). The discrepancy is so large that LHS2924 is for the first authors a bona fide main sequence dwarf of about $0.1 M_\odot$, for the latter it might well be a cooling BD!
Berriman and Reid (1987) have used spectrophotometry of nearby M dwarfs to derive the total flux of these stars, emitted from the ultraviolet to the infrared. They show that the water absorption bands in the near IR depress the continuum, so that the total luminosity is smaller than previously found on the basis of extrapolation at $\lambda \geq 2\mu m$). Further, they derive the T_{eff} by fitting a black-body of the same total flux, normalized at the continuum given by the K band flux.
The values derived for their sample are such that they do not fit the theoretical radii in the whole range $M \leq 0.5 M_\odot$.
So these results -which should be the more accurate we have for VLM stars, do not fit the theory, and are at variance with the other determination from broad band colors only (see the examples above). I assume here a good theoretician point of view (Burrows et al 1989): if the answers are so different passing from an observer to another one, we can be sure that the bars of errors on the observations are much larger than expected.
In other words, we can not derive stringent constraint from the observational HR diagram.

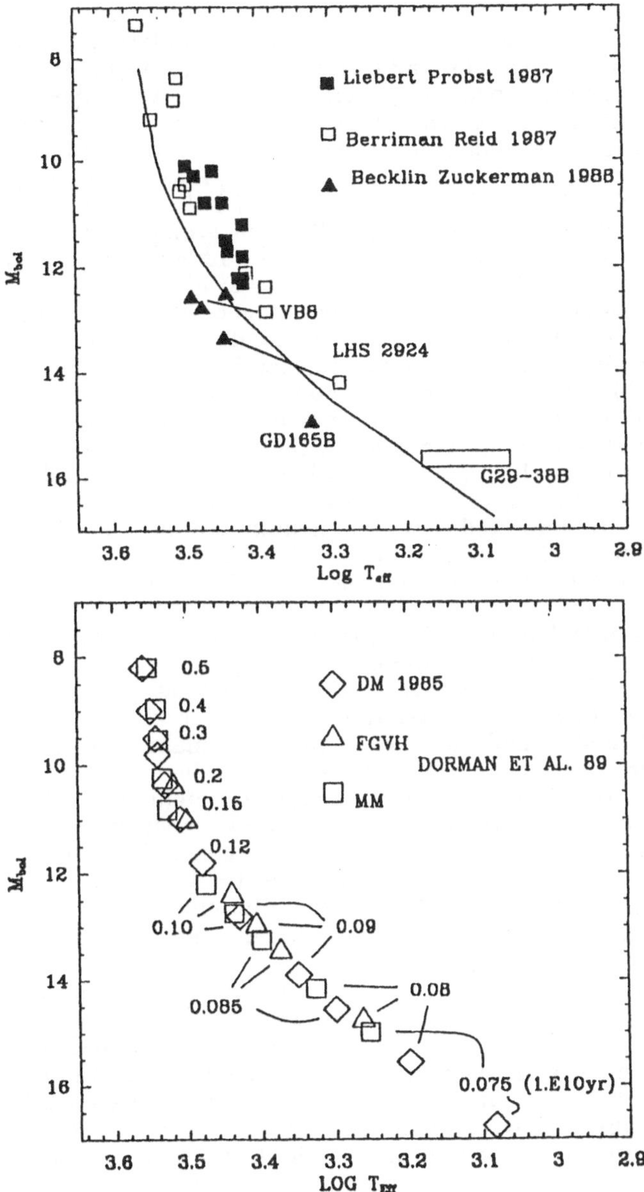

Figure 1: a) shows the HR diagram for low mass stars from recent literature; b) compares the recent theoretical models.

4. Updates of the theory

After D'Antona and Mazzitelli 1985, my study of population II low mass main sequence location (D'Antona 1987) addressed also the problem of the termination of the main sequence in low metallicity stars. The qualitative expectation that the lower opacities in the radiative atmosphere would produce a more luminous end of the main sequence was confirmed by model computation. In fact, the lowest luminosity stars for $Z = 10^{-4}$ are placed at $log\ L/L_{\odot} = -2.9$ ($M_{bol} \sim 12$), and this result fits nicely with the lowest luminosity of halo dwarfs found in a recent analysis of M subdwarfs by Greenstein (1989). Concerning population I, there are mainly two interesting papers on the structure of VLMs and BDs which are going to appear, by Dorman *et al* 1989 and by Burrows *et al.* 1989.
The first paper is an effort to determine the influence of the e.o.s. on the models, and also takes care of adopting the newest Alexander's (1989, unpublished) opacities. The e.o.s. compared are by Magni and Mazzitelli (1979, MM) -also adopted in D'Antona and Mazzitelli 1985- and by Fontaine, Graboske and VanHorn 1977 (FGVH). Although these have been constructed by the same physical approximations (free energy minimization tecnique in the regions of partial ionization and Thomas Fermi model for the regions of full ionization) some details are different, such as the the computation of the internal free energy. Another important difference is that FGVH computed only pure helium and pure hydrogen, while MM considered five different hydrogen contents. This renders interpolation between the tables less uncertain. Further difference are due surely to the numerical tecniques used.
Comparison between D'Antona and Mazzitelli 1985 and the models by Dorman et al. is given in figure 1b. The two different e.o.s. give somewhat different results when $M \leq 0.2M_{\odot}$. This is a further demonstration that good knowledge of the thermodynamics in the region of partial ionization is crucial.
Between the MM models and our 1985 models the main difference (apart from possible other differences in the codes which may also be important) should reside in the atmospheric opacities, as D'Antona and Mazzitelli 1985 adopted the previous Alexander et al. 1983 tables, which probably somewhat overestimated the effect of H_2O. At $M \geq 0.2M_{\odot}$ the differences are insignificant, but they become important for smaller masses, and follow the expected behaviour: models with the smaller opacities (Dorman et al. 1989) are more luminous. Nevertheless, the lowest luminosity of the main sequence is still placed somewhat below $L = 10^{-4}L_{\odot}$, to be compared with our $log\ L/L_{\odot}$=-4.34.
The paper by Burrows et al (1989) also poses interesting questions: in particular the authors consider the effect of inhibition of convection in the envelopes of low mass stars on their luminosity - T_{eff} location. Although it is well known that the envelopes of low mass main sequence stars are scarcely superadiabatic, so that reasonable variation of the parameter $\alpha = l/H_p$ in the range 1 - 2 (to which the theory of superadiabatic convection and the same observations generally constraint it) have scarce effect on the structure, they check the effect of assuming $\alpha = 0.1$, which could represent a zero order approximation of treating the effect of convective inhibition by a magnetic field. As a result, their model *radii* seem to agree much better with the Berriman and Reid's (1987) values from observations in the mass range $M \leq 0.2M_{\odot}$. Whether this is of any significance or not is worth to be discussed only when more reasonable atmospheric parameters will be derived for VLMs.

5. The recent observations

Apart from astrometric measurements and from radial velocity variation measures, which have been widely used also in the past, searches for BDs have followed recently three main lines:

1. Infrared Speckle Interferometry of astrometric pairs (McCarthy and Henry 1987; McCarthy et al. 1987, 1988; Ianna et al. 1988);
2. Infrared photometric survey of $\sim$ 200 white dwarfs (WDs) (Zuckerman and Becklin 1987 and following);
3. Luminosity function, derived through deep star counts towards the South Galactic Pole (Reid and Gilmore 1982, Hawkins and Bessell 1988, Leggett and Hawkins 1988);
4. Infrared observations of open clusters.

Each of these tecniques has given pieces of information.

5.1. INFRARED SPECKLE INTERFEROMETRY

The quoted work of the group McCarthy has produced several interesting determination of masses and luminosities for nearby astrometric binaries, all summarized in the figure 5 in McCarthy et al 1988. As we see, there is no star which can be surely interpreted as BD. Problems in the interpretation of observations versus theory are in any case present (see the quoted papers, and the discussion on Gliese 623 B in the following).

5.2. INFRARED PHOTOMETRIC SURVEY OF 200 WHITE DWARFS

Zuckerman and Becklin have searched for infrared excesses in the bands J,H,K,L,L' around nearby WDs. They have found:

1. an IR excess around G 29-38 (Zuckerman and Becklin 1987a); although this case is still debated, Greenstein (1988) still finds that the BD interpretation is the more appealing;
2. no excess in J, K of a few WDs members of the Hyades or Pleiades supercluster, thus excluding companions of $M \geq 0.03 M_{\odot}$ within 6" for 8 WDs in the Hyades, and of $M \geq 0.015 M_{\odot}$ for a WD in the Pleiades (Zuckerman and Becklin 1987b);
3. an IR excess at 1200 au from GD165, interpreted as an object of T_{eff}=2100K, $L = 7.8 \times 10^{-5} L_{\odot}$;
4. they announce the discovery of *seven* more VLM stars ($M \sim 0.1 M_{\odot}$ companions of other WDs.

We must be very careful in interpreting these results: for instance, as the IR object in G29-38 is not spatially resolved, if indeed it is a BD, it must be very close to the WD, so that, on the one hand it reprocesses the light of the WD itself, and on the other hand it might well have been influenced by the previous giant evolution of the WD; at the end, we might discover that this presumed BD has been formed by accretion on a giant planet of the matter of the giant, in a stage resembling the

pre-cataclysmic binary evolution (Livio and Soaker 1988).
On the other hand, WD cooling is a good signature of the minimum age of this type of system, so that it is also the minimum age of the BD component, if we can assume that the giant evolution has not influenced the BD, like, possibly, in the system GD165 (Becklin and Zuckerman 1988). The WD age of GD165A ($T_{eff} = 12000K$) is $\sim 6 \times 10^8 yr$ (Mazzitelli and D'Antona 1986). Thus GD165B (if it is a true companion of the WD, should have a *minimum mass* of $0.06M_\odot$, otherwhise it would be less luminous. We could refer to GD165B as to the first "transition mass" discovered, unless it will suffer a pitiful death such as for VB8B. But let me suppose that the age really constraints the mass of this object to the given lower limit, and thus that it is at least partially supported by nuclear burning: in some sense we gain a new point on the mass luminosity relation at the lower end, so that we get the "observational" hint that our models correctly predict that the luminosity fast declines with mass around the lower limit for hydrogen ignition.
Unfortunately, an opposite conclusion is reached by examining the results concerning the system Gliese 623, in which binariety, first inferred by astrometric perturbations, has been confirmed by the detection of the companion by speckle interferometry (Mc Carthy and Henry 1987) and by precise radial velocity measurements (Marcy et al. 1986). These observation point to a mass for the companion in the range 0.067 - 0.087 $M_\odot$, while the luminosity is $L \sim 10^{-3} L_\odot$, a factor ten larger than what expected from our theoretical models for an old disk object of such a low mass!
In conclusion, one object (GD 165B) seem to conform to the theoretical drop in the mass - luminosity relation at the end of the main sequence, while another (Gliese 623B) does not agree at all!
While it is clear that deep study of these objects is crucial for our understanding of the structure of VLMs, I also feel obliged to find at least a qualitative explanation for these discordant observations: the luminosity of object in the proximity of the main sequence limit is, as we now well know, very dependent on the opacity in the radiative atmosphere, as it constraints the luminosity output. In cool atmospheres, *the main opacity source is provided by grains, which depend on the relative amounts of different metals (including carbon and oxygen) and may have large variations according to the environment in which the star lives.* Different mixtures of grains may give very different opacities and be the reason for large variations in the mass luminosity relation.

5.3 THE LUMINOSITY FUNCTION FROM DEEP STELLAR COUNTS

The luminosity function derived from the sample of nearby stars is flat or slowly decreasing at $M_v \geq 13$ (Wielen et al. 1983, Dahn et al. 1986), but it is often criticized as it is based on very few stars at these magnitudes. Consideration of luminosity functions based on deep star counts, beginning with the work of Reid and Gilmore (1982, 1984), continued by Hawkins and Bessell (1988) and finishing with Stobie et al 1989, shows that the peak at $M_v = 12$ is becoming sharper and sharper. Although there may be severe, not yet recognized problems in the interpretation of these stellar counts, taken at face value these results indicate the following:

1. the peak at $M_v = 12$ still reflects the flattening of the ML relation at $M = 0.2M_\odot$;
2. a *sharp* peak in the LF such as shown by Stobie et al 1989 cannot be *entirely*

attributed to the shape of the ML relation, preserving a steeply increasing IMF, at least in the framework of existing stellar structure models;

3. the mass function then has to become flatter at $M \leq 0.15 M_\odot$ to fit the data.

If I have to make predictions, probably future observational work will produce "final" LFs with a somewhat flatter shape, which will be consistent with a somewhat increasing IMF, but probably not with a Salpeter type IMF index.

5.4 A SECOND RISE OF THE LF AT VERY LOW LUMINOSITIES

It is also possible that the luminosity function rises again following the maximum at $M_v = 12$. This is indicated mainly in the recent infrared LF by Leggett and Hawkins (1988). Let us concentrate on their results in the J band as the relation $M_J - M_{bol}$ is quite linear. The LF shows a well defined maximum at $M_J \sim 8$ and rises again at $M_J \sim 11$.
The correspondence with bolometric magnitudes, and thus with masses, can be made by means of Berriman and Reid (1987) calibration (see section 4), showing the following result:

Table 1
Mass magnitude calibration

M_J	M_{bol}	$M/M_\odot$(DM)	$M/M_\odot$(Dorman et al)
7	8.65	0.4	
8	9.73	0.25	
9	10.80	0.15	
10	11.9	0.12	0.10
11	13.0	0.10	0.09

Thus this LF suggests two possible interpretations:

1. a drop of the IMF at $M = 0.11 M_\odot$, followed by the first onsight on the cooling of a young BD population. This is the interpretation given by Leggett and Hawkins (1988), based on the very small scale heigth found by Hawkins (1987) for the reddest objects in his samples;

2. an IMF declining at $0.2 \leq M/M_\odot \leq 0.12$ and then rising again at $M = 0.1 M_\odot$ (Reid 1987).

As not only our models (D'Antona and Mazzitelli 1985) but also the most recent ones by Dorman et al predict that the MS is extended down to $M_J = 11$, the second explanation seems to me well possible. The first one implies to postulate a huge (and 'ad hoc') population of BDs below the minmimum mass for hydrogen burning. In any case, extension of the observations down to $M_J = 12$ or 13 may confirm or exclude Leggett and Hawkins interpretation.

In figure 2 I compare the observational LF by Leggett and Hawkins (1988) with the theoretical LFs in M_J. The latter are obtained adopting the ML relation by

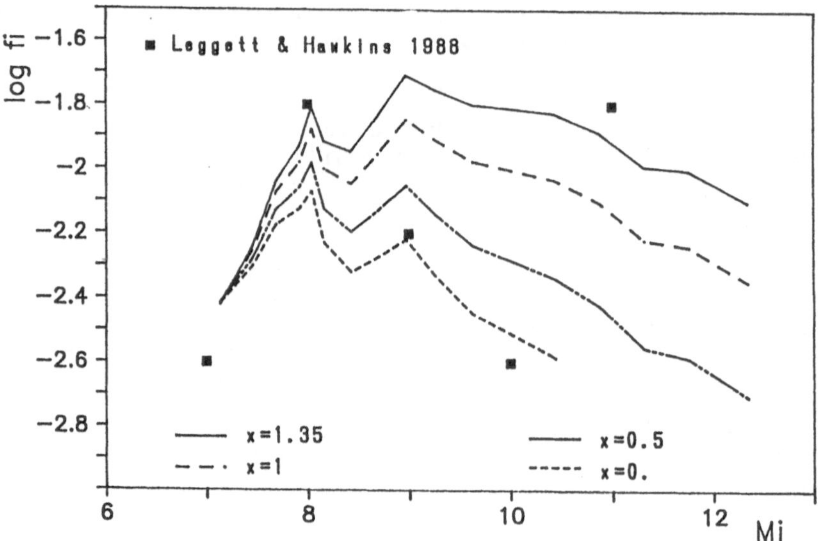

Figure 2: theoretical and observational LFs in the infrared.

D'Antona and Mazzitelli 1985 plus an IMF of the form

$$dN/dM \sim M^{-(1+x)} \tag{3}$$

The first two points ($M_J = 7, 8$) indicate a *large slope* for the IMF (x=1.0 - 1.35), and the point at $M_J = 11$ could represent the *continuation* of this IMF. The theoretical LF flattens at $M_J = 8$ due to the change of slope in the ML relation (as usual). Unfortunately the points at $M_J = 9 - 10$ *can not* be fit by the same power law, and seem to indicate a flaw in the IMF in a small range of masses just before the minimum mass for hydrogen burning.
Thus I have to question whether this interesting work is telling us something exciting about star formation (discontinuity of mass formation around $M \sim 0.15 M_\odot$, or if simply our interpretation of data is still too crude. For instance, the relation $M_J - I - J$ is based on three stars only at $M_J = 11$, but we have seen that the few observations of very low mass objects may indicate that intrinsic scatter exists in the H-R diagram. Reid (1987) gives an interesting discussion on this subject.

5.5 SEARCHES IN OPEN CLUSTERS

This seems to me a promising area of investigation. The only results obtained up today are by Leggett and Hawkins (1989) for the Hyades and by Jameson and Skillen (1989) for the Pleiades. Both find a few candidates BDs, but their results should be confirmed by further observations.

6. Conclusions

In summary, the updates of theory and observations on very low mass stars, let us prospect the following:

1. A huge body of information on stellar fragmentation in Globular Clusters can be derived in the near future by observing the luminosity function of main sequence stars. We need a good definition of the ML relation for the main sequence as a function of the helium and metal abundance. Already, the flattening of the ML relation at increasing luminosity when decreasing the metallicity is confirmed by the observed sharp increase in the LFs, which occurs at smaller luminosities the more metal rich is the cluster.

2. We must solve the observational discrepancy between the luminosity function in the nearby stars sample and in the photometric parallax samples. Present observations are compatible with an increasing mass function at $M \leq 0.2 M_{\odot}$, but the index of the IMF can not be as large as Salpeter's (1.35), and is probable lower than x=1, if the deep star counts are to be taken at face value. This means that brown dwarfs form, but that they are not in such a huge number to provide missing mass in the neighborhood of the Sun.

3. The recent results by Leggett and Hawkins, which give the hint for a rising again LF at very low luminosity, must be carefully explored. Extension of the LF down to $M_J = 12$ could solve the dilemma of whether there is a young population of cooling BDs in the disk, or we are witnessing the low end of the main sequence, at the luminosity predicted by current stellar structure models.

4. There are now several stars close to the limit of the main sequence, but both theory and observations are largely incomplete. We need both good $M_{bol} - T_{eff}$ determinations and good opacities for the regime of low temperature, large density in which the radiative atmospheres of these objects are placed.

5. The scatter in the luminosities of objects close to the end of the main sequence may be intrinsic, and due to a secondary parameter which I tentatively identify with different contributions to the atmospheric opacity given by grains whose composition may differ from star to star.

REFERENCES

Alexander, D.R., Johnson, H.R., and Rypma, R.L. (1983) Ap.J. 272,773.

Bahcall, J.N. (1984), Ap.J. 276,169.

Bienaymè, O., Robin, A.C., Crezè, M. (1987) Astr.Ap. 180,94.

Becklin, E.E., Zuckerman, B. (1989) in preparation.

Berriman, G., Reid, N. (1987) M.N.R.A.S. 227, 315.

Burrows, A., Hubbard, W.B., Lunine, J.I. (1989) Ap.J., in press.

Copeland, H., Jensen, J.O., Jorgensen, H.E. (1970) Astr. Ap. 5, 12.

Dahn, C.C., Liebert, J., Harrington, R.S. (1986) A.J. 91,621.

D'Antona,F., Mazzitelli,I.(1982),Astr. Ap., 113,303.

D'Antona,F., Mazzitelli,I.(1983),Astr. Ap., 127,149.

D'Antona,F., Mazzitelli,I.(1985),Ap.J., 296,502.

D'Antona,F., Mazzitelli,I. (1986), Astr.Ap. 162 ,80.

D'Antona, F. (1987) Ap.J. 320, 653.

Dorman, B, Nelson, L.A., Chau, W.Y. (1989) Ap.J., in press.

Fontaine, G., Graboske, H.C, Van Horn, H.M. (1977) Ap.J. Suppl. 35, 293.

Forrest, W.J., Skrutskie, M.F., Shure, M. (1988) Ap.J. 330, L119.

Greenstein, J.L. (1988) A.J. 95, 1494.

Greenstein, J.L. (1989) preprint

Gould,A. (1989) M.N.R.A.S. in press

Hawkins, M.S.R. (1987) M.N.R.A.S. 234, 533.

Hawkins, M.S.R., Bessell, M.S. (1988) M.N.R.A.S. 234, 177.

Heintz, W.D. (1989) Astr. Ap. 217, 145.

Ianna,P.A., Rohde,J.R., McCarthy,D.W. (1988) A.J. 95, 1226.

Jameson,R.F., Skillen,I. (1989) M.N.R.A.S. 239, 247.

Kuijken, K., Gilmore, G. (1989) M.N.R.A.S. in press.

Latham, D.W., Mazeh, T., Stefanik, R.P., Mayor,M., Burki, G. (1989) Nature 339, 38.

Leggett, S.K., Hawkins, M.S.R. (1988) M.N.R.A.S. 234, 1065.

Leggett, S.K., Hawkins, M.S.R. (1989) M.N.R.A.S., in press.

Liebert, J., Probst, (1987) Ann. Rev. Astr. Ap. 25, 473.

Livio,M., Soaker (1988) Ap.J. 329, 764.

Magni,G., and Mazzitelli,I. 1979, Astr. Ap., 72 , 134.

Marcy, G.W., Lindsay, V., BergengrenJ., Moore,D. (1986) in "Astrophysics of Brown Dwarfs", eds. M.C.Kafatos, R.S.Harrington and S.P.Maran, Cambridge University Press, p. 50.

Marcy, G.W., Moore,D. (1989) Ap.J. 341, 961.

Mazzitelli, I., D'Antona, F. (1986) Ap.J. 308, 706.

McCarthy,D.W.,Probst,R.G., Low,F.J. (1985), Astrophys.J., 290,L9.

McCarthy, D.W., Cobb, M.L., Probst, R.G. (1987) A.J. 93, 1535.

McCarthy, D.W., Henry, T.J. (1987) Ap.J. 319, L93.

McCarthy,D.W.. Henry, T.J., Fleming, T.A., Saffer, R.A., Liebert, J., Christou, J.C. (1988) Ap.J. 333, 943.

McClure, R.D., et al. 1986a, Ap.J., 307 ,L49.

McClure,R.D., et al. 1986b, in IAU Symp. 126: *"Globular Cluster Systems in Galaxies"*, ed. J.E.Grindlay and R.J. Davis (Reidel, Dordrecht)

Mould,J.R., and Hyland,A.R. 1976, Ap.J., 208 , 399.

Neece,G.D. 1984, Ap.J., 277 , 738.

Oort, J. H. (1960) Bull. Astr. Inst. Netherlands, 15, 45.

Perrier,C., Mariotti, J.-M. (1987) Ap.J. 312, L127.

Popper, D.M. (1980) Ann. Rev. Astr. Ap. 18, 115.

Reid, N. (1987) M.N.R.A.S. 225, 873.

Reid, I.N., Gilmore, G. (1982) M.N.R.A.S. 201, 73.

Reid, I.N., Gilmore, G. (1984) M.N.R.A.S. 206, 19.

Renzini,A, Fusi Pecci, F. (1988) Ann. Rev. Astr. Ap. 26, 199

Scalo, J.M. (1986) Fundamentals of Cosmic Physics 11, 1.

Skrutskie, M.F., Forrest, W.J., Shure, M.A. (1987) Ap.J. 312, L55.

Spruit, H.C., Ritter, H. (1983) Astr. Ap. 124, 267.

Stobie,R.S.,Ishida,K.,Peacock,J.A. 1989, M.N.R.A.S. 238,709.

Zuckerman, B., Becklin, E.E. (1987) Nature 330, 138.

Zuckerman, B., Becklin, E.E. (1987) Ap.J. 319, L99.

Zuckerman, B., Becklin, E.E. (1988) Nature 336, 656.

VandenBerg,D.A., and Bell,R.A. 1985, Ap.J.Suppl., 58 , 561.

Wielen,R., Jahreiss,H., and Kruger,R. 1983, in IAU Coll. N.76 'The Nearby Stars and the Stellar Luminosity Function', ed. A.G.Davis Philip and A.R. Upgren (L.Davis Press Inc. Schenectady, New York), p.155.

H.ZINNECKER: A recent paper (Strom et al. 1989, Astron. J. 97, 1451 lists the NIR JHKL magnitudes of most of the known classical and naked T Tauri stars in the Taurus-Auriga T-association. There seems to be a deficiency of objects with faint magnitudes; many objects have K=8 or thereabouts (corresponding to about 1 $L_\odot$ at the distance of 140 pc) but very few objects have K=10 - 13 ($L = 0.1 - 0.01 L_\odot$) which one would expect to be present if the IMF keeps rising towards lower mass stars. Maybe Taurus- Auriga forms only stars with a typical luminosity of around $1 L_\odot$!
A.BOSS (reply to Zinnecker's comment): Recently W.Forrest, M.Skrutskie and colleagues have claimed the detection of half a dozen or so brown dwarfs located close to T Tauri stars, in several cases with proper motions indicating membership in the cluster (abstract submitted to Berkeley A.S.P. meeting). The detections were made using the new University of Rochester IR camera. Hence, perhaps we have some new evidence for brown dwarfs in star forming regions.
F.D'ANTONA: In any case, it is clear that we need a huge amount of observations, in several different fields, before this problem is settled.

H.W.YORKE (reply to Zinnecker's comment): Of course you must remember that stars with masses below 1 $M_{\odot}$ are nowhere near the main sequence. A $0.3 M_{\odot}$ star will need another 10^8yr. These low mass pre-main seuence objects may therefore be overluminous.

MULTIMODALITY OF STAR FORMATION

N. C. RANA
Tata Institute of Fundamental Research
Homi Bhabha Road, Bombay 400 005
INDIA

ABSTRACT. An initial mass function of main sequence stars in the solar neighbourhood is derived from the recent input data on luminosity function, and lifetime of main sequence stars. The luminosity function is corrected for the possible multiplicity of the stellar systems which are counted as single stars in a photographic plate. The mass function shows several humps indicating possibly the multimodality of star formation, and resolves the missing mass problem in a satisfactory manner.

1. Introduction

The initial mass function (IMF) of stars of any given region contains the information on the history of star formation and the physical and dynamical evolution of the stars in and around the region. Salpeter (1955), later Miller and Scalo (1979), and Scalo (1986) did comprehensive analyses of the mass functions of stars in the solar neighbourhood. Salpeter fitted his data on the frequency distribution of stars according to their masses by a power law of the form $\xi(\log m) \propto m^{-\alpha}$, with the value of the index $\alpha = 1.35$, which is known as the Salpeter index. Obviously, this form of IMF is scalefree. However, Salpeter's data were scarce on the low mass stars.

Miller and Scalo (1979) extended it more reliably to the low mass end and suggested a possible turn over of the mass function on the low mass side, which implied a solar mass scale for the formation of stars in the solar neighbourhood. Scalo (1986) in an extensive review of the IMF confirmed this. With more recent input data, Rana (1987) suggested the possibility of existence of multiple scales purely from the observational point of view, as there were signs of several bumps in his derived IMF.

The possible existence of multimodality in the IMF of stars in the solar neighbourhood was theoretically explored by Ferrini (1989, see in this volume). He finds that acoustic-gravitational, magneto-hydrodynamic, rotational and turbulence dominated fragmentation of gas clouds into stars has each got its own preferred scale of mass and a superposition of all these processes might lead to a multimodal IMF. It is also possible that due to low efficiency of the process of star formation in a given giant molecular cloud, the cloud might have to undergo star formation in several steps with increased metallicity, temperature and pressure in successive steps, which might result in a gradual shift of the preferred mass scale.

The present work aims at incorporating further input data to see whether a multimodality is really present or not in the IMF of the stars in the solar neighbourhood.

R. Capuzzo-Dolcetta et al. (eds.), Physical Processes in Fragmentation and Star Formation, 381–386.

2. Modification of the IMF

The luminosity function (LF) for the stars with absolute visual magnitude $M_V < 6$ is taken from Rana (1987), but for $M_V \geq 6$ is taken from Stobie et al (1989). The numbers in the latter paper are usually smaller for any given M_V. M_V – mass (m) calibration curve and the scaleheight distribution $2H(m)$ are both taken from Rana (1987). The main sequence lifetime $T_{ms}(m)$ has been recompiled from various sources (Chiosi 1986, Chiosi and Maeder 1986) which include effects of overshooting, variation in composition, and revised reaction rates. The data are tabulated in Table 1. The IMF is computed assuming further that the age of the disc $T_d = 12$ Gy and that the star formation rate (SFR) is constant and grossly independent of of the stellar masses. The present day mass function (PDMF) and the IMF are plotted in Fig 1.

Table 1

M_V	$\log \Phi_{LF}$	log m	$\frac{dM_V}{d \log m}$	2H in pc	$\log T_{ms}$	$\log \Phi_{ms}$	$\log \xi$
-5	-7.09	1.48	3.2	170			
-4	-6.59	1.25	5.3	180	6.98	-3.61	-0.51
-3	-6.03	1.08	6.9	190	7.28	-2.91	-0.11
-2	-5.45	0.92	5.6	200	7.63	-2.40	0.05
-1	-4.91	0.73	4.9	215	8.03	-1.89	0.16
0	-4.41	0.54	5.6	240	8.49	-1.28	0.31
1	-3.78	0.39	7.8	290	8.88	-0.43	0.77
2	-3.37	0.26	8.5	390	9.21	0.15	1.02
3	-2.98	0.16	10.0	570	9.49	0.78	1.37
4	-2.70	0.06	11.0	940	9.75	1.31	1.64
5	-2.55	-0.02	14.5	1030	10.04	1.62	1.66
6	-2.41	-0.09	14.5	1080	10.34	1.71	1.71
7	-2.34	-0.16	13.2	1100		1.67	1.67
8	-2.35	-0.24	12.0	1100		1.70	1.70
9	-2.28	-0.33	11.4	1100		1.62	1.62
10	-2.34	-0.42	11.1	1100		1.83	1.83
11	-2.12	-0.51	10.1	1100		2.04	2.04
12	-1.87	-0.62	8.6	1100		2.03	2.03
13	-1.81	-0.73	9.5	1100		1.77	1.77
14	-2.11	-0.83	10.4	1100		1.53	1.53
15	-2.39	-0.92	11.4	1100		1.41	1.41
16	-2.55	-1.00	13.5	1100		1.50	1.50
17	-2.53	-1.08	15.3	1100		1.73	1.73
		1.40			6.82	-3.95	-0.69
		1.60			6.67	-4.50	-1.09
		1.80			6.58	-5.00	-1.50
		1.95			6.51	-5.33	-1.76
		2.00			6.49	-6.06	-2.47

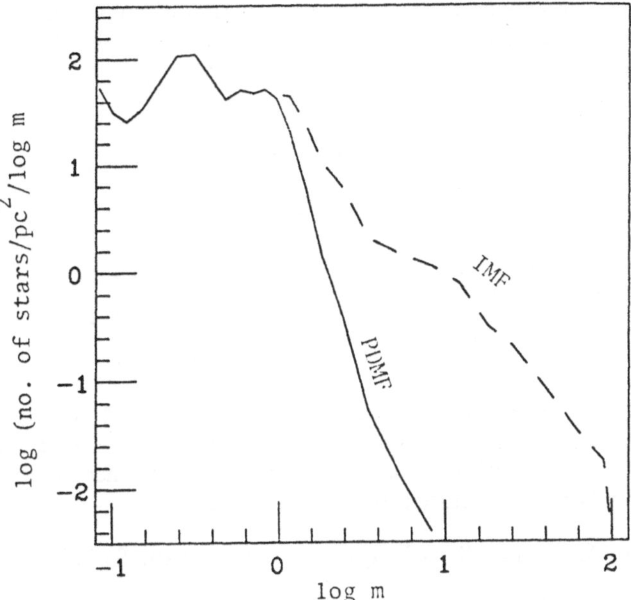

Fig.1: Unmodified mass functions.

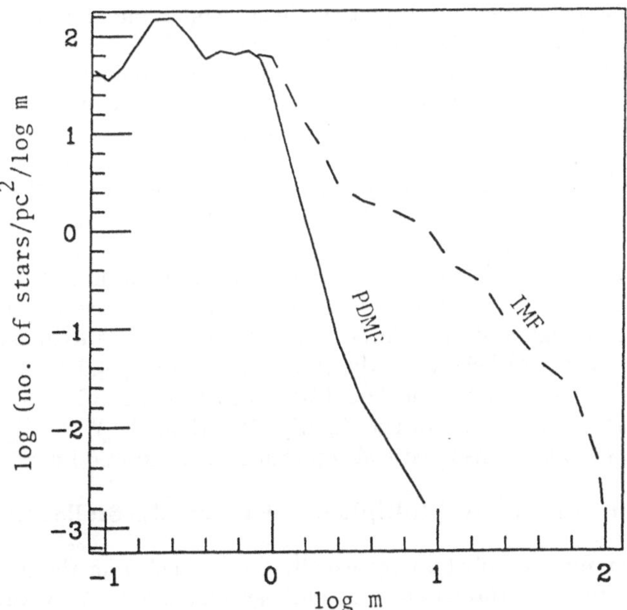

Fig. 2: Modified mass functions.

The IMF does not appear to be a smooth function of stellar masses. The possibility of the existence of a number of scales is apparent with the Wielen dip occurring at the subsolar masses. However, if we wish to fit with a single power law between $m = 1$ and 100 (in units of $M_\odot$), the form the IMF seems to be

$$\xi(\log m) \simeq 46\, m^{-1.73},$$

even though a better fit is possible with the following form

$$\begin{aligned}\xi(\log m) &\simeq 65\, m^{-2.7}, & 1.2 < m < 3.5,\\ &\simeq 5.8\, m^{-0.75}, & 3.5 < m < 12,\\ &\simeq 63\, m^{-1.8}, & 10 < m < 100.\end{aligned}$$

On integrating the PDMF and IMF over the mass range $0.08 < m < 100$, we obtain 65.7 and 72.6 stars/pc^2 respectively, with the constant SFR = 6 stars/pc^2/Gy. The corresponding surface densities of the main sequence stellar masses are given by 26.7 and 44.5 $M_\odot$/pc^2, implying the returned fraction $R = 0.40$.

3. Justification for Large Scaleheights and Constant SFR

We know that stars form from gas and that gas is continually depleted with time as low mass stars once born do not die within 12 Gy. Schmidt's law (1963) suggested that the SFR (ψ in units of $M_\odot$/Gy) is a function of the mass of available gas, that is, if μ is the gas fraction compared to the total baryonic mass in the region, $\psi \propto \mu^k$, $k = 1 - 2$. This would imply a decrease in SFR by at least a factor of 6, if $k = 1$, and the present value of $\mu = 0.15$, compared to unity at the beginning. Rana and Wilkinson (1986, 1988) suggested another form of the law of star formation in which the SFR is to depend on the molecular content of the gas and the molecular fraction is to depend on the metallicity (Z) or grain content in the region, leading to $\psi \propto \mu^{1.4} Z^{1.8}$. If the predisc metallicity is about a factor or 4 smaller compared to its present value (Rana and Wilkinson 1986), the SFR in the solar neighbourhood according to the above formula remains roughly constant.

From a study of the white dwarf luminosity function, Rana (1989) has shown that the rate of formation of white dwarfs in the solar neighbourhood has remained constant. It was also noted that the immediate progenitors of WDs has a local volume density about 2.5 times that of the WDs themselves. From this Rana (1989) argued that the progenitors have an ensemble average lifetime of only 2.5 Gy, compared to average lifetime of WDs (10 – 12 Gy), the scaleheights of WDs have been increased due to dynamical reasons. Since the ensemble average scaleheights of the progenitors correspond to $2H \simeq 450$ pc, the scaleheights of the WD distribution should be about 1 kpc, and hence, the scaleheights of low mass stars which are on an average slightly older than the WDs might be on the same order. Such a large value is also justified from the recent infrared imaging of the local disc.

4. Correction to LF due to Multiplicity of Point-like Stellar Systems

The luminosity function of stars are usually determined from the number count of the point-like stellar images recorded on some photographic plates. Very often the stars occur as multiple systems and are counted as single stars. Within about 5.2 pc of the sun, it is known that there are 32 single stars (of which 5 are WDs), 11 binaries and 2 triples. So 55 main sequence stars will be recorded as 40 stellar images. The luminosity functions based on number counts of point-like stellar images need a suitable correction for this effect, which is not easily determinable from the plates themselves. We introduce an approximate correction to the existing LFs by multiplying the numbers for a given magnitude bin by a factor of 55/40 and by increasing the value of the absolute magnitude by unity (for a double star system of equal masses, the luminosity of individual stars falls by a factor of 2, and for a triple star by 3, with an average of about 2.5 or one magnitude). We do not apply any correction to the mass-magnitude calibration, as masses are usually determined from a sample of known binary systems.

The corrected PDMF and IMF are plotted in Fig 2. The following features are noted.

(1) The behaviour of multimodality and the general shapes of both the PDMF and IMF have not changed significantly.
(2) The integrated numbers correspond to 87 and 95 stars/pc^2 for the PDMF and IMF respectively. The integrated surface densities of main sequence stellar mass are found to be 30 and 46.5 $M_\odot$/pc^2 respectively.
(3) The modified constant SFR becomes 3.9 $M_\odot$/pc^2/Gy, and the returned fraction $R =$ 0.36.
(4) The total number of WDs, obtained by integrating the difference of IMF and PDMF over the mass range of 0.95 to 8 $M_\odot$, is found to be 7.2 stars/pc^2, implying their birth rate in the solar neighbourhood about 0.6 stars/pc^2/Gy. The mass content in the WDs corresponds to 4.2 $M_\odot$/pc^2.
(5) Similarly the number of neutron stars born over the range of progenitor masses 8 to 100 $M_\odot$, turns out to be 0.25 stars/pc^2, or a birth rate of 0.021 stars/pc^2/Gy. This compares well with the observational estimate of the birth rate of pulsars in the solar neighbourhood by Blaauw (1985), which is 0.024 $\pm$ 0.009 pulsars/pc^2/Gy. Neutron stars, dead or alive, contribute to the surface density of mass of the local disc to about 0.4 $M_\odot$/pc^2.
(6) We can now check the account of the seen matter in the solar neighbourhood. The pop I PDMF accounts for 30 $M_\odot$/pc^2, pop I dark remnant (WD + NS) for 4.6 $M_\odot$/pc^2, gas in the ISM for 7 $M_\odot$/pc^2, evolved stars for 0.3 $M_\odot$/pc^2, predisc metal poor stars for 2 $M_\odot$/pc^2. The sum of all these gives 44 $M_\odot$/pc^2, which compares well with the recentmost estimate of the total dynamical mass density in the solar neighbourhood as 46 $\pm$ 9 $M_\odot$/pc^2 (Kuijken and Gilmore 1989).

5. Conclusion

The constancy of star formation rate in the solar neighbourhood is a good approximation. The IMF is possibly multimodal. The LF should be corrected for the multiplicity of the stellar systems which are counted as single stars in photographic plates. Due correction to this effect enhances the mass content in stars in the solar neighbourhood by about 15%. The local dark matter problem seems to be resolved. The corrected IMF also gives the correct number of WDs and pulsars in the solar neighbourhood.

REFERENCES

Blauw, A. (1985) in 'Birth and Evolution of Massive Stars and Stellar Groups', ed. by

W. Boland and H. van Woerden, D. Reidel Publ. Co., p211.
Chiosi, C. (1986) in 'Spectral Evolution of Galaxies', ed. by C. Chiosi and A. Renzini, D. Reidel Publ. Co., p237.
Chiosi, C. and Maeder A. (1986) Ann. Rev. Astr. Astrophys. **24**, 329.
Ferrini, F. (1989) in this volume.
Kuijken, K. and Gilmore, G. (1989) M.N.R.A.S., in press
Miller, G. E. and Scalo, J. M. (1979) Astrophys. J. Suppl. **41**, 513.
Rana, N. C. (1987) Astr. Astrophys. **184**, 197.
Rana, N. C. (1989) in 'White Dwarfs', Proc. IAU Coll. 114, ed. by G. Wegener, Springer-Verlag, p152.
Rana, N. C. and Wilkinson, D. A. (1986) M. N. R. A. S. **218**, 497.
Rana, N. C. and Wilkinson, D. A. (1988) M. N. R. A. S. **231**, 509.
Salpeter, E. E. (1955) Astrophys. J. **121**, 161.
Scalo, J. M. (1986) Fund. Cosmic Phys. **11**, 1.
Schmidt, M. (1963) Astrophys. J. **137**, 758.
Stobie, R. S., Ishida, K., and Peacock, J. A. (1989) M. N. R. A. S. **238**, 709.

DISCUSSION

M. Pérault: Could you please give us details on the physical ideas and empirical assumptions that make you expect undulations and bumps in the IMF?
Rana: Star formation takes place in the ISM under various physical conditions. The clouds collapse on different mass scales depending on whether rotation, and or turbulence, and or magnetic field dominates over the gravo-thermal conditions.
R. B. Larson: You assumed that the star formation rate in galaxies depends only on the molecular gas content. I am not sure that this is supported by the data, since a recent study by Kenney of Virgo cluster galaxies with a wide range of H_2/HI ratios finds that the SFR correlates better with the total gas content than with the molecular content alone.
Rana: Kenney has plotted the integrated molecular content of galaxies as a whole against the integrated SFR derived from H_α light. I am not sure whether H_α emission is the best measure for the SFR, whether CO to H_2 ratio is independent of the galactocentric radii or galaxy type. We studied the correlation on zone to zone basis for the Milky Way only and took average of four independent indicators of SFR and of most of the existing controversial estimates of both HI and H_2, not just one as Kenney has relied upon.
M. Mateo: Does the IMF slope change significantly by considering binaries ?
Rana: No.
M. Mateo: Are the H_2 surface density measurements based on CO observations, and if so, does this not imply that your trend between Σ_{H_2} and metallicity is just a correlation between CO and metallicity which is not unexpected?
Rana: Some people, for example the Durham group, do apply a metallicity correction to the CO to H_2 ratio. We have studied the correlation for both the types of H_2 estimates with the result, as I have shown here (see Fig 1 in Rana and Wilkinson 1988), that the correlation still exists.

5. STAR CLUSTERS: FRAGMENTATION AND STAR FORMATION.

R.B. LARSON - Formation of star clusters.

M. MATEO - The initial mass functions of magellanic cloud star clusters.

R.A.W. ELSON - Are there large variations among IMFs of LMC clusters?

C. CHIOSI - Star clusters in the Large Magellanic Cloud.

P. BATTINELLI, R. CAPUZZO–DOLCETTA - Formation and disruption of open clusters.

FORMATION OF STAR CLUSTERS

RICHARD B. LARSON
Yale Astronomy Department
Box 6666
New Haven, CT 06511, USA

ABSTRACT. Star clusters are the smallest systems in which the processes that generate a spectrum of stellar masses can be studied, and they may be of particular interest as the birth sites of the most massive stars. Observations suggest that bound clusters form in the densest core regions of giant molecular clouds, the best studied example being the Trapezium cluster in Orion. The efficiency with which giant molecular clouds form bound star clusters is estimated to be approximately 2×10^{-3}, about one-tenth of the overall efficiency of star formation. The high local efficiency of star formation that is necessary to form a bound cluster, as well as the apparent preferential formation of massive stars in such environments, may result from protostellar interactions like tidal effects that cause enhanced accretion from residual circumstellar disks onto forming stars. Another effect of interactions may be the formation of binary systems by captures resulting from the gravitational drag produced by protostellar disks; this mechanism might account for a significant fraction of binaries, and for the observed distribution of mass ratios in binaries. Similar gravitational drag effects acting on larger scales may account for the formation of highly condensed clusters of stars with central subgroups of massive stars.

1. Introduction

Star clusters are of fundamental interest for many reasons, among which are that they are good tracers of stellar populations in galaxies; they serve as fossils that can be used to reconstruct the history of stellar systems; and they provide an observational foundation for our current understanding of stellar evolution. In the context of this meeting, they are of interest as the smallest systems in which the processes that give rise to the stellar Initial Mass Function can be studied. It has often been noted that the mass spectra of stars in open clusters are similar, at least for masses above one solar mass, to the IMF of field stars (e.g., Scalo 1986; Mateo 1988, and this volume). It is therefore important to understand the mechanisms that determine the mass spectra of stars in clusters, since they are probably of quite general importance.

Processes occurring in nascent star clusters may be especially important for the formation of the most massive stars, which appear to form in a more spatially confined fashion than low-mass stars; they may in fact form preferentially in clusters, since it is possible that the IMF for the most massive stars declines less steeply with mass in clusters than in the field (Scalo 1986). There is also suggestive evidence that more massive stars

R. Capuzzo-Dolcetta et al. (eds.), Physical Processes in Fragmentation and Star Formation, 389–400.

tend to form in larger groups and in denser environments than less massive stars; for example, in the Taurus clouds, the T Tauri stars have a median mass of only ~ 0.6 $M_\odot$ and are distributed in scattered small groups, whereas in Orion and NGC 2264, the T Tauri stars are mostly more massive than 1 $M_\odot$ and are mostly located in larger, more condensed clusters (Larson 1982). In the Orion cluster, the more massive T Tauri stars appear to be more centrally concentrated than the less massive ones; most strikingly, the Trapezium, which contains the most massive grouping and the most massive single star in the entire Orion region, is located right at the center of the Orion cluster. Infrared array photographs of other regions of star formation have revealed additional examples of dense groupings of young stars with centrally located massive stars, an example being the S106 cluster (Zinnecker, this volume). Scaled-up versions of the Orion cluster with its central multiple system of massive stars are also found in the Galactic HII region NGC 3603 and in the 30 Doradus nebula of the Large Magellanic Cloud, each of which contains a large, centrally condensed young cluster with a central multiple system of very massive stars (Moffat, Seggewiss, and Shara 1985; Baier, Ladebeck, and Weigelt 1985; Weigelt and Baier 1985).

It is probable that the most direct outcome of the fragmentation of molecular clouds is the formation of low-mass stars, and that they form by the collapse of small clumps or cloud cores of comparable mass (Larson 1985). The formation of massive stars may, on the other hand, be a secondary process resulting from the continuing accretion of residual gas by forming stars in favorable locations, such as the cores of large clusters (Larson 1982, 1986). In the following sections, we consider some of the processes that might contribute to the formation of dense bound clusters of young stars, and to the formation of massive stars and binary and multiple systems in them.

2. Observations of Young Clusters and Their Birthplaces

Nearly all stars form in groups or associations of some kind, but only the densest known systems of young stars, which are actually the cores of larger associations, seem likely to survive as bound clusters. The essential requirement for survival as a bound cluster is a high local efficiency of star formation: at least half of the gas initially present must be turned into stars if the residual gas is removed rapidly, although a somewhat lower star formation efficiency of ~ 30 % will suffice if the gas is removed slowly (Wilking and Lada 1985).

In the Taurus clouds, the stellar density is low, and the efficiency of star formation is only a few percent (Jones and Herbig 1979; Myers 1982); therefore the groups of young stars that are observed in the Taurus clouds are very unlikely to survive as bound clusters after the gas is removed. A more promising candidate for a young open cluster is the much denser system of ~ 80 young stars embedded in the core of the ρ Ophiuchus cloud (Wilking, Lada, and Young 1989); however, even in this case the efficiency of star formation is only about 20 %, so it is not clear whether a bound open cluster will survive. The best known candidate for a very young (and still forming) open cluster is probably the exceedingly dense and partly obscured cluster of several hundred young stars around the Trapezium, which forms the core of the much larger Orion association (Herbig and Terndrup 1986). Infrared array photographs have greatly increased the number of known stars in the Trapezium cluster, many of which are invisible optically; this strengthens the conclusion of Herbig and Terndrup that this is the densest known system of young stars (McCaughrean 1989; Zinnecker, this volume). About 500 stars are now known within about 0.4 pc of the Trapezium, and the density of stars in the core region within 0.1 pc of the Trapezium may exceed 10^4 stars per pc^3. The total mass of the stars in the Trapezium cluster is comparable to or larger than the mass of gas in the same region, which is a few hundred $M_\odot$; therefore this cluster has a good chance of surviving as a bound open cluster. The velocity dispersion of about 1.5 km/s measured for the stars near the Trapezium (van Altena et al. 1988) is consistent with this cluster (including gas) already being in virial equilibrium, and implies a total mass of ~ 500 $M_\odot$ within a radius of 0.4 pc. The Trapezium cluster is presently exceptionally compact, but after losing its gas and some massive stars which appear to be escaping from it (van Altena et al. 1988), it will probably expand considerably to become a more typical open cluster.

Infrared array photographs show that the more massive and distant M17 molecular cloud also contains an embedded cluster, larger and even more heavily obscured than the Trapezium cluster (Gatley, De Poy, and Fowler 1988). These examples and others (E.A. Lada, this volume) suggest that the birthplaces of open clusters are the dense core regions of massive molecular clouds like the Orion cloud and the M17 cloud. The most outstanding characteristic of these cores, and the one that seems to be essential for the formation of a bound star cluster, is their combination of high mass and high density: in the OMC1 core region of the Orion cloud, more than 10^3 $M_\odot$ of gas is contained in a region whose average density exceeds 10^4 molecules per cm^3, while in the M17SW core region of the M17 cloud, 10^4 $M_\odot$ of gas is contained in a region of similarly high average density (Larson 1981). A related property of OMC1 and M17SW is that they both have unusually large internal velocities for their sizes (Larson 1981).

How are such massive and dense cloud cores formed? Almost certainly, strong dissipation of the internal motions in at least part of a giant molecular cloud is required; also, loss of some of the magnetic energy contributing to the support of the cloud may be required. The observed cometary shapes of both the ρ Ophiuchus cloud and the Orion A cloud suggest that interaction with a surrounding medium is another effect that exerts an important influence on the structure and evolution of these clouds. In both cases, the dense core containing the cluster of newly formed stars constitutes the head of the "comet", and long filamentary streamers emanate from it in one direction to form the tail. Detailed CO maps of the Orion A cloud give the strong impression that hydrodynamic interaction with a wind or expanding superbubble blowing past the cloud has played a role in generating its intricately filamentary structure, and may also have laterally compressed the filaments (Bally et al. 1987). As noted by these authors, a likely source for such a wind would be the older part of the Orion OB association which lies just beyond the head of the comet. Similar cometary shapes were obtained in a numerical simulation by Woodward (1978) of an interstellar cloud swept over by a shock in the intercloud medium; in this calculation, a dense clump of gas was formed at the head of the cloud, partly by lateral compression.

The Trapezium cluster is located at the center of the most massive and prominent filament in the Orion cloud, a narrow ∫-shaped structure forming the head of the comet, which also contains a string of dense clumps (Batrla et al. 1983; Wilson and Johnston 1989). The large velocity gradient along this filament suggests that rotation is important in it (Bally et al. 1987), and the spiral shape of the filament suggests that gravitational torques may help to redistribute angular momentum within it and thus allow gas to collect at its center and form a cluster of stars there; indeed, the entire structure bears some resemblance to a barred spiral galaxy.

Clearly, many complex processes must have played a role in the formation of the Trapezium cluster, and little is yet understood about how such clusters are formed. One effect that may often be involved is that external compression of a molecular cloud may indirectly trigger the formation of a star cluster by helping to dissipate the turbulent and magnetic energy that support the cloud against collapse (Elmegreen 1989). The filaments in the Orion cloud may represent regions in which such dissipation has been particularly effective.

3. Efficiency of Cluster Formation

Given the complexities involved, an understanding of cluster formation is probably best approached empirically rather than theoretically at present. Accordingly, it is important to try to establish from the available data some of the systematic characteristics of cluster formation, such as the efficiency with which molecular clouds form bound clusters.

The total amount of molecular gas in the Orion A cloud containing the Trapezium cluster is about 10^5 $M_\odot$, and another 10^5 $M_\odot$ of molecular gas is located in the associated Orion B cloud, which contains a number of smaller embedded clusters (E. A. Lada, this volume). Of these clusters, only the Trapezium cluster seems likely to survive for a long time as a bound open cluster, so the surviving cluster mass will probably be a few hundred $M_\odot$. Thus in the Orion region, the efficiency of conversion of molecular gas into bound clusters appears to be of the order of $(1 - 2) \times 10^{-3}$. The M17 cloud is somewhat more massive than the Orion complex, and it may also be forming a somewhat more massive cluster; this would be consistent with its having a similar efficiency of cluster formation. On a much larger scale, the efficiency of cluster formation in the 30 Doradus region may also be of the order of 10^{-3}, since both the total mass of gas in this region and the estimated cluster mass are about two orders of magnitude larger than in Orion (Larson 1988).

We can also use estimates of the cluster formation rate in the solar neighborhood to derive the cluster formation efficiency. According to Elmegreen and Clemens (1985), the mass spectrum of open clusters is approximately a power law with the same slope as the mass spectrum of molecular clouds; this fact is consistent with the hypothesis that, on the average, molecular clouds with different masses convert the same fraction of their mass into bound clusters. Most molecular clouds probably do not form more than one cluster large enough to contain O stars, since the formation of such a cluster rapidly leads to the destruction of the cloud: observations show that molecular clouds are largely dispersed within ~ 5 – 10 Myr after forming a typical open cluster (Leisawitz, Bash, and Thaddeus 1989). The rate of formation of open clusters in the local Galactic disk has been estimated to be roughly 0.4 clusters per kpc^2 per Myr (e.g. Miller and Scalo 1978; Elmegreen and Clemens 1985; Battinelli and Capuzzo-Dolcetta, this volume). This is about equal to the rate of formation of molecular clouds with masses greater than 10^5 $M_\odot$, assuming a typical cloud lifetime of 20 Myr (Larson 1981). Since a typical open cluster mass is a few hundred $M_\odot$, this result suggests a typical cluster formation efficiency of a few times 10^{-3}. As noted by Battinelli and Capuzzo-Dolcetta, however, some of the clusters included in this estimate are likely to be unbound and may not survive for a long time.

Thus, on the basis of somewhat limited evidence, we estimate that the efficiency of formation of bound clusters in molecular clouds is of the order of 2×10^{-3}. This is an order of magnitude smaller than the overall efficiency of star formation, which is typically estimated to be about 0.02 (e.g., Myers et al. 1986). This difference is consistent with earlier estimates that about 10 percent of all stars are formed in open clusters (e.g., von Hoerner 1968; Miller and Scalo 1978). An important implication of this result for the conditions under which stars form is that at least 10 percent of stars must form in regions where the local efficiency of star formation is quite high, i.e. of order 1/2 or more; as we have seen, observations suggest that such efficiencies are achieved only in the very densest parts of giant molecular clouds.

The above considerations apply to open clusters forming in giant molecular clouds like those observed nearby, but they may also apply more generally to massive clusters such as those in the Magellanic Clouds (including the 30 Doradus cluster), and even to globular clusters. In fact, our current knowledge of cluster systems in galaxies does not indicate any sharp distinction between globular and open clusters, so there is no reason to suppose that globular clusters form in a fundamentally different way from open clusters. If a low efficiency of cluster formation such as that inferred above also applies to globular clusters, this would imply that the globular clusters in our Galaxy formed as parts of much larger star-forming systems with masses of 10^8 $M_\odot$ or more, which may have resembled dwarf galaxies and which may have constituted the building blocks of the Galactic halo (Larson 1988).

4. Protostellar Interactions in Dense Clusters

What physical mechanisms might produce a very high efficiency of star formation in dense environments? And how might we understand the apparent preferential formation of massive stars in dense regions? A high efficiency of star formation and larger typical stellar masses could both be explained if the efficiency with which residual cloud gas is accreted by forming stars were strongly enhanced in dense environments. This could result, for example, from the effect of protostellar interactions in redistributing angular momentum in flattened protostellar envelopes or disks, thus causing enhanced accretion onto the central stellar cores (Larson 1982). For example, hydrodynamic interactions might create turbulence in protostellar disks and thereby provide a source of viscosity to drive accretion. Another possibility is that tidal perturbations may trigger episodes of enhanced accretion from such disks (Larson 1982). Tidal effects are, in fact, known to play an important role in driving gas inflows in the disks of spiral galaxies and in triggering bursts of star formation near their centers.

Considerable evidence now exists that at least low-mass young stars often possess residual circumstellar disks with radii of order 100 AU or more and masses of order 10^{-2} to 10^{-1} $M_\odot$ (Strom, Edwards, and Strom 1989; Sargent 1989). Many of the unusual characteristics of the more extreme T Tauri stars may be caused by accretion from such disks, and in fact accretion is the most plausible energy source for the bipolar outflows that are almost ubiquitously present around young stellar objects. Disk accretion would then be an important part of the star formation process itself, adding a significant fraction of the final mass to a forming star. The evidence indicates that gaseous protostellar disks are largely accreted or otherwise removed from around young stars over a time interval that is typically between 10^6 and 10^7 years.

In binary systems containing disks, a possible mechanism for driving an accretion flow is the tidal excitation of spiral shock waves in the disk (Sawada et al. 1987; Spruit et al. 1987; Rozyczka and Spruit 1989). In a forming binary system, this mechanism should be very effective in causing accretion or dispersal of any residual disk gas that experiences strong tidal effects. However, even for a single star with a remnant circumstellar disk, tidal encounters with passing stars can disturb the disk and generate spiral acoustic waves in it, possibly with shocks, just as tidal interactions between galaxies can generate spiral density waves in galactic disks (e.g. Byrd, Saarinen, and Valtonen 1986). For a steady two-armed spiral shock pattern in a disk, the accretion rate has been calculated numerically by Spruit (1987) and from wave theory by Larson (1989); for a typical protostellar disk, the resulting accretion timescale is about 10^6 years at a radius of 1 AU and 10^7 years at a radius of 40 AU. Therefore tidally induced spiral waves of sufficient strength and duration can, at least in principle, disperse protostellar disks in the timescales inferred from observations.

In order to drive an accretion flow, a tidally generated acoustic wave must be strong enough to contain a shock, and this requires that its velocity amplitude exceed about 0.14 times the sound speed (Larson 1989). For a protostellar disk heated by the central star, it can be shown that in order to generate tidal disturbances of this amplitude, a perturbing star of comparable mass must pass within about 5 disk radii. In a dense environment like the Trapezium cluster, such close encounters will be frequent (Herbig 1983; Herbig and Terndrup 1986), especially if some disks have radii exceeding 1000 AU like the disk of HL Tau (Sargent and Beckwith 1987). For example, in the core of the Trapezium cluster, most of the stars will pass within 300 AU of another star at least once during the cluster age of ~ 1 Myr, and about 30 % of the stars will pass within 100 AU of another star; this is close enough to strongly disturb a disk of solar-system size. Thus in such an environment, encounters may well be sufficiently close and frequent to play an important role in the evolution of protostellar disks, and so to increase the rate at which gas is accreted by forming stars. This would increase both the efficiency of star formation and the typical stellar mass, in accordance with the evidence discussed in Sections 1 and 2.

Possible evidence that tidal effects can result in the removal of protostellar disks is provided by the fact that, while the frequency of binaries is normal among the weak-lined T Tauri stars, few close binaries have been found among the strong-lined T Tauri stars whose properties may reflect the presence of disks (Herbig 1989; Mathieu, Walter, and Myers 1989). Thus there may be an anticorrelation between binarity and the presence of disks around young stars, as would be expected if tidal effects in binaries act to disperse disks. Herbig and Terndrup (1986) have also noted that the low-mass stars in the Trapezium cluster appear to be as young as typical T Tauri stars, but few of them actually show strong T Tauri characteristics; a possible explanation would be that circumstellar disks have already been largely removed from these stars by tidal interactions with neighboring stars.

5. Formation of Binaries by Capture

The passage of a star close to or through a protostellar disk surrounding another star will also generate a gravitational drag effect that can cause it to be captured into a bound binary orbit. In effect, the presence of residual disks around young stars greatly increases the cross section for the capture formation of binaries, and may make this an important mechanism of binary formation. To the extent that binary and multiple systems may be viewed as the smallest "star clusters", similar processes may also play a role in the formation of larger condensed stellar systems as well (see Section 6).

The detailed calculations of Shima et al. (1985) show that the deceleration experienced by a body of mass M moving with supersonic velocity V through a uniform medium of density ρ is given to within about 30 percent by

$$dV/dt \sim -12\pi G^2 M\rho / V^2.$$

The resulting decrease in velocity is proportional to the column density of matter traversed; for example, an object passing perpendicularly through a disk of surface density μ has its velocity changed by an amount

$$\Delta V \sim -12\pi G^2 M\mu / V^3.$$

The disk surface density μ can be estimated if we adopt a model for the disk, for example a marginally self-gravitating disk with $Q = 2$, as considered by Larson (1983, 1984). Such a disk contains about one-third of the mass of the central star in a region the size of our Solar System; since this is more mass than is typically observed in protostellar disks, we take as a more representative example a disk with the same radial structure but with only one-tenth the mass of the central star in a region of this size. If a star of mass M_2 passes on a nearly parabolic orbit through such a disk at a distance r from a central star of mass $M_1 > M_2$, its velocity is changed by

$$\Delta V \sim -(M_2/M_1)\, c(r),$$

where $c(r)$ is the sound speed at radius r in the disk. For a disk heated by the central star, as assumed by Larson (1983, 1984), it can then be shown that capture into a bound orbit will occur if the two stars pass within a capture radius that is given approximately, and almost independently of M_1, by

$$r_{cap}(\mathrm{AU}) \sim 350\, M_2(\mathrm{M_\odot})^{4/3}\, V(\mathrm{km/s})^{-8/3},$$

where V here denotes the relative velocity at infinity. This result shows that if V is not too large, capture will often occur when a star passes through a protostellar disk around another star. For disks with the same radial structure but different masses, the capture radius scales as the 4/3 power of disk mass.

The total cross section for captures is dominated for close encounters by the gravitational focusing effect, and is approximately

$$\sigma \sim 2\pi\, G M_1\, r_{cap} / V^2.$$

The capture rate Γ per star, allowing that two stars are involved in each capture, is

$$\Gamma = 2 \int N(V) \sigma(V) V dV$$

where $N(V)$ is the number density of stars as a function of relative velocity V. For a Maxwellian velocity distribution, this integral diverges weakly for small relative velocities and large capture radii, but if we assume a maximum capture radius of order 500 AU, the result is

$$\Gamma \sim 21\, G M_1 N\, r_{cap}(V_0) / V_0$$

where N is the total number density of stars and V_0 is the one-dimensional stellar velocity dispersion. As an example, if we assume that $M_1 = M_2 = \mathrm{M_\odot}$ and adopt $N \sim 10^4$ pc^{-3} and $V_0 = 1.5$ km/s, as appropriate for the core of the Trapezium cluster, we obtain a capture rate per star of

$$\Gamma \sim 0.4\ \mathrm{Myr}^{-1}.$$

This result must be regarded as only an order-of-magnitude estimate, but it does suggest that within the ~ 1 Myr age of the Trapezium cluster, an appreciable fraction (of order 40%) of the stars in the core of this cluster may be captured into binaries. Because of the many uncertainties involved in such an estimate, including the lack of any specific information about disks in dense environments and the lack of a proper treatment of the hydrodynamics of protostellar interactions, it remains unclear whether capture with disks is an important mechanism of binary formation, but this possibility at least seems worth further attention.

Binary systems formed by random captures such as those considered above would typically be quite wide and eccentric, with separations of the order of tens or hundreds of AU; close binaries with separations < 1 AU would not be formed directly in significant numbers by such a process. However, if disks are not immediately destroyed by the encounters and if significant disk material remains present for several orbital periods following a capture, continuing gravitational drag will shrink the orbits of binaries formed by capture; detailed numerical simulations will be needed to study this effect. Another likely possibility is that the stars in a young cluster do not form with random positions and velocities, but in compact subgroups that have relatively small internal motions; such subgroups would provide an especially favorable environment for close encounters and captures, and if the initial stellar velocities are sufficiently small, very close binaries could be formed. Subclustering is, in fact, almost ubiquitously observed in regions of star formation, but a more detailed knowledge of its properties will be needed to evaluate its possible importance for the formation of binary and multiple systems.

Meanwhile, we note that the distribution of mass ratios of binaries formed by capture can be predicted if the radial distribution of surface density in the capturing disks is known. The above approximate equations imply that for any primary mass M_1, the capture rate for stars of mass $M_2 < M_1$ is proportional to the capture radius $r_{\rm cap}$, which in turn is proportional to $M_2^{4/3}$; in fact, a detailed calculation shows that a capture rate more nearly proportional to M_2 is found if a maximum capture radius of order 500 AU is assumed, as before. If stars are captured randomly from a Salpeter mass spectrum with $N(M_2) \propto M_2^{-7/3}$, the total capture rate per unit M_2 is then proportional to $M_2^{-4/3}$. This implies a distribution of mass ratios $q = M_2/M_1$ of the form

$$N(q) \;\propto\; q^{-4/3}.$$

This result is valid if the radial distribution of surface density in the disk is $\mu \propto r^{-7/4}$, as appropriate for a marginally self-gravitating disk; if instead we adopt $\mu \propto r^{-3/2}$, as assumed in some models of the solar nebula (e.g. Nakano 1987), the capture rate becomes approximately proportional to $M_2^{3/2}$, and the predicted distribution of mass ratios becomes

$$N(q) \;\propto\; q^{-5/6}.$$

Both of these predictions are consistent with a recent analysis of the data for a large sample of spectroscopic binaries by Trimble (1989), which yields approximately $N(q) \propto q^{-1}$. Qualitatively, the essential prediction is that the distribution of mass ratios should be flatter than the stellar initial mass spectrum, since the gravitational capture cross section increases with stellar mass.

If gravitational capture is an important mechanism of binary formation, then the fact that the capture rate per star Γ is proportional to both the primary mass M_1 and the density of stars N suggests, assuming similar disk properties, that binary formation should occur most frequently among the most massive stars and in the densest environments. The presence of Trapezium-like multiple systems of massive stars, several of which themselves may be close binaries, at the centers of several young clusters may exemplify both of these effects. It has long been recognized that the most massive stars are rarely found in isolation (e.g. Blaauw 1964), which suggests that they form almost exclusively in binary and multiple systems, although Abt (1983) notes that the frequency of binaries is not a strong function of stellar mass. At present there does not seem to be any evidence that the frequency of binaries depends on environment, but this possibility may be worth further investigation. If no strong dependences are found, a possible reason might be that the binary formation process tends to saturate with the eventual destruction of the capturing protostellar disks.

6. Evolution of Protoclusters

Gravitational drag effects similar to those discussed above may also play a role in the formation of larger condensed systems of stars, including open clusters and even globular clusters. The essential effect of gravitational drag in a system of forming stars is to dissipate orbital kinetic energy, and thus to cause the system to shrink. The capture formation of binaries will itself cause some dynamical cooling and contraction of a forming cluster, since the kinetic energy of relative motion of the encountering pairs is removed from the cluster when binaries form. Moreover, even if capture does not occur, encounters between stars with disks will still be "sticky" and will dissipate some of the kinetic energy of relative motion, causing the whole system to contract. The residual gas need not even be in disks to have this effect, since the gravitational drag experienced by a star depends only on the column density of matter traversed, and not on its spatial distribution.

The gravitational gas drag effect discussed above is essentially similar to the "dynamical friction" effect of stellar dynamics, by which the more massive stars in a cluster are slowed down via gravitational interaction with a background of less massive stars. Both dynamical friction and gas drag result in a deceleration that is proportional to the stellar mass and to the average density of the material causing the drag. Gas drag and dynamical friction will in fact operate together in a forming cluster of stars, and just as in stellar dynamics, the net effect will be strongest for the most massive stars, which as a result will rapidly lose energy and sink toward the center of the cluster. However, unlike dynamical friction, gas drag removes energy from the orbital motions of all of the stars and not just the more massive ones, and this will tend to make the whole system contract. Since the drag effect increases with increasing matter density, the result may be the runaway development of a compact, very dense core (or cores), just as happens in a dynamical collapse. It would then seem almost inescapable that the end result of the processes that have been discussed will be the formation of a highly condensed cluster dominated by a compact central subgroup of massive stars, as is observed in the young clusters mentioned in Section 1.

From the equations given in Section 5, it can be estimated that the timescale for loss of orbital energy due to gravitational drag in the core of the Trapezium cluster is roughly 0.5 Myr for a one-solar-mass star; for more massive stars, this timescale varies inversely with stellar mass. Thus there appears to have been time within the ~ 1 Myr age of the Trapezium cluster for gravitational drag to have produced a much more centrally concentrated mass distribution, especially for the more massive stars.

If star formation typically begins in small dense subclusters, then gravitational drag could also play a role in causing these subsystems to merge into a single larger and more condensed system of stars, much as mergers of small galaxies can play a role in the formation of larger galaxies. Gravitational drag would in fact be much more effective for groups or subsystems of stars than for single stars, because of their much larger masses. It is also possible that such processes could lead to the formation of a hierarchy of subsystems of different sizes; results suggestive of such a hierarchy of subsystems, sometimes dominated by central massive objects, were found in some of the simulations of collapse and fragmentation of Larson (1978). Although these calculations were intended to follow gas dynamics, the formation of condensed objects soon caused gravitational drag effects to become important for the evolution of the system, and to play a significant role in the formation of the binary and multiple systems that frequently resulted.

Although no simulation of the formation of a large cluster of stars has yet been attempted, some of the qualitative features of such a process might resemble the formation of a cluster of galaxies as simulated using standard N-body techniques by West and Richstone (1988). In these calculations, the galaxies are represented by massive particles and the dominant dark matter is represented by a much larger number of low-mass particles; the massive particles then experience a strong dynamical friction effect. Starting from random initial conditions, the calculations show the formation first of a number of subclusters with massive particles concentrated toward their centers; subsequently these subclusters merge into a single relatively symmetrical and highly centrally condensed system in which the more massive particles are strongly concentrated at the center. These results bear a strong resemblance to the distribution of young stars in regions of star formation: as in the simulations, subgroups like the NGC 1999 group in Orion are often seen, and large clusters tend to have compact central subsystems of massive stars and extended halos of low-mass stars, like the flare stars in Orion (Larson 1982). Gas drag would of course further accentuate the formation of compact subsystems beyond what is found in such purely stellar-dynamical simulations. However, continuum hydrodynamics alone, without assistance from the gravitational drag effect that becomes important when stars or groups of stars begin to form, may not suffice to generate such extremely condensed structures. One may then speculate that the interplay between gas dynamics and particle dynamics that results in the sticky stellar dynamics discussed above may be the essential effect responsible for the formation of very dense stellar systems ranging all the way from binary systems through open and globular clusters to the dense cores of large galaxies.

REFERENCES

Abt, H. A., 1983, *Ann. Rev. Astron. Astrophys.*, **21**, 243.

Baier, G., Ladebeck, R., and Weigelt, G., 1985, *Astron. Astrophys.*, **151**, 61.

Bally, J., Langer, W. D., Stark, A. A., and Wilson, R. W., 1987, *Astrophys. J. (Letters)*, **312**, L45.

Batrla, W., Wilson, T. L., Bastien, P., and Ruf, K., 1983, *Astron. Astrophys.*, **128**, 279.

Blaauw, A., 1964, *Ann. Rev. Astron. Astrophys.*, **2**, 213.

Byrd, G. G., Saarinen, S., and Valtonen, M. J., 1986, *Mon. Not. Roy. Astron. Soc.*, **220**, 619.

Elmegreen, B. G., 1989, *Astrophys. J.*, **340**, 786.

Elmegreen, B. G., and Clemens, C., 1985, *Astrophys. J.*, **294**, 523.

Gatley, I., De Poy, D. L., and Fowler, A. M., 1988, *Science*, **242**, 1264 and cover.

Herbig, G. H., 1983, *Highlights of Astron.*, **6**, 15.

Herbig, G. H., 1989, in *The Formation and Evolution of Planetary Systems*, eds. H. A. Weaver and L. Danly, p. 296. Cambridge University Press, Cambridge.

Herbig, G. H., and Terndrup, D. M., 1986, *Astrophys. J.*, **307**, 609.

Jones, B. F., and Herbig, G. H., 1979, *Astron. J.*, **84**, 1872.

Larson, R. B., 1978, *Mon. Not. Roy. Astron. Soc.*, **184**, 69.

Larson, R. B., 1981, *Mon. Not. Roy. Astron. Soc.*, **194**, 809.

Larson, R. B., 1982, *Mon. Not. Roy. Astron. Soc.*, **200**, 159.

Larson, R. B., 1983, *Revista Mexicana Astron. Astrof.*, **7**, 219.

Larson, R. B., 1984, *Mon. Not. Roy. Astron. Soc.*, **206**, 197.

Larson, R. B., 1985, *Mon. Not. Roy. Astron. Soc.*, **214**, 379.

Larson, R. B., 1986, in *Stellar Populations*, eds. C. A. Norman, A. Renzini, and M. Tosi, p. 101. Cambridge University Press, Cambridge.

Larson, R. B., 1988, in *Globular Cluster Systems in Galaxies, IAU Symposium No. 126*, eds. J. E. Grindlay and A. G. D. Philip, p. 311. Kluwer, Dordrecht.

Larson, R. B., 1989, in *The Formation and Evolution of Planetary Systems*, eds. H. A. Weaver and L. Danly, p. 31. Cambridge University Press, Cambridge.
Leisawitz, D., Bash, F. N., and Thaddeus, P., 1989, *Astrophys. J. Suppl.*, **70**, in press.
Mateo, M., 1988, *Astrophys. J.*, **331**, 261.
Mathieu, R. D., Walter, F. M., and Myers, P. C., 1989, *Astron. J.*, in press.
McCaughrean, M. J., 1989, *Bull. Amer. Astron. Soc.*, **21**, 712; see photograph in *Sky and Telescope*, **77**, 352 (1989).
Miller, G. E., and Scalo, J. M., 1978, *Publ. Astron. Soc. Pacific*, **90**, 506.
Moffat, A. F. J., Seggewiss, W., and Shara, M. M., 1985, *Astrophys. J.*, **295**, 109.
Myers, P. C., 1982, *Astrophys. J.*, **257**, 620.
Myers, P. C., Dame, T. M., Thaddeus, P., Cohen, R. S., Silverberg, R. F., Dwek, E., and Hauser, M. G., 1986, *Astrophys. J.*, **301**, 398.
Nakano, T., 1987, *Mon. Not. Roy. Astron. Soc.*, **224**, 107.
Rozyczka, M., and Spruit, H. C., 1989, in The Theory of Accretion Disks, proceedings of a NATO workshop held in Garching, March 1989, in press.
Sargent, A. I., 1989, in *The Formation and Evolution of Planetary Systems*, eds. H. A. Weaver and L. Danly, p. 111. Cambridge University Press, Cambridge.
Sargent, A. I., and Beckwith, S., 1987, *Astrophys. J.*, **323**, 294.
Sawada, K., Matsuda, T., Inoue, M., and Hachisu, I., 1987. *Mon. Not. Roy. Astron. Soc.*, **224**, 307.
Scalo, J. M., 1986, *Fundam. Cosmic Phys.*, **11**, 1.
Shima, E., Matsuda, T., Takeda, H., and Sawada, K., 1985, *Mon. Not. Roy. Astron. Soc.*, **217**, 367.
Spruit, H. C., 1987, *Astron. Astrophys.*, **184**, 173.
Spruit, H. C., Matsuda, T., Inoue, M., and Sawada, K., 1987, *Mon. Not. Roy. Astron. Soc.*, **229**, 517.
Strom, S. E., Edwards, S., and Strom, K. M., 1989, in *The Formation and Evolution of Planetary Systems*, eds. H. A. Weaver and L. Danly, p. 91. Cambridge University Press, Cambridge.
Trimble, V., 1989, *Mon. Not. Roy. Astron. Soc.*, in press.
van Altena, W. F., Lee, J. T., Lee, J.-F., Lu, P. K., and Upgren, A. R., 1988, *Astron. J.*, **95**, 1744.
von Hoerner, S., 1968, in *Interstellar Ionized Hydrogen*, ed. Y. Terzian, p. 101. Benjamin, New York.
Weigelt, G., and Baier, G., 1985, *Astron. Astrophys.*, **150**, L18.
West, M. J., and Richstone, D. O., 1988, *Astrophys. J.*, **335**, 532.
Wilking, B. A., and Lada, C. J., 1985, in *Protostars and Planets II*, eds. D. C. Black and M. S. Matthews, p. 297. University of Arizona Press, Tucson.
Wilking, B. A., Lada, C. J., and Young, E. T., 1989, *Astrophys. J.*, **340**, 823.
Wilson, T. L., and Johnston, K. J., 1989, *Astrophys. J.*, **340**, 894.
Woodward, P. R., 1978, *Ann. Rev. Astron. Astrophys.*, **16**, 555.

DISCUSSION

Zinnecker: I wonder whether the criterion for the formation of bound clusters is as stringent as you indicated, i.e. a star formation efficiency greater than 50%. I am thinking of expanding OB associations whose cores nevertheless appear bound.

Larson: A star formation efficiency of 50% is not a strict requirement, since an efficiency of ~ 30% will suffice if gas removal is gradual, as I indicated in the written version of my talk. This is still much higher than the observed *overall* efficiency of star formation in most associations, so that they cannot remain bound. However, a dense core can survive as a bound cluster if the *local* efficiency of star formation in the core is sufficiently high.

Mouschovias: Even in a dense cluster like the Trapezium cluster, can captures result in anything but very wide binaries? How short a binary period can captures explain, according to your estimate?

Larson: Random captures in an environment like the Trapezium cluster will yield mostly wide binaries with typical separations of the order of 100 AU, assuming that protostellar disks of this size are present. Since the capture cross section is proportional to the distance of closest approach, I would guess that only ~ 1% of directly formed capture binaries would have separations as small as 1 AU. However, if the disk material is not immediately removed, continuing gravitational drag could shrink the orbits of binaries formed by capture. Another, perhaps more promising, way to make close binaries might be in fragmenting clumps or filaments whose angular momentum is small (perhaps because of magnetic braking).

Mateo: What if already existing binaries interact with a third star – will this provide a way to form closer binaries?

Larson: I have not looked into this, but it seems entirely possible. In a star cluster, such effects will tend to shrink the orbits of binaries that are more tightly bound than the cluster; this would in fact be the case for many capture binaries.

THE INITIAL MASS FUNCTIONS OF MAGELLANIC CLOUD STAR CLUSTERS

Mario Mateo
The Observatories of the Carnegie Institution of Washington
813 Santa Barbara Street
Pasadena, CA 91101
U.S.A.

ABSTRACT. Recent attempts to determine the mass functions (MFs) of Magellanic Cloud star clusters are reviewed. A critical comparison of two of these studies – Mateo (1988) and Elson, Fall, and Freeman (1989; EFF) – is presented. Taken at face value, the results of these two investigations suggest that significant variations in the MF slopes exist among young Magellanic Cloud clusters. New observational evidence for one of the clusters studied by EFF – NGC 1866 – is presented, and the issue of whether earlier apparent differences in Magellanic Cloud cluster MF slopes are real or are artifacts of the methods employed in the various studies is addressed. The new determination of the MF slope of NGC 1866 is not consistent with the EFF results, and we suggest a possible explanation for the discrepancy.

1. Introduction

An important constraint for models of star formation and molecular cloud fragmentation is the stellar initial mass function (IMF), but unlike *in situ* observations of star-forming regions, the IMF is a fossil record of the star formation process. Although fossils generally provide more ambiguous information than direct observations of living specimens, astronomers do not have the luxury to observe the complete process involved in forming a group of stars. Thus, the IMF remains our only source of information about the outcome of cloud fragmentation and stellar formation processes.

A number of observational studies have been undertaken to estimate the IMF in a variety of environments. The work of Miller and Scalo (1979) involved the analysis of field stars in the solar neighborhood, and was updated by Scalo (1986). A similar but independent study of the field star IMF has been conducted by Rana (1987, 1989). Determination of the field star IMF has the advantage of including stars that may have formed in a variety of different environments; thus, these studies can legitimately claim to represent the 'average' IMF more reliably than observations of physically distinct subgroups of stars. But field star methods are very sensitive on the assumed star formation rate (SFR): different SFRs can drastically alter the derived form of the IMF for the same set of input data. Field star studies also require subtle but significant corrections to account for effects such as the contribution of evolved stars and changes of scale heights as a function of spectral type and age. It becomes progressively more difficult to determine field star IMFs far from the sun, not only because stars become apparently fainter, but also because the corrections described above grow even more uncertain. Thus, if we want to learn how the IMF changes as a function of environment in our own Galaxy, or if we wish to determine if

R. Capuzzo-Dolcetta et al. (eds.), Physical Processes in Fragmentation and Star Formation, 401–414.

the IMF varies from galaxy to galaxy or as a function of time we are forced to adopt other approaches.

A popular alternative has been to determine the IMF from star clusters. This approach has the advantage of dealing with stars that were probably formed in a single star formation event (but see Eggen and Iben 1988), and therefore have similar metallicities and ages. Further, clusters provide one of the only direct ways of studying the temporal evolution of the IMF. Numerous studies of the luminosity functions of open clusters have been conducted (van den Bergh and Sher 1960; Taff 1974; Burki 1977; Tarrab 1982; Stecklum 1985; Francic 1989). Taken together, these results indicate possible very large variations in the IMF among Galactic open clusters, but because of the sparseness of many of these stellar systems, the statistical significance of the observed variations is not high. In addition, dynamical processes both at the time of the clusters' births (Lada, Marguilis and Dearborn 1984) and later as the clusters interact with other components in the Galaxy (Wielen 1971, 1985) can alter their mass functions (MFs) significantly. Another approach that involves more populous objects – and thus better statistics – is to determine luminosity and mass functions of the stars in massive globular clusters (McClure, *et al.* 1986). Because of their great ages, however, external and internal dynamical effects can be very effective in altering the MFs in globular clusters (Pryor, Smith, and McClure 1986), and only the MF of stars less massive than about $0.7M_\odot$ can be studied in these systems.

Magellanic Cloud (MC) star clusters are in many ways ideal for the purpose of determining the stellar IMF because many are massive systems, and, as a group, they span a large range in age. In addition, the tidal fields of the MCs are probably more benign than in the Galaxy so that external dynamical processes are less effective at altering the MFs of MC clusters than in the Galaxy. As observational techniques and instrumentation have improved, a number of studies have been undertaken in order to determine the MFs of MC clusters – some of the results from these studies are listed in Table 1. Throughout this paper, the mass function will be parameterized as a power-law of the form $\Phi(m) \propto m^{-(x+1)}$. Salpeter's (1955) classic result yielded $x = 1.35$ using this form of $\Phi(m)$ [1]. It can be seen from Table 1 that there is an apparently large range in x for the entire sample of observed clusters. Due to some apparent large discrepancies for individual clusters, it is not at all clear from Table 1 if the large observed range in x is intrinsic to the clusters or due to the various methods used to derive the MFs of the clusters. The purpose of this contributed paper is to a) describe briefly some of the problems involved in determining MC cluster MFs, and b) to present some new results for two Large Magellanic Cloud (LMC) clusters which may help clarify whether the variations seen in Table 1 are real.

2. Determining the Stellar Mass Functions of MC Clusters and Past Results

The primary difficulty encountered in the analysis of typical ground-based imaging data of MC clusters is the extreme image crowding. Central surface densities in these objects can easily exceed 10 stars/arcsec^2 brighter than $V \sim 22$, and one can only hope to measure the brightest stars in the inner regions of most clusters. Even for these bright but crowded

[1] It should be emphasized (see also Taff 1974) that recent studies of solar neighborhood field stars (Rana 1987, 1989; Scalo 1986) find the IMF slope in the range 1-10$M_\odot$ to be about 1.7-2.7: considerably steeper than the classic Salpeter value of 1.35. In addition, a detailed study of eight nearby open clusters with members selected according to proper motion criteria finds an average IMF slope of 2.0 (Francic 1989). The point here is that although the classical Salpeter IMF slope is useful as a fiducial against which other results may be compared, newer studies tend to find a much steeper IMF slope, at least in the solar neighborhood and in the 1-10$M_\odot$ range.

Table 1
Mass Function Determinations for MC Clusters

Cluster	Galaxy	x	Age (Gyr)	[Fe/H]	Reference	Notes
NGC 330	SMC	2.4	0.010	−1.3	M88	1
NGC 1711	LMC	2.5	0.021	0.0		
LW 79	LMC	1.6	1.5	0.0		
NGC 1831	LMC	3.2	0.50	−0.1		
NGC 2010	LMC	2.2	0.063	0.0		
H 4	LMC	2.0	2.5	−0.4		
NGC 1866	LMC	0.0	0.050	0.0:	EFF	2
NGC 2164	LMC	0.8	0.032	0.0:		
NGC 2214	LMC	−0.2	0.032	0.0:		
NGC 2156	LMC	0.5	0.032	0.0:		
NGC 2159	LMC	0.5	0.032	0.0:		
NGC 2172	LMC	0.5	0.032	0.0:		
NGC 1831	LMC	2.5:	0.50	−0.1		
NGC 2070	LMC	1.0	0.002	0.0:	Melnick 1985	
NGC 1835	LMC	1.6	≥ 10	≤ -1.0	Meylan 1988	3
Six Young Clusters	LMC	0.2 – 2.5	≤ 0.05	~ 0.0	Freeman 1977	4
NGC 1866	LMC	1.4 – 3	0.050	0.0:	Chiosi, *et al.* 1989	5
NGC 330	SMC	≥ 1.3	0.010	−1.3	Cayrel, *et al.* 1988	6
NGC 1866	LMC	1.4 – 2.2	0.050	0.0:	This paper	7

Notes to Table 1: 1) Based on CCD observations and automated star counts; 2) Based on photographic observations and visual star counts from the plates; 3) MF slope based on an analysis of the dynamical and structural properties of the cluster; 4) Analysis based on the *same* photographic material and *same* clusters studied by EFF. The range in x refers to the observed range of MF slopes for the sample of six young clusters used by Freeman 1977; 5) The range in x is derived from the text of the Chiosi, *et al.* (1989) paper and their Figure 16; 6) Updated results quoted from a private communication with Dr. Cayrel; 7) An explanation for presenting a range of values for x is given in the text.

stars, the counting completeness may be considerably less than 100% (see Mateo and Hodge 1986). At the other extreme, the outermost regions of clusters merge into the background making it difficult to derive statistically useful cluster-star counts at large radii. Thus, intermediate-radius annuli around clusters provide the best regions to try to count stars to faint magnitudes and yet obtain a statistically useful sample that does not require excessively large corrections for counting incompleteness.

Two recent studies have used this approach to estimate MC cluster MFs. The first was that of Mateo (1988; hereafter M88) who used multi-color CCD frames of 1 SMC and 5 LMC clusters and automatic star finding and counting techniques to determine their present-day MFs. He argued that for the four youngest clusters in the sample the observed MFs are probably very similar to their IMFs. The second study (Elson, Fall and Freeman 1989; hereafter EFF) reported MFs for six young LMC clusters based on photographic observations and visual star counts. Because the objects EFF observed are all young, they also argued that the observed MFs should be very similar to the clusters' IMFs. I will describe these studies in some detail here to try to determine if the apparent MF slope variations listed in Table 1 are real.

The raw data used by M88 and EFF have various relative advantages and disadvantages. First, the photographic approach of EFF covered a very large field so that the background contribution was (in principle) determined with very little contamination from the cluster itself, while the CCD study of M88 necessarily dealt with the smaller field provided by the CTIO prime-focus CCD camera. Second, the CCDs used in M88 have a linear photometric response and their calibration was straightforward, while the photographic data are somewhat more difficult to calibrate due to the non-linearity of photographic emulsions. Finally, the digital nature of the CCD images allowed easy manipulation of these data and various image-processing experiments were readily performed. This was not possible in EFF because their plates had not at that time been scanned and digitized – instead all star counts used by EFF were performed by eye using methods similar to those employed by Da Costa (1982).

The different methodologies used by M88 and EFF are nicely described by Elson (1989) in this volume. The essential points are that M88 derived cluster MFs by direct counts from his CCD images and then correcting the field and cluster counts for incompleteness as a function of magnitude and stellar density using the results of extensive false-star experiments (M88; Mateo and Hodge 1986; Stetson and Harris 1988). Information from two colors was used to explicitly segregate main sequence and evolved stars so that the M88 results refer strictly to main sequence MFs. In contrast, EFF determined their MFs by performing counts on plates of different exposure times, and therefore to different limiting B magnitudes. The completeness corrections employed by EFF came from the empirical relations of King *et al.* (1968) for photographic data and were *not* determined specifically for their own data. Additional (small) corrections for the presence of evolved stars were required to determine the main sequence MFs from EFF's data. In both cases, mass functions were determined by transforming the observed luminosity functions (differential in the case of M88 and integrated in the case of EFF) to mass functions using standard stellar evolutionary models (*e.g.*, VandenBerg 1985; Bertelli, *et al.* 1986; Bressan, *et al.* 1986).

It is apparent from Table 1 that there are significant differences in the mean slopes derived by EFF and M88 from their cluster samples, although it is important to note that there was only one cluster in common to the two studies. In the case of that one cluster, NGC 1831, EFF note that they data available to them were not ideal for the determination of the cluster mass function. Nevertheless, their results for this cluster agree with those of M88 over a limited magnitude range, and their analysis of the NGC 1831 data from M88 yields a MF slope of 2.5, in good agreement with the M88 result given a typical uncertainty (in the best cases) of 0.5 in the m ss function slope. The latter point suggests

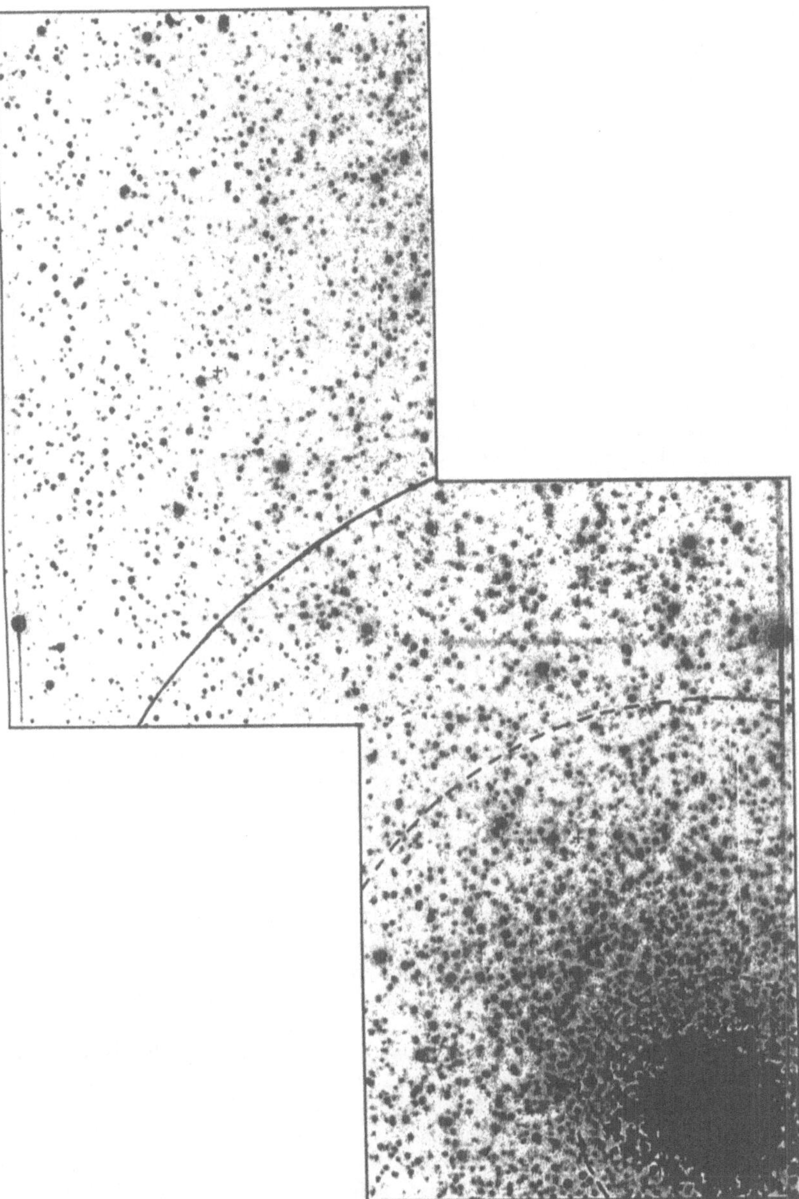

Figure 1 – The CCD fields used for the present study of the mass function of NGC 1866. The solid line is located 5 arcmin from the cluster center and denotes the adopted field/cluster boundary – only stars located further from the center of the cluster were used to determine the field star luminosity function. The dashed lines line show the annulus used to extract cluster stars ($R_{in} = 1'$ and $R_{out} = 3'$).

that the adopted mass-luminosity relations or the manner in which they were applied to the data does not explain the different mean mass function slopes measured by M88 and EFF. In this volume, Elson (1989) argues that the apparent differences between the M88 and EFF results are not due to errors in completeness corrections or contamination from evolved stars, and I agree with this conclusion based on my investigation of how these factors effect the MF determinations. Taken together, the M88 and EFF studies seem to provide evidence that genuine differences in the initial mass functions exist among young MC clusters.

Two independent studies have appeared since EFF was published that challenge this conclusion. The first is the study of Chiosi, *et al.* (1989) who find that the slope of the mass function of NGC 1866 is consistent with $x = 1.35$, although their Figure 16 suggests the main sequence MF slope of this cluster may be as large as 2-3. These values are all considerably steeper than the slope quoted by EFF for the same cluster. More recently, Cayrel, Tarrab, and Richtler (1988, and private communication) have found that the MF slope of the SMC cluster NGC 330 is larger than about 1.3 – possibly considerably larger once counting completeness corrections are applied to the observed counts. This is at least not inconsistent with the M88 results for this system and suggests that the possible differences between the EFF and M88 results may not be intrinsic to the clusters after all, but may be simply an artifact of the methods used to derive the mass functions.

3. New Results for the Young LMC Cluster NGC 1866

With this possibility in mind, I obtained a new set of deep photometric observations of MC clusters using the prime-focus CCD camera at the CTIO 4m telescope in Dec 1987. Among the program clusters was NGC 1866 which was also studied in detail by EFF. Although the basic procedures used to determine the NGC 1866 MF using these new data were similar to those employed in M88, there were some important differences. For example, because NGC 1866 is a particularly large cluster (Elson, Fall, and Freeman 1987), a single CCD frame is not large enough to observe both the cluster and field stars simultaneously. Thus, a dedicated CCD field was obtained which overlapped slightly with the cluster frame in order to determine the background star contribution without danger of contamination from the cluster itself (Figure 1). In addition, significant changes in the false-star procedure were implemented in order to take into account more explicitly the effects described in M88 and by Drukier, Fahlman, and Richer (1988), and more false-star experiments (using the photometry program DoPHOT (Mateo and Schechter 1989)) were conducted than in M88 in order to improve the statistical accuracy of the derived completeness corrections.

The deep color-magnitude diagram (CMD) of NGC 1866 is presented in Figure 2a for stars in the inner annulus shown Figure 1. Figure 2b shows the deep field star CMD for stars located beyond the outer boundary illustrated in Figure 1. The lack of bright main sequence stars and of large numbers of evolved giants in Figure 2b compared to Figure 2a confirm that the cluster contribution to the field region is slight.

Because short and long BV exposures were obtained of NGC 1866 and its field, it is possible to compare corrected luminosity functions derived from two independent sets of CCD frames. Such a comparison for the NGC 1866 field stars is shown in Figure 3 where it can be seen that the luminosity functions derived from the short and long exposure frames agree very well, and never differ by even 1σ. This is true even for the bin at $V = 21.75$ for which the shallow and deep frame completeness corrections differ by a factor of nearly two, and indicates that the completeness corrections have been reliably determined.

At the present time, completeness corrections have been evaluated for only the shallow NGC 1866 cluster frames, but this includes data for about 4.5 magnitudes of the upper

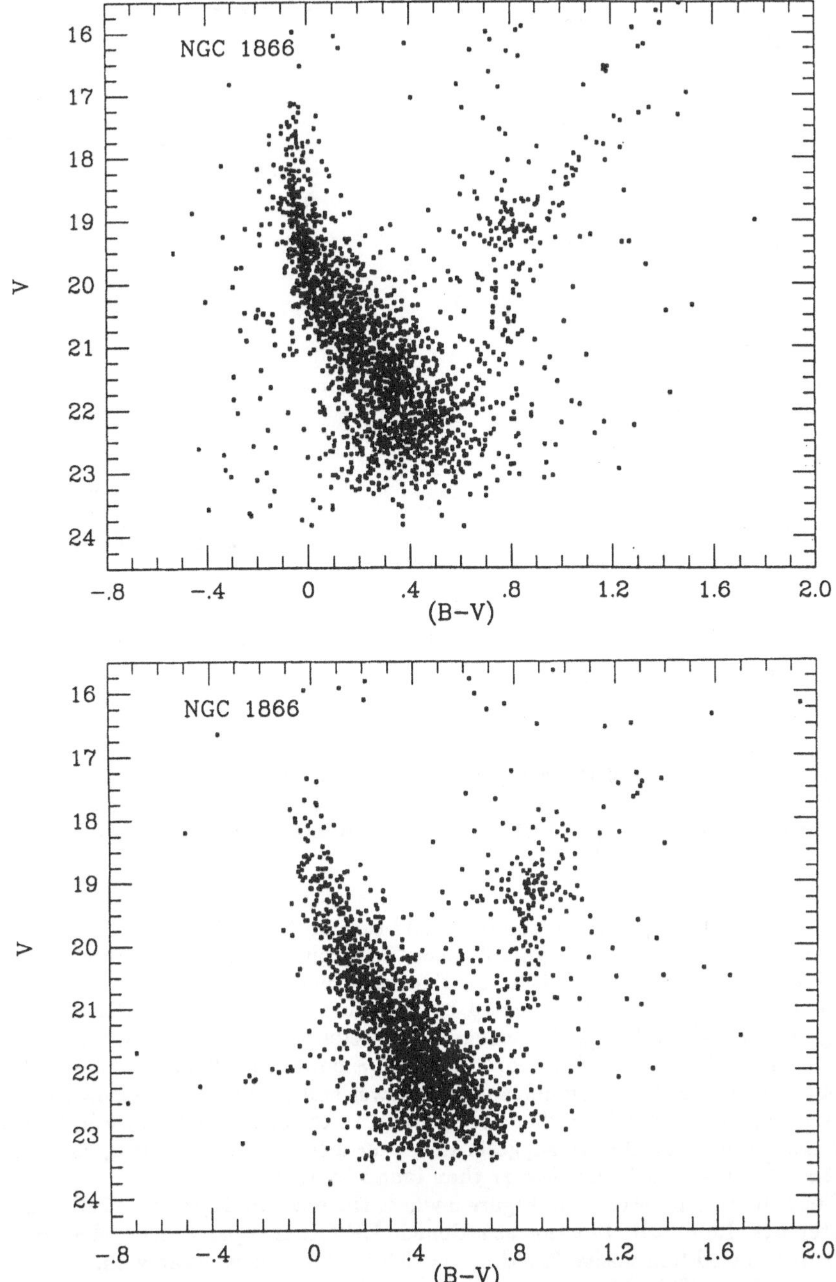

Figure 2 – a) ***top panel***; **The color-magnitude diagram of stars in the inner annulus shown in Figure 1. b)** ***bottom panel***; **The color-magnitude diagram of stars located in the field region shown in Figure 1.**

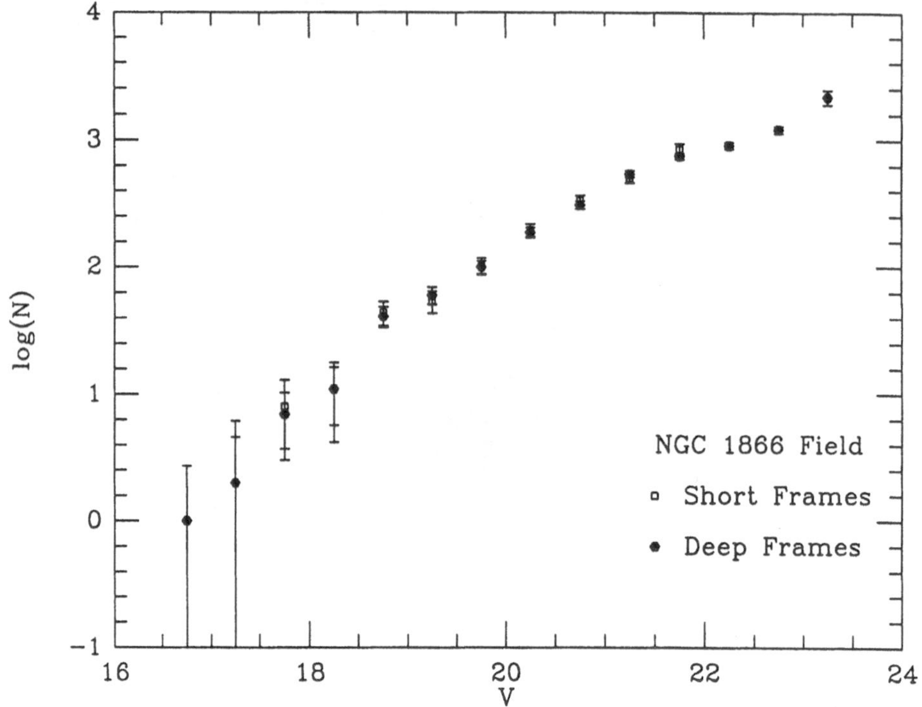

Figure 3 – A comparison of the completeness-corrected field star luminosity functions for NGC 1866 derived from short and long exposure BV CCD frame pairs. Note the excellent agreement between the two data sets over the entire magnitude range of overlap.

main sequence of the cluster. The resulting field-star subtracted cluster luminosity function is shown in Figure 4a where the symbols refer to two slightly different methods of applying the completeness corrections. This luminosity function was transformed to a mass function using the 63 Myr isochrone constructed by M88 for NGC 2010 from the models of VandenBerg (1985) and Becker (1981) using the procedures described in M88. The resulting MF is shown in Figure 4b and is consistent with a power law of slope of $x \sim 2.2$ for $m \gtrsim 2M_{\odot}$, but appears to flatten significantly for lower stellar masses. Alternatively, a power law with a slope of 1.4 can be rather poorly fitted to the entire MF shown in Figure 4b. In either case, the new CCD data suggests that over a large range of mass, the slope of the NGC 1866 MF is considerably steeper than found by EFF.

This is further illustrated in Figure 5 where the new results for NGC 1866 are plotted as an integrated luminosity function as is done in EFF. It is apparent from this diagram that the EFF integrated luminosity function for NGC 1866 is considerably shallower than the one obtained here. This difference persists *even if no completeness corrections whatsoever are applied to the new CCD data*, and leads one to the conclusion that the MF slope derived by EFF for this cluster is significantly too low.

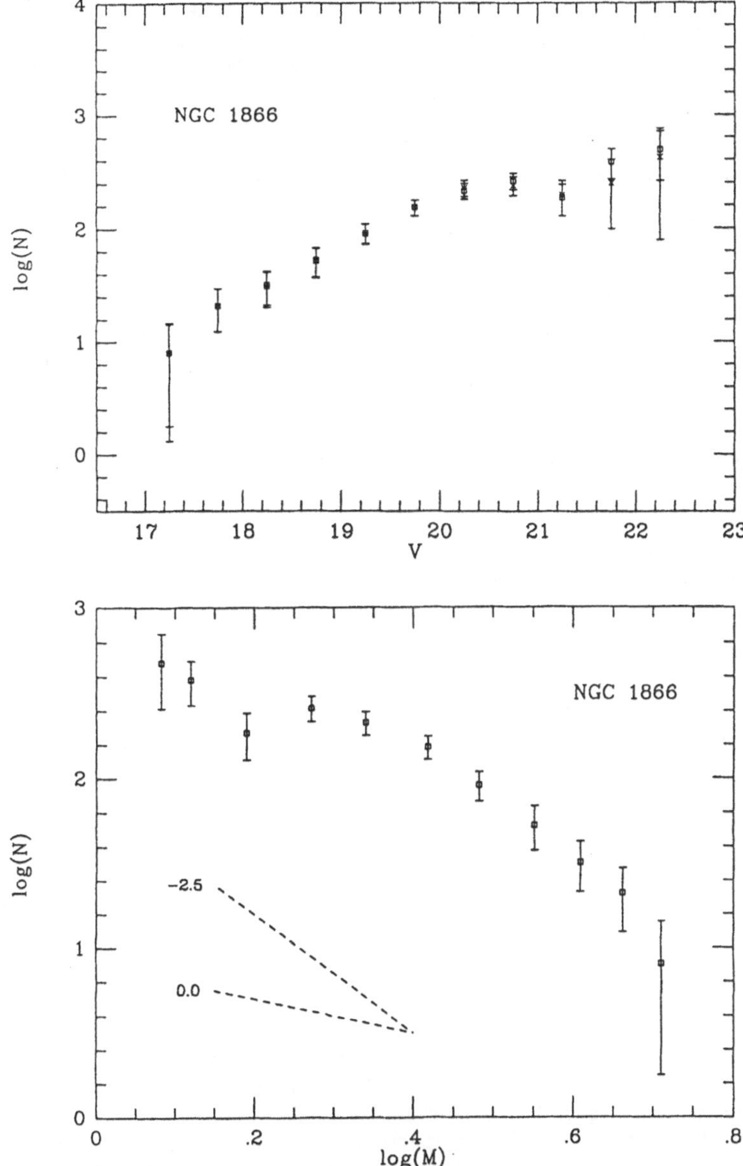

Figure 4 – a) ***top panel***; The differential luminosity function of NGC 1866 corrected for incompleteness and field star contamination. This diagram is based only on the short exposure NGC 1866 CCD frames. b) ***bottom panel***; The differential mass function of NGC 1866 based on the data in Figure 4a, and converted to masses using the 63 Myr isochrone constructed in M88 using the stellar evolutionary models of Becker (1981) and VandenBerg (1985).

How could such an error occur? As pointed out by EFF and Elson (1989), uncertainties in the limiting magnitudes of the individual plates of up to 0.5 magnitudes would not alone explain such a large difference in the derived mass function slopes. In addition, even though the completeness corrections used by EFF were not necessarily optimal for their data, even errors as large as a factor of two would not account for the differences seen in Figure 5. A more subtle error *may* affect the photographic counts in such a way that they yield systematically shallow mass functions. Specifically, a critical factor in successfully applying the star count method employed by EFF is to maintain a constant limiting magnitude as a function of position on the plate (Da Costa, private communication). Even a relatively small decrease in the limiting magnitude with decreasing radius can result

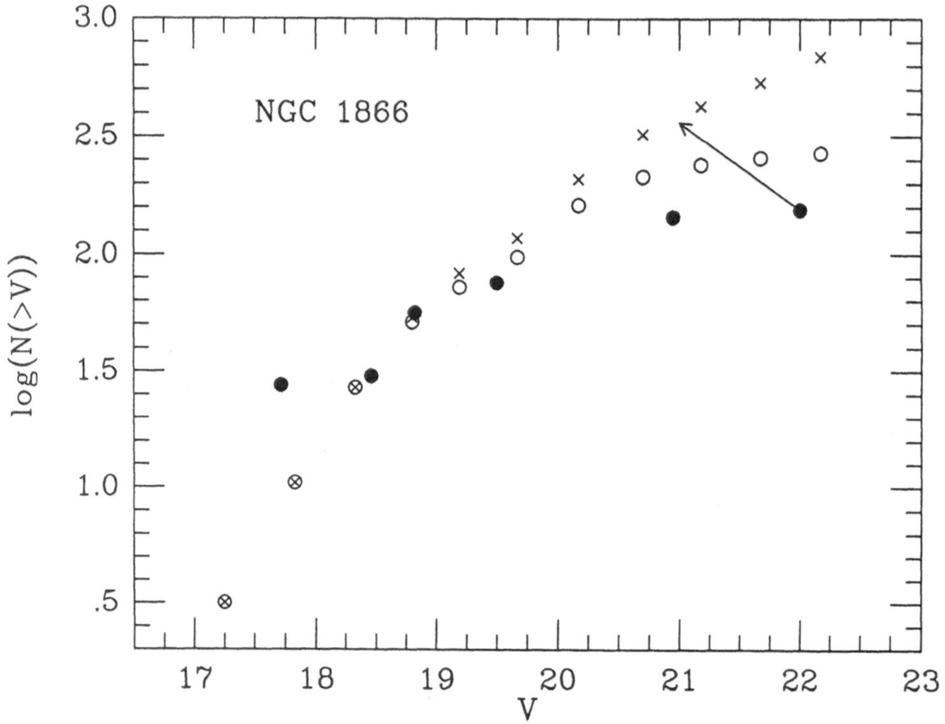

Figure 5 – A comparison of the present NGC 1866 results and those of EFF. The differential luminosity function in Figure 4a has been replotted (crosses) as an integrated V luminosity function in order to facilitate the comparison. The last three points from the present study are marked with the field and cluster completeness factors used to derive them. The correction factors for the cluster counts in the last two bins are in larger than a factor of four. All brighter bins required corrections smaller than a factor of two. The EFF results (closed circles) have been converted to the V band by taking the mean color of the main sequence as a function of magnitude from Figure 2b. The open circles show the mass function that results from the present data if *no corrections except for field star subtraction are applied.* The EFF results have been normalized to the present luminosity functions at $V \sim 18.8$. The arrow shows the change in the last EFF bin if the systematic error described in the text affects the EFF counts.

in a significant underestimate in the cluster counts because a) the limiting magnitude is inadvertently increased, and b) an excess number of field stars are subtracted since they were counted to a deeper limiting magnitude. The effect of overestimating m_{lim} by 0.5 magnitude and having m_{lim} differ by 0.5 magnitude between the cluster and field so that the limiting magnitude is brighter in the cluster annulus than in the field, is shown for the last point in Figure 5. This sort of error should be smaller for plates with brighter limiting magnitudes – thus decrease with decreasing magnitude – and result in a luminosity and mass function that appear shallower than they really are.

Whether this explains the difference between the present results for NGC 1866 and those of EFF is not certain, but it suggests a possible solution to the problem. Nor does this argument suggest that the CCD photometry and analyses performed by M88 and in this paper are free of systematic errors. For example, when adding scaled point-spread-functions ('false stars') to a CCD image, the crowding of the frame is necessarily altered, and the false stars are almost sure to be recovered with a slightly different success rate than true stars are. The subtle ways in which true and false stars can interact in crowded fields using relatively complex reduction routines such as DoPHOT and DAOPHOT is very difficult to predict or model and it would be surprising if even the most careful analyses of CCD data are free of all systematic errors when it comes to determining reliable luminosity and mass functions.

However, if the present results for NGC 1866 are taken at face value (and this analysis has addressed many of the identifiable sources of possible systematic error described by M88), then this result suggests that the MF slopes derived by EFF for the other clusters in their sample may be underestimated by a significant amount. The Chiosi, *et al.* (1989) study of the luminosity and mass function of NGC 1866 tends to confirm the findings of this paper, but clearly further observations of other clusters in the M88 and EFF samples by other workers, possibly using different techniques, are needed to settle the issue in these other cases.

4. Magellanic Cloud Clusters as Constraints on Star Formation Theories: Not Quite Yet

Because of the important practical difficulties involved, the full potential of using MC cluster observations to constrain models of star formation and cloud fragmentation through definitive and statistically robust studies of the IMF of these stellar systems clearly has yet to be realized. This paper has summarized the various areas of disagreement in the detailed results of the past few years in this field and has highlighted the problem by comparing the methods and results of two recent studies (M88 and EFF).

There does seem to be hope that in the near future the sample of clusters for which luminosity and mass functions are available will increase substantially both as observational techniques are refined and the quality of imaging data improves using telescopes and instrumentation designed to improve ground-based seeing substantially. Moreover, new computers and more automated reduction codes are already providing faster and more accurate reductions and false-star experiments to be conducted routinely[2]. Motivated by the poor agreement among earlier studies, more attention is being paid to confirming and reproducing earlier results such as this paper in which we reobserved NGC 1866 as a check of the EFF results. Likewise, Elson (1989) reports that the EFF plates are being scanned

[2] As an example, the reductions in M88 were conducted using DAOPHOT running on a CRAY X-MP supercomputer, while the present study is based on DoPHOT reductions on a microVAX-3 computer.

to determine if an automated method of performing the star counts will significantly alter their results.

In conclusion, it is unfortunately probably still too early to discuss how MC cluster observations constrain physical models of star formation. Planned HST observations of MC clusters and accompanying field regions may help by providing good, high spatial resolution data with which to measure cluster MFs reliably and over a large range of radius. However, because of the high demand on HST, it cannot be expected to survey a large set of MC clusters quickly, and additional studies from the ground in good seeing conditions with larger format CCDs will remain particularly valuable for the foreseeable future.

It is a pleasure to thank Drs. Da Costa, Chiosi, and Cayrel for useful discussions. DoPHOT was developed by Paul Schechter under grant AST 83-18504 from the U. S. National Science Foundation, and I thank him for permission to use the code for this work.

References

Becker, S. A. 1981, *Ap. J. Suppl.*, **45**, 475.

Bertelli, G., Bressan, A., Chiosi, C., and Angerer, K. 1986, *Astr. Ap. Suppl.*, **66**, 191.

Bressan, A., Bertelli, G., and Chiosi, C. 1986, *Mem. Soc. Astr. Ital.*, **57**, 427.

Burki, G. 1977, *Astr. Ap.*, **57**, 135.

Cayrel, R., Tarrab, I., and Richtler, T. 1988, *ESO Messenger*, **54**, 29.

Chiosi, C., Bertelli, G., Meylan, G., and Ortolani, S. 1989, *Astr. Ap.*, **219**, 167.

Da Costa, G. S. 1982, *A. J.*, **87**, 990.

Drukier, G. A., Fahlman, G. G., and Richer, H. B., and VandenBerg, D. A., 1988, *A. J.*, **95**, 1415.

Eggen, O. J., and Iben, I. 1988, *A. J.*, **96**, 635.

Elson, R. A. W. 1989, this volume.

Elson, R. A. W., Fall, S. M., and Freeman, K. C. 1987, *Ap. J.*, **323**, 54.

Elson, R. A. W., Fall, S. M., and Freeman, K. C. 1989, *Ap. J.*, **336**, 734. (EFF)

Francic, S. P. 1989, *A. J.*, **98**, 888.

Freeman, K. C. 1977, in *The Evolution of Galaxies and Stellar Populations*, eds. B. M. Tinsley, and R. B. Larson, (New Haven: Yale Univ. Press), p. 133.

King, I., Hedemann, E., Hodge, S. M., and White, R. E. 1968, *A. J.*, **73**, 456.

Lada, C. J., Marguilis, M., and Dearborn, D. 1984, *Ap. J.*, **285**, 141.

McClure, R. D., *et al.* 1986, *Ap. J. Lett.*, **307**, L49.

Mateo, M. 1988, *Ap. J.*, **331**, 261. (M88)

Mateo, M. and Hodge, P. 1986, *Ap. J. Suppl.*, **60**, 893.

Mateo, M., and Schechter, P. 1989, in *The First ESO Data Analysis Workshop*, in press.

Melnick, J. 1985, in *Star Forming Dwarf Galaxies and Related Objects*, eds. D. Kunth, T. X. Thuan, and J. T. T. Van, (Paris: Editions Frontieres), p. 171.

Meylan, G. 1988, *Ap. J.*, **331**, 718.
Miller, G. E., and Scalo, J. M. 1979, *Ap. J. Suppl.*, **41**, 513.
Pryor, C., Smith, G. H., and McClure, R. D. 1986, *A. J.*, **92**, 1358.
Rana, N. C. 1989, this volume.
Rana, N. C. 1987, *Astr. Ap.*, **184**, 104.
Salpeter, E. E. 1955, *Ap. J.*, **121**, 161.
Scalo, J. M. 1986, *Fund. Cosmic Phys.*, **11**, 1.
Stecklum, B. 1985, *Astr. Nach.*, **306**, 45.
Stetson, P. B., and Harris, W. E. 1988, *A. J.*, **96**, 909.
Taff, L. G. 1974, *A. J.*, **79**, 1280.
Tarrab, I. 1982, *Astr. Ap.*, **109**, 285.
van den Bergh, S., and Sher, D. 1960, *Publ. D. D. O.*, **2**, # 7, 203.
VandenBerg, D. A. 1985, *Ap. J. Suppl.*, **58**, 711.
Wielen, R. 1971, *Astr. Ap.*, **13**, 309.
Wielen, R. 1985, in *Dynamics of Star Clusters*, eds. J. Goodman and P. Hut, (Dordrecht: Reidel), p. 449.

DISCUSSION

ZINNECKER: I notice with interest that the average IMF slope of your six LMC cluster is $x = 2.5$; *i.e.* steeper than that of the solar neighborhood clusters ($x \sim 1.3$, reviewed by Scalo 1986). There have been suggestions in the literature that extragalactic HII regions/clusters have IMFs with a slope flattening with decreasing metallicity (Terlevich and Melnick 1988). Clearly, your results do not confirm this proposed trend of x *vs.* metallicity.

MATEO: There have been some recent studies on solar neighborhood clusters not reviewed by Scalo (1986) but mentioned in the written text of my talk which indicate that, in fact, the mass function slopes of nearby clusters are relatively steep (see Stecklum (1985) and Francic (1989)). The latter study, in particular, used proper motion information to weed out field stars and found an average mass function slope of $x \sim 2.0$ for eight nearby, bright clusters. Thus, I don't think your first assertion is necessarily correct, and the differences between the mass function slopes of nearby open clusters and Magellanic Cloud clusters may not be so large after all. If any real differences can be established, it remains to be seen if they are due to real variations in the IMF or due to dynamical effects.

All I can say regarding your second point is that the Terlevich and Melnick trend deals with the most massive stars in a given cluster associated with an HII region (*i.e.*, greater than about $10M_{\odot}$). None of my results refer to stars this massive, so it is not really possible to use my findings to make any statements about how the IMF slope varies with metallicity in the way proposed by those authors.

CAPUZZO-DOLCETA: Do you think that at least part of the disagreement between yours and EFF's results on the mass function of NGC 1866 can be due to the different total number of stars in the samples?

MATEO: No. Referring to Figure 5, both studies used samples that involved hundreds of stars by the time the bins fainter than $V \sim 19$ are reached. Thus, counting errors alone would generate error bars smaller than 0.04 in log N which does not explain the observed difference. Systematic errors in the completeness corrections used in one or both of the studies are more likely to be possible culprits, but it seems unlikely that even this would explain the discrepancy completely (see the contribution by Elson in this volume).

ARE THERE LARGE VARIATIONS AMONG IMFS OF LMC CLUSTERS?

REBECCA A. W. ELSON
Institute for Advanced Study
Princeton, N.J. 08540, U.S.A.

ABSTRACT. Two studies of initial mass functions (IMFs) in Magellanic Cloud clusters have appeared recently in the literature: Mateo (1988), and Elson, Fall and Freeman (1989). The first reports exceptionally steep IMFs ($x \approx 2.5$) for six clusters, while the second reports exceptionally flat IMFs ($x \approx 0$) for a different sample of six clusters, drawn from roughly the same population. A brief comparison of these studies addresses the question of whether the differences between the results reflect real cluster-to-cluster variations in the IMFs, or simply differences in the method for deriving them. Possible sources of a discrepancy are the difficulty in accounting for incompleteness in crowded fields, and the disparate treatment of evolved stars in the two studies. It is not clear that either of these can account for the full difference between the results.

1. Introduction

Two studies of initial mass functions (IMFs) in Magellanic Cloud clusters have been published recently. One is for a sample of six clusters in the Large Magellanic Cloud (LMC) with ages $\sim 10^7$ yr: NGC 1866, NGC 2156, NGC 2159, NGC 2164, NGC 2172, and NGC 2214 (Elson, Fall, and Freeman 1989; hereafter EFF). The other is for a sample including one SMC cluster and five LMC clusters, with ages $1 \times 10^7 - 2.5 \times 10^9$ yr: NGC 330, NGC 1711, NGC 1831, NGC 2010, LW 79, and H4 (Mateo 1989). EFF find a typical IMF slope for their clusters that is exceptionally flat ($-0.2 \leq x \leq 0.8$) compared to the Salpeter value ($x = 1.35$), while Mateo finds an exceptionally steep IMF ($x \approx 2.5$). (A power law IMF of the form $\Phi \propto m^{-(1+x)}$ is used throughout.) Both EFF and Mateo used the same stellar mass-luminosity relations for converting luminosity functions to IMFs. Neither Mateo nor EFF finds evidence for cluster-to-cluster variations in IMF slope within their samples greater than $\Delta x \approx \pm 0.5$, so the striking difference between their results is surprising. Clearly it is important to determine whether it reflects real differences in the IMFs, or simply a difference in the methods used to derive them.

2. Comparison of Methods and Results

EFF determined luminosity functions by counting stars on photographic plates with increasingly faint limiting magnitudes. Between four and seven plates in the B passband with $17.5 \leq B_{lim} \leq 22.4$, were counted for each cluster. (At the distance of the LMC, this

R. Capuzzo-Dolcetta et al. (eds.), Physical Processes in Fragmentation and Star Formation, 415–419.

magnitude range corresponds to a mass range of $\sim 1 - 6M_\odot$.) Figure 1 shows a typical (shallow) color-magnitude diagram (CMD) with limiting magnitudes for six plates, labeled $A, B1 \ldots B5$. (CMDs have been published for all the clusters in EFF's sample; see references in EFF). For each cluster, the shallowest plate is selected to have a limiting magnitude (B1) *fainter* than all the evolved stars in the cluster. This plate then contains *all* the stars with mass $\geq$ the mass of a main-sequence star with magnitude B1. If a plate with limiting magnitude A had been included, the number of stars on that plate would not have been representative of the total number of stars with mass greater than the mass of a main-sequence star with magnitude A, since some of the evolved stars would have been fainter than the plate limit.

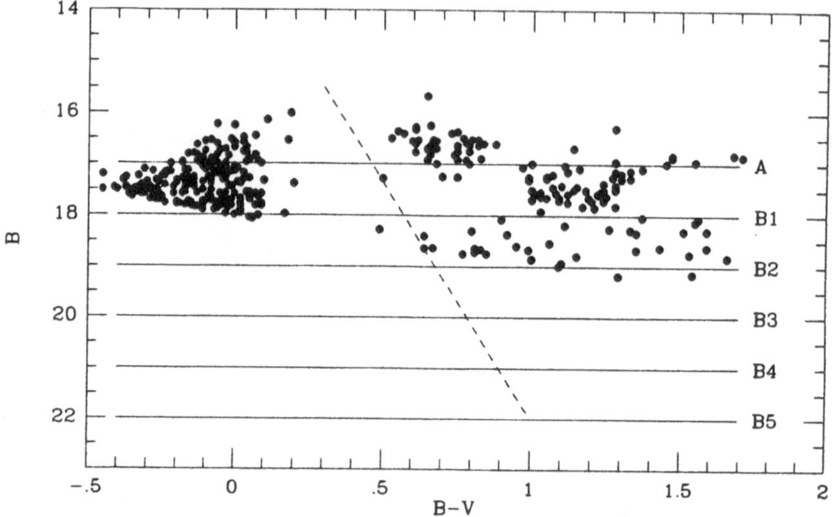

Figure 1. Typical CMD for a young LMC cluster illustrating the plates used $(B1 - B5)$ and not used (A) by EFF to determine luminosity functions. The dashed line illustrates a typical 'demarcation line' used by Mateo to eliminate evolved stars.

Figure 1 also illustrates Mateo's approach. Mateo determined luminosity functions from B and V CCD frames, and used color information to eliminate evolved stars with a 'demarcation line' (dashed line in Fig. 1). Eliminating these stars produces a slightly steeper luminosity function, and the open symbols in Fig. 2 illustrate this difference. The open squares are a cumulative luminosity function for NGC 1831, derived directly from Table 7 of Mateo (1988) and do not include evolved stars. The open circles are from the same data, but with the evolved stars included (numbers provided by Mateo). The effect is in the right sense but too small to account for the full difference between the results of EFF and Mateo.

Another possible source of systematic difference between the two results is a difference in the radial range within the clusters from which the IMFs are drawn. The angular size of the CCD used by Mateo is comparable to that of the clusters; many CCD frames would be required for the same coverage of a cluster and surrounding field afforded by a single photographic plate. In order to include enough cluster stars to minimize $\sqrt{N}$ uncertainties, one is forced with a CCD to work in the inner regions of the clusters. Thus, Mateo derives luminosity functions in a radial range $\sim 0.5 - 1.3$ arcmin, whereas EFF study annuli at $\sim 1 - 4$ arcmin. Any mass segregation within the clusters could therefore produce different IMFs, however the difference would probably be in the opposite sense. Two-body relaxation would cause the more massive stars to sink towards the centers of the clusters, thereby producing a *flatter* IMF at smaller radii. In any case, the clusters with ages $\leq 10^8$ yr are too young to have undergone significant two-body relaxation (EFF).

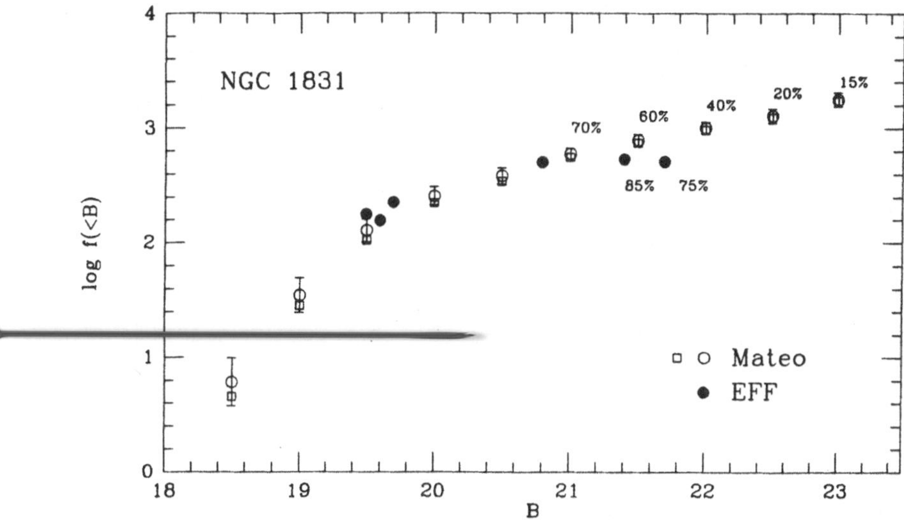

Figure 2: Cumulative luminosity functions for NGC 1831 constructed from Mateo's data calculated with evolved stars (open circles), without evolved stars (open squares), and from the starcounts of Elson, Fall and Freeman (1987) (filled circles). Approximate completeness fractions are indicated (Mateo, private communication; EFF Table 7).

Completeness corrections may be a major source of uncertainty in determining luminosity functions in dense star clusters. EFF used the prescription from King *et al.* (1968) for estimating corrections to compensate for incompleteness in the starcounts due to the crowding of stellar images. These are from an empirical relation between plate scale and the size and denisty of the faintest stellar image on the plate, derived by King *et al.* by comparing plates taken of the same fields with different angular scales, limiting magnitudes, and seeing conditions. In practice, the completeness in a given annulus within a cluster depends mainly on the limiting magnitude of the plate, and the radius of the annulus. This is illustrated in Fig. 3 for NGC 1866 (filled symbols). The completeness fractions are from Table 7 of Elson *et al.* (1987), and are typical for the clusters in EFF's sample; they range from $0.5-0.7$ for the deepest plate ($B_{lim} = 22.4$), and from $0.5-1.0$ for the shallower plate ($B_{lim} = 21.2$). Since no crowding corrections are entirely trustworthy, EFF only used starcounts with completeness fraction ≥ 0.5 in determining IMFs.

Mateo estimated completeness corrections by adding artificial stars to his CCD frames, and determining what fraction were recovered by DAOPHOT's star-finding software. For NGC 2010, completeness at limiting magnitudes B_{lim} = 21.2 and 22.4 are plotted as the open symbols in Fig. 3. These values are from Fig. 7 of Mateo (1988), and range from 0.05 to 0.55. The difference in the determinination and application of completeness corrections probably contributes to the difference between the IMFs found by EFF and Mateo. The false star method possible with CCD frames is more direct, and the corrections may be more reliable, however because the CCDs give less spatial coverage, one must work in regions at small radii where crowding is worse and the corrections are larger.

Unfortunately there are no clusters common to the samples of EFF and Mateo. However, Elson *et al.* (1987) have published starcounts for one of Mateo's clusters, NGC 1831, and EFF derived a luminosity function for this cluster for comparison with Mateo's data. This is plotted as the filled circles in Fig. 2. Approximate completeness fractions are indicated for the fainter points (Mateo 1989, private communication; Table 7 of Elson *et al.* 1987). The agreement between the two luminosity functions is good to $B \approx 21$, although fainter than this they begin to diverge.

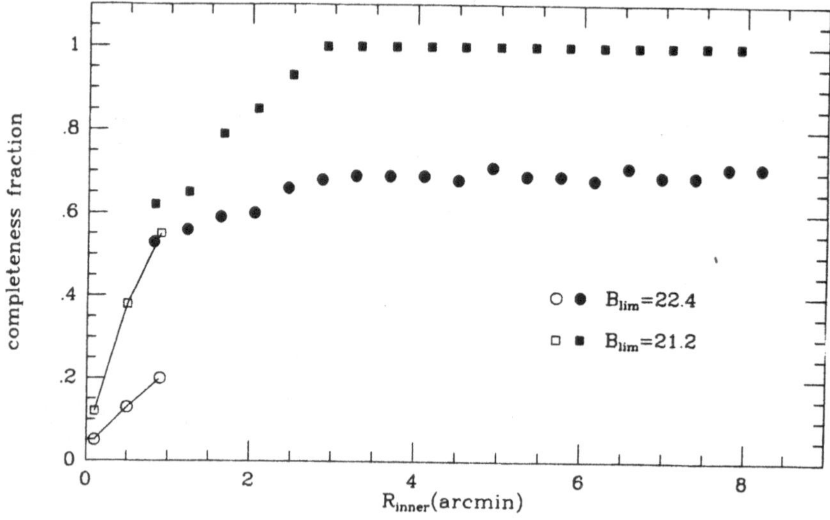

Figure 3: Completeness as a function of radius at limiting magnitudes of B = 21.2 and 22.4, for two plates of NGC 1866 (filled symbols; see Table 7 of Elson, Fall and Freeman, 1987), and for NGC 2010 (open symbols symbols; see Fig.7 of Mateo 1988).

3. Conclusions

It is likely that the difference between the flat IMFs found by EFF and the steep IMFs found by Mateo is due at least in part to the methods used to determine them. In particular, the treatment of evolved stars produces a difference in the right sense, but too small to account for the full difference observed. Completeness corrections in both studies are a source of uncertainty, although there is no a priori reason why any systematic discrepancy would be in the right sense to account for the difference in the derived IMFs.

Cluster-to-cluster variations in x of ± 0.5 are consistent with the results of both Mateo and EFF, but more work is needed to determine whether there is real evidence for larger variations. To this end, Mateo is using CCD data to derive an IMF for NGC 1866, one of the clusters from EFF's sample, and Elson is re-deriving IMFs for several of the clusters in EFF's sample using the Cambridge Automatic Plate Measuring machine.

REFERENCES

Elson, R.A.W., Fall, S.M. and Freeman, K.C. 1987, *Ap. J.* **323**, 54.
Elson, R.A.W., Fall, S.M. and Freeman, K.C. 1989, *Ap. J.* **336**, 734 (EFF).
King, I., Hedemann, E., Hodge, S.M. and White, R.E. 1968, *A. J.* **73**, 456.
Mateo, M. 1988, *Ap. J.* **331**, 261.

STAR CLUSTERS IN THE LARGE MAGELLANIC CLOUD

Cesare Chiosi
Department of Astronomy.
University of Padova.
Italy

ABSTRACT
In this paper, we discuss the main properties of the family of star clusters in the Large Magellanic Cloud (LMC). To this aim. we first review the stellar models for low and intermediate mass stars in occurrence of convective overshoot, and then describe an algorithm suitable for a theoretical approach to cluster photometry (colour magnitude (C-M) diagrams, luminosity functions (LF), integrated magnitudes and colours, and total masses). The Johnson BV CCD photometry of the stellar content of NGC 1866 is presented and used to test the theoretical models for intermediate mass stars with convective overshoot. The conclusion of this analysis is that these models fit the C-M diagram and LF of NGC 1866 much better than any other type of model in literature. In addition to this. the Johnson BV CCD photometry of NGC 1831, another populous cluster of LMC somewhat older than NGC 1866, is presented and analyzed by means of the same stellar models. In particular, we discuss the effect of unresolved binary stars on alleviating the discrepancy between the age determined from the main sequence stars and the age determined from the red giant stars. Following an idea advanced by Eggen and Iben (1988, 1989) to explain the existence of blue stragglers in young disk aggregates, we explore the possibility that delayed star formation has taken place in NGC 1831. The final part of the review deals with the integrated photometry of star clusters. Theoretical calibrations between integrated colours and age as a function of the chemical composition are briefly discussed. The aim is to derive the age distribution function of LMC cluster and interpret the physical cause of their bimodal distribution in (B-V). The main result is that bimodality simply originates from the kind of age distribution that LMC clusters happen to possess, whereas all other explanations put forward in literature play a marginal role. Furthermore, it is argued that the age distribution function show two peaks of major activity of cluster formation, whose ages are centered at about 1.5×10^{8} and 2×10^{9} yr ago. We also present new estimates of the total mass of many clusters derived from their integrated BV magnitudes. Finally a few words are addressed to the issue of the history of star formation in the field stars of LMC.

R. Capuzzo-Dolcetta et al. (eds.), Physical Processes in Fragmentation and Star Formation, 421–439.

1. INTRODUCTION

In this paper we report on a series of studies aimed at interpreting in a coherent fashion the properties of star clusters in LMC. To this purpose, on one hand we have undertaken an observational program to determine the Johnson BVRI CCD photometry for a large sample of clusters spanning the whole range of ages and chemical composition, on the other hand we have calculated many evolutionary tracks in the mass range 0.6 to 9 $M_{\odot}$ and with different chemical compositions, from the main sequence till the latest evolutionary phases. The observational material is used to derive for each cluster the C-M diagram, the luminosity functions, the surface brightness photometry, and the integrated magnitudes and colours. The theoretical models provide an homogeneous background from which the above observational properties can be derived as a function of the age, chemical composition, initial mass function, etc. The paper is organized in three parts. In the first one, the main properties of recent stellar models incorporating the effect of convective overshoot are outlined, and our theoretical approch to cluster photometry is described. The second part deals with the stellar content of two clusters of LMC, namely NGC 1866 and NGC 1831. The analysis of NGC 1866 indicates that stellar models with convective overshoot ought to be preferred, whereas the study on NGC 1831 brings into evidence the important role played by unresolved binary stars, and also suggests the possibility that star formation has occurred either on a relatively long time scale or in a series of episodes, not excluding however the possibilty that the cluster itself might be the result of merging of smaller subunits. In the third part, we address the subject of the integrated photometry (magnidutes and colours) and its dependence on age and chemical composition. New calibrating relationships are presented, which in principle enable us to estimate the age of a cluster from its integrated colours. We also discuss the nature of the bimodal distribution in (B-V) exhibited by LMC clusters, pointing out that the age distribution function (the final result of the cluster formation and disruption histories) is the main parameter determing the observed (B-V) colour distribution. In addition to this, we also discuss the possibility that the age distribution function shows evidences of two epochs of major activity of cluster formation. Furthermore, we present preliminary results on the total mass of clusters derived from their integrated magnitudes. Finally, we briefly report on a study on the law of star formation underlying the C-M diagrams and luminosity function of stellar fields in LMC. In principle, if we are able to determine the efficiency of star (and likely cluster) formation in LMC, subtracting it from the age distribution, we may be also able to infer something on the efficiency of cluster disruption by various dynamical processes all over the galaxian lifetime.

2. STELLAR EVOLUTION WITH CONVECTIVE OVERSHOOT

In the classical view of stellar evolution, semiconvection caused by the high radiation pressure and electron scattering opacity affects the mass extention of convective cores of massive stars, during both core H and He-burning phase, whereas a similar phenomenon, driven by the opacity difference between carbon-rich and carbon poor-mixtures affects those of intermediate and low mass stars, however limited to the core He-burning phase. These topics have been the subject of many theoretical studies (see the review article by Chiosi, 1986) and are still a matter of a vivid debate. However, in addition to the problem of semiconvection, there is a more fundamental question concerning the mass extension of convective cores, i.e. the amount of convective overshoot from their borders. This, in fact, is always expected to occur at the boundary of a convective core, where the kinetic energy of convective elements carries them a finite distance into layers which are formally stable against convection. Though simply motivated from the standpoint of physical arguments, great controversy exists on the exact amount of convective overshoot. According to the theoretical formulations of this phenomenon proposed by different authors, the overshoot distance at the edge of the formally convective core goes from zero to about 2 pressure scale heights (see Chiosi, 1986 for a review of the subject and exhaustive referencing). In any case, convective overshoot gives rise in stellar models to differences in structure and evolution that are much more substantial than those given by semiconvection. In addition to the above arguments of mere theoretical nature, there are many observational facts requiring a deep revision of the basic physics, such as convective cores significantly bigger than usually assumed (see the discussion by Bertelli et al., 1985, 1986a,b; Bressan et al., 1986; Maeder and Meynet, 1988).

2.1 Stellar Models with Convective Overshoot

Among the possible formulations of convective overshoot, we adopt the method developed by Bressan et al. (1981), which stands on a non local view of convection, uses the mixing length theory, and takes the mean free path of convective elements as a parameter to be fixed by comparing model results with the observational properties of star clusters. The mean free path is given by $l = \lambda Hp$, where Hp is the local pressure scale height and λ is a free parameter. The evolutionary models we are going to describe are for $\lambda = 1$, which has been found to give the most interesting results (Bertelli et al., 1985, 1986a,b: Bressan et al., 1986).

i) The Core H-Burning Phase

Stars whose initial mass is high enough to develop a convective core on the main sequence ($M > 1.1\ M_{\odot}$), are strongly affected by convective overshoot. In fact, owing to their more massive convective cores, they run at higher luminosities and live longer than classical models. They also extend the main sequence band over a wider range of effective temperatures, this trend increasing with the stellar mass. Massive stars ($M > 40\ M_{\odot}$) would spread all across the colour-magnitude (CM) diagram, were it not for the contrasting effect of mass loss (see Chiosi and Maeder, 1986).

ii) The Core He-Burning Phase in Massive and Intermediate Mass Stars

The overluminosity caused by overshoot during the core H-burning phase still remains during the shell H- and core He-burning phases. The mass of the H-exhausted core, M_{He}, and the mass of the C-O rich, He-burning convective core, are increased by overshoot. However, as a consequence of the higher luminosity, the lifetime of the He-burning phase (t_{He}) gets shorter in spite of the increase in the core mass. This, combined with the longer H-burning lifetime, t_H, makes the ratio t_{He}/t_H fairly low (from 0.12 to 0.06 when the stellar mass varies from 1.6 to 9 $M_\odot$). The location of the core He-burning models in the CM diagram can be schematically summarized as follows. In massive stars, where mass loss all over the stellar life dominates, the He-burning phase takes place entirely in the blue for M > 60 $M_\odot$, partly in the blue and partly in the red for stars in the range 60 to 20 $M_\odot$ or thereabouts, the lifetime of the two sub-phases being controlled by the mass loss rate. The blue loops of intermediate mass stars ($M < M_{HeF}$) are much less extended, whereas their dependence on chemical composition and nuclear reaction rates is the same as in classical models. This implies that the blue band of core He-burning models is less inclined in the HR diagram and, therefore, for a given chemical composition the intersection with the instability strip occurs at different luminosities, hence masses.

iii) The Core He-Burning Phase of Low Mass Stars

HB stars while burning helium in the centre, possess a convective core whose mass extension may be affected by convective overshoot. Models of HB stars have been calculated according to the Bressan et al. (1981) algorithm with λ = 1, core mass equal to 0.475 $M_\odot$ and total mass in the range 0.6 to 1.4 $M_\odot$ (Bressan et al., 1986). The most important result of these calculations is that in presence of convective overshoot, semiconvection and/or breathing pulses of convection (see Castellani et al., 1985 for details) never develop. In particular, these models predict lifetime ratios in RGB, HB and AGB that are in closer agreement with the observational ones derived from star counts in globular clusters (see Buonanno et al. 1985, for a discussion of this topic). A detailed description of these models and a comparison with the observational data for globular clusters are given in Bressan et al. (1986) to whom we refer.

iv) The Critical Masses M_{mas}, M_{up} and M_{HeF}

By virtue of the larger masses of the He and C-O core left over at the end of core H and He-burning phases respectively, the relationships between the initial mass and the mass of the He and C-O core, which define the critical masses in question are different in models with convective overshoot. The most important result is that now both M_{up} and M_{HeF} are significantly lower than in classical models. They turn out to be about 1.5 smaller than in classical models. This means that no AGB and no prolonged RGB phases are expected for stars of initial mass above the new M_{up} and M_{HeF}, respectively. The impact of this result on the observational front is straightforward and of a paramount importance.

v) The AGB Phase

Over the past few years, the evolution of AGB stars has been the subject of a great deal of theoretical studies recently reviewed by Lattanzio (1988a) aimed at understanding why carbon stars can be formed at lower luminosities than predicted by the classical models of Iben (1975a,b, 1976; Becker, 1981; Becker and Iben, 1979). Since the evolution along the thermally pulsing AGB phase is essentially governed by the linear relationship between the H-exhausted core mass, or equivalently the C-O core mass, and the luminosity (see Iben and Renzini, 1983 for all details), this is equivalent to say that the third dredge up must operate at lower core masses than suggested by the current theory. The series of recent models by Lattanzio (1986, 1987a, 1987b, 1988a,b, 1989), Boothroyd and Sackmann (1988a-d) in the mass range 1-3 $M_\odot$, in which revised algorithms for semiconvection and breathing convection, the use of better opacities, the inclusion of mass loss by stellar wind, and the estimate of the dependence of the C-O core mass at the first thermal pulse on the initial chemical composition and initial mass of the star, concur to succeed in producing models of carbon stars of quite low luminosity. Models incorporating convective overshoot (see Chiosi et al., 1987) have not yet been evolved into the thermally pulsing regime, however we may foresee that they should give similar results, since they share many features in common with models calculated with semiconvection and breathing convection. Among other things, since they predict that M_{up} is much lower than the classical value, this would help removing all very bright AGB stars as indicated by the observational data (very few AGB stars brighter than $M_b = -6$).

3. THE THEORETICAL APPROACH TO THE CLUSTER PHOTOMETRY

The models described in the previous section have been adopted to set up a code suited to study the properties of star clusters from many points of view (Chiosi et al., 1988; 1989a,b: Vallenari et al., 1989).

i) **C-M Diagrams**. Synthetic C-M diagrams are constructed, in which stars are distributed along a given isochrone according to evolutionary lifetimes and initial mass function (IMF). This is the Salpeter (1955) law in which the number of stars dN in the mass interval dM is given by

$$dN = \phi(M)\ dM, \tag{1}$$

where $\phi(M) = A\ M^{-(1+x)}$. The constant A is fixed by the normalization condition

$$N_T = A \int_{M_L}^{M_U} M^{-(1+x)}\ dM\ . \tag{2}$$

where N_T is the total number of stars in a cluster, whereas M_L and M_U are the minimum and maximum mass of stars allowed to form in each stellar generation. Star formation is supposed to take place either instantaneously or over a finite time period, thus introducing an age spread among the stars present in a cluster of any age. A gaussian distribution of width σ is assumed for the age dispersion. Finally, at any given age only stars up to a maximum mass M(t) are let exist along the isochrone, which is supposed to include stars from the main sequence till the latest observable stage, namely before either the final explosion and/or the formation of a collapsed remnant, depending on the initial mass (see Renzini 1981, and Renzini and Buzzoni, 1986 for more details). This makes the total number $N_T(t)$ of living stars in a cluster a decreasing function of the age. In practice a "Monte Carlo" technique is employed, which randomly distributes stars of fixed age and age dispersion along all possible evolutionary stages. All values of the stellar mass in the range M_L to M(t) are allowed, each of which weighed on the IMF. The procedure is repeated until the number of stars has grown to $N_T(t)$. Since under any plausible IMF, main sequence stars are by far more numerous than those in post main sequence stages and M_L is fairly small (0.1 $M_\odot$ or thereabouts), this way of proceeding is not very convenient, as an enormous number of trials would be required. On the other hand, observations of real clusters with nowadays instrumentation customarily provide data only for stars of the upper main sequence and later evolutionary stages. Even worse, in many cases only the brightest evolved stars can be seen. In the light of this, we prefer to replace condition (2) with the equivalent one that a fixed number of post main sequence stars N_{PMS} (blue and red giants and AGB's) is matched. The total number of living stars in a synthetic cluster follows from integrating relation (1) from M(t) down to M_L.

ii) **Integrated fluxes and colours**. If the M_V, M_B and M_U magnitudes of each star of the synthetic cluster are known, the integrated magnitudes and colours immediately follow from summing up the fluxes of the component stars. The integrated flux in a given passband is expressed by

$$F_{\Delta\lambda} = \sum_i \phi(M_i)\, 10^{-0.4\, M_{\Delta\lambda}(M_i,t)}\, \Delta M \qquad (3)$$

where $M_{\Delta\lambda}(M_i,t)$ is the magnitude of a star of mass M_i at time t. All normalization factors entering the definition of $\phi(M)$ and F_λ are not of interest here. The integrated colours are computed with the relations (see Searle et al., 1973)

$$(B-V)_o = 2.5\, \mathrm{Log}(F_V/F_B) \quad \text{and} \quad (U-B)_o = 2.5\, \mathrm{Log}(F_B/F_U). \qquad (4)$$

Similar expressions hold good in the case of an isochrone with a continuous distribution of stars (masses), provided the summation is replaced by an integral and $M_{\Delta\lambda}(M_i,t)$ is considered as the magnitude in the pass band $\Delta\lambda$ of the particular stage along the isochrone to which a mass M_i corresponds.

iii) **Luminosity functions for the main sequence and the evolved stars**. These turn out to be very very useful tools for discriminating among the various scenarios for the evolution of stars of any mass that have been proposed during the recent years (see section 1).

3.1 The Conversion to a Colour-Magnitude Diagram

The theoretical luminosities and effective temperatures are translated into M_V, M_B, M_U magnitudes and $(B-V)_o$, $(U-B)_o$ colours using tables of conversion, where colours and bolometric corrections are given as a function of the effective temperature (T_{eff}) and gravity (g). There are two main sets of Tables, i.e. the ones elaborated by Chiosi et al. (1988), which are based on model atmospheres calculatedby several authors (Bell and Gustafsson, 1978; Buser and Kurucz, 1978; Gustafsson et al., 1975) and extend to the UBV passbands, and those contained in the revised Yale isochrones (Green et al., 1987) which extends to the UBVRI passbands. Though similar, they differ in several details which may bear some importance in the comparison with the observational data.

4. THE STELLAR CONTENT OF YOUNG LMC CLUSTERS

4.1 NGC 1866: A Test for Convective Overshoot

Although the stellar models calculated with semiconvection and breathing convection look similar to those calculated with convective overshoot, the two types of model deeply differ in many quantitative details. The characterizing features are the core H and He-burning lifetimes and evolutionary brightening from the main sequence up to the most advanced stages. The difference in lifetimes immediately reflect onto the morphology of cluster C-M diagrams, i.e. the number of evolved to main sequence stars and the luminosity functions. It goes without saying that the ideal laboratory for testing theories of stellar structure are the rich young clusters of the Magellanic Clouds. These in fact by virtue of their large number of stars allow statistically meaningful comparisons between theory and observations in samples of about coeval and chemically homogeneous objects even for the shortest lived evolutionary stages. A classical test for stellar evolution theories is the stellar content of the LMC cluster NGC 1866. To this purpose, Chiosi et al. (1989a,b) have taken accurate Johnson BV CCD photometry of 1517 stars in the cluster, and 640 stars in a nearby field with the RCA #5 CCD at 2.2m ESO telescope, and reduced the data using the ESO version of DAOPHOT. The resulting C-M diagram and main sequence star luminosity function have been compared with the theoretical predictions based on both classical models and models with convective overshoot. To single out the sample of stars suffering from the least degree of uncertainty by photometric incompleteness and contamination by field stars, the following procedure is adopted. First, the surface brightness profile of the cluster is generated in a way similar to that by Meylan and Djorgovski (1987) and compared to the theoretical density profiles calculated from multi-mass, King-Michie models (Meylan, 1987a,b; 1988) in the case of the Salpeter law with slope x = 2.35 and isotropical velocity dispersion. This allows to determine the following parameters: concentration $c = \log r_t/r_c = 1.5$, core radius $r_c = 6.''69$, and tidal radius $r_t = 211.''6$. Second, the surface brightness profile is reconstructed from the photometric data (V and/or B magnitudes) of individual stars falling within circles of increasing radius from the cluster centre. The simultaneous comparison of the surface brightness profile, the density profile, and the surface brightness profile reconstructed from individual stars, shows that the ring confined between 0.'52 and 0.'98 is the one suffering from the least degree of incompleteness by crowding and other causes, and it can be taken to represent the cluster population. Finally, this sample is analysed for photometric incompleteness by means of the classical technique of the artificial stars. After correction for photometric completeness (crowding) and contamination by field stars by means of the zapping technique of Mateo and Hodge (1986a,b) the total number of stars amounts to 1020 and it is composed of 954 main sequence stars (MS), 41 bright red giants (BRGS), and 25 faint red giants (FRGS). The magnitude separating BRGS from FRGS is V = 17. The group of BRGS likely represents the population of evolved stars belonging to the cluster, i.e. genetically linked to the populations of MS. whereas FGRS are likely field stars. NGC 1866 contains also several Cepheids: 11 well established objects (Walker, 1987. and references therein) and about 10 more recent candidates (Storm et al., 1988), which are located in the middle of BRGS. The sample of BRGS can be split into three subgroups (labeled A, B and C). Group A and B contain almost equal numbers of stars and are located at the blue and red side of the instability strip. whereas group C is populated by stars (very few indeed) of the same luminosity but of much redder colour. Likely. these stars do not belong to the same population as those of groups A and B, but probably they are genetically related to FRGS. The luminosity function for the MS stars is given by the integrated number of stars counted from the tip of the main sequence band down to the current magnitude bin and normalized to the number of BRGS. The advantage of this type of luminosity function is that it can be simply related to the ratio of core H- to core He-burning lifetime. The comparison between theory and observations based on several fiducial characteristics of the the C-M diagram, such as the mean location of the main sequence band, its terminal luminosity, the mean luminosity of the blue and red giant stars. the tip of the blue loop (if present), and the location of the reddest stars (Hayashi limit) allows to derive the age and the chemical composition and to estimate the distance modulus in the two alternatives for the stellar evolution background, namely classical models and models with overshoot. The simultaneous fit of the above constraints indicates that the appropriate chemical composition is Y = 0.28 and Z = 0.02 (or slightly less, say Z = 0.010-0.015), the true distance modulus is (m-M)o = 18.6 mag in good agreement with the value determined from the near infrared photometry of Cepheid stars (Welch et al., 1987), and the age is 200×10^{6} yr for models with overshoot, or 70×10^{6} yr for classical models. However, with these latter the fit of the red giant stars is rather poor, as they turn out to be fainter than the mean luminosity of the loop and bluer than the Hayashi limit. Other combinations of age and chemical composition can be neglected as they would lead either to a very poor fit of the global properties of the C-M diagram or to unacceptably low distance moduli. While the C-M diagram does not clearly indicate the underlying evolutionary scheme, this is possible looking at the integrated main sequence luminosity function normalized to the number of post main sequence stars (N_{PMS}). The comparison with the observational luminosity function strongly favours models with convective overshoot. In other words, star counts in the clusters suggest that the ratio of core H to

He-burning lifetimes and the range of luminosity spanned by main sequence stars, whose masses are compatible with those of the evolved stars, must be smaller and wider, respectively, than given by the classical models. This is possible only if more efficient mixing occurs all over a star's evolutionary history and not only in stages beyond the main sequence phase. It is worth emphasizing that other mixing processes like semiconvection and/or breathing convection, would not be able to satisfy the observational demand, since they tend to increase the number of evolved stars with respect to those of the main sequence. Agreement with the classical models is marginally possible only by increasing the slope of the initial mass function above 4.

4.2 NGC 1831: Binary Stars or Delayed Star Formation ?

In order to test the effects of convective overhoot on stars in a lower range of mass, Vallenari et al. (1989) have studied the stellar content of the older cluster NGC 1831, using the same instrumentation, detectors and reduction technique as for NGC 1866. In total about 1800 stars in the cluster and of 500 stars in a nearby field have been measured. The concentration, the core radius and tidal radius of the cluster are found to be $c = \log r_t/r_c = 1.2$, $r_c = 11.''8$ and $r_t = 187.''0$, respectively. whereas the annulus with radii 36."7 and 1.'32 is seen to contain the stars best representing the cluster population. The C-M diagram shows a well defined main sequence band terminating at about V = 18.4, a red giant clump centered at about half of a magnitude below the termination of the main sequence band and spanning the colour range $0.8 < (B-V) < 0.9$, a sprinkle of stars of the same luminosity but of bluer colour scattered toward the main sequence band, and finally a scarce population of bright red stars. With the aid of the method developed by Da Costa et al. (1984), the metallicity of the cluster is estimated to be about [Fe/H] = -0.6 or equivalently Z = 0.004. The colour excess $E_{B-V} = 0.04$ is taken from Burstein and Heiles (1982), whereas the intrinsic distance modulus $(m-M)_o = 18.6$ is assumed. With the adopted true distance modulus and extinction, the termination magnitude of the main sequence band is $M_V = -0.32$, whereas the luminosity at the bottom of the clump of red stars is $M_V = 0.28$. These two critical luminosities can be used to set a preliminary estimate of the age (Chiosi et al., 1988). It turns out that, independently of the adopted chemical composition and conversion relations, the age estimated from the main sequence termination magnitude is in contrast with the age derived from the clump of red stars. In fact with the metallicity Z = 0.004, the observational and theoretical main sequence bands agree both in luminosity and colour, but the predicted red giants, though of about the same colour, are about 0.5 mag brighter than the observed ones. The age derived from the main sequence stars is approximately 8×10^8 yr, whereas that obtained from the red giants is obout 5×10^8 older. There are several causes for the disagreement between the expected and observed location of red giants with respect to the termination magnitude of the main sequence band, among which we recall: i) inadequacy of the conversion relationships; ii) inadequacy of the relationships converting luminosities and effective temperatures into visual magnitudes and (B-V) colours; iii) underestimate of the visual extinction for the red giant stars, in the sense that A_V should be much higher than simply given by $A_V = 3.1 E_{(B-V)}$ or the colour excess assigned to the cluster should be greater than $E_{(B-V)} = 0.04$; iv) inadequacy of the theoretical models for intermediate mass stars (the turn-off mass here is about 2.5 $M_\odot$) in the sense that the luminosity of the core He-burning band is too bright with respect to the main sequence tip; v) the observational termination magnitude does not correspond to the stage of core H-exhaustion of single stars, but to a more complex situation where the occurrence of unresolved binaries of suitable mass ratio and separation extends the main sequence band to brighter magnitudes. Points (i) through (iii) are found to yield a marginal effect, whereas points (iv) and (v) may be more important. Nevertheless, the analysis by Vallenari et al. (1989) has clarified that the effect of point (iv) cannot be the cause for the disagreement, whereas point (v) may potentially lead to a very satisfactory agreement between theory and observations.

The Effect of Binary Stars. Binary stars of suitable mass ratios spread the main sequence band to brighter magnitudes and alter the luminosity function. If the mass ratio is close to one, a second sequence about 0.7 magnitude brighter than the sequence of single stars is generated (Maeder, 1974). This means, in particular, that the main sequence tip is shifted to 0.7 brighter magnitudes, and therefore the real termination magnitude fixing the age must be taken 0.7 mag below the observed one. For Z = 0.004 this corresponds to an age of 1.04×10^9 yr. In Vallenari's et al. (1989) simulations, the fraction of binary stars is taken to amount to 30% and the mass ratio $Q = M_1/M_2$ is assumed in the range 0.8 to 1.25. With the inclusion of the binary stars, the apparent main sequence termination magnitude is as bright as $M_V = -0.3$ in agreement with the observational determination, the red giant stars are older hence fainter than simply determined by the apparent termination magnitude, and finally the red stars are scattered toward bluer colours as it occurs in the observational C-M diagram. However even in this case, although the magnitude difference between the apparent tip of the main sequence and the red stars has almost entirely disappeared and the age discrepancy ruled out, the bulk giants tend to be about 0.3 mag brighter than the observed ones. More important, when binary stars are considered, the apparent tip of the main sequence can no longer be taken as a firm age indicator.

The Main Sequence Star Luminosity Function. The comparison of the theoretical with the observational ILF, based on the assumed distance modulus and extinction, reveals that the case with age of 1.04×10^9 yr. the Salpeter law, and 30% of unresolved binaries fits the data very well. Although Vallenari et al. (1989) conclude that the age of 1.04×10^9 yr and metallicity Z = 0.004 are the appropriate parameters for the cluster, in that the termination magnitude of the main sequence and the luminosity function of these stars are reproduced, there still remains the disagreement between observed and expected location of the red giants. Considering the entity of the discrepancy in question, they argue that suitable adjustments of the colour equation, conversion relations, extinction, and adopted stellar models could entirely account for the above discrepancy. Nonetheless, among the many factors affecting the appearence of a star cluster's CM diagram, dispersion in age, if substantial, can easily contribute to scatter stars about the tip of the main sequence and in the region of red giants.

Prolonged Activity of Star Formation ? Interpreting the C-M diagram of a cluster the assumption is customarily made that all stars are formed in a single episode of activity spanning a narrow age range, so that all stars are essentially coeval. However, the possibility has been suggested that star formation in stellar aggregates may take place over extended periods of time, often in recurrent bursts of activity (Eggen and Iben, 1988, 1989). Exploiting further Eggen' and Iben' (1988, 1989) suggestion, Vallenari et al. (1989) have tried to investigate whether this idea could be applied to NGC 1831 to ameliorate the interpretation of its C-M diagram and luminosity function. To this aim, they have simulated many C-M diagrams at varying law of star formation. and imposing that the termination magnitude and the mean colour of the main sequence band, together with the luminosities and colours of the red giant stars are matched at a time. Furthermore, the effects of binary stars and photometrical errors are also included. It is found that the C-M diagram and main sequence star luminosity function of NGC 1831 are compatible with ages in the range 0.8-1.4x10^9 yr and metallicitiy Z = 0.004. Binary stars are assumed to amount to 30% of the total, and their presence is meant to account for the spread of red giants toward bluer colours seen in the observational C-M diagram. As for the law of star formation, the many experiments performed to this purpose show that assuming it either constant or declining toward the present by a factor of two to three with respect to the initial value, yield simulations that almost perfectly match the C-M diagram and ILF of main sequence stars of NGC 1831. The possibility of several bursts of activity cannot be excluded as they would lead to similar results provided they have occurred within the time interval of about 6x10^8 yr. As a result of this analysis, if in NGC 1831 either star formation has continued or several bursts have occured during a time scale of about 6x10^8 yr, the age of the cluster should be indicated by its oldest component, i.e. 1.4 x10^9 yr. Previous episodes of star formation, if they exist, cannot be identified with the present observational data. The duration of the star forming activity we have found with our simulations does constitute an upper limit. It could in fact be shorter if the other causes of the discrepacy between main sequence and red giant stars are alleviated. Whether the agreement we have found by allowing for a prolonged activity of star formation is a mere coincidence and other explanations ought to be put forward we cannot say and, sharing Eggen's and Iben's (1988, 1989) concern, leave the subject to future studies.

5. PROPERTIES FROM INTEGRATED COLOURS AND MAGNITUDES

In recent years, integrated properties (colours and magnitudes) as a function of the age and chemical compositions have been derived for both Pop I and Pop II stars. Among others, we recall the studies of Searle et al. (1973), Barbaro and Bertelli (1977), Barbaro (1981, 1982). Rabin

(1982), Bruzual (1983), and Arimoto and Yoshii (1986). Even if the main properties of real clusters have already been accounted for by those calibrations, still their use is hampered by many uncertainties. First of all, they are often limited to narrow ranges of age and composition, which may generate systematic errors when different relationships are used to date clusters all over the range of possible ages. Secondly, not always all evolutionary phases implied by the types of star expected to occur in a cluster of given age are taken into account in the derivation of the integrated fluxes. In most cases, only the core H and He-burning phases are considered. Thirdly, evolutionary sequences calculated by different authors are amalgamated to derive the evolutionary scenario to start with. This also can be a source of systematic errors. Lastly, they rest on rather old evolutionary calculations, while, on the contrary, much progress has been made in this context over the recent years. In addition to this, there are properties of the star clusters that have not yet been fully understood. In particular, the clusters of LMC and SMC show two distinct groupings in the plane of integrated (B-V) colours and V magnitudes (van den Bergh, 1981). In fact there are clusters either bluer than about 0.4 or redder than about 0.6 and very few in between, independently of the total V magnitude. This is also visible in the correlation between the (B-V) colour and the cluster type SWB according to the classification proposed by Searle et al. (1980), where it appears that all SWB I, II and III are blue, all V, VI and VII are red, and that the blue to red transition occurs within the cluster type IV. Furthermore, the Searle et al. (1980) classification has not yet been translated into a well defined age and/or composition ranking, due to the lack of sufficient theoretical modeling of integrated properties as a function of these two parameters. In the following we will summarize the studies by Chiosi et al. (1988) and Alongi and Chiosi (1989a,b), in which the above topics have been analysed in a coherent fashion.

5.1 The Theoretical Colour-Age Relationships

The colour-age relationships at varying chemical composition and efficiency of mass loss in the RGB and AGB phases have been calculated by Chiosi et al. (1988) to whom we refer for the discussion below. Remarkable features of the $(B-V)_o$ versus age relations are:

i) In general the $(B-V)_o$ colour gets redder at increasing age. However, under suitable combinations of the mass loss rate and chemical composition (metallicity), the colour tends to become blue again as the cluster gets older than a few billion years. This is caused by the presence in the cluster of very blue He-burning stars (mostly in HB). As expected, this effect occurs at younger ages as the mass loss rate increases.

ii) Looking at the separate contribution to $(B-V)_o$ colour from stars on the main sequence, in RGB and/or core He-burning, and later stages, we notice that in clusters of high metallicity the dominant trend is dictated by the main sequence stars with major deviations (in the sense that the colour gets much redder) at the onset of RGB and core He-burning in a red HB. In clusters of low metal content the main sequence trend, which remains also after the inclusion of core He-burning stars with mass greater than M_{HeF}, is first destroyed by the appearance of early and late AGB stars in this mass range, and then even reversed by the effect of very blue HB stars for initial masses smaller than M_{HeF}.

iii) The relative contribution of AGB stars mainly depends on the metallicity via the effect of mass loss. The rate of mass loss has been taken either from Reimers (1975) or Fusi-Pecci and Renzini (1976). In clusters of high metal content and almost independently of the particular algorithm adopted for the mass loss rate, AGB stars are expected to contribute slightly to the integrated $(B-V)_o$ colour, whereas a much greater effect is expected in clusters of low metal content.

iv) No sudden reddening in the $(B-V)_o$ colour is found at the particular ages at which the RGB and AGB phases develop for the first time in a cluster (see Renzini and Buzzoni, 1986 for a detailed discussion of this point). The reddening in the colour prior to the onset of the AGB phase is caused by those stars more massive than M_{up} that spent an appreciable fraction of their lifetime, a few 10^5 years, along the Hayashi line in stages following central He-exhaustion and prior to core C-ignition in non degenerate conditions.

v) There is an age range (approximately from $3x10^8$ to $1x10^9$ years) in which clusters of very different metallicity possess the same $(B-V)_o$ colour.

5.2 Determining the Age of Clusters from their Integrated Photometry
In principle, despite the uncertainty generated by the stochastic nature of the IMF (see the discussion by Chiosi et al., 1988 on the subject), the colour-age relationships discussed in the previous section could be used to date clusters of various chemical composition when only the integrated colours are known. Prior to this, the colour-age relationships must be tested against clusters for which both colours and independent estimates of the age are available. While the integrated colours are obtained by direct observational measurements, ages can be derived from one of the following methods: i) Main Sequence turn-off and termination magnitudes, ii) The red giant star luminosity, and iii) The maximum AGB luminosity. Since all the details relative to the methods in question are given in Chiosi et al. (1988), they will not be repeated here. Using one of the above methods, ages have been derived for a sample of selected clusters (Chiosi et al., 1988). Although the chemical abundances used to derive the theoretical calibrations may not be fully suited to the clusters in question, the agreement between theory and observations is fairly good, thus confirming the validity of calibrations. Better comparison will be possible only when accurate and homogeneous determinations of chemical abundances for individual clusters will be obtained, and the corresponding calibrating relationships calculated. However, only a small number of clusters out of the sample for which integrated colours have been measured, possess all the information required to date them on the basis of colour

magnitude diagrams (turn off and main sequence termination luminosities, red giant clump luminosity, brightest AGB stars luminosities). On the other hand, it may be worth of interest to rank clusters as a function of the age even in those (more numerous) cases, in which only integrated colours and magnitudes are available. To this aim, two different methods can be adopted. The first one follows Elson and Fall (1985) and makes use of parameter S, analogous to the SWB types, defined by the location of a cluster in the (U-B) versus (B-V) plane (see Elson and Fall, 1985, for more details). With the aid of those clusters for which both age and S are known, Chiosi et al. (1988) got the following relation

$$\mathrm{Log}\ t = 0.062\ S + 6.99, \tag{5}$$

where t is the age in years, and in turn a relation between SWB and S or t. Relations (5) is somewhat different from the correspondent one by Elson and Fall (1985). This arises only from the different ages assigned to the calibrating clusters in virtue of the novel isochrones from models with convective overshoot. The second method proposed by Alongi and Chiosi (1989a) can determine the age and the metallicity of a cluster from its integrated colours at a time. Adopting for each cluster the parameter S of Elson and Fall (1985), it leads to a S(t) relation similar to that of equation (5).

5.3 The Colour Gap: Phase Transition or Cluster Formation History ?

It has been known for long time that the integrated magnitude V versus (B-V) diagram of LMC (as well as SMC) clusters shows a gap in the colour range $0.4 < (B-V) < 0.6$, in the sense that very few clusters of any total luminosity fall in this region (van den Bergh, 1981). A similar gap seems to occur also in the plane (B-V) versus cluster type (SWB) of Searle et al. (1980), the transition being centered at type IV. Several explanations have been advanced: age gap and/or effects of cluster disruption (van den Bergh, 1981); RGB and/or AGB phase transitions, in that the sudden appearance of those phases would drastically redden the colour on a very short time scale (Gascoigne, 1980; Renzini, 1981; Renzini and Buzzoni, 1983); transition of the core He-burning phase from loop to clump. All these possibilities have been recently discussed by Renzini and Buzzoni (1986). However, none of these has been yet quantitatively tested on the basis of detailed colour evolution models. Although the preference goes to the RGB and/or AGB hypotheses, the calibrating relationships presented in so far show little evidence of this. In fact, although these phase transitions are included in the derivation of the integrated fluxes, there is no visible effect in the integrated $(B-V)_o$ and $(U-B)_o$ colours. Chiosi et al. (1988) suggest that the colour gap originates from other causes, among which are the effect of a period of star formation inactivity in LMC, the result of the rapid disruption and/or fading of the young blue clusters as suggested by van den Bergh (1981), or the superposition of these two.

5.4 The Age Distribution of LMC Clusters

In order to test the hypothesis of periods of inactivity in the cluster formation history, Chiosi et al. (1988) derive the age distribution function and check whether the observational data are consistent with the idea that the cluster formation proceeded in bursts of activity. To this aim, they follow the study of Elson and Fall (1985). who determined the age distribution (dN/dt) for a mass limited sample of LMC clusters sorted out from the van den Bergh (1981) list. The cluster mass is limited by means of theoretical fading lines in the B versus S plane. The fading line is locus along which a cluster of given total number of stars hence mass would evolve at increasing age. The number of clusters falling in each S or age bin is then corrected for photometric completeness. Adapting Elson's and Fall' (1985) precedure to the adopted ages in, Chiosi et al. (1988) get the following relation

$$\Delta N/\Delta t = 2.39\ 10^{-7}\ \Delta N'\ /\ (\ 10^{0.062\ S}) \qquad (6)$$

The resulting age distribution (dN/dt) is steeper than in Elson and Fall (1985), and the difference arises from the different S(t) relation. However, looking at the age distribution by Chiosi et al. (1989) in more detail, it seems that two major peaks, centered at about 2×10^9 yr and 1.5×10^8 yr, are likely to exist, which are not present in Elson and Fall (1985). In principle they could be attributed to periods of enhanced activity in the rate of cluster formation. However, since no great statistical significance can be given to those glitches owing to the large error bars (the square root of N), in agreement to what already claimed out by Elson and Fall (1985), Chiosi et al. (1989) safely conclude that there is no convincing evidence for strong discontinuities in the rate of formation of LMC clusters, even though they cannot be definitely ruled out. Different conclusion has been reached by Alongi and Chiosi (1989b), who by means of the ages obtained by Alongi and Chiosi (1989a) suggest that the two peaks in the age distribution are likely to be real, thus indicating two major episodes of activity.

5.5 The Colour Distribution of LMC Clusters

As already recalled, the number frequency distribution of (B-V) colours of LMC clusters is bimodal with two peaks centered at about 0.1 and 0.6 and a gap ranging from 0.2 to 0.5. The theoretical colour distribution function f[(B-V)o] can be expressed by

$$f[(B-V)o] = (dN/dt)\ /\ (d(B-V)_o/dt) \qquad (7)$$

where (dN/dt) is the observational age distribution function and $d(B-V)_o/dt$ is the colour speed. This latter can be easily derived from any colour-age relationship. To make the colour evolution of the cluster population closer to reality, Chiosi et al. (1988) also considered the possibility that some degree of metal enrichment over the galaxian lifetime have taken place. Such an hypothesis is certainly not ad hoc, as many arguments exist in literature supporting this fact (see Chiosi and Matteucci, 1983; Cohen 1982; Hartwick and Cowley, 1981; Searle and Smith, 1981). To this aim, a linear relation has been taken, assuming Z to be zero 15×10^9 years ago and equal to 0.02 at the present epoch (see Chiosi et al., 1988, for more details). The careful analysis of the dependence of the colour distribution on the various factors contributing to it clarifies that it is primarily reflective of the age distribution, which determines the ratio of the blue to red clusters, whereas the colour speed fixes the location and width of the gap. In, particular it has been pointed out that the colour gap cannot be simply attributed to the increase in the mean metallicity of the gas with galaxian age as suggested by Frenk and Fall (1982). In fact, even including the increase in metallicity, the right colour distribution is obtained only with the observational age distribution. Chiosi et al. (1988) analysis together with the studies by Alongi and Chiosi (199a,b) seem to favour the idea that the colour distribution mainly reflect the periods of enhanced activity in the cluster formation history that occurred in LMC.

5.6 Determining the Mass of Clusters from their Integrated Photometry
In principle, with the aid of eq.(1) and eq. (2) together with a suitable expression for the mass of stars in the cluster as function of the age, it is possible to get information on the mass of a cluster out of its integrated magnitudes and colours. The procedure works as follows. The normalization constant A is given by

$$\log A = 0.4\ (M_{\Delta\lambda t} - M_{\Delta\lambda o} + DM + A_{\Delta\lambda}) \qquad (8)$$

where $M_{\Delta\lambda t}$ and $M_{\Delta\lambda o}$ are the theoretical and observed integrated magnitudes in a given passband (V for instance), DM is the distance modulus and $A_{\Delta\lambda}$ is the reddening in that passband. Furtheremore, the total mass of stars in the cluster can be expressed as

$$M_{tot} = A\int_{M_L}^{M_U(t)} M\ \Phi(M)\ dM + A\int_{M_U(t)}^{M_U} M_x(M)\ \Phi(M)\ dM \qquad (9)$$

where $M_U(t)$ is the maximum mass of living stars in the cluster at that particular age, whereas M_x is the mass of remnants. M_U and M_L have been already defined in section 2. Having an estimate of the age and metallicity, we may infer $M_{\Delta\lambda t}$ from theory, fix the normalization constant A, and finally determine the mass of the cluster. It goes without saying that this will depend on the slope of the initial mass function and the assumed lower limit M_L for main sequence stars. The results obtained by Alongi and Chiosi (1989a) assuming x = 1.35 M_L = 0.05 $M_\odot$ and M_U = 100 $M_\odot$, indicate that most of the clusters have masses lower than 10^5 $M_\odot$ with a few up to 7×10^5 $M_\odot$.

6. STAR FORMATION IN THE FIELDS OF LMC
A recent study of stellar fields in LMC by Mateo et al. (1989) has shown that the distribution of stars in the C-M diagrams V versus (B-V) is compatible with a law of star formation schematically approximated by a long period of lower efficiency followed by a global enhancement that began about 3 to 5 Gyr ago. The ratio of the past to recent level of star formation, called the enhancement factor α, that yields a good agreement between theoretical simulations and observational C-M

diagrams, is about 10. This finding somewhat supports what already indicated by the age distribution function of the star clusters discussed in the previous sections.

ACKNOWLEDGEMENTS

This work has been financially supported by the Ministry of Public Education (Funds 40% and 60% MPI), and the Italian Space Agency (ASI).

REFERENCES

Alongi, M., Chiosi, C.:1989a. Astrophys. J. submitted
Alongi, M., Chiosi, C.:1989b. Preprint
Arimoto, N., Yoshii, Y.:1986. Astron. Astrophys. 164, 260
Barbaro, G.:1981. Astrophys. Sp. Sci. 77. 23
Barbaro, G.:1982. Astrophys. Sp. Sci. 83, 143
Barbaro, G., Bertelli, G.:1977. Astron. Astrophys. 54. 243
Becker, S.A.:1981. Astrophys. J. Suppl. 45, 478
Becker, S. A., Iben, I. Jr.:1979. Astrophys. J. 232. 831
Becker, S. A., Mathews, G. J.:1983. Astrophys. J. 270, 155
Bell, R. A., Gustafsson, B.:1978. Astron. Astrophys. Suppl. 34, 229
van den Bergh, S.:1981. Astron. Astrophys. Suppl. 46, 79
Bertelli, G., Bressan, A., Chiosi, C.:1985. Astron. Astrophys. 150. 33
Bertelli, G., Bressan, A., Chiosi, C., Angerer, K.:1986a. Astron. Astrophys. Suppl. Ser. 66, 191
Bertelli, G., Bressan, A., Chiosi, C., Angerer, K.:1986b. In "The Age of Star Clusters",ed. F. Caputo, Mem. Soc. Astron. It. 57.427
Boothroyd, A. I., Sackmann, I. J.:1988a. Astrophys. J. 328, 632
Boothroyd, A. I., Sackmann, I. J.:1988b. Astrophys. J. 328, 641
Boothroyd, A. I., Sackmann, I. J.:1988c. Astrophys. J. 328, 653
Boothroyd, A. I., Sackmann, I. J.:1988d. Astrophys. J. 328, 671
Bressan, A., Bertelli, G., Chiosi, C.:1981. Astron. Astrophys. 102, 25
Bressan, A., Bertelli, G., Chiosi, C. :1986. In "The Age of Star Clusters", ed. F. Caputo, Mem. Soc. Astron. It. 57, 411
Bruzual, A. G.:1983. Astrophys. J. 273, 105
Buonanno, R., Corsi, C. E., Fusi-Pecci, F.:1985. Astron. Astrophys. 145, 97
Burstein, D. Heiles, C.:1982. Astron. J. 87, 1165
Buser, R., Kurucz, R. L.:1978. Astron. Astrophys. 70, 555
Castellani, V., Chieffi, A., Pulone, L., Tornambe', A.:1985. Astron. Astrophys. 296, 204
Chiosi, C.:1986. In "Nucleosynthesis and Stellar Evolution", 16th Saas-Fee Course, ed. B. Hauck et al., p. 199, Geneva Observatory
Chiosi, C., Bertelli, G., Bressan, A.:1987. In "Late Stages of Stellar Evolution", ed. S. Kwok and S. R. Pottasch, p. 239, Dordrecht: Reidel
Chiosi, C., Bertelli, G., Bressan, A.:1988. Astron. Astrophys. 196, 84
Chiosi, C., Bertelli, G., Meylan, G., Ortolani, S.:1989a. Astron. Astrophys. Suppl. Ser. 78, 89
Chiosi, C., Bertelli, G., Meylan, G., Ortolani, S. (1989b). Astron. Astrophys. 219, 167
Chiosi, C., Maeder, A.:1986. Ann. Rev. Astron. Astrophys. 24, 329

Chiosi. C.. Matteucci, F.:1983. Astron. Astrophys. 105, 140
Cohen, J. G.:1982. Astrophys. J. 258, 143
Da Costa, G. S., Mould, J., Crawford, M. D.:1984. Astrophys. J. 280, 595
Eggen, O.J., Iben, I.Jr.:1988. Astron. J. 96, 635
Eggen, O.J., Iben, I.Jr.:1989. Astron. J. 97, 431
Elson, R. A. W., Fall S. M.:1985. Astrophys. J. 299, 211
Frenk, C.S., Fall, S.M.:1982. Mon. Not. R. Astr. Soc. 199, 565
Fusi-Pecci, F., Renzini, A.:1976. Astron. Astrophys. 46, 447
Gascoigne. S. C. B.:1980. In "Star Clusters", ed. J. Hesser, p. 305. Dordrecht: Reidel
Green, E. M., Demarque P., King, C. R.:1987. "The Revised Yale Isochrones and Luminosity Functions (Yale University, Observatory, New Haven
Gustafsson, B., Bell, R. A., Eriksson, K., Nordlund, A.:1975. Astron. Astrophys. 42, 407
Hartwick F. D. A., Cowley, A.:1981. In "Astrophysical Parameters for Globular Clusters" eds. A. G. D. Philip & D. S. Hayes, L. Davis Press Inc., New York, p. 205
Heckman, T. M.:1976. Astrophys. J. Suppl. 36, 451
Iben. I. Jr.:1975a. Astrophys. J. 196, 525
Iben, I. Jr.:1975b. Astrophys. J. 196, 549
Iben. I. Jr.:1976. Astrophys. J. 208. 165
Iben, I.Jr., Renzini, A.:1983. Ann. Rev. Astron. Astrophys. 21, 271
Lattanzio, J. C.:1986. Astrophys. J. 311. 708
Lattanzio, J. C.:1987a. In "Late Stages of Stellar Evolution", ed. S. Kwok and S. R. Pottasch, p. 235, Dordrecht: Reidel
Lattanzio, J. C.:1987b. Astrophys. J. Lett. 313, L15
Lattanzio. J. C.:1988a. In "Evolution of Peculiar Red Giant Stars". In press
Lattanzio, J. C.:1988b. In "Origin and Distribution of the Elements", ed. G. J. Mathews, p. 398, Singapore: World Scientific
Lattanzio, J.C.:1989. UCRL-100238 preprint
Maeder, A.:1974. Astron. Astrophys. 32, 177
Maeder, A., Maynet, G.:1989. Astron. Astrophys. 210, 155
Mateo, M., Bertelli, G., Chiosi, C.:1989. Preprint
Mateo, M., Hodge, P.:1986a. Astrophys. J. 311, 113
Mateo, M., Hodge, P.:1986b. Astrophys. J. Suppl. 60. 893
Mateo, M., Hodge P. W.:1987. Private Communication
Meylan, G.:1987a. Astron. Astrophys. 184, 144
Meylan, G.:1987b. In "Stellar Evolution and Dynamics in the Outer Halo of the Galaxy, eds. M. Azzopardi, F. Matteucci, Garching, ESO, p. 665
Meylan, G.:1988. Astron. Astrophys. 191, 215
Meylan, G., Djorgovski, S.:1987. Astrophys. J. Lett. 322, L94
Persson, S. E., Aaronson, M., Cohen, J. G., Frogel, J. A., Matthews, K.: 1983. Astrophys. J. 266, 105
Rabin, D.:1982. Astrophys. J. 261, 85
Reimers, D.:1975. Mem. Soc. Roy. Sci. Liege 6 serie, tome 8,p. 369
Renzini, A.:1981. Ann. Phys. Fr. 6, 87
Renzini, A., Buzzoni, A.:1983. Mem. Soc. Astron. It. 54, 739

Renzini, A., Buzzoni, A.:1986. In "Spectral Evolution of Galaxies", ed. C. Chiosi, A. Renzini, p. 195. Dordrecht: Reidel
Salpeter, E. E.:1955. Astrophys. J. 121, 161
Searle, L., Sargent, W. L. W., Bagnuolo, W. G.:1973). Astrophys. J. 179, 427
Searle, L., Smith, H.:1981. In "Astrophysical Parameters for Globular Clusters", eds. A. G. D. Philip and D. S. Hayes. L. Davis Press Inc., New York, p. 201
Searle, L., Wilkinson, A., Bagnuolo, W. G.:1980. Astrophys. J. 239, 803
Storm, J., Andersen, J., Blecha, A., Walker, M. F.:1988. Astron. Astrophys. 190. L18
Vallenari, A., Chiosi, C., Bertelli, G., Meylan, G., Ortolani, S.: 1989. Astron. J. submitted
Walker, A. R.:1987. Mon. Not. R. Astr. Soc. 225, 627
Welch, D. L., McLaren, R. A., Madore, B. F., McAlary, C. W.:1987. Astrophys.J. 321, 162
Wielen, R.:1971. Astron. Astrophys. 13, 309

Discussion

Capuzzo-Dolcetta (comment): With regard to the synthetic properties of clusters in the Magellanic Clouds I would like to remind the works by Capuzzo-Dolcetta (1986, in 'The age of stellar systems', Mem. Soc. Astr. Italiana, v. 57, n. 3, p. 469) and Battinelli and Capuzzo-Dolcetta (1989, Astrophys. J., **347**, 794). In particular, in these papers can be found a detailed discussion of the problem of the colour gap in the HR diagram of the MC clusters and is proposed and quantitatively checked an explanation on the basis of evolution of synthetic model clusters obtained with classical stellar evolutionary tracks.

FORMATION AND DISRUPTION OF OPEN CLUSTERS

P. BATTINELLI
Osservatorio Astronomico di Roma
viale del Parco Mellini 84
I-00136, Roma, Italy

R. CAPUZZO-DOLCETTA
Istituto Astronomico Universitá 'La Sapienza'
via G.M. Lancisi 29
I-00161, Roma, Italy

ABSTRACT. The fifth edition of the Lynga 'Catalogue of open cluster data'(1987) is used to obtain a complete sample of galactic open clusters, which represents a suitable data base to deduce information on the system of open clusters in our Galaxy, like, for example, the formation and disruption rates. The latter parameters, together with the characteristic time $t_{1/2}$ (which is representative of the lifetime of the clusters), show a slight dependence on the necessary assumptions made on the IMF of the open cluster system. The short lifetimes obtained confirm that a large fraction of clusters which are commonly referred as open clusters are indeed gravitationally unbound systems.

1. THE OPEN CLUSTER SYSTEM.

1.1 GENERALITIES.

The present characteristics of the open cluster system in our Galaxy depend on its formation history and its following time evolution.

The main parameter which we can identify as describing the formation history is the formation rate R giving the number of clusters formed per unit time and unit area on the disc. Two kinds of time evolution are relevant to determine the observed characteristics of any sample of open clusters: i) the dynamical evolution, which in some case can lead to the destruction of a cluster; ii) the luminosity evolution of stars in the cluster, which can make a cluster so faint to be difficult to be detected. Consequently, in order to interpret the main features of the open cluster system it is necessary to try to separate the substantially different effects of the two types of time evolution, in such a way to allow to distinguish the luminosity fading from the dynamical effects. To accomplish with reliability this task it is needed the disposal of a large and unbiased data base, which can be obtained by the rich 'Catalogue of Open Cluster Data' (Lynga, 1987) in a way similar to that described in Battinelli and Capuzzo-Dolcetta (1989a) and briefly summarized in the following Section 1.2.

1.2 THE COMPLETE SAMPLE OF GALACTIC OPEN CLUSTERS.

Any theoretical interpretation of structural properties of the ensemble of the

R. Capuzzo-Dolcetta et al. (eds.), Physical Processes in Fragmentation and Star Formation, 441–449.

open clusters should rely on a data base which is reasonably statistically complete. Very often, the determination of a sample which is complete with respect to a chosen variable (for instance age, $B-V$, etc.) is done assuming that the spatial distribution of the variable is (statistically) the same in different circular shells centred on the Sun, hypothesis which is not always valid. We preferred to select a sample basing on the simple assumption of a uniform disc number density ($cluster/pc^2$), as, for instance, supported (at least up to $2kpc$ from the Sun) by the map of clusters in the galactic plane (see for example Figure 7a from Janes and Adler,1982). In this hypothesis, if all open clusters were known, the function $N(d)$, i.e. the number of clusters having projected distance from the Sun less than d, would be, in a bi-logarithmic plane, a straight line with slope 2. Figure 1 shows the Log N vs Log d observed relation for six samples with different integrated V cut–off magnitudes; the richest possible complete sample is given by $M_V \leq -4.5$ and $d \leq 2kpc$. It is worth noting, that the needed assumption of uniform disc number density is somehow supported by the fact that the observed points get really closer to the theoretical line of completeness ($N \propto d^2$) when decreasing the cut–off magnitude of the sample.

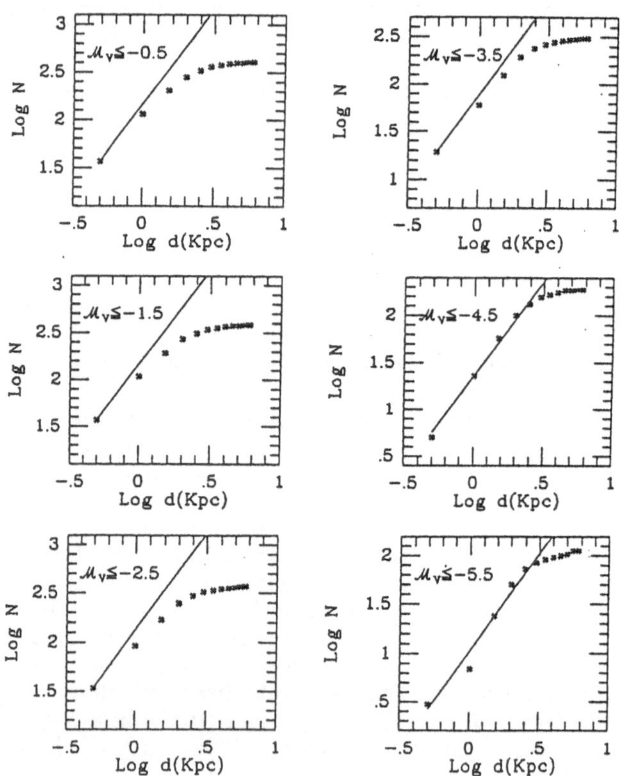

Figure 1. Comparison between uniform disc number density (straight lines) and observed number density of open cluster (points) for samples characterized by six different values of the absolute magnitude cut–off.

1.3 THE AGE DISTRIBUTION OF CLUSTERS IN THE COMPLETE SAMPLE.

A first important result deduced from data relative to the complete sample previously discussed is the distribution of the ages of the clusters in the sample. The discussion of the age distribution and of how the conclusions derived from incomplete samples may be misleading, is given in Battinelli and Capuzzo–Dolcetta (1989). Figure 2 shows the behaviour of the age frequency $\tilde{\nu}(t) = \nu(t)/n_{tot} = (1/n_{tot})dn(t)/dt$ (where $n(t)$ is the number of clusters with age t) for clusters in our sample compared to the ones obtained by Wielen (1971) and by Pandey and Mahra (1986). This comparison has the aim to show the difference in the age distribution of samples whose completeness has been assessed in different ways. The main feature in Figure 2 is the overabundance of old clusters (Log $t \geq 7.6$, hereafter t is in years) in the samples of Pandey and Mahra, and Wielen with respect to ours; this is due to two facts: i) Pandey and Mahra, and Wielen samples contain all clusters with $d \leq 1kpc$ including also the faint ($M_V \geq -4.5$) and consequently old (at least in the average) objects which we excluded for completeness reasons; ii) our sample extends up to $2kpc$, thus containing a large number of young spiral arm clusters (see Battinelli and Capuzzo–Dolcetta, 1989a).

2. SOME STRUCTURAL PROPERTIES OF THE OPEN CLUSTER SYSTEM.

2.1 METHOD.

The age distribution shown in Section 1.3 can be used in order to obtain information about the history of the open cluster system. In fact, it can be said that the present shape of the age distribution has been moulded by three different time functions reflecting the birth (formation rate, $\rho(t)$) and death (due to disruption, $D(t)$, and to luminosity fading, $F(t)$) of clusters, so that:

$$\nu dt = F(t)D(t)\rho(t)dt. \tag{1}$$

In eq. 1 we wrote the presently observed age distribution as the factorization of the number of clusters, $\rho(t)dt$, formed between t and $t + dt$ years ago, with $F(t)D(t)$ which represents the fraction of clusters neither faded below the cut–off magnitude nor disrupted. $\rho(t)$ differs from the usual definition of formation rate R just by a trasformation of the time axis such that: $\rho(t) = R(\tau - t)$, where τ is the age of the galactic disc.

2.1.1 THE FADING FACTOR F.

As it is well known, the luminosity of a star cluster is a decreasing function of time so that sooner or later each cluster becomes (if it is not disrupted earlier) too faint to be included in our complete sample simply because its absolute V–magnitude gets fainter than the cut–off magnitude −4.5. Therefore, at any given age t, only a fraction $F(t)$ of all clusters is included in our sample. $F(t)$ is obtained from the cluster luminosity function $\Phi(M_V, t)$ and it results to be:

$$F(t) = \int_{-\infty}^{-4.5} \Phi(M_V, t)dM_V = \int_{m_{lim}(t)}^{\infty} \Psi(m, t)dm. \tag{2}$$

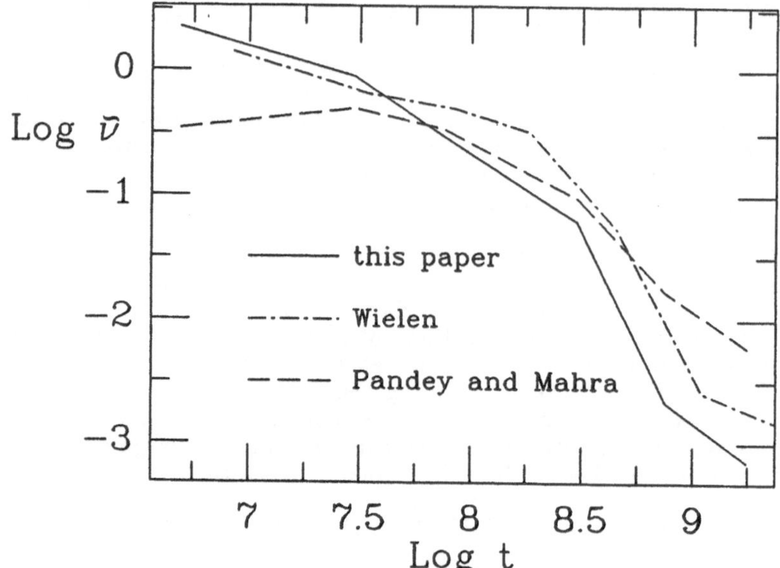

Figure 2. Comparison of the age frequency as obtained in this paper and by Pandey and Mahra (1986) and Wielen (1971).

where the mass function $\Psi(m,t)$ is normalized to one, $m_{lim}(t)$ represents the mass of the lightest cluster still visible at the age t. In this paper we actually chose a time–independent mass function in the simple form:

$$\Psi(m,t) = \begin{cases} 0, & m \leq m_1; \\ km_2^{-s}(m-m_1)/\Delta m, & m_1 \leq m \leq m_2; \\ km^{-s}, & m_2 \leq m \leq m_3; \\ 0, & m > m_3. \end{cases} \tag{3}$$

where $\Delta m = m_2 - m_1$; the dependence of our results on the assumed values for the free parameters $m_1, \Delta m, m_3$ and s is discussed in Section 2.2. As far as the lower limit of the integral in eq. 2 is concerned, the time function $m_{lim}(t)$ is given by:

$$m_{lim}(t) = A(t)10^{-0.4(-4.5-M_V^{\odot})} \tag{4}$$

where $A(t)$ is the time dependent mass to V–luminosity ratio. For 38 galactic open clusters for which homogeneous age, mass and luminosity determinations are available, a least square fit in the bi-logarithmic plane results to be:

$$\text{Log } A(t) = 0.39\text{Log } t - 3.9. \tag{5}$$

A theoretical evaluation of the function $A(t)$ can also be done on the basis of photometric synthetic models (see for instance Battinelli and Capuzzo–Dolcetta, 1989b, for classical models, or Chiosi *et al.*, 1988, for models with convective overshooting). In the present work only the observational fit (5) is used.

2.1.2 THE DISRUPTION FUNCTION D.

Another important reason why some clusters are not detectable is that star clusters undergo to several disruptive processes due to both external (e.g. close encounters with massive objects, tidal force, dynamical friction etc.) and internal causes (mass loss: stellar wind, stellar evaporation and sudden gas removal by supernova explosion). All these (and any other possible) phenomena are accounted for in eq. 1 by the disruption function $D(t)$ defined as the fraction of the visible clusters (brighter than the cut–off magnitude -4.5), with age t, that have not been destroyed yet. Of course, this treatment cannot tell apart, among all the possible disruptive mechanisms, which ones are really working and which ones are neglectable. Anyway, eq. 1 , is surely an improvement with respect to the equation used first by Wielen (1971) and later by Pandey and Mahra (1986) where the link between the observed age distribution and the true age distribution (in our notation $\nu_{true}dt$ is $\rho(t)dt$) was described by a single time function making impossible for them to distinguish death by disruption from luminosity fading of open clusters.

Although our aim is to obtain the function $D(t)$ from eq. 1, we had to make an assumption about it. In fact, in order to obtain an estimate of the formation rate of open clusters, it is needed (as we will soon see in the next Section 2.1.3) to assume that no disruptive process is effective for very young clusters (Log $t < 6.5$).

2.1.3 RATE OF FORMATION OF OPEN CLUSTERS.

In line of principle, the use of eq. 1 to obtain the disruption function for the galactic open clusters requires us to know how the formation rate has changed with time. The assumption $D(t) = 1$ for Log $t < 6.5$ allows us to estimate the present–day formation rate for the galactic open clusters. Anyway, there is no evidence for a strong time dependence of the formation rate so that the common assumption of a constant formation rate seems, especially for the open cluster system, quite reasonable. Indeed, for open clusters, the constancy of the rate of formation can be restricted to the last few billion years (age of the oldest clusters).

2.2 RESULTS.

The present–day cluster formation rate $R(\tau)$ was determined, from the age distribution of clusters in our sample, using eq. 1 once that $F(t)$ for Log $t < 6.5$ (which is the fading factor of the newly formed open clusters, F_0) is computed by eq. 2. It is important to recall that the $R(\tau)$ evaluation is presumably underestimating the true formation rate because not for all clusters age and magnitude are known. Anyway, our evaluation of $R(\tau)$ relies on the observed number of very young and luminous clusters, for which this problem is less important. Table 1 gives our evaluation of the cluster formation rate compared with others available in literature. The ± 4 variation in our estimate of $R(\tau)$ is due to the variation of parameters in the mass function (3) (hereafter masses are in solar units):

$$40 \leq m_1 \leq 120; \;\; 250 \leq \Delta m \leq 900; \;\; 10^4 \leq m_3 \leq 3 \cdot 10^4; \;\; 1 \leq s \leq 3.5. \qquad (6)$$

The value of F_0 is in all the cases under consideration $0.84 \leq F_0 \leq 1$: this means that the neglecting of fading in the determination of $R(\tau)$ (which is an implicit

Table 1. The present day cluster formation rate $R(\tau)$ in units 10^{-8}*clusters* kpc^{-2} yr^{-1} as obtained by various authors using different catalogues. d_{max} is the cut–off projected distance (in kpc) from the Sun of the sample.

Reference	$R(\tau)$	Catalogue	d_{max}
Wielen (1971)	16	[1], [2]	1
van den Bergh (1981)	50	[3]	0.75
Elmegreen and Clemens (1985)	25 ± 10	[1], [4]	1
Pandey and Mahra (1986)	18	[5]	1
Janes *et al.* (1988)	25	[6]	–
this paper	45 ± 4	[6]	2

Notes: The key to the references quoted in the catalogue column is: [1] Becker and Fenkart (1971), [2] Lindoff (1968), [3] Mermilliod (1980), [4] Lynga (1981), [5] Lynga (1983), [6] Lynga (1987).

assumption of the authors whose $R(\tau)$ is given in Table 1) is not the main cause of the difference between ours and other's $R(\tau)$ estimate. The difference is rather due to:

(i) different sources of data;
(ii) different intervals of age used for the $R(\tau)$ determination;
(iii) neglecting of the problem of the data completeness.

For example, Pandey and Mahra obtained their $R(\tau) = 18$ dividing the number of open clusters of any magnitude with $d \leq 1kpc$ and age $7 \leq \text{Log}\, t \leq 7.8$. This value of $R(\tau)$ may be uncorrect because at these ages the neglecting of fading and disruption is not allowed. The neglecting of fading and disruption leads to a significant underestimate of the true $R(\tau)$.

In the range of ages of cluster in our complete sample and of the parameters (6), the functions $F(t)$ and $D(t)$ depend essentially on Δm and s, in the way shown in Figures 3, and 4. It is clear from their definition (1) that F and D are not independent, $F(t)D(t)$ being (in our assumption of constant $\rho(t)$) proportional to the observed age distribution; this explains the inverse dependence of F and D on the free parameter in the IMF. It is worthy noted from Figures 3a and 4a that the significant variation of F occurs at ages larger than those of clusters we used to determine $R(\tau)$, thus confirming the reliability of our determination of the present–day cluster formation rate.

A rough indication of the cluster lifetimes may be deduced by Figures 3b and 4b. This indication may be quantified through the definition of a parameter, $t_{1/2}$, as the age at which D is halved. By this definition, 50% of the unfaded ($M_V \leq -4.5$) clusters are actually destroyed. The range of variation of $t_{1/2}$ is $6.9 \lesssim \text{Log}\, t_{1/2} \lesssim 7.2$. In our opinion, such very short "lifetimes" imply so efficient destructive external forces to make more plausible that a large fraction of the newly formed clusters is in the form of intrinsically unstable systems, i.e. they have initial positive mechanical energy, or they may dissolve due to energy input by SN in their primordial stage, when a large fraction of the total mass is still in gas (see for instance Margulis *et al.*, 1985; and Terlevich, 1987).

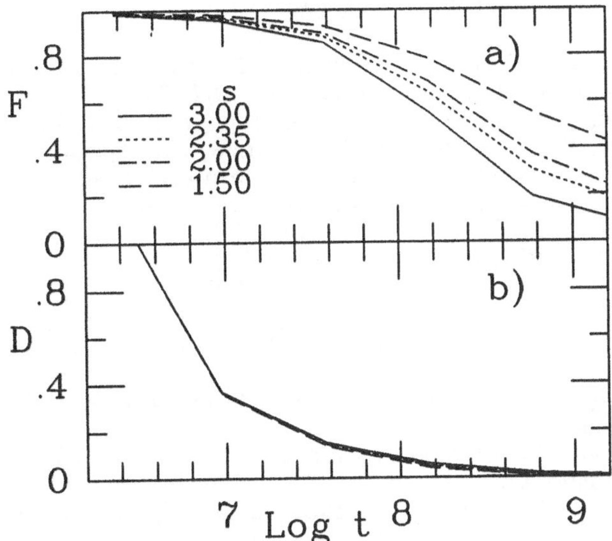

Figure 3. Dependence on time of F (panel a) and D (panel b) at varying the exponent s in the mass function, and other parameters set to: $m_1 = 40$, $\Delta m = 900$, $m_3 = 2 \cdot 10^4$

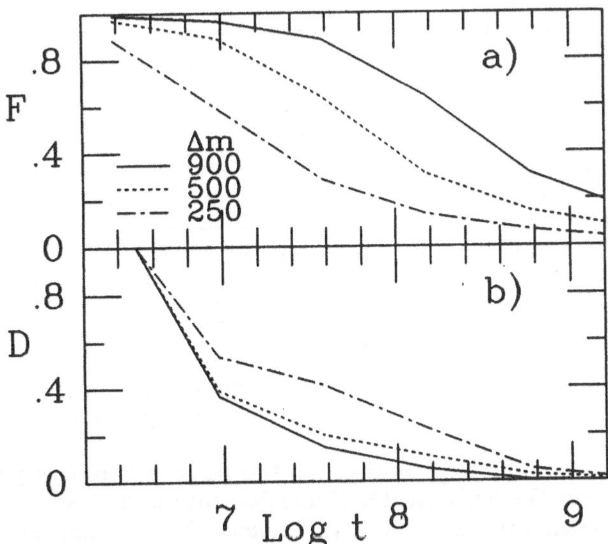

Figure 4. Dependence on time of F (panel a) and D (panel b) at varying the parameter Δm in the mass function, and other parameters set to: $m_1 = 40$, $m_3 = 2 \cdot 10^4$, $s = 2.35$

REFERENCES

Battinelli, P. and Capuzzo–Dolcetta, R.: 1989a, in *Evolutionary Phenomena in Galaxies*, J. Beckman ed., *Astrophys. Space Sci.*, **157**, 75.
Battinelli, P. and Capuzzo–Dolcetta, R.: 1989b, *Astrophys. J*, **347**, 794.
Becker, W. and Fenkart, R.: 1970, in IAU Symp. n.38, *The Spiral Structure of Our Galaxy*, W. Becker, G. Contopoulos eds., (Dordrecth:Reidel), p.205.
Chiosi, C., Bertelli, G. and Bressan, A.: 1988, *Astr. Ap.* **196**, 84.
Elmegreen, B.G. and Clemens, C.: 1985, *Astrophys. J.* **294**, 523.
Janes, K.A. and Adler, D.: 1982, *Astrophys. J. Suppl.* **49**, 425.
Janes, K.A., Tilley, C. and Lynga,G.: 1988, *Astron. J.* **95**, 771.
Lindoff, U.: 1968, *Ark. Astr.* **5**, 1.
Lynga, G.: 1981, *Catalogue of Open Clusters*, Stellar Data Center, Observatoire de Strasbourg.
Lynga, G.: 1983, *Catalogue of Open Cluster Data*, 3rd ed., Stellar Data Center, Observatoire de Strasbourg.
Lynga, G.: 1987, *Catalogue of Open Cluster Data*, 5th edition, Stellar Data Center, Observatoire de Strasbourg.
Margulis, M., Lada, C.J. and Dearborn, D.: 1985, in IAU Symp. 113, *Dynamics of Star Clusters*, J. Goodman and P. Hut eds., p.463.
Mermilliod, J.C.: 1980, in IAU Symp. 85, *Star Clusters*, J.E. Hesser ed. (Dordrecht:Reidel), p. 129.
Pandey, A.K. and Mahra, H.S.: 1986, *Astrophys. Space Sci.* **126**, 167.
Terlevich, E.: 1987, *Monthly Notices Roy. Astron. Soc.* **224**, 193.
van den Bergh, S.: 1981, *Publ. Astron. Soc. Pacific* **93**, 712.
Wielen, R.: 1971, *Astr. Ap.* **13**, 309.

Discussion

Mouschovias: Does your sample of open clusters include associations?

Battinelli: Our source of data is the 5th edition of the Lynga's 'Catalogue of Open Cluster Data'; however it cannot be excluded, also at the light of our results, that some of the objects in the Catalogue should be betterly named star associations, according to the usual definition of young, unbound systems.

Cayrel: Can you tell us how your conclusions will be altered if you had included associations as well as clusters in your study?

Battinelli: Qualitatively, smaller values of the parameter $t_{1/2}$ would have been obtained. It is harder to give a quantitative answer to your question, because often it is difficult to share loose and very young open clusters from associations.

Mateo: Do you have a possible explanation for the large differences you found between your own and Wielen's cluster formation rate and $t_{1/2}$?

Battinelli: There are many reasons for these differences. We obtained shorter $t_{1/2}$ (which is, however, not exactly the same parameter used by Wielen) due to the completeness of our sample, which excludes very old low luminosity clusters. Moreover, we extend our sample up to $2kpc$ in the projected distance from the Sun, thus including young spiral arm open clusters. A further, and even more important, source of discrepancy is the Wielen's assumption of no fading and disruption for clusters younger than $\approx 5 \cdot 10^7$ yr while we computed the fading factor and made just the less restrictive assumption of no disruption for ages less than $\approx 3 \cdot 10^6$ yr. The last two points also explain our larger value for the cluster formation rate.

6. DISCUSSION ON THE LOW MASS END OF THE IMF AND CONCLUDING REMARKS

DISCUSSION ON THE LOW MASS END OF THE IMF.

E. KRÜGEL - Radio–observation perspectives and star formation.

R. B. LARSON - Concluding remarks.

DISCUSSION ON THE LOW MASS END OF THE IMF.

Larson: Several key questions that arise in studies of the IMF of low mass stars are:

(1) Does the hydrogen–burning main sequence end at $M_V \sim +12$?

My impression is that stellar models are on firm grounds at this mass ($\sim 0.2 M_\odot$), and they exclude such an interpretation of the peak in the luminosity function at this magnitude.

(2) What is the mass–luminosity relation for low–mass stars?

This is important to know accurately, but it is not known well enough to constrain strongly the shape at the lower IMF. Nevertheless, it is my impression that the data exclude a change in slope as large as that suggested by Hawkins.

(3) Does the IMF clearly depart from a power law at low masses?

If not, the IMF is a scale–free function with no preferred mass, but if there is a departure from a power law, then there is a mass scale or characteristic mass that must be understood. Different published determinations of the IMF below $0.2\ M_\odot$ disagree on whether it rises or falls with decreasing mass, but agree that its slope is not as steep as the Salpeter law. In this case, there is a characteristic stellar mass of order one solar mass, in the sense that the IMF contains comparable amounts of mass in stars above and below one solar mass.

(4) Are there large numbers of brown dwarfs, and, if so, where are they?

Many efforts to find brown dwarfs have yielded only a few candidates, and most of them have turned out to be spurious. It seems clear that if brown dwarfs exist in large numbers, they mostly avoid associating with more massive hydrogen–burning stars in binary systems. They would therefore probably have to be a separate population of objects not closely related to normal stars.

(5) Can we understand theoretically how brown dwarfs might form?

I think that a reasonable theoretical understanding of the formation of solar–mass stars is beginning to emerge, as discussed by Lizano, but it does not in any very natural way account for the formation of brown dwarfs; to do this, one would have to assume that the star formation process is somehow aborted shortly after it begins.

(6) Do we need brown dwarfs?

For many years it has been thought that the solar neighborhood contains substantial unseen mass, but recently this conclusion has been questioned. If it should turn out that there is no unseen mass, there is no need for large numbers of brown dwarfs.

R. Capuzzo-Dolcetta et al. (eds.), Physical Processes in Fragmentation and Star Formation, 453–457.

comment related to Larson's points by an unidentified person:
(1) I want to point out that the radial velocity surveys by Campbell and Waltur (≈ 20 stars) and by the CfA group (≈ 70 stars) were aimed at finding planets, and not brown dwarfs. So they mostly looked at G and K stars. Because double stars having periods less than 100 days tend to have similar masses one should use late M primaries to look for brown dwarfs with the best chances.
(2) It seems to me that the crucial point to decide if one has reached non–hydrogen burning objects along the observational sequence is that the main sequence does not exist any more for such objects (stars of different ages have no more the same luminosity so there is no mass–luminosity relationship for brown dwarfs).

Rana: With regard to the low mass end of the IMF I can say that invisible low mass companions of M dwarfs within 5.2 *pc* off the solar neighborhood may be as numerous as 7 within the mass range 0.01 to 0.1 $M_{\odot}$. So the possibility of a rise in the LF in the extreme low mass range cannot be ruled out.

Another comment is that the luminosity functions are constructed out of the observed counts of the systems of stars, not of the single stars. The total number of stars compared to the total number of systems bears a ratio of 4:3 in the solar neighborhood. So the number should increase by the factor 4/3, but the luminosity of the individual stars being almost half will shift in magnitude by about 1. Such a luminosity function is constructed by me and shows that the integrated IMF has about 10% more mass than otherwise.

Palla: As for the presence of a characteristic mass in the IMF, I would like to stress the relevance of the recent developments in the investigation of young embedded stellar clusters made possible by the use of IR array cameras, as it has been shown by Zinnecker and Lada during their presentations.
So far, the direct investigation of the IMF has been limited to either the solar neighborhood or to few open clusters, in both case old systems where the star formation process has long been exhausted, and where evolutionary effects might have influenced the shaping of the IMF. By direct imaging of regions of current star formation, it has now become possible to probe "in situ" the process of assembling of the IMF. Luminosity functions limited to the K–band have been presented for several regions, down to a completeness limit of about 15 mag., showing a decrease in the number counts after a peak whose position varies from cloud to cloud.
In the case of the IR cluster in the ρ Ophiuchi core, Rieke *et al.* (1989, Ap. J. Lett., in press) have implemented their data at 2 μm with the long wavelength measurements presented by Wilking and collaborators (1989, Ap. J., in press) to determine the total bolometric luminosities, and assuming that all the detected stars are pre–main–sequence objects, they have been able to derive an IMF, the first of this kind. Although several important criticism should be raised to such a result (completeness problems, proper treatment of local extinction, the distinction between protostars and PMS stars, and so on) the important point is that, in principle, reliable luminosity functions can be constructed for star forming regions, and therefore the question of the characteristic mass can be fully and, hopefully, unambiguously answered. However, it is fair to say that it may take a considerable amount of time before the transformation from the LF to the IMF will be possible, this requiring a lot of theoretical work and a better understanding of the processes that determine the mass of a star.

Capuzzo–Dolcetta: With regard to the investigation of young embedded stellar clusters Francesco referred to, it is worth mentioning the good work by Leisawitz *et al.* (1989, Ap. J. Suppl., in press) who made a CO survey of regions around 34 open clusters and found that very young clusters (age < 5 *Myr*) have associated with them molecular clouds heavier than 10^4 $M_\odot$, while clusters older than 10 *Myr* do not have associated with clouds more massive than few times 10^3 $M_\odot$. Moreover, it seems that molecular clouds are receding from young clusters, and evidence of ongoing star formation is noted in the clouds associated with young clusters. They obtain for the ratio of stellar mass to total around the youngest clusters 1% – –2%. I like to note that these observations may be put in relation with the results described by Battinelli (this volume) concerning the evolution of the open cluster system of our Galaxy, which show that a significant fraction of open clusters is destroyed in the first ≈ 3.5 *Myr*, suggesting that many of them are unbound systems. Indeed, this fits with the picture of a formation of clusters from a massive molecular cloud with low star formation efficiency and subsequent lost of surrounding gas taking away a large part of the binding energy.

Di Fazio: I would like to call your attention on two theoretical points related to the low–mass end of the IMF, and on two observative ones.

(1) In making evolutionary models of protostructures (such as protogalaxies or protoclouds where stars form, etc.) many authors, for the sake of simplicity, have either neglected the dynamical or the thermodynamical evolution of the system, calculating, under some hypotheses, only one of the two evolutions, and assuming some give mathematical time behaviour for the other one. Unfortunately, this method can be useful only to test some new model–phernomena, but not to predict the actual protocloud evolutions. In particular, in trying to model the IMF at low masses for evolving self–gravitating objects, often a given mathematical law is assumed for the time evolution of the gas density (e.g. Palla, Salpeter, and Stahler, 1983, Ap. J. **271**, 632: they assume an exponentially decreasing time–dependence). Even thoguh a detailed description of the chemical reactions and of the radiative properties of matter is often included, the very strong assumptions on the gas density time–behaviour (and completely independent from the thermodynamics) invalidate any overall predictions on the studied Jeans mass time dependence. In the well–known example mentioned above, for example, the authors study the formation of molecular hydrogen, and show a very efficient cooling function due to the presence of the H_2 molecule. The application of this study to a model where the gas density is assumed to be ever–increasing exponentially (a collapse of the cloud) yields a fast dropping Jeans mass. The attained low values induce the authors to conclude that the cloud forms low–mass stars from primordial matter. Nevertheless, if we examine the process of fragmentation in such a self–gravitating cloud, we should note the following: i) the ever–increasing gas density hypothesis is not acceptable, as the initially gaseous system turns very fast (in about one free–fall time) into a fragment phase–dominated one (similar to an N–body system), which of course stops collapsing and bounces back up, as its equation of state is "stiff" to compression. This bounce of the fragment phase decrease drastically the active gravitational mass "seen" by the gas, which then bounces back, too. The gas density then, in a time slightly larger than the free–fall time, stops increasing and starts to decrease, causing the Jeans mass to increase; ii) a part from the latter gravitational–dynamical consideration, a sec-

ond effect sends the gas density back down. In fact, even before the bounce of the system, a very efficient star formation (or any fragmentation) subtracts mass from the environment gas, and, sooner or later, it can depauperate the gas phase to the point when the gas density starts to drop notwithstanding the system's collapse. The gas exhaustion due to fragmentation and star formation can thus be a further powerful mechanism that stops the initially increasing trend of the gas density, and thus sends the Jeans mass back up. The above discussed example shows that, in attempting to predict the time behaviour of the Jeans mass (and thus the shape of the low mass end of the IMF) we can make all the simplifications we want, but without introducing artificial, non self–consistent separations between the dynamics and the thermodynamics of the object.

(2) An effort should be made, in the models, not to restrict the investigation to star formation alone, but to extend it to fragmentation in general, as probably many different self–gravitating astronomical objects (such as stars, open and globular clusters, gas clouds, etc.) have had common birth mechanisms (even if modulated by several different processes). Also from the observative data analysis point of view, the hypothesis mentioned above (already formulated by Reddish, 1978) should be checked. Moreover, the comparison of the mass functions of different objects in dimensionless coordinates can yield precious information on the formation mechanisms, and, sometimes, on the early evolution of the examined systems.

(3) For what regards the long–lasting discussion, in the literature, about the low–mass end of the stellar IMF in the solar neighborhood, and, in particular, about the existence and location of the turn–over, I would like to make the following remarks. The assessment of the extension and of the quantitative relevance of the selection effects on the low–mass end of the stellar IMF has been and still is a very difficult task, to the point that no conclusive statement can archive the problem as solved, for the time being. Nevertheless, after: i) the conclusion of Miller and Scalo (1979, Ap. J. Suppl. **41**, 513), that "the turn–over or flattening" at low masses corresponding to a visual magnitude of ≈ 15 "appears to be a real effect"; ii) the IR search for very low–mass stars of Probst and O'Connell (1982, Ap. J. Lett. **252**, L69) designed to be complete down to $M_V \approx 20.7$, that also confirms a turn–over in the same region; iii) the IMF of the single stars of the solar neighborhood by Scalo (1985, in 'Protostars and Planets', Univ. of Arizona Press, p. 201), that shows a peak around 0.1 $M_\odot$, we can consider the determination of the turn–over point as being not so far away as it was. The recent discussions on the contribution of the brown dwarfs to the displacement at lower masses of the discussed IMF peak should, on the other hand, take into account that the brown dwarfs appear to members of bynary systems, in their great majority. This means that, due to the fact that binary systems very likely originate from a single cloud, in the star counts for the statistics needed in order to build the IMF, we should not count the brown dwarfs separately, but together with the mass of their companion star (i.e., use the total mass M_1+M_2 in the statistics for the binary systems). This, of course, results practically in a non–modification of the IMF that has been evaluated without brown dwarfs.

(4) In trying to evaluate the IMF for stars of the solar neighborhood, we know that some spurious information (or entropy) is introduced by using the luminosity

function and the theoretical evolution models (the M/L ratio plus the MS lifetime) in order to convert the present–day mass function into an initial mass function. But, the bynary star systems are obviously born through the same processes as the single stars (and only subsequently split), and thus their IMF will have the same shape as that of the single stars. Moreover, a great number of visual and spectro–photometrical systems (where the mass can be dynamically measured) are now in the available catalogues. so, a fairly natural suggestion is to build the mass function for unevolved bynary systems (counting the total mass for each system) and then compare its shape with the IMF of the single stars. The sense of this is, of course, that the binary IMF is more directly determined, and does not rely on the model–dependent M/L ratio, as does an IMF deduced from the LF.

Mateo: But what if -as we heard the first day of conference and as Larson argued during his talk on formation of star clusters- captures in dense star formation regions form a significant number of binaries–they are not then born as a physical unit? Note that Latham and collaborators have indicated that a 'typical' binary separation is not $\sim 30\ AU$ but hundreds, perhaps thousands of AU's.

Di Fazio: Of course, my proposal is not to use binaries in "dense star formation" regions, but, so to speak, in "clear" regions, where it is also easier to measure the orbital parameters. As a matter of fact, the binaries for which in the available catalogues it is possible to have both the mass measurements are often not embedded in overcrowded regions.
Furthermore, the average interstellar distance in the galaxy is $D \approx 500,000\ AU$, and taking the values you mentioned for the 'typical' binary separation d, we get that d is only $\approx 0.002D$. This value is small enough to make capture or disturbance by another system far less probable than the "splitting" hypothesis for the birth of a binary system from a single mother cloud (e.g. in the fashion of Norman, Wilson, and Barton, Ap. J. **239**, 968).

Capuzzo–Dolcetta: I would try to move the discussion to globular clusters of our galaxy. Actually I think that it is more worthy for the physical understanding of the low mass end of the IMF to refer to those environments, like globular clusters, which are more easily theoretically modeled, because they are geometrically well defined, composed by coeval stars of almost the same metallicity, with well known relaxation times, etc.. In my opinion, the solution of more difficult tasks should be postponed to the solution of the easier ones. What is your feeling as an observer, Roberto Buonanno?

Buonanno: Globular clusters are a useful tool to study the PDMF. According to the work I am doing together with H. Richer, G. Fahlman and F. Fusi Pecci, we can observe masses as small as $0.18\ M_{\odot}$ in globular clusters, and determine LF's down to these faint limits. We already determined the LF's in two fields of the globular clusters M4 and NGC 6752.

RADIO–OBSERVATION PERSPECTIVES AND STAR FORMATION

E. KRÜGEL
Max Planck Institut für Radioastronomie
Auf dem Hügel 69
53 Bonn, FRG

Instead of trying to put the results and new ideas on star formation presented during this meeting into a global perspective, I will restrict myself to one comment. It is prompted by my background as an astronomer working among radio observers.

In the recent past, a number of new radio telescopes have become operational and others will be added in the next two or three years. They are all aiming at the millimeter/sybmillimiter region, the highest radio frequencies accessible from the ground. Among them are the IRAM $30m$ telescope in Andalucia, the Maxwell-telescope in Hawaii, the Swedish-ESO submillimeter telescope in Chile and the Submillimeter telescope in Arizona; the latter will be the most accurate large dish and the first to allow ground-based observations down to $0.3mm$.

As the costs for these new instriments are of the order of ten million US $, in each case a sound scientific justification was required to justify the spending. One of the strongest arguments was always the expectation to unravel with these new telescopes the riddles of star formation: dense, gravitationally unstable condensations buried inside molecular clouds behind large amounts of visual extinction are best studied in the it mm/sub*mm* region, photometrically as well as spectroscopically.

The arguments seem simple and convincing. For instance, at the very early, low luminosity stage in protostellar evolution, continuum emission from dust has to occur mainly at sub*mm* wavelengths and it reaches us almost unattenuated. As it is unclear today to what degree, at the relevant high densities and low temperatures, molecules freeze out on the grains and are thus observable at all, sub*mm* dust emission may turn out to be the best way to detect early protostars. On the other hand, sub*mm*/*mm* line emission from molecules and atoms provides (for most combinations of temperature and density which prevail in the various stages of star formation) a probe of the interstellar medium. CO is the most common example: in the rotational transitions of this molecule one samples with increasing quantum number progressively warmer and denser regions. Observations of other fairly abundant molecules $(CS, NH_3, HCN, H_2O, OH, ...)$, of selected fine structure lines and of recombination lines can complement the study.

Furthermore, the new telescopes are large and offer a diffraction limited spatial resolution of $\approx 10"$. In the nearest clouds they thus allow to study structures less than 0.01 *pc* across and the existing and forthcoming *mm*/sub*mm* arrays, like the

R. Capuzzo-Dolcetta et al. (eds.), Physical Processes in Fragmentation and Star Formation, 459–460.

IRAM array at Plateau de Bure, can do even better.

Now it struck me that submillimeter astronomy (even if we include the adjacent wavelength bands) played a surprisingly modest role in the contributions during this meeting compared to what it claims it can achieve. One might have expected suggestions for future submillimeter experiments or a more frequent reference to existing data. Of course, some of the questions addressed here may be better tackled, say, with optical methods (determination of the luminosity function); others are of a very theoretical nature (collapse calculations) and cannot at present be subjected to observational tests. Nevertheless, either radio astronomers are too favorably inclined towards their own field and overestimate its possibilities, or its potential is not sufficiently appreciated by non–radio astronomers. Whatever the personal view, I encourage everyone to think over, in the light of the new telescopes, what experiments can be done to enhance our understanding of star formation.

CONCLUDING REMARKS

RICHARD B. LARSON
Yale Astronomy Department
Box 6666
New Haven, CT 06511, USA

At this meeting, we have heard about many interesting new developments in the study of fragmentation and star formation; yet it must be admitted that most of the processes involved are still poorly understood, and a fully quantitative understanding of star formation remains a distant goal. Perhaps, however, we can at least begin to formulate more clearly some of the questions that need to be addressed. Also, we have seen that new techniques, particularly in the observational study of regions of star formation, promise to yield rapid progress in answering certain of these questions. Therefore, rather than attempting to summarize what has been learned, I have chosen to conclude with a list of some of the outstanding questions which seem to me to be raised but not yet answered by the work that has been presented; I hope that these issues will be the subject of much continuing research.

The Initial Mass Function. A major motivation for observational studies of the IMF has always been to advance our understanding of star formation, but what constraints, exactly, does the observed IMF place on a theoretical understanding of star formation? A fundamental question is whether star formation tends to make stars with a characteristic mass; if so, this mass is a basic quantity that theory should attempt to explain. My impression is that one solar mass is indeed a characteristic stellar mass, but this question will not be completely settled until the IMF for very low-mass stars is reliably determined. A well-defined and universal slope for the upper IMF would be another parameter that theory should attempt to explain. At present, observations do not closely constrain the slope of the upper IMF, and the question of its possible variability remains open; however, observations are gradually placing better constraints on such variability.

Cloud Structure. Several contributions to this meeting addressed the structure of star forming clouds, but left unanswered the question of what structural studies can teach us about the physics of these clouds. Do filamentary shapes, fractal geometries, etc., contain important clues to the origin and evolution of interstellar clouds? As with terrestrial clouds, one might expect to see structural evidence for hierarchical processes such as turbulence, but whether this is true in interstellar clouds remains unclear. By contrast, the cometary shapes of some star forming clouds seem to me to provide fairly compelling evidence for hydrodynamic interaction with a surrounding medium, a phenomenon that has so far received relatively little theoretical attention.

R. Capuzzo-Dolcetta et al. (eds.), Physical Processes in Fragmentation and Star Formation, 461–463.

Cloud Dynamics. A major theme of this meeting has been the internal dynamics of star forming clouds; some new approaches to this subject were suggested here, but this remains perhaps the most poorly understood aspect of star formation, owing to its obvious complexity. It has been clear for many years that large molecular clouds possess supersonic internal motions which have been loosely described as "turbulence", but their nature and origin have remained quite unclear – are they best described in terms of eddies, clump motions, stochastic shock waves, magnetohydrodynamic waves, or some combination of all of these? Almost certainly, a description involving only classical eddy turbulence is not adequate, and magnetohydrodynamic waves have recently become more popular. Several formalisms for describing chaotic motions in molecular clouds have been proposed here, based on some of the above ideas, but they do not yet appear to have much predictive power or to make much contact with observations. I believe that efforts in this direction will be most fruitful if they focus on producing observationally testable predictions, and if theoretical work maintains close contact with observations.

Fragmentation. Molecular clouds are observed to be very clumpy, which is not surprising given that they contain many Jeans masses, but what processes or forces actually generate the clumps? It is worth noting that, while gravity and magnetic fields may be the dominant forces in molecular clouds, gravity can only amplify existing density enhancements and magnetic fields can only resist this tendency; but neither gravity nor magnetic fields can create density enhancements *ab initio*. However, purely hydrodynamic phenomena such as turbulence or shocks can create density fluctuations which can then be amplified by gravity. Thus, understanding the internal dynamics of star forming clouds is probably an essential aspect of understanding fragmentation.

Protostellar Collapse. For the formation of low-mass stars, a generally accepted picture seems to be emerging that involves the initial slow development of cloud cores via ambipolar diffusion, followed by a dynamical collapse that forms a small stellar core which then continues to gain mass by accretion. The predicted final properties of the stellar core are in good agreement with the observed properties of T Tauri stars, and circumstellar disks are also predicted when rotation is included. We have heard, however, that the core can show surprisingly complex behavior characterized by very large-amplitude oscillations before settling into hydrostatic equilibrium. Only a few percent of the total mass is involved in these oscillations, so it is not clear whether the final properties of the stellar core at the end of the accretion phase are much altered; it will be interesting to pursue the calculations further to answer this question.

Binary Formation. The formation of binary stars is not yet well understood, but two possible mechanisms for making binaries were discussed at this meeting, namely formation by fragmentation and formation by capture due to gravitational drag with disks. It may well be that binaries form in both ways, or even that fragmentation and gravitational drag work together in many cases. It may also be that the final result is not very directly related to the initial conditions, and that quasi-chaotic effects are important. Again, it will be important to try to produce testable predictions that can be compared with the available data on binary stars.

Cluster Formation. It seems clear that many very complex processes are involved in the formation of star clusters; indeed, the formation of a cluster of stars may be the culminating event in the history of a star forming cloud, involving all of the effects that have been discussed. Rather than attempting to develop models for such a complex sequence of events, the best hope for understanding cluster formation is probably to rely strongly on observations such as detailed molecular maps and infrared array photographs to suggest the effects that are likely to be important; theoretical work can then focus on trying to understand these effects one by one, an example being the role of interactions in protoclusters. The consequences, as well as the mechanisms, of cluster formation will also be important to study, since they are bound to have major effects on the subsequent development of star forming regions.

Galaxy Evolution. A long-term goal of studies of star formation is to apply our knowledge of star formation toward building up a better understanding of the formation and evolution of galaxies. Possibly the smallest galaxies can be viewed as resembling large star clusters, but clearly most galaxies are far more complex yet than star clusters. Consequently, we must rely even more heavily on observational input. A new feature is that not only do we have to understand star formation as it is observed locally, but we also have to understand star formation in different and sometimes much more extreme conditions, for example in starbursts. As a first step in this direction, it will be important to study star formation in nearby galaxies differing in some respects from our own, such as the Magellanic Clouds. Can we understand, for example, why the Magellanic Clouds seem to form larger clusters than our own Galaxy?

The Interstellar Medium. Finally, since star formation is intimately connected with the evolution of the interstellar medium and is one of the many processes by which matter is cycled between different forms in galaxies, we must eventually understand all of the interactions and exchanges that occur between stars, star forming clouds, and the interstellar medium generally. How do stars, once formed, react back on the interstellar medium? How do the various possible feedback effects influence subsequent star formation? And how do the elements produced in evolving stars eventually become incorporated into later generations of stars? Answers to such questions are required for a better understanding of both star formation and galactic evolution, and will require much further study of many processes of interstellar physics.

DISCUSSION

Mouschovias: On the issue of fragmentation, I agree that gravity only enhances perturbations while magnetic fields restrict them, but we have shown definitely, both analytically and numerically, that gravity and magnetic fields together set a length scale and therefore a mass scale which is essentially the product of the Alfven velocity in the neutrals and the ion-neutral collision time. We are far from predicting a mass spectrum, but we have demonstrated, I believe, that magnetic fields do introduce a mass scale of about 1 $M_\odot$ at typical molecular cloud densities.

Larson. Clearly the question of a characteristic mass scale is one of the most fundamental in the theory of star formation, and I believe it is a good one for both theory and observation to focus on in an effort to attain a more quantitative understanding of the subject.

KEY WORD INDEX

Page numbers are those of the title page of the article in which the subject is mentioned.